BONDING, STRUCTURE AND SOLID-STATE CHEMISTRY

Bonding, Structure and Solid-State Chemistry

Mark Ladd

Formerly Head of Chemical Physics, University of Surrey, UK

One is almost tempted to say ... at last I can almost see a bond. But that will never be, for a bond does not really exist at all: it is a most convenient fiction which, as we have seen, is convenient both to experimental and theoretical chemists.

Charles A Coulson

OXFORD

UNIVERSITY PRESS

OXFORD

UNIVERSITY PRESS

Great Clarendon Street, Oxford, OX2 6DP,
United Kingdom

Oxford University Press is a department of the University of Oxford.
It furthers the University's objective of excellence in research, scholarship,
and education by publishing worldwide. Oxford is a registered trade mark of
Oxford University Press in the UK and in certain other countries

© Mark Ladd 2016

The moral rights of the author have been asserted

First Edition published in 2016

Impression: 1

Published in the United States of America by Oxford University Press
198 Madison Avenue, New York, NY 10016, United States of America

British Library Cataloguing in Publication Data

Data available

Library of Congress Control Number: 2015944565

ISBN 978–0–19–872994–5 (hbk.)
ISBN 978–0–19–872995–2 (pbk.)

Printed and bound by
CPI Group (UK) Ltd, Croydon, CR0 4YY

Foreword

It is not an easy decision to decide to write a scientific book. What is the target audience? How will the content interface with the web and with computer-literate readers? Above all, is there a need for it? A common response is a local one. The author gives a course for which no really compatible text exists, so he or she writes one. But there is a higher level response also; several independent courses are given but need to be brought to a unity, to be assembled into a single coherent unit. Here, the common solution is to follow a single, all-embracing text. But this imposes its own structure; what if this is not acceptable?

In this book, Mark Ladd has bravely addressed this situation. He is well qualified. He is a well-established book author, not only in science, as well as a researcher and Journal Editor—and so experienced in critical assessment. *Bonding, Structure and Solid-State Chemistry* finds a basis in, and draws from, books that the author has written on *Physical Chemistry*, the *Symmetry of Crystals and Molecules*, *Symmetry and Group Theory in Chemistry*, *Chemical Bonding in Solids and Fluids* and *Structure Determination by X-ray Crystallography*. In doing so, it brings together much of contemporary chemistry and its background.

The author does not shrink from exploring the mathematical models which provide the backbone of the subject, but he does so recognizing that not all readers will find this the easiest of approaches. Further, that there will be wide variations in the depth and approach to the material that has already been encountered by different readers. Careful definition and explanation are the keys. The availability of programs, specifically written and available online, to accompany the text provides additional areas of support, covering not only specific topics (electron-in-a-box) but also more general subjects (angular and radial wavefunctions, Gaussian quadrature, Hückel).

The freshness and originality of the book is evident in some unexpected arrangements. So, paramagnetism is covered in the chapter on Nanoscience and Nanotechnology, as well as being associated with ligand field theory (which is part of the chapter on Covalent Compounds). Although much of the content can be regarded as a modern view of traditional material, the contemporary is not neglected. So, vibrational bonding is covered and you will find 'recent advances' mentioned in contexts such as desalination, superconductivity, fuel cells, Li–S batteries and perovskite solar cells. To meet the phrase 'at the time of writing', which we do, is always encouraging.

January 2016

S. F. A. Kettle
Professorial Fellow, University of East Anglia

Preface

This book derives from lecture courses given over a number of years to undergraduate students of both chemistry and those subjects in which chemistry forms a significant part. It is not set at any particular academic level and should find a place in any year of the normal span of degree studies in the physical and biochemical sciences. It presumes a background in science and mathematics of approximately A-level standard; mathematical treatments herein that are above that level are discussed in Appendices so as not to perturb the development of the textual matter.

Each chapter has been provided with a set of problems of varying degrees of difficulty, which the reader is encouraged to solve so as to gain familiarity with the subject matter and to test its application to new situations. Some recommendations for the solution of numerical problems are given. It can be only too easy to rely on a 'device' for numerical computation; a feel for the correctness of a calculation is desirable, and careful attention to units and their balance in an expression can often assist in this acquisition.

The Système International (SI) units are employed in the text. However, there are situations in practice where other units are still in use, and for good reason: thus, the electron-volt eV for energy, the Ångström Å for interatomic distance and the cm^{-1} for spectroscopic 'frequency', all enshrined in the scientific literature, remain as practical units in everyday scientific use. Competency in more than one system of units can enhance ability in a subject and in the understanding of its literature; to this end, the physical constants that are not one of the seven basic SI units are expressed also in k_m_s symbols on page xxi.

Notation is standardized to common usage as far as possible, and is tabulated on pages xxii–xxiv. The naming of equations and figures follows standard practice; in addition Eq. (A2.3) is equation 3 in Appendix A2, Fig. P1.2 means figure 2 in the problems section of Chapter 1, and Fig. S2.1 refers to Fig. 1 in the Solutions for Chapter 2; A–B indicates a bond between A and B and A–B–C a bond angle at atom B with arms BA and BC. Frequently, and particularly in matrices and character tables, a negative sign is placed above the parameter to which it refers; thus, for example, $\bar{1}$ is used for −1, with a consequent neatness in presentation.

Point-group symbols and symmetry elements are written in italic font, as in D_{3h} or σ, using the Schönflies symmetry notation; operators, including symmetry operators, are given in bold italic, as in $\boldsymbol{C_2}$ (a twofold symmetry operator), whereas vectors and matrices are expressed in bold font, as in $\mathbf{C_2}$ (a twofold symmetry matrix) or $\boldsymbol{\mu}$ (a dipole moment vector).

The study of chemical structure, whether at the molecular or crystal level, requires an appreciation of relationships in three dimensions. The acquisition of this important ability is aided by stereoscopic illustrations. Many molecular and crystal structure diagrams are presented as stereoviews, most of which have been drawn with the program PLUTO, by courtesy of Dr W. D. S. Motherwell, University of Cambridge; instructions for correct stereoviewing and for constructing a stereoviewer, are given in an Appendix.

The main chapters of the book are devoted to a study of structure and solids under a four-point classification, dependent upon the main type of cohesive force acting in the solid state of a substance, namely, covalent, molecular (aka van der Waals), ionic and metallic. Important applications of the topics studied are interspersed at appropriate points within the text.

Computational aids form an important adjunct to both the understanding of the textual material chapters, and the solving of exercises and problems based on them. Thus, a set of computer programs has been devised for these applications and accompanies the book: it can be accessed from the publisher's website at http://www.oup.co.uk/companion/ladd. The programs are very straightforward in operation, and of each of them is mostly self-explanatory. A short chapter provides some additional information and practice in their execution, and the reader may wish to peruse this chapter at an early stage. Interactive computing is employed in certain aspects of the work, and the necessary instructions are provided. It should be noted that *two complete sets* of programs are provided in order to meet the needs of both 32-bit and 64-bit operating systems.

The sets of problems are supplemented by detailed tutorial solutions, in some of which additional relevant material, not treated explicitly in the book, is given where a useful extension of a topic is justified.

Chemical bonding and solid-state chemistry are such wide-ranging topics that not all of their aspects and applications can be included, but enough has been provided, it is hoped, to form a sound basis for wider explorations of these subjects. The chapter on nanoscience explores an important topic that is the subject of much intensive current research and technological application.

I would like to express my thanks to those publishers and authors for permission to reproduce the diagrams that carry appropriate credits. In particular, several illustrations have been gleaned from the following sources: (a) *Structure Determination by X-ray Crystallography*, 5th ed. Springer Science+Business Media, 2013; (b) *Crystal Structures: Lattices and Solids in Stereoview*, Ellis Horwood Ltd. UK, 1998 / Woodhead Publishing, UK; (c) *Introduction to Physical Chemistry*, 3rd ed. Cambridge University Press, UK 1998, and the courtesy of their publishers is greatly appreciated.

It is my great pleasure to thank Professor Anthony Ladd of the University of Florida for the Python graph and contour-plotting programs, Professor Neil Ward of the University of Surrey for reading the draft manuscript and for timely comments, and Dr Andrew Coulson, of the Institute of Local Government, University of Birmingham for permission to reproduce the quote on the title page.

Also, thanks go to Professor Sidney Kettle of the University of East Anglia for writing the Foreword. I am indebted to Sönke Adlung, Senior Commissioning Editor, Physical Science, Oxford University Press for his encouragement; to Ania Wronski, Assistant Commissioning Editor, Physical Sciences; and Maegan Reed, Production Editor, Law and Academic Sciences & Medicine, of Oxford University Press, for their advice and technical assistance throughout the preparation of this book.

Bramshott   Mark Ladd
January 2016

Disclaimer

Every effort has been made to ensure the correct functioning of the software associated with this book. However, the reader planning to use the software should note that, from a legal point of view, there is no warranty, expressed or implied, that the programs are free from error or will prove suitable for a particular application. By using the software the reader accepts full responsibility for all the results produced, and the author and publisher disclaim all liability from any consequences arising from the use of the software. The software should not be relied upon for solving a problem, the incorrect solution of which could lead to injury to a person or loss of property. If you do use the programs in such a manner, it is at your own risk. The author and publisher disclaim all liability for direct or consequential damages resulting from your use of the programs.

Contents

Tutorial Solutions

Web Program Suite (accessed from http://www.oup.co.uk/companion/ladd)

Physical Data, Notation and Online Materials

Physical constants

The physical constants tabulated hereunder are CODATA (2010)[a] recommended values; the majority of these constants are available with precisions greater than those indicated here. Where a unit is expressed other than in any of the seven basic SI units (kg, m, s, A, K, mol, cd) the corresponding kg_m_s components (and some useful equivalents) are added in parentheses.

A current review of the SI system is under consideration in which exact values will be set for the basic units, excluding the Kelvin, but including the hyperfine splitting frequency for ^{133}Cs; the Kelvin will be set by defining an exact value for the Boltzmann constant. Such changes, if agreed, will not come into force before 2018 and would not affect the use of the values of these constants employed in this book.

Atomic mass unit	m_u	1.66054×10^{-27} kg
Avogadro constant	L	6.02214×10^{23} mol^{-1}
Bohr magneton[b]	μ_B	9.27401×10^{-24} J T^{-1} (kg m^2 s$^{-2} \equiv$ C V) (kg s^{-2} A^{-1})
Bohr radius for hydrogen[c]	a_0	5.29177×10^{-11} m
Boltzmann constant	k	1.38065×10^{-23} J K^{-1} (kg m^2 s^{-2})
Boltzmann constant (eV)	k' (k/e)	8.61733×10^{-5} eV K^{-1}
Electric constant[d]	ε_0	8.85419×10^{-12} F m^{-1} (kg^{-1} m^{-2} s^4 A$^2 \equiv$ A s V$^{-1} \equiv$ C V^{-1})
Electron mass	m_e	9.10938×10^{-31} kg
Electron radius	r_e	2.81792×10^{-15} m
Elementary charge	e	1.60218×10^{-19} C (A s)
Faraday	F	9.648535×10^4 C mol^{-1} (A s mol^{-1})
Magnetic constant[e]	μ_0 (exact)[f]	$4\pi \times 10^{-7}$ N A^{-2} (kg m s$^{-2} \equiv$ J m^{-1})
Molar gas constant	R	8.31446 J K^{-1} mol^{-1} (kg m^2 s^{-2})
Molar mass of carbon-12	$M(^{12}C)$ (exact)	12×10^{-3} kg mol^{-1}
Muon mass	m_μ	1.88353×10^{-28} kg
Neutron mass	m_n	1.67493×10^{-27} kg

Planck constant	h	6.62607×10^{-34} J s (kg m^2 s^{-2})
Planck constant over 2π	$\hbar$	1.05457×10^{-34} J s (kg m^2 s^{-2})
Proton mass	m_p	1.67262×10^{-27} kg
Rydberg constant	R_∞	1.09737×10^{7} m^{-1}
Speed of light in a vacuum	c	2.99792×10^{8} m s^{-1}

[a] http://physics.nist.gov/cgi-bin/cuu/Category?view=html.

[b] $\mu_B = \dfrac{e\hbar}{2m_e}$

[c] $a_0 = \dfrac{4\pi \varepsilon_0 \hbar^2}{m_e e^2}$

[d] aka permittivity of a vacuum

[e] aka permeability of a vacuum

[f] Dependent numerically on the expression of π.

Prefixes to units[a]

atto	femto	pico	nano	micro	milli	centi	deci	kilo	mega	giga	tera	peta	exa
a	f	p	n	μ	m	c	d	K	M	G	T	P	E
10^{-18}	10^{-15}	10^{-12}	10^{-9}	10^{-6}	10^{-3}	10^{-2}	10^{-1}	10^{3}	10^{6}	10^{9}	10^{12}	10^{15}	10^{18}

[a]The prefixes deca (10^1) and hecto (10^2) are used only rarely.

Notation

These notes indicate the main symbols used throughout the book. Inevitably, some of them have more than one application, partly from general usage, and partly from a desire to preserve a mnemonic character in the notation wherever possible. Two or more uses of one and the same symbol are separated by a semicolon in the presentation hereunder.

Å	Ångström unit (1 Å = 10^{-10} m)
a, b, c	Unit cell edge lengths parallel to the x, y and z axes, respectively
a, b, c	Unit cell translation vectors along the x, y and z axes, respectively
a_0	Bohr radius
B	Magnetic field strength vector (some sources use **H**)
C	Capacitance
c	Speed of light
D	Density
d	Interplanar spacing of the *hkl* family of planes
E	Energy; electric field strength
e	Electron charge

e, exp	Exponential function
esd	Estimated standard deviation
(hkl)	Miller indices of planes associated with the x, y and z axes, respectively
hkl	Reciprocal lattice point corresponding to the (hkl) family of planes
h	Miller index parallel to the x-axis; Planck constant; order of a group
$\hbar$	Cross-$h = h/2\pi = 1.05457 \times 10^{-34}$ J s
I	Moment of inertia
$\mathbf{i}, \mathbf{j}, \mathbf{k}$	Unit vectors parallel to x, y and z axes, respectively
i	$\sqrt{-1}$
k	Boltzmann constant; $\lvert \mathbf{k} \rvert$ (in $\mathbf{k}$-space)
l	Angular momentum quantum number
L	Avogadro constant
M_r	Relative molecular mass
$M(^{12}\mathrm{C})$	Molar mass of carbon (^{12}C isotope)
m_e	Mass of electron
m_n	Mass of neutron
m_p	Mass of proton
m_l	Magnetic quantum number
m_s	Spin quantum number
m_u	Atomic mass unit
n	Principal quantum number; refractive index
P	Polarization; total bond order
p	Pressure; π-bond order
R	Symmetry element; rotation axis of degree R; gas constant
R_∞	Rydberg constant
$\mathbf{R}$	Symmetry operator or operation
$\mathbf{R}$	Matrix representing symmetry operation R
$\mathbf{r}$	Vector distance, as in $\mathbf{r} = \mathbf{i}x + \mathbf{j}y + \mathbf{k}z$, for example
r	Spherical coordinate; general distance
r_e	Equilibrium interatomic distance; radius of electron
rms	Root mean square
$\mathbf{v}$	Velocity
v	Speed (magnitude of $\mathbf{v}$)
$V_\mathrm{c} V_\mathrm{c}$	Volume of a crystal unit cell
x, y, z	Labels for reference axes
Z	Atomic number; number of formula entities of relative molecular mass M_r per unit cell
Z_eff	Effective atomic number ($Z - \sigma$)
α	Polarizability; coefficient of linear thermal expansivity
α'	Volume polarizability
β	Delocalization energy; coefficient of volume expansivity
ε_0	Electric constant (aka permittivity of a vacuum)
ε	Permittivity; energy parameter in the Lennard-Jones potential

ε_r	Relative permittivity (aka dielectric constant)
Φ	Molecular wavefunction
ϕ	Spherical coordinate
γ	Surface tension; Sommerfeld constant
Γ	Gamma function
κ	Coefficient of isothermal compressibility; thermal conductivity
λ	Wavelength
$\boldsymbol{\mu}/\mu$	Dipole moment vector / dipole moment magnitude
μ_0	Magnetic constant (aka permeability of a vacuum)
μ_B	Bohr magneton
ν	Frequency (distinguish from v velocity)
π	Pi (ratio of the circumference of a circle to its diameter); π-bond
θ	Spherical coordinate; Bragg angle
ρ	Electron density; (also density in some contexts); electrical resistivity; $2Zr/na_0$
σ	σ-bond; surface charge; quantum mechanical screening constant; electrical conductivity
ψ	Atomic wavefunction
Ψ	Molecular wavefunction
$\circ$	Superscript: degree, as in $90°$
$\ominus$	Superscript: standard thermodynamic property, as in $S^{\ominus}$
$<>$	Average value, as in $< X >$
[....]	Polyatomic ion, as in $[\text{Fe(CN)}_6]^{4-}$

Energy conversion table

An energy term carrying a unit in column X is converted to its equivalent in a unit of row Y by multiplying it by the table entry at the intersection of column X and row Y: for example, $1 \text{ cm}^{-1} \equiv 1.98645 \times 10^{-23}$ J.

X \ Y	eV	cm^{-1}	kJ mol^{-1}	K	J	Hz
eV	1	8.06554×10^3	9.64853×10^1	1.16045×10^4	1.60218×10^{-19}	2.41799×10^{14}
cm^{-1}	1.23984×10^{-4}	1	1.19627×10^{-2}	1.43878	1.98645×10^{-23}	2.99792×10^{10}
kJ mol^{-1}	1.03643×10^{-2}	8.35935×10^1	1	1.20272×10^2	1.66054×10^{-21}	2.50607×10^{12}
K	8.61733×10^{-5}	6.95035×10^{-1}	8.31447×10^{-3}	1	1.38065×10^{-23}	2.08366×10^{10}
J	6.24151×10^{18}	5.03412×10^{22}	6.02214×10^{20}	7.24297×10^{22}	1	1.50919×10^{33}
Hz	4.13567×10^{-15}	3.33564×10^{-11}	3.99031×10^{-13}	4.79924×10^{-11}	6.62607×10^{-34}	1

Online materials

Computer programs relevant to the text and the problems have been devised and are available via the publisher's website http://www.oup.co.uk/companion/ladd.

Preamble

"Wherever we look, the work of the chemist has raised the level of civilization and has increased the productive capacity of our nation."

President Calvin Coolidge

1.1 Introduction

It has been suggested that a contribution to the failure of Napoleon's 1812 Russian campaign was linked to buttons, tin buttons that fastened the greatcoats and trousers of Napoleon's officers and foot soldiers. When the temperature falls below 13 °C, 'white' β-tin begins to transform to a crumbling powder of brittle, 'grey' α-tin, a process that proceeds rapidly at –40 °C, a temperature not unknown in a Russian winter:

$$\beta - \text{Sn} \underset{}{\overset{13\,°C}{\rightleftharpoons}} \alpha - \text{Sn} \tag{1.1}$$

Were the soldiers of the Grande Armée fatally weakened by cold, because the buttons of their uniforms degraded to powder and their clothing fell apart (Fig. 1.1)?

Fig. 1.1 Napoleon's retreat from Moscow, October 1812. [Adolf Northern, German painter, 1828–1876.]

Bonding, Structure and Solid-State Chemistry. First Edition. Mark Ladd.
© Mark Ladd 2016. Published in 2016 by Oxford University Press.

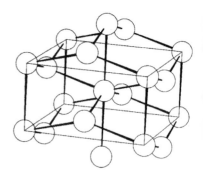

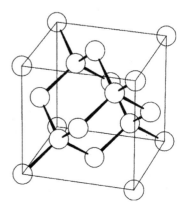

Fig. 1.2 Stereoview of the unit cell and environs of the crystal structure of β-tin, showing the distorted octahedral arrangement of six tin atoms about any other tin atom.

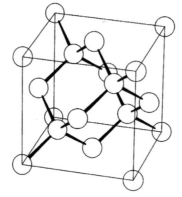

Fig. 1.3 Stereoview of the unit cell and environs of the crystal structure of α-tin, showing the regular tetrahedral arrangement of four tin atoms about any other tin atom; the cubic structure type is the same as that of the diamond form of carbon.

How different might the world have been if tin did not disintegrate at low temperatures and the French had continued their eastward expansion—but the veracity of this intriguing story of failure is not without challenge.

The crystal structures of these two allotropes of tin are illustrated by the stereoviews of Figs. 1.2 and 1.3. The silvery β-tin is the better known and exhibits the properties of a metal, whereas α-tin is a grey, powdery, non-metallic substance. The crystal structure of α-tin is similar in type to that of the diamond form of carbon, a cubic crystal with coordination number four, whereas β-tin forms an irregular six-coordinated solid; the transformation Eq. (1.1) will be considered again later. Appendix A1 provides information on stereoviewers and stereoviewing.

1.2   Atomic nature of matter

Notwithstanding many subatomic particles are now known, the fundamental particles of interest here are electrons, protons and neutrons. Atoms may be regarded as spherical in shape, with diameters of the order of 0.1 nm. The atomic nucleus is 1–10 fm in size[1] and contains positively charged protons and uncharged

[1] 1 fm = 10^{-15} m.

neutrons. These particles constitute most of the mass of an atom, being approximately 1.67×10^{-27} kg each. The nucleus is surrounded by negatively charged electrons, each of mass approximately 9.11×10^{-31} kg; the number of electrons is equal to that of the protons, the atomic number, thus ensuring electrical neutrality of the atom as a whole.

Electrons may be considered as particles located mostly within regions of space surrounding the atom known as *atomic orbitals*; they are associated with specific energy states. In terms of an electron density around the atom, the term atomic orbital implies a *one-electron wavefunction* ψ describing the electron. The product $|\psi\psi^*|$ may be regarded as an electron density,[2] such that $|\psi\psi^*|\mathrm{d}\tau$ is a measure of the probability of finding an electron in a volume element $\mathrm{d}\tau$ (see Appendix A6) in a region around the nucleus, as will be discussed further in the next chapter. The chemistry of a substance is determined by its electrons, particularly those that are furthest from the nucleus.

Atoms in combination exist in four different states.[3] Sodium chloride, for example, is encountered usually as a colourless, crystalline, ionic solid. If pure, it is a non-conductor of electricity, but at 1077 K it melts to form a colourless liquid that conducts electricity by ion transport. At the higher temperature of 1686 K the liquid boils and the vapour[4] consists almost entirely of NaCl molecules in which the atoms are mainly covalently bonded, in strong contrast to the ionic nature of the substance in the solid state.

The changes solid $\rightarrow$ liquid $\rightarrow$ gas occur for most substances, and at widely different ranges of temperature.

$$\text{ice (s)} \xrightleftharpoons{273\,\text{K}} \text{water (l)} \xrightleftharpoons{373\,\text{K}} \text{water vapour (g)}$$

$$\text{gallium (s)} \xrightleftharpoons{29.76\,\text{K}} \text{gallium (l)} \xrightleftharpoons{2477\,\text{K}} \text{gallium (g)} \tag{1.2}$$

$$\textbf{Two solid} \rightleftharpoons \textbf{liquid} \rightleftharpoons \textbf{gas transitions}$$

In the first of these transition sequences, the covalency of the water molecule is preserved in all three phases.

1.3 States of matter

The state, solid, liquid or gas, of any substance at a given temperature is determined by the result of a competition between the interatomic forces acting on its components and their thermal energy. At equilibrium, there exists a balance between forces of attraction and repulsion that is dependent upon temperature.

Gases may be characterized by their large volume changes consequent upon variations in the temperature or pressure to which they are subjected, and by their ability to flow into the space available to them. Gases are fixed in neither shape nor volume; they are miscible one with the other in all proportions. They

[2] ψ^* is the complex conjugate of ψ; $\psi\psi^* \equiv |\psi|^2$ if the electron density function is real.

[3] Plasma, the fourth state of matter, is not discussed herein [1].

[4] A vapour is a gas below its critical temperature.

differ significantly from liquids and solids, in that many of their properties are independent of their nature and respond to general laws.

A liquid is fixed in volume at a given temperature but, like a gas, has no definite form: it assumes the shape of its bounding container. Unlike a gas, the molecular species in a liquid are close enough one to the other for interatomic forces to influence their relative spatial distribution and cause local small clusters of molecules to form.

The *radial distribution function* $g(r)$ of a liquid measures the probability of finding molecules of the liquid within a spherical shell of width δr at a distance r from a reference molecule acting as an origin (see also Section 3.9.2). The following diagram, obtained by x-ray diffraction from liquid water, shows a strong peak at approximately 0.28 nm. It derives from multiple polar interactions at this distance arising from intermolecular hydrogen bonding; an average intermolecular distance in the absence of polarity is *ca.* 0.37 nm. However, the thermal energy of a liquid is commensurate with its intermolecular bonding energy, and order does not exist over more than a few atomic dimensions, often termed *short-range order.*

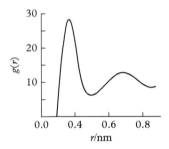

Radial distribution function for water based on x-ray diffraction measurements

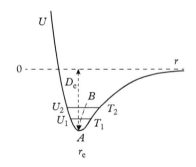

Fig. 1.4 Variation in potential energy U with interatomic distance r for a pair of atoms. The curve is anharmonic and U rises steeply as r decreases below the equilibrium value r_e. The energies U_1 and U_2 may be taken to correspond with temperatures T_1 and T_2.

A solid is a material of fixed volume and shape at a given temperature. In atomic terms, the mean, or equilibrium, positions of the atoms are invariant with time at a given temperature. However, the atoms are not static; they are vibrating about their mean positions, and their vibrational energy makes a major contribution to the heat capacity of the solid. The vibrations themselves are *anharmonic*: Fig. 1.4 illustrates the variation of the energy U of a pair of atoms as a function of their interatomic distance r. At absolute zero, the theoretical potential (dissociation) energy D_e at the equilibrium distance r_e is represented by the point A. As the temperature is increased from T_1 to T_2, the potential energy of the system decreases (becomes less negative) in moving from U_1 to U_2. Owing to the anharmonicity of the atomic vibrations, the potential energy variation is asymmetric, and the mean interatomic distance moves along the curved path $A \rightarrow B$ as the temperature is increased.

Anharmonicity arises because at any given value of the energy U, a movement $\pm\Delta r$ from the equilibrium value r_e produces the steeper energy change for $-\Delta r$ because of electron–electron repulsion; hence, solids have very small values of compressibility.

In an assembly of atoms there are added complications. The problem must be considered in terms of free energy, which introduces an entropy factor, and it may be shown [2] that even if the atomic vibrations were harmonic an increase in free energy of the system would lead to an increase in its volume. Nevertheless, the simple potential energy curve provides a useful qualitative picture that assists in understanding physical properties of solids.

The invariance of mean atomic position with time needs further consideration. In some solids, certain groups of atoms may behave as though their symmetry were higher than that apparent from their structure.

Potassium cyanide at room temperature has the sodium chloride structure type (Fig. 1.5); although the cyanide ion is linear, it behaves as though it possessed spherical symmetry. This situation may arise in one of two ways. On the one hand the $(C–N)^-$ group may exhibit *dynamic disorder*, performing *free rotation* about its mean position so that, although linear, its envelope of motion over a period of time is spherical in shape. On the other hand the group may exhibit *static*, or *orientational*, disorder, with the multiplicity of the positional distributions of the anionic groups over many unit cells again simulating a sphere. In the case of potassium cyanide experimental evidence supports the orientationally-disordered model, with the axes of the cyanide ions lying mostly normal to planes of the types (100) and (111) in the cubic unit cell.[5] Thus, the mean positions *averaged over time* are constant at a given temperature.

At 233 K, potassium cyanide transforms to a polymorphic structure (Fig. 1.6) in which the cyanide ions are orientated regularly in the unit cell, lending support for the preference of orientational disorder at the lower temperatures.

[5] Miller indices (hkl) are discussed in Appendix A2.

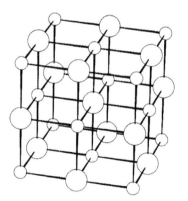 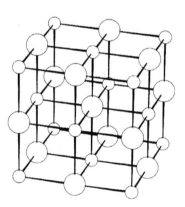

Fig. 1.5 Stereoview of the unit cell and environs of the face-centred sodium chloride crystal structure type for potassium cyanide at ambient temperature; circles in decreasing order of size represent the $(CN)^-$ and K^+ species. Random orientations of the cyanide anions about their mean positions account for the effective spherical symmetry of the anions in the cubic unit cell.

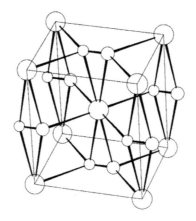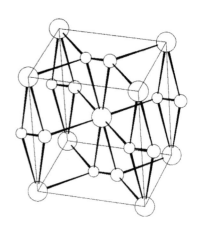

Fig. 1.6 Stereoview of the unit cell and environs of the body-centred orthorhombic crystal structure of potassium cyanide below 233 K; circles in decreasing order of size represent the K, C and N species.

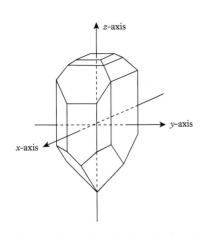

Fig. 1.7 Crystal of the hydrated silicate mineral hemimorphite, $Zn_4Si_2O_7(OH)_2.H_2O$. The crystallographic right-handed reference axes are shown in their conventional orientation, with y and z in the plane of the diagram and x directed forwards.

1.4   Crystalline and amorphous solids

Crystals, sometimes referred to now as classical crystals, are characterized by a *three-dimensional periodicity* of their structural units, atoms or groups of atoms, a feature termed *long-range order*. The regular appearance of a well-formed crystal (Fig. 1.7) is a manifestation of its ordered internal arrangement.

Lower degrees of crystallinity are recognized: certain stretched polymer sheets exhibit two-dimensional order in the planes of the sheets, and many natural and synthetic fibres exhibit long-range order in only one dimension, that of the fibre direction. Mammalian hair contains the fibrous protein α-keratin, the structural unit of which is illustrated here:

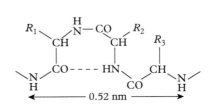

Repeating unit of the α-keratin fibrous protein

The R_i groups are amino acid residues and other chemical groups. α-Keratin shows a periodicity to x-rays of approximately 0.52 nm along its fibre axis, but if the hair is extended in steam by 100% of its length, β-keratin is obtained, with a periodicity of 0.35 nm along its fibre axis:

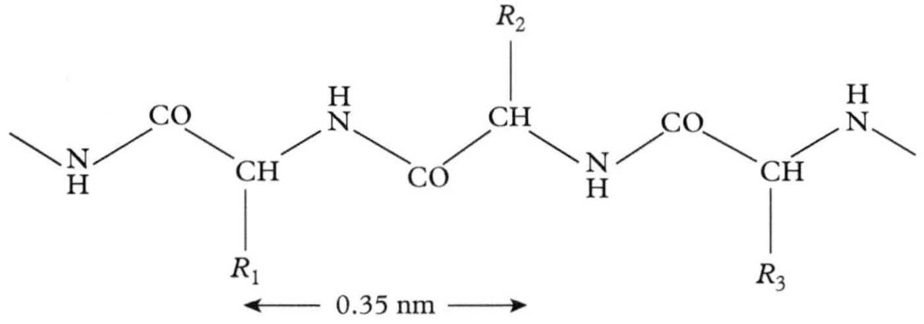

Repeating unit of the β-keratin fibrous protein

In steam the $>C=O---H-N<$ hydrogen bonds in α-keratin are broken, allowing the polymer chain to extend and so develop the smaller periodicity. The $\alpha \rightarrow \beta$ transformation is reversible, but if the stretched hair is held in the steam for some time the reversibility is lost, a feature that is involved in the 'permanent waving' of hair.

Most crystalline materials are anisotropic in their physical properties, that is, the magnitude of a physical property varies with direction in the material. Stretched fibres often exhibit optical anisotropy under polarized light. However, this may not indicate crystallinity because optical anisotropy can be induced by stress in an unannealed, drawn fibre; further evidence must be sought by x-ray diffraction. A fibre of ethylene–polypropylene–diene (co)polymer when subjected to x-ray

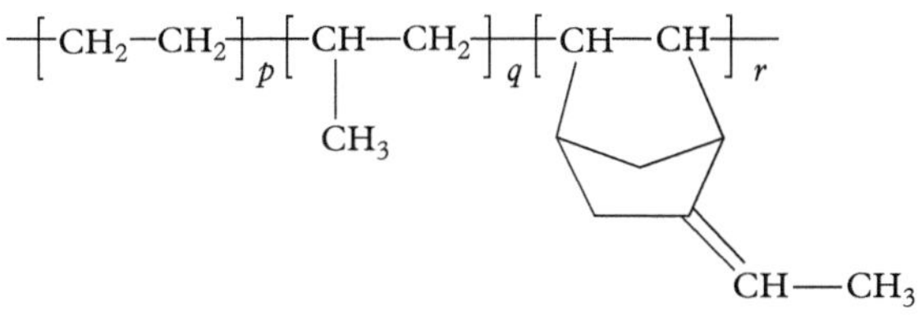

A repeating unit of ethylene–polypropylene–diene (co)polymer

irradiation produced the diffraction diagrams in Fig. 1.8. They indicate an increase in the crystallinity of the polymer from 10% in (a) to 70% in (b), corresponding to an increase in the extension of the fibre.

Amorphous solids, like liquids, exhibit short-range order; Fig. 1.9 shows a schematic diagram for experimental x-ray diffraction from a powdered material. The sharply defined rings characteristic of a crystalline material, of which only two are shown on the diagram, arise from an aluminium sample. The diffuse

Fig. 1.8 X-ray diffraction patterns from an ethylene–polypropylene-diene (co)polymer: (a) zero extension, 10% crystallinity; (b) 220% extension, 70% crystallinity. The increasing crystallinity from (a) to (b) is indicated by a beginning of the development of a spot pattern, a characteristic of all crystalline materials [Dr EJ Wheeler. Personal communication, 1979.]

Fig. 1.9 Schematic diagram for the diffraction of x-rays from a powder specimen: X, incident x-ray beam; S, powder specimen; F, photographic film; L, lead trap for transmitted beam. The two sharp diffraction rings arise from the (111) and (200) planes in aluminium, and the diffuse band is the strongest diffraction spectrum from a Sellotape film enclosing the powder. The dashed lines are the traces in the plane of the 200 'cone' of diffracted rays. In practice, a strip of film encompassing the central portion of the film F is sufficient, as shown in Fig. 5.35.

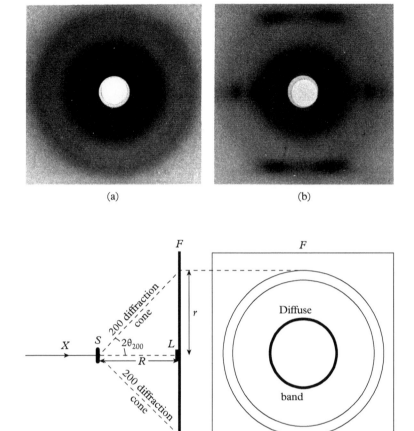

(a)   (b)

band is from the amorphous Sellotape that is used to enclose the powder in the specimen holder. From Fig. 1.9,

$$r/R = \tan 2\theta_{hkl} \tag{1.3}$$

where θ_{hkl} is the Bragg angle for dffraction from (hkl) planes in the aluminium crystals; Appendix A3 shows that

$$\sin \theta_{hkl} = (\lambda/2a)(h^2 + k^2 + l^2)^{1/2} \tag{1.4}$$

where λ is the wavelength of the x-rays and a is the dimensions of the cubic unit cell, of aluminium in this discussion. Hence,

$$a = \lambda(h^2 + k^2 + l^2)^{1/2}/\{2 \sin[\tfrac{1}{2} \tan^{-1}(r/R)]\} \tag{1.5}$$

In an actual experiment using Cu $K\alpha$ x-radiation of wavelength 0.15418 nm and with $R = 30.00$ mm, measurements on the film gave $r_{200} = 29.5$ mm; thus, from Eq. (1.5), the unit cell dimension a evaluates as 0.4070 nm.

Example 1.1

The spacings d of the (200), (220), etc. planes in aluminium, which is cubic with unit cell of sides a (= $b = c$) = 0.4070 nm, follow from the discussions in Appendices A2 and A3:

$$d_{hkl} = a/\sqrt{h^2 + k^2 + l^2}$$

Hence, d_{200} evaluates to 0.2035 nm, d_{220} to 0.1439 nm, and so on.

The diffuse band cannot be examined in this manner; it is the strongest part of the scattered radiation from the Sellotape, a part of its radial distribution, and it arises from a prominent interatomic spacing D given by [3]:

$$D = 0.61\lambda/\sin\theta \qquad (1.6)$$

By experiment, the mean radius of the diffuse ring was 11.5 mm; hence, from Eqs. (1.3) and (1.6), $D = 0.61 \times (0.15418\,\text{nm})/\sin\{\frac{1}{2}\tan^{-1}[(11.5\,\text{mm})/(30.00\,\text{mm})]\}$, which evaluates to 0.517 nm. Sellotape is a polysaccharide and a model of a 1,4-glucosidic monomer fragment, constructed from standard molecular geometry, shows an overall length of the major structural unit of 0.52 nm. Discussions on crystal geometry and x-ray diffraction may be found in the literature [4].

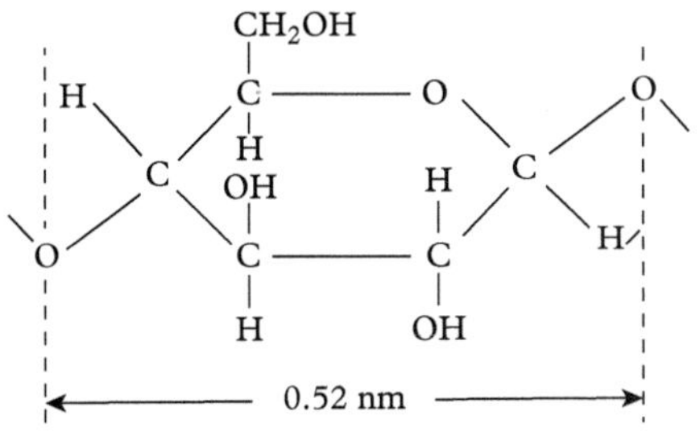

Repeating unit of a 1,4-glucoside monomer

As well as their different responses to x-ray diffraction, crystalline solids may be differentiated from amorphous solids by their behaviour on melting. Pure crystalline solids generally have a clearly defined melting point, with a melting range of 0.5–1 degree, whereas amorphous solids soften over a wider range of temperature before becoming fully liquid.

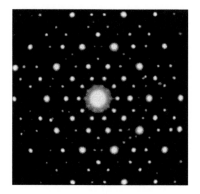

Fig. 1.10 Transmission electron micrograph of an Al_6Mn alloy surface; the apparent tenfold symmetry pattern indicates a crystalline nature for the specimen, but without the periodicity normally associated with a crystal. The true symmetry is fivefold; x-ray diffraction introduces a centre of symmetry into the pattern [4]. [Professor Daniel Shechtman, personal communication, 2013.]

Table 1.1 *Axial ratios of selected isomorphous sulphates and selenates*

	a/b	b	c/b
K_2SO_4	0.573	1	0.742
Rb_2SO_4	0.572	1	0.749
Cs_2SO_4	0.571	1	0.753
Tl_2SO_4	0.564	1	0.732
K_2SeO_4	0.573	1	0.732
Rb_2SeO_4	0.571	1	0.739
Cs_2SeO_4	0.570	1	0.742
Tl_2SeO_4	0.555	1	0.724

The materials normally understood as crystals (classical crystals) possess an ideally infinite three-dimensional periodic arrangement of atoms, and exhibit rotational symmetry degrees R of 1, 2, 3, 4 or 6: a rotation of $(360/R)°$ about the rotation axis brings the crystal into a position that is indistinguishable from that prior to the rotation.

In 1982, Shechtman, experimenting with an alloy of composition Al_6Mn, obtained transmission electron microscopy photographs (Section 2.5.3) showing apparent tenfold symmetry (Fig. 1.10); however, x-ray diffraction introduces a centre of symmetry into an x-ray spectral record. Further experimentation established unequivocally the presence of *fivefold* symmetry in the crystalline material [5]. The crystal structure of the alloy material exhibits long-range order and is space-filling in an *aperiodic* manner, that is, without the three-dimensional periodicity that is characteristic of normal crystals. The aluminium–manganese alloy is one of a class of substances termed *quasicrystals* [6].

The concept of aperiodic space-filling material was introduced by Schrödinger in 1944 [7]. He sought to explain how hereditary information is stored. Molecules were deemed to be too small, and amorphous solids were plainly chaotic; so it had to be a kind of crystal. Since the periodic structure of a crystal could not encode information it had to be pattern of another type, a new type of periodicity, namely aperiodicity.

Schrödinger proposed that the blueprint of life would be found in a compound with components arranged in a long and irregular sequence, but which carried information in the form of a genetic code embedded within its chemical structure. A protein was the obvious candidate for an aperiodic crystal, with the amino acid sequence providing the code. Later, the structure of DNA (deoxyribonucleic acid) was discovered and shown to possess properties similar to those predicted by Schrödinger—a regular but aperiodic structure. An on-line account of the DNA story [8] summarizes briefly events over the period 1869, when Miescher extracted DNA from white blood cell material, to 1953, when the structure of DNA was finally determined.

1.5   Isomorphism and polymorphism

The correspondence between similar crystal shape and chemical composition was demonstrated first by Mitscherlich in 1819. His *law of isomorphism* may be illustrated with data on the axial ratios a/b and c/b of sulphates and selenates of a number of singly-charged cations listed in Table 1.1. The results show that the components of the different crystals pack with Rb^+ taking the place of the K^+ cation, Se^{2-} taking the place of the S^{2-} anion and so forth.

On the basis of the totality of structural information now available, a wider definition of isomorphism is recognized. Chemical resemblance is not wholly essential; more significant is a similarity in the nature, size and shape of atoms and groups of atoms comprising a structure. Thus, KCl and RbCl are isomorphous in

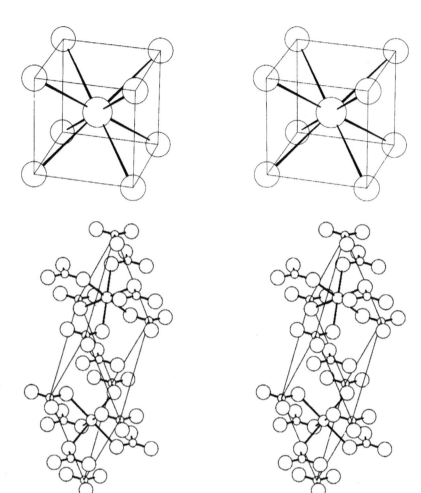

Fig. 1.11 Stereoview of the unit cell and environs of the cesium chloride crystal structure type; circles in decreasing order of size represent the Cs and Cl species. The unit cell is primitive, not body-centred.

Fig. 1.12 Stereoview of the unit cell and environs of the calcite $CaCO_3$ crystal structure type; circles in decreasing order of size represent the O, Ca and C species; this structure is isomorphous with that of sodium nitrate $NaNO_3$.

the sodium chloride structure type (Fig. 1.5), but CsCl has a different structure (Fig. 1.11), on account of the larger radius of the Cs^+ cation.

Again, KNO_3 and $NaNO_3$ are not isomorphous, but $NaNO_3$ is isomorphous with the calcite form of $CaCO_3$ (Fig. 1.12). Here, the similarity of ionic radii is the controlling factor:

Ion	Na^+	Ca^{2+}	K^+
r/nm	0.112	0.118	0.144

In these examples, the shape and size of the *structural units* in the crystals are important factors, as well as the ionic radii themselves (see Table 4.9); both

Fig. 1.13 Stereoview of the unit cell and environs of the aragonite CaCO$_3$ crystal structure type; circles in decreasing order of size represent the O, Ca and C species; this structure is isomorphous with that of potassium nitrate, KNO$_3$.

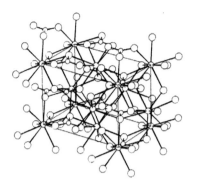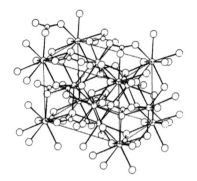

the (NO$_3$)$^-$ and (CO$_3$)$^{2-}$ anions are trigonal planar in shape, with bond lengths N–O ≈ C–O ≈ 0.13 nm. The potassium cation is too large to be accommodated in the calcite structure type, but potassium nitrate is isomorphous with calcium carbonate in the aragonite structure type (Fig. 1.13). An octahedral coordination is present in the crystal structures of both KNO$_3$ and NaNO$_3$, but energetics determine the patterns of crystallization.

Solids that possess more than one structure type are said to exhibit *polymorphism*, which is often referred to as *allotropy* when the species involved are elements. The different structures may involve quite different interatomic forces and, therefore, show quite different physical properties, as with elemental tin (Section 1.1). Another example of allotropy is afforded by the graphite and diamond structures of carbon (see Figs. 2.49 and 2.50); graphene, a two-dimensional allotropic form carbon, is discussed in Section 6.3.4ff.

1.6 Solid-state transitions

Generally, polymorphs are stable over a particular range of temperature and pressure and transformations between them may take place rapidly or over a period of time. For example, at atmospheric pressure α-tin is stable below 286 K and β-tin is stable above this temperature; the $\beta \rightarrow \alpha$ transformation becomes more rapid as the temperature is further decreased below the transition temperature of 13 °C.

Much of the available data on the solid state refers to ambient conditions, 293–298 K and 1 atm.[6] The temperature range of availability of solids is from –272.2 K (He, s) to 4488 K (tantalum hafnium carbide, Ta$_4$HfC$_5$), above which latter temperature all known solids melt, vaporize or decompose.

Conditions of temperature and pressure are of importance in the study of polymorphic transitions. For example, above 500 K potassium nitrate transforms from the aragonite structure type (Fig. 1.13) to that of calcite (Fig. 1.12), so that potassium nitrate and sodium nitrate then become isomorphous under these conditions. Transitions may be classed as sharp or gradual and each type possesses specific characteristics.

[6] 1 atm = 101,325 Pa (N m^{-2}).

1.6.1 Sharp transitions

A *sharp transition* occurs at a precise temperature and pressure, and the Clapeyron equation (1.7) is obeyed. This equation expresses the rate of variation of vapour pressure p with temperature T in terms of the changes in both enthalpy ΔH_t and volume ΔV_t at the transition temperature [9]:

$$\frac{\mathrm{d}p}{\mathrm{d}T} = \frac{\Delta H_t}{T \Delta V_t} \tag{1.7}$$

Many physical properties, such as density, heat capacity and entropy, exhibit a *discontinuity* at the transition point that is characteristic of a first-order transition, one in which the discontinuity is in the parameter itself. The variation of the density of white phosphorus with temperature is shown in Fig. 1.14; a sharp, first-order transition occurs at 317 K, when the density falls markedly but then continues to decrease at approximately the same rate as before the transition.

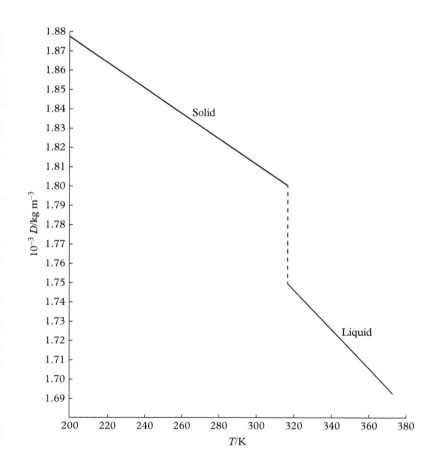

Fig. 1.14 Variation of density with temperature for white phosphorus P_4; the sharp transition at 317 K is accompanied by a discontinuity in the density.

When a solid undergoes a transition, the packing of its component species changes significantly and abruptly. The resultant molar volume, at constant mass, is reflected in the change in density. Generally, the solid phase of a substance is denser than its liquid form; ice is a notable exception.

1.6.2 Gradual transitions

In *gradual transitions*, the temperature of transformation is not always clearly defined and the change may take place over a considerable range of temperature. Discontinuities in physical properties are not observed, but maxima or minima occur in the temperature or pressure variations of properties such as heat capacity and compressibility. The Clapeyron equation is not obeyed, and *hysteresis* may occur in a physical property, such as that in the variation of heat capacity with temperature for cesium (Fig. 1.15). The gradual transition is of second order, that is, one in which the discontinuity is in the first derivative of the parameter.

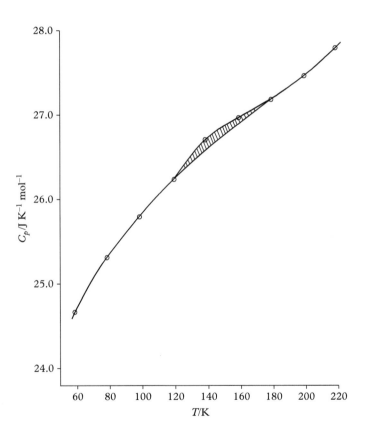

Fig. 1.15 Variation of molar heat capacity with temperature for elemental cesium; hysteresis occurs between 110 K and 170 K, the lower curve indicating the path followed on cooling. The area of the hysteresis loop represents the heat content retained on cooling. Similar effects in other substances have been attributed to crystal imperfections and to surface entropy changes arising from migration of surface atoms from lattice sites to a random distribution.

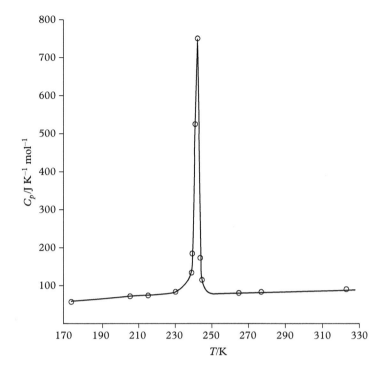

Fig. 1.16 Variation of molar heat capacity with temperature for ammonium chloride; a Λ-type transition occurs at 242.5 K, from a non-centrosymmetric cubic crystal structure to the cesium chloride structure type, Fig 1.12; there is further transition to the sodium chloride structure type at 457.5 K.

The change in heat capacity with temperature for ammonium chloride exhibits a Λ-*point transition* (Fig. 1.16), so-called because of the mnemonic resemblance of the heat capacity versus temperature graph to the Greek letter *lambda*, Λ.

Ammonim chloride has the cesium chloride structure type (Fig. 1.11); at low temperatures, the N–H bonds in the $(NH_4)^+$ ion are directed tetrahedrally to the same four corners of the unit cells throughout the crystal. As the temperature is increased a new structure arises, now of twice the size in order to accommodate the tetrahedral ammonium cation in both possible orientations in the unit cells. Further heating leads to the Λ-point transition in the heat capacity at 242 K. A new structure is formed having the volume of the low temperature form, but with the $(NH_4)^+$ ions now in twofold orientational disorder. The two possible positions of the cation are distributed randomly throughout the structure, so that the cation assumes statistically the spherical envelope required by the cesium chloride structure type; the thermodynamic properties of this transition are discussed in Section 1.6.3.

Another well-known example of a Λ-point transition is in crystalline quartz, for which the low temperature form α-quartz transforms at 846 K to the high temperature β-quartz (see Fig. 2.53). The α → β transition is reversible and is accompanied by a linear expansion of *ca.* 0.5%; on account of this type of change, silica-containing ceramic ware must be cooled very slowly after firing in order to avoid cracking of the material.

1.6.3 Entropy of transition and the Boltzmann equation

The molar entropy change of ammonium chloride at the transition point is known from experiment to be $4.6\,\mathrm{J\,K^{-1}\,mol^{-1}}$. Entropy S may be defined statistically as a measure of the *disorder* of a system: for example, a liquid has a higher value of entropy (greater disorder) than that of its solid phase because the components of the liquid phase exhibit only short-range order.

Entropy is related to disorder through the *Boltzmann equation* [9]:

$$S = k \ln W \tag{1.8}$$

where W is a measure of the disorder, or probability of the system. A molar entropy change between states 1 and 2 is then

$$\Delta S_\mathrm{m} = Lk \ln(W_2/W_1) = R \ln(W_2/W_1) \tag{1.9}$$

For ammonium chloride (W_2/W_1) is 2 for the disordering of the structure;[7] the nitrogen atom is the central species in the diagrams of the two structures of ammonium chloride:

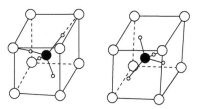

Disordered forms of NH_4Cl

Hence, $\Delta S_\mathrm{m} = (8.3145\,\mathrm{J\,K^{-1}\,mol^{-1}} \times 0.6931/1)$, or $5.76\,\mathrm{J\,K^{-1}\,mol^{-1}}$. The difference of *ca.* $1\,\mathrm{J\,K^{-1}\,mol^{-1}}$ from the experimental value indicates that the disorder is probably not total throughout the crystal.

The *Boltzmann distribution of energies* in a system of molecules at equilibrium relates the number n_ε of molecules of energy greater than a given value ε to the energy ε by the equation

$$n_\varepsilon = n_0 \exp(-\varepsilon/kT) \tag{1.10}$$

where n_0 is a constant, equal to the number of molecules in the lowest energy state. In molar terms, Eq. (1.10) may be written as

$$N_\varepsilon = N_0 \exp(-E/RT) \tag{1.11}$$

A simple derivation of this important distribution equation, used in later chapters, is given in Appendix A4.

[7] W_1 represents the state of no disorder.

As the temperature of any solid is increased, the thermal motion imparted to its components increases. The consequent changes in the structure and properties of the substance depend upon the balance between this thermal energy and the internal bonding energy. Increased thermal vibrations may bring about polymorphic transformations: melting, in KCN at 233 K and in NaCN at 371 K; sublimation, in I_2 at 458 K;[8] decomposition, in $Pb(NO_3)_2$ at 743 K. In each example the entropy change ΔS for the process is positive, indicating an increase in the degree of disorder in the substance arising from the transformation.

According to the *third law* of thermodynamics, the entropy of an infinite crystal of a pure substance is zero at 0 K, indicating a state of perfect order. This law does not preclude complete order at a higher temperature. Indeed, it is possible that it exists in a superconductor as, for example, in elemental molybdenum at 0.9 K. The ordering process near absolute zero may be more complex than that simply of progressive reduction in entropy by cooling based on the third law and may also involve magnetic interactions.

As the temperature of a solid is decreased, its atomic vibrations decrease in amplitude and rotational motion ceases until, finally at absolute zero, the species remains with only the zero-point energy of vibrational motion. The changes in atomic vibrations are accompanied by a decrease in heat capacity which, from the third law of thermodynamics, tends to zero as the temperature approaches 0 K. The loss of vibrational modes can be detected by a decrease in intensity of infrared spectra, and finally its absence.

Electrical resistivities of solids change markedly with changes in temperature: the resistivities of ionic solids tend to high values whereas those of metals tend to zero at 0 K. These effects will be discussed further after a study of ionic and metallic solids in the later chapters.

The effect of pressure on solids, though important, is often less dramatic than that of temperature. Generally, an increase in pressure on a solid produces a change in its physical properties akin to that occasioned by a decrease in temperature. On application of external pressure, a crystal may undergo a polymorphic transition that produces a closer packed structure. This process is usually accompanied by both a change in the coordination number, and a weakening in the intensity of infrared spectra occasioned by the decreased atomic vibrations. Cesium chloride, for example (Fig. 1.11), transforms to the sodium chloride structure type (Fig. 1.5) at 743 K but also at ambient temperature under a pressure of 10 GPa.

The variations in electrical resistivity differ considerably among solids within the four classifications chosen; Figs. 1.17 and 1.18 illustrate the changes in resistivity for cesium and selenium respectively as a function of applied pressure. The discontinuity in the curve for cesium occurs at a pressure of *ca.* 24 kbar (2.4 GPa) and corresponds to a structural change from the body-centred cubic structure, coordination number 8, to the face-centred cubic structure, coordination

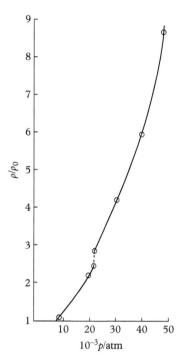

Fig. 1.17 Variation of relative electrical resistivity with pressure for elemental cesium at ambient pressure; ρ_0 is the resistivity at zero pressure. A discontinuity accompanies the polymorphic transition occurring at a pressure of 23.4×10^3 atm (23.7 kbar).

[8] At a partial pressure less than 90 mmHg (*ca.* 12,000 Pa).

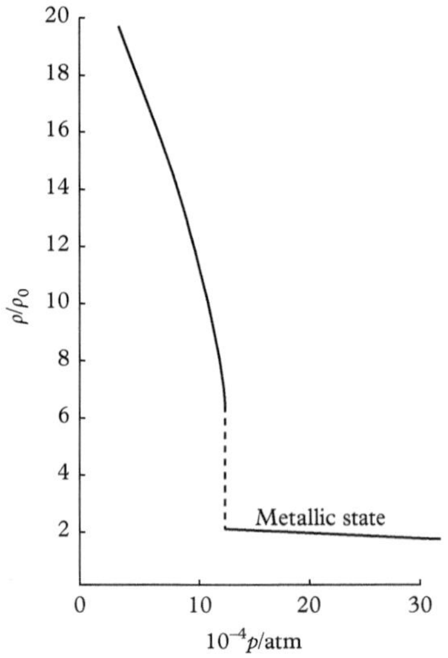

Fig. 1.18 Variation of relative electrical resistivity with pressure for elemental selenium at ambient temperature; ρ_0 is the resistivity at zero pressure. The discontinuity at *ca.* 120,000 atm initiates metallic characteristics for selenium.

number 12. In the case of selenium, a metallic structure develops at a pressure of *ca.*120 kbar.

1.6.4   Thermodynamic properties at transition points

Thermodynamic data on a selection of substances at their transition points are listed in Table 1.2. All ΔH_t values are positive: liquids are in a state of higher enthalpy (and energy) than are the corresponding solids at their melting points; gases are of higher enthalpy than the vapours of their liquids at the boiling points, as Table 1.2 shows.

Entropy changes at transition points are positive; the degree of randomness increases in the order solid → liquid → gas. With the exception of helium, enthalpies of fusion range from *ca.* 0.1 to 46 kJ mol^{-1}, and the corresponding entropy changes from *ca.* 6 to 44 JK^{-1} mol^{-1}. For the vaporization transitions, the enthalpy change ranges from *ca.* 1 to 300 kJ mol^{-1}, but the ΔS_t values cluster around 100 JK^{-1} mol^{-1} to within 20%, excluding hydrogen and helium.

Table 1.2 *Selected thermodynamic data at transition points*

	Mp/K	Bp/K	ΔH_t/kJ mol^{-1} Fusion	ΔH_t/kJ mol^{-1} Vaporization	ΔS_t/J K^{-1} mol^{-1} Fusion	ΔS_t/J K^{-1} mol^{-1} Vaporization[a]
NaCl	1074	1738	28.5	171	26.5	98
KCl	1043	1680	25.5	162	24.4	96
BeCl$_2$	678	793	12.6	105	18.6	132
MgCl$_2$	985	1691	43.1	137	43.9	81
H$_2$O	273	373	5.86	47.3	21.5	127
CH$_4$	91	112	0.96	9.20	10.5	82
Hg	234	630	2.43	64.9	10.4	103
Ge	1210	3103	34.7	285	28.7	92
Si	1683	2953	46.4	297	27.6	101
Na	371	1165	2.64	103	7.1	88
H$_2$	14.0	20.0	0.13	0.92	9.3	46
He	1.0[b]	4.2	0.021	0.084	6.3	21

[a] From Trouton's rule.
[b] At 26 atm.

The thermodynamic free energy change ΔG expresses the tendency for a reaction to take place spontaneously:

$$\Delta G = \Delta H - T\Delta S \qquad (1.12)$$

A negative value for ΔG in a given reaction $A \rightarrow B$ indicates a spontaneous reaction in the forward direction; a free energy change represents a compromise between the normally opposing enthalpic and entropic terms in the equation. The term 'spontaneous' in this context does not mean necessarily that the reaction is immediate; kinetic and catalytic factors may control the rates of reactions. For example, a mixture of hydrogen and oxygen confined in darkness will not react although the free energy change is $-237.1\,\text{kJ}\,\text{mol}^{-1}$ for the reaction:

$$H_2(g) + \tfrac{1}{2}O_2(g) \rightarrow H_2O(l)$$

However, in the presence of sunlight or other intense radiation the reaction takes place at an explosive rate.

In the liquid $\rightarrow$ gas transition, because of the much greater randomness of a gas compared to its liquid state, the entropy change of the reaction is dominated by the entropy of the gas. From the Avogadro law, gases have approximately equal molar volumes; hence, the entropy of vaporization tends to a constant value. This result is embodied in *Trouton's rule*, which states that the entropy of vaporization at the transition point, given by

$$\Delta H_t / T_{bp} = \Delta S_t \qquad (1.13)$$

is approximately $88\,\text{J}\,\text{K}^{-1}\,\text{mol}^{-1}$ for a range of different liquids. However, it is evident from Table 1.2 that Trouton's rule is only approximate; other factors, such as hydrogen bonding or dimerization, modify the Trouton value because they influence the degree of randomness in the liquid and vapour phases.

Hydrogen-bonded acetic acid dimer: Trouton constant $62.4\,\text{J}\,\text{K}^{-1}\,\text{mol}^{-1}$

1.7 Liquid crystals

Certain organic substances when heated pass into a state that is intermediate between those of solid and liquid. Physically, they flow like liquids, yet have some properties of crystalline solids. *Liquid crystals* were discovered in 1888 by the Austrian botanist Friedrich Reinitzer. If cholesteryl benzoate $C_{34}H_{50}O_2$ is heated, it melts sharply at 419 K forming a *cholesteric* type, opaque liquid crystal, and at 452 K there is a sudden change to an isotropic liquid.

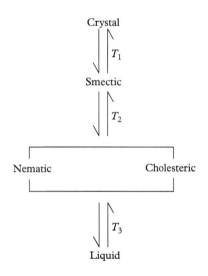

Cholesteryl benzoate

Liquid crystals may be considered as crystalline substances that have lost some or all of their positional order, while maintaining full orientational order. They consist usually of large elongated molecules, at least 1.3 nm in length, not severely branched or angular and possessed of polar groups such as $-NH_2$, $-OH$ or $>CO$ and low melting points. The best materials tend to be structurally rigid and anisotropic in shape, so that the more useful liquid crystals are frequently based on benzene derivatives.

In the crystalline state, the molecules of liquid crystals are aligned parallel one to the other and bonding takes place between the polar groups and also through van der Waals (London) forces of attraction. On heating, the weaker van der Waals forces are overcome first by the thermal energy supplied to the crystal, and relative movement of the molecules can then occur. Further heating breaks the dipolar linkages and the substance passes into the true liquid state.

Several phases (mesophases) of liquid crystal are recognized (Fig. 1.19), but not all liquid crystals necessarily exhibit each phase. As well as the cholesteric phase, there is the *nematic* phase, taking ammonium oleate $CH_3(CH_2)_7CH=CH(CH_2)_7CO_2^-NH_4^+$ as an example,

Ammonium oleate

Fig. 1.19 Liquid crystal mesophases; T_1, T_2 and T_3 are transition temperatures; not all liquid crystals exhibit all mesophases.

in which elongated molecules are arranged parallel one to the other but without periodicity, rather like an army of descending parachutists. The *smectic* phase, exemplified by *p*-azoxyanisole $C_{14}H_{14}N_2O_3$, has its molecules arranged on equally spaced planes but without lateral periodicity,

$$H_3CO \text{—} \bigcirc \text{—} N \overset{O^-}{\underset{N^+}{\text{=}}} \text{—} \bigcirc \text{—} OCH_3$$

***p*-Azoxyanisole**

rather like a crowd of shoppers in a department store. The transitions between mesophases are reversible and occur at definite temperatures which vary according to the Clapeyron equation (1.7).

- Liquid crystals have numerous applications in science, technology and engineering on account of both their thermochromic sensitivity and their reactions to electrical and magnetic fields. Applications of these materials, which are still being developed, have already provided effective solutions to many different problems.

- Liquid crystal displays (LCDs) find uses in watches, calculators and computer screens. An LCD consists of an array of tiny segments, or *pixels*, of a material that can be manipulated to present information to an observer or a detector device (LCDs are discussed further in a later chapter).

- Liquid crystal thermometers using nematic or cholesteric mesophases that reflect light with a colour that depends upon temperature find applications in thermometric devices over a wide range of temperatures.

- Optical imaging and recording technology make use of a liquid crystal between two layers of photoconducting material. On irradiation by light the conductivity of the material is increased, which develops an electric field in the crystal of a magnitude dependent on the light intensity. The electric pattern is then converted to an image which can be digitized and recorded.

- Other applications of liquid crystals include *inter alia* non-destructive stress analysis of materials, medical applications, computer-aided design and even the somewhat less scientific mood ring (Fig. 1.20), which changes colour in response to the body temperature and is alleged to indicate the emotional state of the wearer.

Fig. 1.20 Mood rings: a 'stone' is a glass or quartz shell containing a strip of thermochromic liquid crystal material. A change in temperature causes the crystalline material to deform and so bring about a change in structure which modifies the wavelength of light interacting with the crystal. [Reproduced by courtesy of Bestmoodrings.com]

A recent innovative application in the field of nematic liquid-crystal technology incorporates strongly-emitting inorganic cluster species into a nematic liquid, in particular, the $Cs_2Mo_6Br_{14}$ cluster combined with a polyanionic crown ether (CE) species. The Cs^+ cation and the CE species are brought together to form a nematic mesophase $(2CE_9 \cdot Cs)_2Mo_6Br_{14}$. It preserves the strong red–NIR luminescent properties of the functional inorganic cluster species and the resulting liquid crystal material has important applications in temperature sensors and liquid crystal displays [10].

1.8 Classification of solids

No scheme of classification of solids or structures is free from some degree of ambiguity, yet it is desirable that the vast body of available structural information be discussed over a framework that groups solids according to a few chosen parameters. The method selected herein is based on the type of bonding that is mainly responsible for cohesion in the solid state. Four classes evolve; they are introduced briefly in the following subsections and will be discussed in detail in the following chapters. The subject of nanoscience has been accorded a chapter outside this classification since the materials involved therein embrace all four classes, and belong to a rapidly developing subject of academic and technological importance.

1.8.1 Covalent solids

Covalent bonding involves an *electron sharing* mechanism by overlap of the atomic orbitals of atoms whereby a stable electron configuration is obtained. By virtue of the nature of atomic orbitals the covalent bond is strongly directional. A necessary repulsion energy, balancing the attraction derived from electron sharing, arises mainly from the inner electron shells. A pair of electrons shared by two atoms may remain localized to the orbitals of these atoms or they may be delocalized over the complete molecular entity.

The diamond form of carbon is an excellent example of a *covalent solid*. Crystals of organic molecules in which the atoms are bonded covalently are sometimes quoted incorrectly as covalent solids, but the bonding responsible for their cohesion derives from dipolar and van der Waals forces, and so they belong to the 'molecular' class.

1.8.2 Molecular solids

While van der Waals forces make a small contribution to bonding in most solids, they are solely responsible for the cohesion between electrically neutral species, such as the inert gases, methane, benzene and other non-polar compounds in the solid state, the *molecular solids*. The cohesive energy in these substances arises through dipolar or induced dipolar interactions. The close-packed cubic structure of krypton, typical of the inert gases in the solid state, is illustrated by Fig. 1.21. Its unit cell dimension a is 0.5648 nm at 4 K; since the krypton atoms are in close contact along the face diagonal of the cubic unit cell of length $a\sqrt{2}$, $r = a\sqrt{2}/4$, or 0.1997 nm, which is a measure of the van der Waals radius for krypton at 4 K.

1.8.2.1 Hydrogen-bonded solids

Hydrogen bonding is discussed conveniently within the molecular group of solids; it is an important interaction in the solid and liquid states. Bonded hydrogen

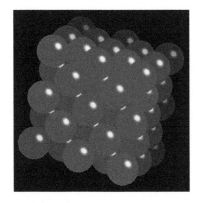

Fig. 1.21 Unit cell and environs of the crystal structure of krypton; the krypton atoms form a close-packed cubic structure which is representative of the solid inert gases and many elemental metals; eight unit cells are shown in the figure. [Reproduced from http://www.webelements.com/krypton/crystal_structure.html by courtesy of Professor Mark Winter, University of Sheffield, UK.]

Table 1.3 *Melting temperature/K of hydrides of periodic table groups[a] 15 (V A), 16 (VI A) and 17 (VII A)*

H_3N	195	H_2O	273	HF	190
H_3P	140	H_2S	190	HCl	159
H_3As	157	H_2Se	207	HBr	186
H_3Sb	185	H_2Te	224	HI	222

[a] Earlier group notation is given in parentheses.

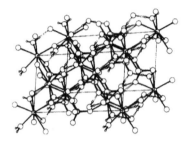 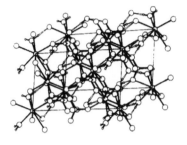

Fig. 1.22 Stereoview of the unit cell and environs of gypsum $CaSO_4.2H_2O$; circles in decreasing order of size represent the O, Ca, S and H species. The hydrogen bonds, shown by double lines, are responsible for cohesion along one direction in the crystal.

atoms are able to form additional links between two atoms, most strongly with fluorine, oxygen and nitrogen, which enhance the total bonding energy. Hydrogen bonding can be either *intramolecular*, between atoms of one and the same molecule, or *intermolecular*, between atoms in different, adjacent molecules in a condensed phase. Clear evidence of hydrogen bonding can be found among the hydrides of groups 15, 16 and 17 of the periodic table, as may be judged from the example of Table 1.3. Normally, melting temperatures increase with an increase in molecular mass; exceptions may be seen along the first row of the table. Hydrogen bonds increase the attractive energy in both ionic and molecular (van der Waals) compounds. In gypsum $CaSO_4.2H_2O$, for example, the cohesion in one direction is governed by hydrogen bonds (Fig. 1.22). As well as hydrogen-bonded compounds, it will be convenient to include clathrate compounds, charge transfer structures, π-electron overlap compounds as well as other structure-types in Chapter 3.

1.8.3 Ionic solids

Ionic bonding involves an *electron donor–acceptor* mechanism among the participating atoms. A structure is formed in which ions are attracted one to the other by forces that are predominantly coulombic in nature. They are balanced by repulsion forces that increase markedly with a decrease in the interionic distance below the equilibrium value for the solid. Potassium fluoride, which has the sodium chloride structure type (Fig. 1.5), is a good example of an ionically bonded solid.

1.8.4 Metallic solids

In a molecule such as benzene, the valence electrons are *delocalized*, or distributed, over the whole molecule. In metals this delocalization is extended to the *complete crystal*. A regular array of positive ions is bound by an all-pervading distribution of electron density, sometimes referred to as a 'sea' of electrons. Because of the mobility of the outermost, *conduction electrons*, metals have high electrical and thermal conductivities. Gold is a good example of a metallic solid and it has the same close-packed cubic structure as that shown by Fig. 1.21.

1.8.5 Comments on the classification of solids

A schematic illustration of the four types of bonding by which solids have been classified is shown by Fig. 1.23a–d. With the exception of the crystalline forms of the inert gases, solids cohere by an amalgam of bonding forces, frequently with a given compound exhibiting properties that are associated mainly with one type of idealized bonding.

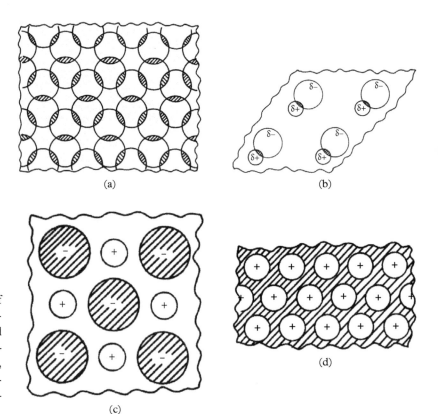

Fig. 1.23 Pictorial representations of the four main bonding forces: (a) co-valent, (b) molecular, (c) ionic and (d) metallic. [*Crystal Structures: Lattices and Solids in Stereoview*, 1999, Ellis Horwood Limited, UK; reproduced by courtesy of Woodhead Publishing, UK.]

Table 1.4 *Structural and physical properties of solids*

Bonding class	COVALENT		MOLECULAR	
Structure class	I[a]	II	I	II
Structural units	Atoms or groups bonded covalently in three dimensions.	As for structure class I	Atoms.	Molecules.
Close packed or nearly close packed	Diamond, Si, Ge; compounds of groups 14–16 elements among themselves. Generally, hard, high mp crystals; low thermal and electrical conductivity.	Compounds of groups 14–16 metals with P, As; NiAs; pyrite.	Inert gases. Low mp; poor thermal and electrical conductivity.	Molecular gases, e.g. N_2, O_2, HCl; S; organic compounds, e.g. methane, sucrose, phenol; low to moderately high mp.
Chain				Rubber, cellulose; fibrous proteins.
Layer				Graphite, graphene.
Framework	Quartz, cristobalite.			Globular proteins.

Bonding class	IONIC			METALLIC	
Structure class	I $EBS^b < \lvert z_X \rvert /2$	II $EBS = \lvert z_X \rvert /2$	III $EBS > \lvert z_X \rvert /2$	I	II
Structural units	Positive and negative ions or groups, and covalently-bonded groups carrying + or − charge.			Atoms.	Atoms.
Close packed or nearly close packed	Halides and oxides of metals: MX, MX_2 types; spinels; perovskites.	Borates, silicates, germanates.	Salts of inorganic oxy-acids.	True metals and their alloys.	Zn, Cd, Sn; alloys of more metallic group 14–16 elements one with the other.
	Hard, brittle crystals. Poor electrical and thermal conductivity in solid; conduct electricity in the melt.			Soft to hard crystals; low to very high mp; excellent thermal and electrical conductivity.	
Chain		Pyroxenes, amphiboes.		Na, Mg, Al	Se, Te, Sb_2S_3
Layer		Mica	Gypsum		As, Sb, Bi; MoS_2
Framework		Felspars, zeolites.		Interstitial compounds.	

[a] I = Class bond type predominates; II = Overlap with other areas of the classification.

[b] EBS (electrostatic bond strength) is the oxidation number of a species divided by the coordination number, and z_X is the change on an anionic species: NaCl, $EBS_{(Na-Cl)} = 1/6$; $\lvert z_x \rvert = \frac{1}{2}$.

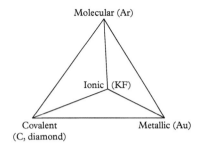

Molecular (Ar)

Ionic (KF)

Covalent
(C, diamond)

Metallic (Au)

Fig. 1.24 Schematic representation of bond type, with typical representatives in parentheses.

In general, it may be considered that a given bond type is disposed, conceptually, somewhere within a tetrahedron (Fig. 1.24), with each apex indicating a good example of its principal bond type. An actual bond can have the character of more than one of the four extreme types while remaining a unique chemical bond; Table 1.4 summarizes, with examples, the classification described. Fuller discussions appear in the following chapters of the book which will reveal more clearly the nature of interatomic bonding in the chosen classes.

....

REFERENCES 1

[1] Eliezer S and Eliezer Y. *The Fourth State of Matter: An Introduction to Plasma Science*, 2**nd** ed. Taylor and Francis, 2001.

[2] Fowler RH and Guggenheim EA. *Statistical Mechanics*, 2nd ed. Cambridge University Press, 1960.

[3] James RW. *The Crystalline State. Vol. 2: The Optical Principles of the Diffraction of X-rays from Crystals*. G. Bell and Sons Limited, 1948.

[4] Ladd M and Palmer R. *Structure Determination by X-ray Crystallography*, 5th ed. Springer Science+Business Media, 2013.

[5] Shechtman D *et al. Phys. Rev. Lett.* 1984; 53: 1951.

[6] Senechal M. *Quasicrystals and Geometry*. Cambridge University Press, 1996.

[7] Schrödinger E. *What Is Life*. Cambridge University Press, 1944.

[8] The DNA story. *Chem. World*, 2003; http://www.rsc.org/chemistryworld/Issues/2003/April/story.asp

[9] Atkins PW and de Paula J. *Atkins' Physical Chemistry*, 9th ed. Oxford University Press, 2009.

[10] Nayak SK *et al. Chem. Commun.* 2015; 51: 3774,

[11] Wyckoff RWG. *Crystal Structures*, Vol. I, 2nd ed. Wiley, 1963.

[12] Wyckoff RWG. *The Structure of Crystals*, Reinhold, 1935; on-line as *The Structure of Crystals*. https://archive.org/details/structureofcryst030914mbp

Problems 1

1.1. At 420 K, ammonium nitrate has been found to exhibit the cesium chloride structure type. What structural implication may be associated with this observation?

1.2. In an x-ray diffraction experiment, similar to that described in Section 1.4, but with ammonium nitrate at 420 K, the first two sharp diffraction rings had diameters of 22.1 mm and 32.7 mm respectively. The indices of these spectral rings were 100 and 110, and the specimen to film distance was

30.00 mm. If the x-ray wavelength was 0.15418 nm, calculate an average value for the unit cell dimension a.

1.3. Which of the following nine pairs of substances illustrate structural isomorphism?

$NaCl$	KCl	$BaSO_4$	$PbSeO_4$	CaF_2	$\beta - PbF_2$
$RbCl$	$CsCl$	$SrSO_4$	$CaSO_4$	$NaBr$	MgO
KNO_3	$CaCO_3$ (calcite)	CaF_2	MgF_2	CaO	BeO

Publications by R. W. G. Wyckoff may be helpful in this problem [11, 12].

1.4. The entropy of vaporization of $BeCl_2$ is 36 J K^{-1} mol^{-1} greater than that for KCl. What does this result suggest about the liquid state of $BeCl_2$?

1.5. The enthalpies of combustion of orthorhombic α-sulphur (the thermodynamically stable form) and monoclinic β-sulphur to form SO_2 at 298.15 K are -297.0 kJ mol^{-1} and -297.3 kJ mol^{-1} respectively. Draw an enthalpy-level diagram to illustrate the changes involved. Assuming that the enthalpies of α-sulphur and β-sulphur do not change between 298 K and 386 K, calculate the entropy of β-sulphur at the transition temperature of 386 K given that $S_\alpha = 31.73$ J K^{-1} mol^{-1}.

1.6. Classify the following 12 substances according to the principal type of bonding type responsible for cohesion in the solid state.

RbF	CO_2	C_6H_6	Cu_3Au	P_4	AlN
Ne	Na_2SO_4	P_2Cl_{10}	Pb	$KClO_3$	SiC

1.7. What are the Miller indices of crystal planes that make the following intercepts on the x, y and z crystallographic axes?

(i)	$a/2$	b	$-c/3$
(ii)	a	$\parallel$ to b	$2c/3$
(iii)	$-a/3$	$\parallel$ to b	$\parallel$ to c
(iv)	a	$-2b/3$	$3c/4$
(v)	$\parallel$ to a	$b/2$	$-3c/4$
(vi)	$\parallel$ to a	$-b$	$2c$

1.8. Consider the close-packed, cubic structure of aluminium (the same structure type as that in Fig. 1.21). Using the information in Section 1.4, what is the diameter of the first sharp diffraction ring on Fig. 1.9?

1.9. Iron transforms from a body-centred cubic structure to a face-centred cubic structure at $812\,°C$. If the atoms of iron behave as hard spheres of radius r, what is the percentage change in packing fraction accompanying the transition? (The packing fraction is the volume of an atom divided by the volume occupied by that atom in the structure.)

1.10. The unit cell dimension for krypton (Fig. 1.21) at 4 K is 0.5648 nm. What is its density at 4 K?

Covalent Compounds

2

Let there be brothers first, then there will be brotherhood, and only then will there be a fair sharing of goods among brothers.

Fyodor Dostoyevsky

2.1 Introduction

The classification of solids that forms the main substance of this book was introduced in the first chapter. That to be studied now in detail is concerned with the *covalent bond* between atoms. The term *covalence* was used first by Langmuir, writing in 1919 about the arrangement of electrons in atoms and molecules, but the idea of covalent bonding as the *sharing* of pairs of electrons was put forward by G. N. Lewis in 1916. The Lewis notation uses dots to indicate valence (outer shell) electrons between atoms. Two dots located between a pair of atom symbols indicate a covalent *single* bond between those atoms. Two or three pairs of dots similarly placed represent *double* or *triple* bonds respectively, all forming an outer, *valence octet* of electrons around the atoms of any bonded species that has sufficient electrons so to do. Later, the two dots representing a single bond were replaced by a single line between the two bonded atoms, two and three lines becoming the obvious extension for double and triple bonds, as the following diagrams of ethane, ethene (aka ethylene) and ethyne (aka acetylene) show:

$$
\begin{array}{ccc}
\text{H } \text{H} & \text{H } \text{H} & \\
\text{H:C:C:H} & \text{C::C} & \text{H:C:::C:H} \\
\text{H } \text{H} & \text{H } \text{H} &
\end{array}
$$

$$
\begin{array}{ccc}
\text{H} \quad \text{H} & \text{H} \quad \text{H} & \\
| \quad | & | \quad | & \\
\text{H–C–C–H} & \text{C==C} & \text{H–C}\equiv\text{C–H} \\
| \quad | & | \quad | & \\
\text{H} \quad \text{H} & \text{H} \quad \text{H} &
\end{array}
$$

| **Ethane** | **Ethene** | **Ethyne** |

Such diagrams are still of use today, but they provide no evidence of true molecular shape. For this and other properties, models based on mathematical arguments must be employed for the understanding of bonding in chemical species.

Bonding, Structure and Solid-State Chemistry. First Edition. Mark Ladd.
© Mark Ladd 2016. Published in 2016 by Oxford University Press.

The classical mechanical principles described by Newton in his publication of 1687 [1] allowed particles to be determined exactly in both position and momentum, and for their internal energy, translational, vibrational and rotational, to be excited to any value by the application of appropriate external stimuli, such as heat or radiation. While Newtonian mechanics remains applicable to matter in bulk, it fails to explain satisfactorily the properties of electrons and atoms. It will transpire also that a significant modification of the Maxwellian theory of light is required, in order to account fully for the findings in experiments with light radiation.

2.2   Black-body radiation

When objects are heated they emit radiation. As the temperature of the radiator is increased, the frequency of the emitted radiation increases from the infrared region of the electromagnetic spectrum to its ultraviolet region and beyond. This radiation is *black-body radiation* and its colour is a measure of the temperature of the radiator. An ideal black body may be defined as an object that is capable of absorbing and emitting radiation of all frequencies in a continuous and uniform manner. An approximation to an ideal black body is a pinhole in one wall of an empty container at a constant temperature. Radiation emitted from the hole is in thermal equilibrium with the walls of the container because it has been absorbed and re-emitted continuously within the container.

The energy distribution $E(v)$ of radiation as a function of its frequency v at several temperatures is shown by Fig. 2.1. Experiments on black-body radiation conducted by Wien in 1894 showed that

$$T \times \lambda_{max} = b \tag{2.1}$$

When Eq. (2.1) is written in terms of wavelength, as here, this equation is known as the *Wien displacement law*. The parameter b is a constant and T is the absolute temperature of the black body; v_{max}, equal to c/λ_{max}, is the maximum in the frequency of the radiation emitted from it at that temperature.

On the basis of the equipartition of energy (Appendix A5), Rayleigh proposed that radiation was emitted by molecular oscillators of all possible frequencies inside the radiator. He determined the number $N(v)$ of oscillators in a cubical enclosure of side length c/v, where c is the speed of light *in vacuo* and v the frequency of the radiation. His result was modified later by Jeans and is given now as:

$$N(v) = 8\pi v^2 / c^3 \tag{2.2}$$

From Appendix A5, the mean energy of a classical oscillator at temperature T is kT, where k is the Boltzmann constant. Hence, the energy distribution, or energy

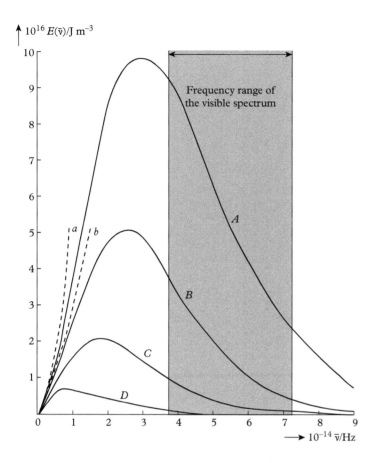

Fig. 2.1 Energy distribution $E(\nu)$ from a black body as a function of frequency ν and temperature. Rayleigh–Jeans equation (broken lines): a, 5000 K; b, 4000 K. Planck equation (full lines): A, 5000 K; B, 4000 K; C, 3000 K; D, 2000 K.

density, between the frequencies ν and $\nu + \mathrm{d}\nu$ is the Rayleigh–Jeans equation, given by

$$E(\nu)\mathrm{d}\nu = N(\nu)kT\mathrm{d}\nu = 8\pi\nu^2(kT/c^3)\,\mathrm{d}\nu \qquad (2.3)$$

or

$$E(\nu) = 8\pi\nu^2 kT/c^3 \qquad (2.4)$$

Evidently, this equation does not agree with Wien's observation given in Eq. (2.1). Furthermore, the Rayleigh–Jeans equation predicts a very large distribution of energy in the very high frequency region of the spectrum, leading to the so-called ultraviolet catastrophe. In fact, it would imply that even cool objects would radiate in the visible region of the electromagnetic spectrum and that objects would glow in the dark—if there was any darkness left.

2.3 Planck's quantum theory

In 1900, Planck proposed a major revision to the classical physics of radiation. He showed that the experimental results on black-body radiation were explicable if the energy of an electromagnetic oscillator of frequency ν were limited to multiples of a discrete amount, a *quantum of* energy $h\nu$, where h is the *Planck constant*. Thus:

$$E(\nu) = nh\nu \tag{2.5}$$

where n is an integer; the curves A to D may be compared with a and b, in Fig. 2.1.

The *Planck equation* [2] for black-body radiation is the Rayleigh–Jeans equation modified by replacing the classical mean energy kT of an oscillator by the quantum value $\frac{h\nu}{[\exp(h\nu/kT)-1]}$:

$$E(\nu) = \frac{8\pi h\nu^3}{c^3} \frac{1}{[\exp(h\nu/kT) - 1]} \tag{2.6}$$

Classical conditions correspond to no quantization, that is, to the limit of $E(\nu)$ as $h\nu/kT$ tends to zero. Expanding the exponential term:

$$E(\nu) = \frac{8\pi h\nu^3}{c^3} [1 + (h\nu/kT) + (h\nu/kT)^2/2! + (h\nu/kT)^3/3! + \ldots - 1]^{-1}$$

or

$$E(\nu) = \frac{8\pi h\nu^3}{c^3} \frac{kT}{h\nu} \frac{1}{[1 + (h\nu/kT)/2! + (h\nu/kT)^2/3! + \ldots]}$$

Hence, on cancelling h and taking the limit:

$$\lim_{(h\nu/kT)\to 0} E(\nu) = 8\pi \nu^2 kT/c^3 \tag{2.7}$$

which is the Rayleigh–Jeans equation. On the one hand, classical conditions are attained for $(h\nu/kT) << 1$, that is, at low frequencies and high temperatures, typically $\nu \approx 10^{11}$ Hz and $T \approx 3000$ K, while on the other hand, at high frequencies and low temperatures, $(h\nu/kT) >> 1$, and Eq. (2.6) tends to:

$$E(\nu) = (8\pi \nu^3 kT/c^3) \exp(-h\nu/kT) \tag{2.8}$$

Differentiating $E(\nu)$ with respect to ν and setting the derivative to zero for a maximum (Fig. 2.1):

$$T/\nu_{max} = h/(3k) \tag{2.9}$$

Thus, Wien's observations reflect quantum behaviour, whereas the Rayleigh–Jeans equation applies over a small frequency range under classical conditions. The value of the constant b in Eq. (2.1), the Wien displacement law constant b, was determined experimentally as 4.8×10^{-3} m K, and since Planck had determined a value for k during his work on black-body radiation as 1.346×10^{-23} J K^{-1}, it followed that h was evaluated as 6.46×10^{-34} J s, an early value for the Planck constant.

A mechanical oscillator in the form of a simple pendulum of length 248.4 mm would have a frequency of 1 Hz. Thus, the separation of its energy levels would be of the order of h, or *ca.* 6.6×10^{-34} J, which would present a *continuum* with respect to experimental methods for their detection. For optical frequencies, lying in the region of 5×10^{14} Hz, an energy quantum is 3.3×10^{-19} J, or 199 kJ mol^{-1}, a significant and measurable quantity.

At any temperature T, an oscillator can be excited by absorption of one quantum of energy $h\nu$. In classical terms, when the mean thermal energy is kT, oscillators would be excited continuously and even the oscillators of highest frequency would contribute to black-body radiation. One of the successes of the Planck theory is the provision of the necessary damping of the high frequencies through the exponential term in Eq. (2.6), so that an ultraviolet catastrophe cannot occur.

2.4 Heat capacity

Heat capacity raised another problem for the classical physical theory. The *Dulong and Petit law*, based on experimental studies, stated that the molar heat capacity of all monatomic solids was *ca.* 25 J K^{-1} mol^{-1}. An atomic vibration can be resolved into components of vibration in three mutually perpendicular directions, as indicated by the dotted lines x, y and z in the diagram below. Each such vibration constitutes a *degree of freedom*, represented here by a quadratic term, that is, a term involving a square of velocity or momentum.

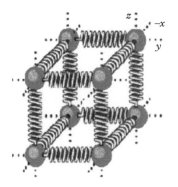

Atomic vibration directions x, y and z

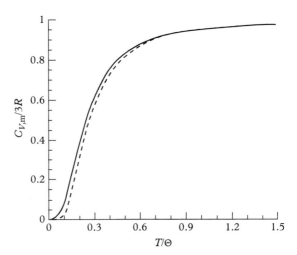

Fig. 2.2 Constant volume molar heat capacity $C_{V,m}$ normalized by $1/(3R)$ to a maximum of unity as a function of temperature T (normalized by the Debye temperature Θ): solid line, Debye theory; broken line, Einstein theory. The Debye temperature is that temperature of a crystalline solid in its highest normal mode of vibration ($\Theta = h\nu_{max}/k$). The Dulong and Petit limit (unity on this scale) is reached at the higher temperatures.

According to the *equipartition theorem*, discussed in Appendix A5, each vibrational quadratic term contributes kT to the energy of the system, or $3kT$ in total. Hence, the molar internal energy U is $3LkT$, or $3RT$. The molar heat capacity at constant volume is $(\partial U/\partial T)_V$, which is $3R$ or 24.94 J K^{-1} mol^{-1}. The Dulong and Petit law is obeyed well above the value $T/\Theta \approx 0.7$ but then begins to decrease rapidly with a decrease in temperature as T approaches absolute zero (Fig. 2.2). The quantum approach to heat capacity will be discussed further in a later chapter.

2.5   Wave–particle duality

It is necessary to consider next the experimental evidence that indicates a dual behaviour of both radiation and matter, according to the given experimental conditions.

2.5.1   Photoelectric effect

The diagram in Fig. 2.3 indicates how the photoelectric effect may be detected and measured. A monochromatic light source incident upon the metal photocathode *in vacuo* results in the ejection of electrons from the metal, provided that the applied potential V is at least a certain minimum value V_0, which depends upon the nature of the metal. Several conditions were found to exist, of which the following are the more significant:

- No electrons were emitted unless the frequency of the incident light exceeded a certain minimum value.

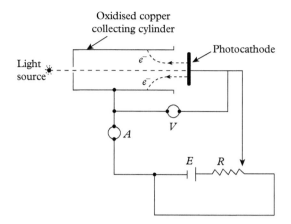

Fig. 2.3 Schematic circuit diagram for observing and measuring a photo-electric effect; the photocathode and collecting cylinder are *in vacuo*: *V*, voltmeter; *A*, galvanometer; *E*, source of electromotive force; *R*, variable resistor. The potential difference between the cathode and cylinder is arranged for a positive potential at the cylinder. Electrons are expelled from the photocathode and travel to the cylinder; the photocurrent is measured by the galvanometer.

- The mean kinetic energy of the electrons emitted was proportional to the frequency of the incident light and independent of its intensity.
- At low light intensity, electrons were still emitted once the light frequency was sufficiently great.

If the light source comprises photons, that is, quanta each of energy $h\nu$, then by conservation of energy the kinetic energy of the electrons emitted would be given by:[1]

$$\tfrac{1}{2}mv^2 = h\nu - \phi_M \tag{2.10}$$

Once the incident radiant energy $h\nu$ transferred to an electron in the metal cathode exceeds ϕ_M, an electron is ejected from the metal with a kinetic energy equal to the excess of $h\nu$ over ϕ_M. The parameter ϕ_M is the *work function* of the species M, and is characteristic of it, and the three observations listed above are fully explained through Eq. (2.10).

This theory has implications for the interpretation of the nature of light. Maxwellian theory propounds its *wave nature*, which explains optical phenomena such as refraction and diffraction, whereas the photoelectric effect confers a *particulate nature* on it. If this be the case, then light particles, and by implication photons, should have a momentum p given by:

$$p = mc \tag{2.11}$$

where m is the mass associated with the photon. Using the Einstein equation that relates mass and energy, namely,

$$E = mc^2 \tag{2.12}$$

---

[1] Be careful to distinguish between v (velocity) and ν (frequency).

together with Eq. (2.5), taking n equal to 1:

$$p = E/c = h\nu/c(= h/\lambda) \qquad (2.13)$$

Example 2.1

One mole of water is raised in temperature from 293 K to 373 K; the heat energy needed is $0.01801 \, \text{kg} \times 4186 \, \text{J kg}^{-1} \, \text{K}^{-1} \times 80 \, \text{K} = 6031.2 \, \text{J}$. If the heating equipment were an infrared lamp of frequency 3×10^{14} Hz, the number of quanta needed would be $6031.2 \, \text{J}/(6.6261 \times 10^{-34} \, \text{J s} \times 3 \times 10^{14} \, \text{Hz}) \approx 3 \times 10^{22}$; specifically, 3.034×10^{22}, or 0.05 mol of quanta of photons.

2.5.2 De Broglie equation

An electron diffraction experiment carried out by Davisson and Germer in 1925 on a crystal of metallic nickel showed that the electrons were diffracted by the material, a process that is a property of waves. Earlier, experiments by Thomson on cathode rays (electrons) had demonstrated their particulate nature.

The two aspects of electrons and, indeed, of radiation were coordinated by the *de Broglie equation*, which was proposed in 1924:

$$p = h/\lambda \qquad (2.14)$$

The particulate property of momentum is linked inversely to the wavelength property of radiation by the Planck constant. Matter in bulk has such high momentum that its wavelength is far too small to be detected by experimental means: a mass of 1 kg moving at 1 m s^{-1} has an associated wavelength of *ca.* 6.6×10^{-34} m.

Example 2.2

The de Broglie equation enables the wavelength of accelerated electrons to be determined. The kinetic energy E_K of an electron is $\frac{1}{2}m_e v^2$ or $p^2/2m_e$, and at the end of the acceleration is wholly potential energy. If the accelerating potential difference is 150 V, then the electron wavelength is given by:

$$\lambda = 6.6261 \times 10^{-34} \, \text{J s}/(2 \times 9.1094 \times 10^{-31} \, \text{kg} \times 1.6022 \times 10^{-19} \, \text{C}$$
$$\times 150 \, \text{V})^{1/2} = 1.001 \times 10^{-10} \, \text{m, or } ca. \, 0.1 \, \text{nm}$$

(Note: $\text{J} \equiv \text{kg m}^2 \, \text{s}^{-2} \equiv \text{C V}$)

2.5.3   Electron microscopy

The electron wavelength determined in Example 2.2 is of the order of interatomic distance. Visible light used with an optical microscope can resolve, at best, objects at a separation of *ca.* 250 nm, which is approximately 10^3 times greater than interatomic spacings. However, electron beams because of their small de Broglie wavelength can achieve nanometre-size resolution. In the technique of *transmission electron microscopy*, a beam of electrons is passed through a very thin specimen. The electrons interact with it, and the transmitted beams form an image which is magnified and recorded on a sensitive film or other sensor such as a CCD collector.[2] The resolution is then of the order of 1000 times better than that of the optical microscopy under its best conditions. The photograph in Fig. 1.10, which revealed the existence of fivefold symmetry in crystalline material, was obtained by this technique. The recently developed technique of *super-resolution light microscopy*, work whose authors were awarded the 2014 Nobel Prize in Chemistry, has reached atomic resolution and has enabled the study of molecular processes in real time.

2.6   Atomic spectra

The Balmer series in the spectrum of atomic hydrogen is shown in Fig. 2.4, from which it is evident that energy is emitted at *discrete* frequencies. This result would not occur if energy were absorbed or emitted as a continuum. If the energy of an atom has discrete values E_1 and E_2, then a transition between these levels would be accompanied by an energy change ΔE, given by $(E_2 - E_1)$, which, from Eq. (2.5), has the discrete value $nh\Delta\nu$. The Balmer spectral 'frequencies' fit the equation:

$$\bar{\nu} = R_H(1/n_1^2 - 1/n_2^2) \tag{2.15}$$

where R_H is the *Rydberg constant* for hydrogen and is known with high precision from spectroscopic measurements as 1.0967758×10^7 m^{-1}; in this spectral series $n_1 = 2$ and n_2 is another integer, greater than 2.

[2] CCD: charge-coupled device [8]

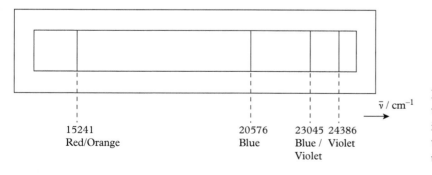

Fig. 2.4 Balmer series in the spectrum of atomic hydrogen corresponding to discrete energy levels; the spectral frequencies are listed in cm^{-1}, as is usual in spectroscopy; $\bar{\nu} = 1/\lambda = \nu/c$.

The wavenumber of any spectral line is proportional to the difference between two terms, as in Eq. (2.15), which is a statement of the *Ritz combination principle* (1908). The *Bohr theory* of the atom explains the spectra of atomic hydrogen in terms of transitions between two energy levels: an electron moving from a level E_2 to a level of lower energy E_1 emitted radiation of frequency v given by:

$$v = (E_2 - E_1)/h = c\bar{v} \tag{2.16}$$

The energy levels were predicted correctly as:

$$E_n = -\mu e^4/(8h^2\varepsilon_0^2 n^2) \; [= -13.6\,\text{eV}/n^2] \tag{2.17}$$

where μ is here the *reduced mass*[3] of the system of one electron and one proton (Appendix A8) and ε_0 is the electric constant. Evaluating $[\mu e^4/(8h^2\varepsilon_0^2)]/hc$ gives 1.0967757×10^7 m^{-1} for the value of R_H. The Rydberg constant itself R_∞ is given by $(m_\text{e}/\mu_\text{H})R_\text{H}$, that is, by the replacement of μ by m_e as in Eq. (2.17), leading to $1.0973731.6 \times 10^7$ m^{-1}.

The Bohr theory was based on classical mechanics, but with quantum conditions imposed in order to fit the experimental results. It was a patchwork theory, assuming that electrons moved around the nucleus like planets in the solar system, and it could not explain atomic spectra other than that of atomic hydrogen.

Two further, fundamental objections to the Bohr theory are the need to define both the position and momentum of an electron, and to follow its orbit. The uncertainty principle, discussed in the next section, shows that these two requirements cannot be achieved simultaneously. That it can be done with planets and not with electrons depends on the very different sizes of these bodies in relation to the methods used for their observation.

2.7   Heisenberg's uncertainty principle

It can be shown by the following experiment that the simultaneous determination of both position and momentum of an electron is feasible only within a certain limit.

Let a beam of electrons pass through a narrow slit and the resulting pattern be recorded on a photographic film or other detector (Fig. 2.5). If the width of the slit is δx then every electron that is recorded must have passed through the slit, and so has a positional uncertainty δx in the direction normal to the incident beam. If the slit width δx is decreased sufficiently, a diffraction pattern is obtained, which may be interpreted to mean that the electron receives a certain momentum δp_x as it passes through the slit.

If the momentum of an electron is p and its angle of scatter α, the component of momentum along the direction of δx is:

[3] $\mu = 9.10442 \times 10^{-31}$ kg.

$$\delta p_x = p \sin \alpha \tag{2.18}$$

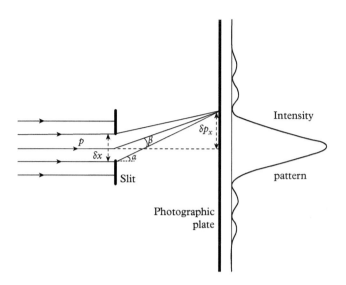

Fig. 2.5 Experimental arrangement for the diffraction of electrons at a slit of width δx; α is the angle of scatter.

The angular width of the diffraction pattern is given by:

$$\delta x = \lambda / \sin \beta \qquad (2.19)$$

Using the de Broglie equation (2.14):

$$\delta p_x \delta x = p\lambda \sin \alpha / \sin \beta = h \sin \alpha / \sin \beta \qquad (2.20)$$

Since $\alpha \approx \beta$, it follows that:

$$\delta p_x \delta x \approx h \qquad (2.21)$$

which is one representation of the *Heisenberg uncertainty principle*, expressing a limit to the simultaneous exact knowledge of both momentum and position of an electron. A more precise statement of the uncertainty principle is $\delta p_x \delta x \geq \hbar/2$, where $\hbar = h/(2\pi) = 1.054572 \times 10^{-34}$ J s.

To use an analogy in this context, the fact that the location of London (UK) cannot be defined to better than a region of several miles is not a feature of the precision of the surveying instrument used, but rather the fact that the city does not have a precise location.

Example 2.3

If the speed of an object of mass 0.5 g has an uncertainty in speed δv of 2×10^{-6} m s^{-1}, the uncertainty δp in its momentum is given by $m\delta v$, or 0.5×10^{-3} kg $\times$ 2×10^{-6} m s^{-1}. Hence, the uncertainty δx in its position is at least 6.6261×10^{-34} J s$/(0.5 \times 10^{-3}$ kg $\times 2 \times 10^{-6}$ m s$^{-1}) \approx 6.6 \times 10^{-25}$ m.

If, then, the required success in explaining the properties of the electron is not to be obtained by following descriptions of its particulate behaviour, it is reasonable to consider next theories based on its wave nature.

2.8   Wave mechanics of particles

The *wave mechanical equation* for a particle (or electron) cannot be derived, but may be presented through certain basic ideas and justified *a posteriori* by the success of results based on it. The relevant equation is the *Schrödinger equation* which, although a postulate, can be seen to be plausible by the following argument.

The classical wave equation has special solutions known as standing (time-independent) waves which have positions of zero amplitudes termed nodes. In the case of a string stretched between two fixed points, these standing waves are the fundamental and overtone vibrations of the string and can be labelled $0, 1, 2, \ldots,$ depending on the number of nodes in the string between its fixed ends. The integer term n in Eq. (2.17) arises from the standing-wave solution of a wave equation for the electron, as in the hydrogen atom.

A harmonic standing wave ψ in one dimension x can be expressed as:

$$\psi = A \sin 2\pi (x/\lambda - vt) \sin 2\pi vt \tag{2.22}$$

where A is amplitude, t is time and other symbols have meanings as before. Differentiating Eq. (2.22) twice with respect to x:

$$d^2\psi/dx^2 = -(4\pi^2/\lambda^2)\psi \tag{2.23}$$

which, using the de Broglie equation, becomes:

$$d^2\psi/dx^2 = -(4\pi^2 p^2/h^2)\psi \tag{2.24}$$

The classical law of conservation of energy is:

$$E = T + V = \tfrac{1}{2}mv^2 + V = p^2/(2m) + V \tag{2.25}$$

where T and V are, respectively, the kinetic and potential energy components of the total energy E.

Substituting for p^2 in Eq. (2.24) and rearranging:

$$-\frac{h^2}{2m_e}d^2\psi/dx^2 - (E - V)\psi = 0 \tag{2.26}$$

which is a one-dimensional analogue of the *time-independent* Schrödinger equation.

In three dimensions, the corresponding wave equation for an electron is given by Schrödinger as:

$$-\frac{\hbar^2}{2m_e}\nabla^2\psi + V\psi = E\psi \tag{2.27}$$

where ∇^2 is the *Laplacian operator* (aka '*del squared*'; aka '*nabla squared*'), $\partial^2/\partial x^2 + \partial^2/\partial y^2 + \partial^2/\partial z^2$. More concisely, Eq. (2.27) may be written as:

$$\mathcal{H}\psi = E\psi \tag{2.28}$$

where $\mathcal{H}$ is the time-independent *Hamiltonian operator*: $\mathcal{H} = -(\hbar^2/2m)\nabla^2 + V$. Thus, the equation of motion (2.25) has been converted into the corresponding wave equation by the replacement:

$$\mathbf{p} \rightarrow -i\hbar\nabla \tag{2.29}$$

2.9 Born's interpretation of the wavefunction

The *Born interpretation* of the wavefunction for a particle focuses on its location and states that the product $|\psi|^2\,dx$, or $|\psi\psi^*|\,dx$ if the wavefunction is complex, represents the probability of finding a particle in a one-dimensional system within the region x and $x + dx$ from a given origin. The wavefunction for an electron has the same probability interpretation and in three dimensions the function $|\psi|^2\,d\tau$ represents the probability of finding the electron, or electron density, in the infinitesimal volume element[4] $d\tau$. In general, $|\psi|^2$ is real and non-negative for electrons, and is the physically significant feature of the wavefunction. The wavefunction ψ itself may be positive, negative or complex, which is of significance in atomic bonding situations.

Example 2.4

The Born interpretation can be examined by calculating the probability P of an electron in a region of volume 1 pm^3 at given distances r from the nucleus of an atom. Let the wavefunction be proportional to $\exp(-r/a_0)$, where a_0 is the Bohr radius. The probabilities $P(r)$ are listed below for the constant volume element of 1 pm^3 and with r in units of a_0:

r	0	$a_0/2$	a_0	$2a_0$	$5a_0$
$P(r)$	1	0.37	0.14	1.8×10^{-2}	4.5×10^{-5}

continued

---

[4] $d\tau$ is the volume element $r\sin\theta\,dr\,d\theta\,d\phi$, equivalent to $dx\,dy\,dz$ in Cartesian space (see Appendix A6).

Example 2.4 *continued*

It is clear that the probability falls off rapidly with increasing distance from the nucleus, as expected from the nature of the wavefunction. Acceptable wavefunctions must be continuous, single-valued, continuous in the first derivative and quadratically integrable which is to say that $-\infty < \int_{-\infty}^{\infty} |\psi(r)^2| \, dr < +\infty$.

2.9.1 Normalization

For any solution $\psi(r)$ of a wavefunction, $C\psi(r)$ is also a solution where C is a constant, as can be shown by differentiating the function twice with respect to the variable r. Hence, any wavefunction can be normalized as follows, since the integral of the probability over all space must be unity. If N is the *normalization constant* for a wavefunction ψ, then the equation of normalization is:

$$N^2 \int_{-\infty}^{\infty} |\psi|^2 d\tau = 1 \tag{2.30}$$

or using $|\psi\psi^*|$ if the wavefunction is complex. Transforming to spherical polar coordinates according to Appendix A6 and adopting the wavefunction in Example 2.4, Eq. (2.30) becomes:

$$N^2 \int_0^{\infty} r^2 \exp(-2r/a_0) dr \int_0^{\pi} \sin\theta d\theta \int_0^{2\pi} d\phi = 1 \tag{2.31}$$

The integrals over θ and ϕ may be shown readily to be 2 and 2π, respectively; the integral over r may be solved along the lines of Example 7.1 in Appendix A7. Making the substitution $t = 2r/a_0$ leads to the value of $(a_0^3/8)\Gamma(3)$, or $a_0^3/4$, for the integral. Thus, $4\pi N^2 \frac{a_0^3}{4} = 1$, so that $N = 1/\sqrt{\pi a_0^3}$ and $\psi = \frac{1}{\sqrt{\pi a_0^3}} \exp(-r/a_0)$.

It will transpire shortly that this wavefunction represents the lowest energy state for the hydrogen atom; it would allow the results for $P(r)$ in Example 2.4 to be calculated exactly.

In addition to the Born interpretation, the wavefunction may be neither infinite nor multivalued: the first of these conditions would lead to Eq. (2.30) being infinite, so that $N = 0$, whereas the second condition would imply more than one probability value for a given set of conditions.

2.9.2 Orthogonality and orthonormality

Two coordinate axes or two functions are termed *orthogonal* when there exists no component of one on the other. Thus, mutually perpendicular x, y and z axes are orthogonal. Normalized functions have equal weight, frequently unity. The term *orthonormal* combines the terms orthogonal and normal, as the next example shows.

Example 2.5

In order to determine whether or not the following two functions $f_1 = \cos \psi$ and $f_2 = \sin \psi$ are orthonormal over the range $\pm \pi$, the normalization condition is first set up with N as the normalizing constant:

$$N_1^2 \int_{-\pi}^{\pi} \cos^2 \psi \, d\psi = N_1^2 \int_{-\pi}^{\pi} \tfrac{1}{2}(1 + \cos 2\psi) \, d\psi = N_1^2 \pi = 1$$

from which it follows that $N_1 = 1/\sqrt{\pi}$. A similar result in N_2 is obtained for f_2. Furthermore, for the product of f_1 and f_2:

$$N_1 N_2 \int_{-\pi}^{\pi} \cos \psi \sin \psi \, d\psi = N_1 N_2 \int_{-\pi}^{\pi} \tfrac{1}{2} \sin 2\psi \, d\psi = 0$$

Thus, the functions f_1 and f_2 are orthogonal and become orthonormal if each is multiplied by $1/\sqrt{\pi}$.

2.10 Particle-in-a-box: quantization of translational energy

Let a particle of mass m be constrained to linear motion in a one-dimensional box of length a. The path of the particle is terminated at both ends by potential barriers and these walls of the box are of infinite height (Fig. 2.6). Then, the potential energy V of the system is zero, which leads to the boundary conditions for acceptable wavefunctions:

- ψ is zero for $0 \leq x \leq a$;
- ψ is infinite for $0 > x > a$.

The wave equation is Eq. (2.26) with $V = 0$:

$$d^2\psi/dx^2 + (8\pi^2 m/h^2)E\psi = 0 \tag{2.32}$$

which is of the form

$$d^2y/dx^2 + \kappa^2 y = 0 \tag{2.33}$$

and its solution, from Appendix 9, is:

$$\psi = A \exp(i\kappa x) + B \exp(-i\kappa x) \tag{2.34}$$

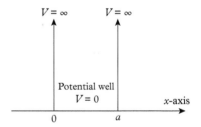

Fig. 2.6 Electron in a one-dimensional box. Potential well of infinite height; $V = 0$ within the walls.

where A and B are constants and $\kappa = (2mE/\hbar^2)$. Applying Euler's theorem[5] and summing:

$$\psi = (A + B)\cos\kappa x + i(A - B)\sin\kappa x$$

Absorbing the numerical factors into new constants C and D:

$$\psi = C\cos(\kappa x) + D\sin(\kappa x) \tag{2.35}$$

The boundary conditions given are satisfied if $C = 0$ and $\kappa = n\pi/a$; then the wavefunction takes the form

$$\psi_n = D\sin(n\pi x/a) \tag{2.36}$$

so that the allowed energies are given through Eq. (2.32) by

$$E_n = n^2 h^2/8ma^2 \tag{2.37}$$

Equation (2.37) introduces the quantization of energy through the integral *quantum number n*. It follows that the lighter the mass m of the particle and the closer together the walls become, the greater is the separation of successive energy levels. This fact can be seen by writing an energy level difference ΔE for successive values of *n*:

$$\Delta E = E_{n+1} - E_n = (2n + 1)h^2/(8ma^2) \tag{2.38}$$

which makes clear the dependence of ΔE on both m and a.

The probability of finding the particle between x and $x + \mathrm{d}x$ within the region 0 to a is unity. Hence, from Section 2.9.2:

$$\int_0^a D^2\sin^2(n\pi x/a)\mathrm{d}x = 1 \tag{2.39}$$

whence the normalization constant D is given by

$$D = (2/a)^{1/2} \tag{2.40}$$

The wavefunctions representing *stationary states*[6] are known as *eigenfunctions* (Ger. *eigen* = particular, characteristic) and the corresponding energies as *eigenvalues* (aka eigenenergies). In wave mechanical terminology, and referring to Eq. (2.28), $\mathcal{H}$ is an operator, the function ψ an eigenfunction, and the constant E an eigenvalue. The solutions of Eq. (2.36) for $n = 1$ to $n = 5$ are shown in Fig. 2.7; the number of *nodes*[7] in a wavefunction ψ_n is $n - 1$. The forms of the amplitudes are similar to those of the fundamental ($n = 1$) and first four overtone ($n = 2$ to 5) vibrations of a stretched string.

---

[5] $\exp(\pm i\theta) = \cos\theta \pm i\sin\theta$.
[6] States with probability functions $|\psi|^2$ that are time-independent.
[7] Regions of zero amplitude *between* 0 and a.

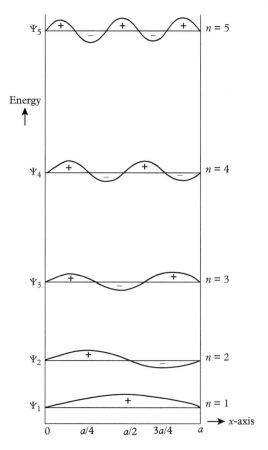

Fig. 2.7 Eigenfunctions of the first five solutions of an electron in a one-dimensional box; the number of nodes is $n - 1$; the positions $\psi_n = 0$ at $x = 0$ and a do not constitute nodes (see also Section 2.10, program BOXS).

The lowest energy state corresponds to wavefunction ψ_1, and has an energy E_1 that is the *zero-point energy* of the system; it is kinetic energy since $V = 0$. Even in the lowest energy state, the particle is in motion, a property that is entirely wave mechanical, because if h were zero the minimum energy would be zero. The existence of zero-point energy is in accord with the uncertainty principle. The particle must possess kinetic energy, being confined to motion within the region 0 to a. Since ψ is non-zero except at the walls, the particle must follow a curved trajectory, which itself implies kinetic energy; the second differential of $\psi \propto (E - V) = T$. Also, if the kinetic energy of the particle were zero, its momentum would be known exactly, zero; thus, $\delta p_x \delta x$ also would be zero, again flouting the uncertainty principle.

From Eq. (2.38), the value of ΔE becomes greater as a is made smaller, but the limit as a tends to zero would have no physical meaning as there would then be no box.

As a is made larger, the energy difference $E_{n+1} - E_n$ becomes smaller, and in the limit as a tends to infinity all values of E are allowed and the particle is free. A completely free particle has unquantized translational energy, for which reason atoms and molecules involved in laboratory experiments behave as though of unquantized energy.

The program BOXS in the Web Program Suite allows the calculation of wave-functions and energies for the particle-in-a-box procedure with $n = 1–15$. The program produces also a file of $\psi\ v.\ X/a$ data for each value of n, which can be plotted separately and in summation (see Section 7.2.2).

The state ψ of the particle may be described as a superposition of eigenfunctions given by the harmonic function $\psi\ =\ \sum_{n} c_n \exp(ikx)$ or the sum of its sine and cosine components. Interference occurs, and a wave packet results that progresses in time with a centre of gravity corresponding to the centre of mass of the equivalent particle (see also Appendix 21, particularly Figs. A21.1 and A21.2). The particle is then said to be localized, within the confines of the Heisenberg uncertainty principle; superposition of wavefunctions can be carried out by *Fourier transformation* [3], which involves a sum of sine and cosine terms. In Fig. 2.8a wave packets are illustrated, formed from two significantly differing numbers n of waves. Where the particles are electrons, which will be the concern in this book, an energy continuum can be observed in an atomic spectrum (Fig. 2.8b) when the electrons are free at a sufficiently high energy.

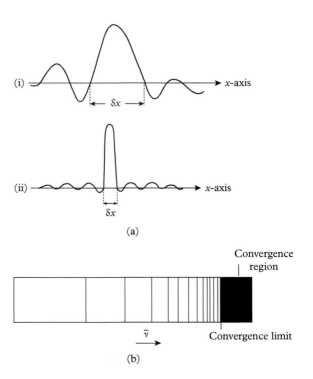

Fig. 2.8 (a) Wave packets formed by the superposition of one-dimensional eigenfunctions: (i) a small number of wavefunctions; (ii) a large number of wavefunctions (see also Appendix A21). The most probable location of an electron is within the region δx in each case. Comparable situations are encountered with the location of atoms in x-ray electron density syntheses. (b) Free electrons produce an energy continuum. The schematic diagram represents an absorption spectrum of sodium; the convergence limit represents the energy level as the quantum number n tends to infinity, and a continuum is then developed in the spectrum.

Example 2.6

The system of alternating single and double bonds in β-carotene $C_{40}H_{56}$ can be approximated to a one-dimensional box of twenty-two carbon atoms in length (not including terminal CH_3 groups). Each C–C bond length is *ca.* 0.140 nm, and each carbon atom contributes one *p* electron to π-bonding in the molecule. In its lowest energy state, these electrons will occupy the first eleven energy levels in pairs. The energy required to promote an electron from the ground state to the next level above it may be calculated from Eq. (2.38).

β-Carotene molecule

Thus, $\Delta E_{11 \to 12} = 23 \times (6.6261 \times 10^{-34}\,\mathrm{J\,s})^2/(8 \times 9.1094 \times 10^{-31}\,\mathrm{kg} \times (21 \times 0.14 \times 10^{-9}\,\mathrm{m})^2 \approx (1/7) \times 10^{-18}\,\mathrm{J}$, or specifically $1.603 \times 10^{-19}\,\mathrm{J}$. When the promoted electron falls back to the ground state level, the frequency v of the radiation emitted is, from Eq. (2.16), $\Delta E_{12 \to 11}/h$, so that $v = 2.419 \times 10^{14}\,\mathrm{Hz}$ ($\lambda = 1239\,\mathrm{nm}$). Experimentally, $v = 6.030 \times 10^{14}\,\mathrm{Hz}$, so that $\lambda = 497$ nm; the result highlights the approximate nature of the calculation with this large number of double bonds. So, what colour are carrots?

2.10.1 Tunnelling

If a tennis ball be locked in a well constructed metal safe, then Newtonian mechanics predicts that the probability of the ball being outside the box would be zero. Not surprisingly, the apparently similar electron-in-a-box problem when treated by wave mechanics reveals a somewhat different picture. As long as the potential rises to infinity at the barrier E is less than V in Eq. (2.26), and the solution of the wave equation is given by Eq. (2.36).

However, if the potential barrier in the box is of a finite height and additionally E is less than V, then the wavefunction does not decrease abruptly to zero at the barrier. Instead, it continues to oscillate within the barrier walls, decreasing exponentially throughout the wall thickness then continuing to oscillate beyond the barrier itself. Thus, the electron may be found outside the box, contrary to

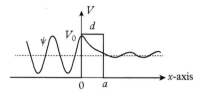

Fig. 2.9 Tunnelling: electron in a one-dimensional box; the potential well is of finite height V_0 and the wall thickness is d.

the prediction of classical mechanics. Such penetration through a potential barrier is known as *tunnelling* (Fig. 2.9).

A wave mechanical analysis [4] shows that the transmission probability P_T for a wall thickness d is given by:

$$P_T = \left\{1 + \frac{[\exp(\kappa'd) - \exp(-\kappa'd)]^2}{16\alpha(1-\alpha)}\right\}^{-1} \tag{2.41}$$

where κ' is the wavenumber in the barrier and is equal to $[2m_e(V-E)]^{1/2}/\hbar$ and $\alpha = E/V$. For a high barrier and a box of long dimension, that is, when $\kappa'd \gg 1$, Eq. (2.41) reduces to:

$$P_T = 16\alpha(1-\alpha)\exp(-2\kappa'a) \tag{2.42}$$

Thus, the transmission probability decreases with $\sqrt{m_e}$ and exponentially with the thickness of the barrier.

Tunnelling is essentially a quantum mechanical property. In classical mechanics a particle cannot have a potential energy greater than its total energy, but in quantum mechanics this impossibility is turned into an improbability.

Example 2.7

A characteristic barrier $(V-E)$ for elemental gold is of the order of 5 eV. The decay length for an electron travelling in a one-dimensional space through the barrier can be determined from the above discussion. Since $\kappa' = \sqrt{2m_e(V-E)}/\hbar$, the decay length $\frac{1}{2\kappa'}$ is given by $\hbar/\sqrt{8m_e(V-E)}$. Inserting the appropriate data gives a decay length of *ca.* 0.044 nm, a value commensurate with atomic sizes.

2.10.1.1 *Scanning probe microscopy*

Wave mechanical tunnelling finds an important application in *scanning probe (tunnelling) microscopy* (STM). If a needle-point of a platinum or tungsten probe held by a cantilever system is brought to approximately 1 nm distance from another conducting or semiconducting material and a bias voltage applied, a tunnelling current will flow between the tip of the probe and the surface of the material under examination. If the tunnelling current is monitored and the probe to surface distance maintained at a correct value, the topography of the surface can be analysed at atomic scale.

As an example of the application of STM, Fig. 2.10a illustrates an STM image of a zigzag chain of cesium atoms (red) deposited on to a gallium arsenide (110) surface (blue); the image size is approximately 7 nm square. The deposited atomic 'wires' form naturally when cesium atoms are deposited on to the gallium arsenide surface. A self-assembly of wires results from cesium atoms diffusing on to the surface, and attaching preferentially to the ends of already existing cesium atom chains [5]; Fig. 2.10b shows gallium atoms highlighted on a (110) surface of gallium arsenide.

2.10.1.2 *Tunnelling in electron transfer reactions*

A form of tunnelling occurs in an *electron transfer* reaction, which describes the process of movement of an electron from one chemical species to another. It is embodied in the term *redox*; reduction of one species implies oxidation of another and two mechanisms may be involved.

In *outer sphere transfer* electrons are moved from donor to acceptor, a process of oxidation. In the following example, the coordination spheres of the reacting species remain intact, and electron transfer is rapid from one complex to the other; the molar rate constant for the reaction is *ca.* 10^4 s^{-1}:

$$[Fe^{II}(CN)_6]^{4-} + [Ir^{IV}Cl_6]^{2-} \rightarrow [Fe^{III}(CN)_6]^{3-} + [Ir^{III}Cl_6]^{3-}$$

An *inner sphere electron transfer* involves covalent bonding between the donor and acceptor, as in the sharing of a ligand across two species in their coordination spheres. The electron transfer takes place across the bridging ligand; it is exemplified by the Taube reaction:

$$[Co^{III}(NH_3)_5Cl]^{2+} + [Cr^{II}(H_2O)_6]^{2+} \xrightarrow[1M\,HClO_4]{} [Co^{II}(NH_3)_5H_2O]^{2+}$$
$$+ [Cr^{III}(H_2O)_5Cl]^{2+}$$

An intermediary in the form of a chloride ion bridge serves as a temporary bridging ligand:

$$[Co^{III}(NH_3)_{5-}(\mu\text{-}Cl^-)\text{-}Cr^{II}(H_2O)_6]^{4+}$$

where μ indicates that Cl$^-$ is more closely linked to the Cr species at this stage.

Following the electron transfer, or tunnelling, across the chloride bridge, Cl$^-$ is exchanged with an H$_2$O molecule to accommodate the change in charges on the central cations of the ligands.

The electron transfer process, known as *light harvesting*, has many applications in biochemistry. In plant photosynthesis, for example, electron transfer catalyzed by radiation of visible wavelengths plays an important role wherein carbon dioxide is reduced to glucose with simultaneous oxidation of water to oxygen; it is, overall, the reverse of the glycolic and Krebs phases in the oxidation of glucose.

$$6\,CO_2 + 6\,H_2O \xrightarrow{\lambda_{600\,nm}} C_6H_{12}O_6 + 6\,O_2$$

(a)

(b)

Fig. 2.10 (a) Self-assembly of 'wires' imaged by scanning tunnelling (probe) microscopy (STM): a single-atom zigzag chain of cesium atoms (red) is shown deposited on a (110) surface of gallium arsenide, $a = 0.400$ nm; the array is *ca.* 7 nm square. [Reproduced from Whitman LJ *et al. Phys. Rev. Lett.* 1991; 66: 1338, by permission of the American Physical Society.] (b) STM photograph of gallium arsenide GaAs, highlighting the gallium atoms on a (110) surface.

2.10.2 Boxes of higher dimensions

It is not difficult to extend the analysis in Section 2.10.1 to a two-dimensional box. The wave equation analogous to Eq. (2.32) is

$$\partial^2\psi/\partial x^2 + \partial^2\psi/\partial y^2 + (8\pi^2 m/h^2)E\psi = 0 \qquad (2.43)$$

with the box extending from 0 to a and 0 to b along the x and y directions respectively. The conditions $0 \leq x \leq a$ and $0 \leq y \leq b$ apply, and the solutions depend now on two quantum numbers. The wave equation is separable, that is, $\psi_{n_1,n_2} = \psi_{n_1}\psi_{n_2}$ and the two resulting one-dimensional equations can be treated separately. Following the procedure already described, two solutions result:

$$\psi_{n_1} = (2/a)^{1/2}\sin(n_1\pi x/a)$$
$$\psi_{n_2} = (2/b)^{1/2}\sin(n_2\pi y/b) \qquad (2.44)$$

where the quantum numbers n_1 and n_2 refer to the directions x and y, respectively. Since $\psi_{n_1,n_2} = \psi_{n_1}\psi_{n_2}$ and $E_{n_1,n_2} = E_{n_1} + E_{n_2}$, the complete solution of the wave equation is

$$\psi_{n_1,n_2} = 2/(ab)^{1/2}\sin(n_1\pi x/a)\sin(n_2\pi y/b) \qquad (2.45)$$

and the energy is given by:

$$E_{n_1,n_2} = (h^2/8m_e)\left(\frac{n_1^2}{a^2} + \frac{n_2^2}{b^2}\right) \qquad (2.46)$$

The plot of the function of energy E_{n_1,n_2} against the ratio a/b for two energy states just above that of the lowest state ($E_{1,1}$) shows that when $a = b$, the energy is *doubly degenerate* (Fig. 2.11).

In three dimensions, the energy levels are characterized by three integers, and it is not difficult to show that an electron in a rectangular box of sides a, b and c has energies given by:

$$E_{n_1,n_2,n_3} = (h^2/8m_e)\left(\frac{n_1^2}{a^2} + \frac{n_2^2}{b^2} + \frac{n_3^2}{c^2}\right) \qquad (2.47)$$

If the box is cubic $a = b = c$, with a consequent simplification of the equation. The three-dimensional solution of the wavefunction can be studied conveniently with reference to the hydrogen atom, after a consideration of two other aspects of quantization, those of vibrational and rotational motion.

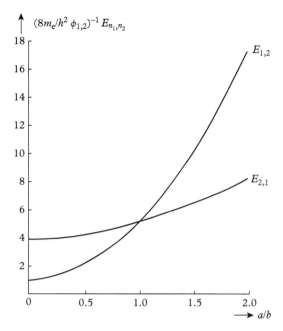

Fig. 2.11 Plot of a function of energy E_{n_1,n_2} against the ratio a/b for an electron in a two-dimensional box; a and b have arbitrary, different values and $\phi_{1,2} = \left(\frac{n_1^2}{a^2} + \frac{n_2^2}{b^2}\right)$. For a square box $a/b = 1$, and at this value the two curves intersect; the equality of $E_{1,2}$ and $E_{2,1}$ for this a/b ratio indicates a twofold degeneracy in energy.

2.10.3 Vibrational motion

Let a particle of mass m suspended at the end of a stretched spring undergo simple harmonic motion about the position $x = 0$, where x is measured along the axis of the spring. It experiences a restoring force $-k_f x$ that is proportional to a distance x, where k_f is a force constant (or *bond force constant* for atomic vibrations) and $\frac{1}{2}k_f x^2$ is the potential energy V of the particle. The Schrödinger equation for the *vibrational motion* takes the form:

$$-[\hbar^2/(2m)]\frac{\partial^2 \psi}{\partial x^2} + \frac{1}{2}k_f x^2 \psi = E\psi \tag{2.48}$$

The solution of this equation leads to the permitted values for the energy; quantization arises from the boundary condition $\psi = 0$ for $x = \pm\infty$, since vibration cannot be infinite [4]. To a first approximation and expressing 'energy' ε in terms of $\bar{v}$ (cm^{-1})

(i) $\varepsilon_v = (v + \frac{1}{2})\bar{v}$

and in order to allow for anharmonicity of vibration: (2.49)

(ii) $\varepsilon_v = (v + \frac{1}{2})\bar{v} - (v + \frac{1}{2})^2 x_a\bar{v} + (v + \frac{1}{2})^3 y_a\bar{v}\dots$

where the vibrational level v takes the values 0, 1, 2, etc. and $\bar{v} = 1/(2\pi c)\sqrt{k_f/\mu}$; x_a and y_a are the first and second *anharmonic constants* and μ is the effective (reduced) mass.

The separation $\Delta\varepsilon_v$ of successive vibrational levels is unity in this notation (or hc in real terms) for all values of v, provided that the motion remains harmonic. From the expression for $\bar{v}$ it follows that $\Delta\varepsilon_v$ is negligibly small for objects of large mass, but very significant with those of atomic size. The value of zero for v shows that the minimum energy $\varepsilon_0 = \bar{v}/2$, or in real terms:

$$E_0 = \tfrac{1}{2}hc\bar{v} \tag{2.50}$$

also written as $\tfrac{1}{2}h\omega$ where $\omega = 2\pi c\bar{v}$, and is the zero-point energy of vibration. The effect of the anharmonicity correction evolves from the next example.

Example 2.8

The reduced mass of $^1H^{35}Cl$ is 1.627×10^{-27} kg and by experiment it shows a fundamental spectral band at 2884.9 cm^{-1} and an overtone band at 5668.1 cm^{-1}. These values can be used to find the anharmonicity constant x_a for HCl. From Eq. (2.49, ii), ignoring the second anharmonicity constant y_a:

$$(\varepsilon_2 - \varepsilon_0) = 5668.1 \text{ cm}^{-1} = 2\bar{v} - 6\bar{v}x_a = 2\bar{v}(1 - 3x_a)$$

$$(\varepsilon_1 - \varepsilon_0) = 2885.9 \text{ cm}^{-1} = \bar{v} - 2\bar{v}x_a = \bar{v}(1 - 2x_a)$$

Solving these equations gives $x_a = 0.01734$ and $\bar{v} = 2989.6$ cm^{-1}. The bond force constant f is $\mu \times (2\pi c\bar{v})^2$, and on introducing data from the above discussion (recall the conversion, cm$^{-1} \rightarrow 100$ m^{-1}), $f = 516.0$ N m^{-1}.

The general solutions of Eq. (2.48) are complex, but for the ground state ($v = 0$), the result takes the form:

$$\psi_0(x) = N_0 \exp(-x^2/2\alpha^2) \tag{2.51}$$

where $\alpha = (\hbar^2/mk)^{1/4}$; hence, the probability density $|\psi_0(x)|^2$ has the form of a Gaussian distribution. Classical mechanics, however, allows the possibility for the particle on a spring to be at rest with E_0 equal to zero. But if this particle were at rest, its position and momentum would be known precisely, which would be at variance with the uncertainty principle. An important contrast between quantization and the mass of a particle is shown by the following example.

Example 2.9

A given mechanical spring has a frequency of 1 Hz, which corresponds to an energy of approximately 6.6×10^{-34} J. A vibrating molecule of hydrogen chloride has a frequency of *ca.* 10^{14} Hz, which corresponds to energy of *ca.* 6.6×10^{-20} J. On the one hand, the energy width for the spring is negligibly small and energy is transferred in an apparently continuous manner. On the other hand, in hydrogen chloride, the energy width is approximately 40 kJ mol^{-1}, which is an experimentally significant quantity; the zero-point energy for the hydrogen chloride vibrator is thus *ca.* 20 kJ mol^{-1}.

2.10.4 Rotational motion

A particle of mass m moving on a circular path of radius r has an *angular momentum $\mathcal{J}$* equal to pr, where p is the tangential linear momentum at any point on the circle. The expression $p^2/(2m)$ is replaced by $\mathcal{J}^2/(2I)$, where I is the moment of inertia mr^2 of the system; thus, the kinetic energy E of the system is now:

$$E = \mathcal{J}^2/(2I) \tag{2.52}$$

Since only certain values for the wavelength λ are permitted, E is quantized, and a number n of wavelengths must fit exactly the circumference of the circular path. This closure leads to a cyclic boundary condition illustrated by Fig. 2.12, such that λ has only the values $2\pi r/n$ $(n = 0, 1, 2, \ldots)$, which implies that $p = h/\lambda = n\hbar/r$. Hence, $\mathcal{J} = n\hbar$, and the permitted kinetic energies are given by:

$$E = n^2\hbar^2/(2I) \tag{2.53}$$

Furthermore, the momentum p may be directed in one of two ways, corresponding to an anticlockwise or a clockwise rotation of the particle. By convention, angular momentum is indicated by a vector along the z-axis, perpendicular to the plane of rotation, and the angular momentum is quantized in units of $\hbar$:

$$\mathcal{J}_z = m_l\hbar(m_l = 0, \pm 1, \pm 2, \ldots) \tag{2.54}$$

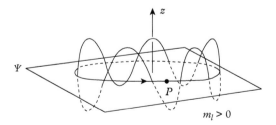

Fig. 2.12 Particle P on a ring: the z-axis corresponds to the direction of ψ for the wavefunction and to the direction of the $\mathcal{J}_z$ angular momentum vector. For the direction of rotation indicated, the quantum number m_l is positive, by convention.

where m_l is the *angular momentum quantum number*. Positive values of m_l correspond to anticlockwise rotation, as viewed along the direction $-z$ (Fig. 2.12). The energies of the particle are given better as:

$$E = m_l^2 \hbar^2 / (2I) \tag{2.55}$$

from which it is evident that the rotational energy is, not surprisingly, independent of the direction of rotation. These results are confirmed by solving the appropriate time-independent Schrödinger equation

$$-\left(\frac{\hbar^2}{2mr^2}\right) \frac{\partial^2 \psi}{\partial \phi^2} = E\psi \tag{2.56}$$

in polar coordinates [6, 7], the solutions for which are the travelling waves:

$$\psi_{m_l} = N \exp(im_l \phi)\,(m_l = 0, \pm 1, \pm 2, \ldots) \tag{2.57}$$

It is left as an exercise for the reader to determine the normalization constant N (recall that ψ_{m_l} is complex) and to show that the quantized energies derived through Eq. (2.56) and Eq. (2.57) are as given by Eq. (2.55). Since ψ must fit the circumference of the circle in order to satisfy the boundary condition, it must be the same at ϕ and $\phi + 2\pi$, which leads to integral values of m_l.

In three dimensions, the particle is free to move anywhere on the surface of a sphere of radius r. The wavefunction must match along any path that is a great circle on the sphere, that is, the equator and the trace on the sphere of any other plane passing through the poles. The kinetic energy and angular momentum are determined by rotation about the three Cartesian x, y and z axes, and the further *quantum number l* is needed to govern the new boundary condition. The Schrödinger equation for this rotation is:

$$-[\hbar^2/(2mr^2)]\frac{1}{\sin\theta}\frac{\partial\psi}{\partial\theta}\left(\sin\theta\frac{\partial\psi}{\partial\theta}\right) = E\psi \tag{2.58}$$

The mathematics of the solution of this equation [6, 7] leads to the following three results. These solutions are discussed in more detail in the next sections.

- The energy E of the particle is quantized to the values:

$$E_l = l(l+1)\hbar^2/(2I)\,(l = 0, 1, 2, \ldots) \tag{2.59}$$

  where l is the orbital angular momentum quantum number, and is non-negative; for any value l, there are $2l + 1$ permitted values of the magnetic quantum number m_l; thus, for $l = 1$, m_l has the three values $0, \pm 1$.

- The energy of the particle is related classically to the angular momentum given by Eq. (2.52) so that By comparison of Eq. (2.59) with Eq. (2.52), it follows that the total angular momentum is also quantized:

$$\mathcal{J} = \sqrt{l(l+1)}\hbar \tag{2.60}$$

- The component of angular momentum along the z-axis is given by Eq. (2.54), and quantization follows from Eq. (2.59), The range of value of m_l depends on l. Since this quantum number can be positive, negative or zero, it has a total of $2l + 1$ values for any value of l; hence, Eq. (2.54) is written as:

$$\mathcal{J}_z = m_l \hbar \ [m_l = l, l-1, \ldots 0, \ldots -(l-1), -l] \tag{2.61}$$

2.11 The hydrogen atom

The hydrogen atom was the subject occupying the first part of series of papers entitled *Quantizierung als Eigenwertproblem* published by Schrödinger in 1926 that produced the equation[8] $i\hbar\Psi = H\Psi$, which may be found on his tombstone at Alpbach in the Tyrol.

The equation was introduced here in its usual form in Section 2.8 and the applications of interest herein will not involve potential energy terms V that are dependent upon time. Thus, a time-independent Hamiltonian operator $\mathcal{H}$ for the hydrogen atom consists of terms involving $1/M_n$, where M_n is the mass of a nucleus, terms involving $1/m_e$, where m_e is the mass of an electron, and terms involving interparticle attractions.

Except for the hydrogen atom, the Schrödinger equation cannot be solved exactly. The ratio M_n/m_e is approximately 1836, and in the *Born–Oppenheimer approximation* the nuclei are assumed to move so slowly with respect to the electrons that they may be treated as stationary in fixed locations, whereupon the Schrödinger equation is solved for the electrons alone. Thus, the operator used here in the Schrödinger equation may be termed the *electronic Hamiltonian* $\mathcal{H}_e$, which takes the form:

$$-\frac{\hbar^2}{2\mu}\nabla^2 - Ze^2/(4\pi\varepsilon_0 r) \tag{2.62}$$

where ε_0 is the electric constant and the other terms have the meanings as before; Z is the atomic number of the species in question, unity for the hydrogen atom.

It is desirable to transform from Cartesian to the spherical polar coordinates described in Appendix A6, and it has been shown [6, 7] that the Schrödinger equation (2.28), $-\frac{\hbar^2}{2m_e}\nabla^2\psi + V\psi = E\psi$, in spherical polar coordinates takes the form

$$\frac{1}{r^2}\left\{\frac{\partial}{\partial r}\left(r^2\frac{\partial\psi}{\partial r}\right) + \frac{1}{\sin\theta}\frac{\partial}{\partial\theta}\left(\sin\theta\frac{\partial\psi}{\partial\theta}\right) + \frac{1}{\sin^2\theta}\frac{\partial^2\psi}{\partial\phi^2}\right\} + \frac{2\mu}{\hbar^2}\left[E + \frac{e^2}{4\pi\varepsilon_0 r}\right]\psi = 0 \tag{2.63}$$

and the full solution of this equation may be found under the same references.

[8] *Ann. Phys.* 1926; 79: 3ff.

A significant feature of this equation is that it is *separable*: as Eq. (2.43) was separable to give Eq. (2.44), so each term in Eq. (2.63) depends upon one of the variables. The radial term is governed by r, and the angular term by θ and ϕ, hence, the wavefunction may be written as:

$$\psi(r, \theta, \phi) = R(r) Y(\theta, \phi) \tag{2.64}$$

Since Y does not depend on r, and R does not depend on the angular variables it follows from Eq. (2.64) that

$$\frac{\partial \psi}{\partial r} = \frac{\partial (RY)}{\partial r} = Y \frac{dR}{dr} \tag{2.65}$$

with similar results in θ and ϕ. Thus, using Eq. (2.65) in Eq. (2.63) and rearranging:

$$\frac{Y}{r^2} \frac{d}{dr}\left(r^2 \frac{dR}{dr}\right) + \frac{2\mu}{\hbar^2}\left[E + \frac{e^2}{4\pi\varepsilon_0 r}\right]RY = -\frac{R}{r^2 \sin\theta}\frac{\partial}{\partial\theta}\left(\sin\theta \frac{dY}{d\theta}\right) - \frac{R}{r^2 \sin^2\theta}\frac{\partial^2 Y}{d\phi^2} = 0$$

Multiplying throughout by r^2/RY:

$$\frac{1}{R}\left\{\frac{d}{dr}\left(r^2 \frac{dR}{dr}\right) + \frac{2\mu r^2}{\hbar^2}\left[E + \frac{e^2}{4\pi\varepsilon_0 r}\right]\psi = -\frac{1}{Y \sin\theta}\frac{\partial}{\partial\theta}\left(\sin\theta \frac{\partial Y}{\partial\theta}\right)\right.$$
$$\left. - \frac{1}{Y \sin^2\theta}\frac{\partial^2 Y}{\partial\phi^2}\right\} = 0 \tag{2.66}$$

As r, θ and ϕ are independent variables, both sides of Eq. (2.66) must be equal to a constant ξ.

In order to balance each other at all points in space, the terms in R and Y take the same values of the separation constant ξ but with opposing signs. Hence, the radial and angular equations may be written as:

$$\frac{d}{dr}\left(r^2 \frac{dR}{dr}\right) + \frac{2\mu r^2}{\hbar^2}\left[E + \frac{e^2}{4\pi\varepsilon_0 r}\right]R - \xi R = 0 \tag{2.67}$$

and

$$\frac{1}{\sin\theta}\frac{\partial}{\partial\theta}\left(\sin\theta \frac{\partial Y}{\partial\theta}\right) + \frac{1}{\sin^2\theta}\frac{\partial^2 Y}{\partial\phi^2} + \xi Y = 0 \tag{2.68}$$

where the complete derivatives in r in Eq. (2.67) indicate its independence from Y.

Furthermore, the angular term may be separated as:

$$Y(\theta, \phi) = \Theta(\theta)\Phi(\phi) \tag{2.69}$$

Following the argument as before, Y is replaced and the differentials modified simultaneously in Eq. (2.68):

$$\frac{\Phi}{\sin\theta}\frac{\partial}{\partial\theta}\left(\sin\theta\frac{\partial\Theta}{\partial\theta}\right) + \frac{\Theta}{\sin^2\theta}\frac{\partial^2\Phi}{\partial\phi^2} + \xi\Theta\Phi = 0 \tag{2.70}$$

Multiplying by $\sin^2\theta/\Theta\Phi$ and rearranging gives:

$$-\frac{1}{\Phi}\frac{\partial^2\Phi}{\partial\phi^2} + \frac{\sin\theta}{\Theta}\frac{\partial}{\partial\theta}\left(\sin\theta\frac{\partial\Theta}{\partial\theta}\right) + +\xi\sin^2\theta = 0 \tag{2.71}$$

2.11.1 The angular equations

The separation of $\Phi(\phi)$ and $\Theta(\theta)$ is accomplished by introducing a second separation constant ζ with opposing signs for the two parts, as before.

The Φ-*equation* from Eq. (2.71) becomes

$$\frac{1}{\Phi}\frac{d^2\Phi}{d\phi^2} + \zeta = 0$$

or

$$\frac{d^2\Phi}{d\phi^2} + \zeta\Phi = 0$$

Conveniently, the constant ζ is set equal to m_l^2, which may be allowed positive or negative values, and the solutions, following Appendix A9, are:

$$\Phi(\phi) = C\exp(i\zeta\phi) + D\exp(-i\zeta\phi) \tag{2.72}$$

or

$$\Phi(\phi) = C'\exp(im_l\phi) \tag{2.73}$$

where m_l takes the integral values $0, \pm1, \pm2, \ldots$, because $\Phi(\phi + 2\pi)$ and $\Phi(\phi)$ must match, as discussed in Section 2.10.4. The conjugate equation is $\Phi^*(\phi) = C'\exp(-im_l\phi)$, so that from Section 2.9.1 the normalizing constant C' is $1/\sqrt{2\pi}$.

The Θ-*equation* from Eq. (2.71) is written as

$$\frac{\sin\theta}{\Theta}\frac{d}{d\theta}\left(\sin\theta\frac{d\Theta}{d\theta}\right) + \xi\sin^2\theta - \zeta = 0$$

Since ζ has been shown to be equal to m_l^2, division by $\sin^2\theta$ leads to:

$$\frac{1}{\sin\theta}\frac{d}{d\theta}\left(\sin\theta\frac{d\Theta}{d\theta}\right) + \left(\xi - \frac{m_l^2}{\sin^2\theta}\right)\Theta = 0 \tag{2.74}$$

Table 2.1 *Normalized surface harmonics Y_{l,m_l} for $l = 0$ to 2*

l	m_l	Y_{l,m_l}	Orbital type
0	0	$\sqrt{1/4\pi}$	s
1	0	$\sqrt{3/4\pi}\,\cos\theta$	p_0
1	+1	$-\sqrt{3/8\pi}\,\sin\theta\,\exp(i\phi)$	p_1
1	−1	$\sqrt{3/8\pi}\,\sin\theta\,\exp(-i\phi)$	p_{-1}
2	0	$\sqrt{5/16\pi}\,(3\cos^2\theta - 1)$	d_0
2	+1	$-\sqrt{15/8\pi}\,\cos\theta\,\sin\theta\,\exp(i\phi)$	d_1
2	−1	$\sqrt{15/8\pi}\,\cos\theta\,\sin\theta\,\exp(-i\phi)$	d_{-1}
2	+2	$-\sqrt{15/32\pi}\,\sin^2\theta\,\exp(i2\phi)$	d_2
2	−2	$\sqrt{15/32\pi}\,\sin^2\theta\,\exp(-i2\phi)$	d_{-2}

By making the substitutions $\cos\theta = z$ and $\Theta(\theta) = P(z)$, Eq. (2.74) evolves to:

$$\frac{\mathrm{d}}{\mathrm{d}z}\left[(1-z^2)\frac{\mathrm{d}P}{\mathrm{d}z}\right] + \left[\xi - \frac{m_l}{(1-z^2)}\right]P = 0 \tag{2.75}$$

The coefficients in this equation depend on z, and its solutions are the associated Legendre polynomials [6, 7] which converge if the constant ξ is set equal to $l(l+1)$, where l is another constant. The complete equation for $Y(\theta,\phi)$ can be written in the form:

$$Y_{l,m_l}(\theta,\phi) = (-1)^{(m_l+|m_l|)/2}\left[\left(\frac{2l+1}{4\pi}\right)\frac{(l-|m_l|)!}{(l+|m_l|)!}\right]^{1/2}P_l^{m_l}(\cos\theta)\,\exp(im_l\phi) \tag{2.76}$$

Values of $P_l^{m_l}$ are available from the literature references already given, and the first few results for $P_l^{m_l}(\cos\theta)$ are given below:

$$P_0^0 = 1 \quad P_1^0 = \cos\theta \quad P_1^1 = -\sin\theta \quad P_1^{-1} = \sin\theta \quad P_2^0 = \tfrac{1}{2}(3\cos^2\theta - 1)$$

The normalized surface harmonics (aka spherical harmonics) are listed in Table 2.1 for values of l from 0 to 2.

2.11.2 The radial equation

The *radial (R) equation* develops from Eq. (2.67) as

$$r^2\frac{\mathrm{d}^2R}{\mathrm{d}r^2} + 2r\frac{\mathrm{d}R}{\mathrm{d}r} + \frac{2\mu r^2}{\hbar^2}\left[E + \frac{e^2}{4\pi\varepsilon_0 r}\right]R - \xi R = 0$$

and division by r^2 leads to

$$\frac{d^2R}{dr^2} + \frac{2}{r}\frac{dR}{dr} + \frac{2\mu}{\hbar^2}\left[E + \frac{e^2}{4\pi\varepsilon_0 r}\right]R - \frac{\xi R}{r^2} = 0$$

or, since ξ has been defined already as $l(l+1)$, in the form

$$\frac{d^2R}{dr^2} + \frac{2}{r}\frac{dR}{dr} + \frac{2\mu}{\hbar^2}\left[E + \frac{e^2}{4\pi\varepsilon_0 r}\right]R - \frac{l(l+1)R}{r^2} = 0 \qquad (2.77)$$

The detailed solution of this equation has been discussed in the literature [6, 7]. However, a change of variable $\sigma = Rr$ allows Eq. (2.77) to be expressed, after some manipulation, as

$$\frac{\hbar^2}{2\mu}\frac{d^2\sigma}{dr^2} + \left\{E - \left[\frac{\hbar^2}{2\mu}\frac{l(l+1)}{r^2} - \frac{e^2}{4\pi\varepsilon_0 r}\right]\right\}\sigma = 0 \qquad (2.78)$$

and comparison with Eq. (2.63) in a Cartesian form, such as Eq. (2.62), shows that the potential energy term is:

$$V = \frac{\hbar^2}{2\mu}\frac{l(l+1)}{r^2} - \frac{e^2}{4\pi\varepsilon_0 r} \qquad (2.79)$$

For $l = 0$, the potential energy lies solely in the coulombic term, decreasing rapidly towards $-\infty$ as r tends to zero, but rising asymptotically to zero with increase in r; the electron has zero angular momentum when l is zero. When l is greater than zero, the term in l opposes the coulombic energy, and a minimum exists in V for small values of r, as shown by the radial function R_{2s} in Fig. 2.13, for example. The solution of the radial equation reveals that the coexistence of the two types of behaviour demands integral values of the quantum number n, and the radial equation can be written in the form:

$$R_{n,l}(r) = N\rho^l \mathcal{L}_{n+1}^{2l+1}(\rho)\exp(-\rho/2) \qquad (2.80)$$

where N is a normalizing constant and ρ is a function of r given by:

$$\rho = 2Zr/(na_0) \qquad (2.81)$$

where a_0 is the Bohr radius of value 5.29177×10^{-11} m, obtained from the equation:

$$a_0 = 4\pi\varepsilon_0\hbar^2/(m_e e^2) \qquad (2.82)$$

Table 2.2 *Normalized hydrogenic radial wavefunctions $R_{n,l}$ for $n = 1$ to 3; $\rho = 2Zr/na_0$*

Orbital	n	l	$R_{n,l}(r)$
1s	1	0	$2(Z/a_0)^{3/2}\exp(-\rho/2)$
2s	2	0	$(Z/a_0)^{3/2}(1/\sqrt{8})(2-\rho)\exp(-\rho/2)$
2p	2	1	$(Z/a_0)^{3/2}(1/\sqrt{24})\rho\exp(-\rho/2)$
3s	3	0	$(Z/a_0)^{3/2}(1/\sqrt{243})(6-6\rho+\rho^2)\exp(-\rho/2)$
3p	3	1	$(Z/a_0)^{3/2}(1/\sqrt{486})(4\rho-\rho^2)\exp(-\rho/2)$
3d	3	2	$(Z/a_0)^{3/2}(1/\sqrt{2430})\rho^2\exp(-\rho/2)$

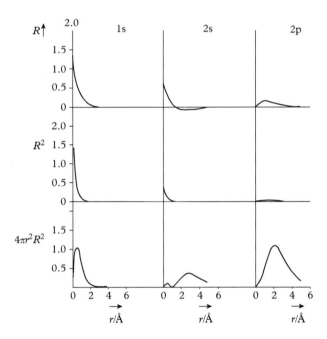

Fig. 2.13 Radial 1s, 2s and 2p wavefunctions for hydrogen-like species as a function of (r/a_0). The maximum in the function $4\pi r^2 R^2$ occurs at $r = a_0$ for the 1s function, which may be confirmed by differentiating this function with respect to r. Where along the abscissae are the maxima for the 2s and 2p radial wavefunctions? (See also Plotting exercise 2.1, Section 2.11.2.)

The $\mathcal{L}$-functions are associated *Laguerre polynomials*, and may be found in the literature references already quoted; the normalized radial wavefunctions are listed in Table 2.2 for values of l from 0 to 2.

Certain of the $R(r)$ functions in Fig. 2.13 exhibit non-trivial nodes. In general, there are $n - l - 1$ non-trivial nodes: thus 2s has one such node and 2p has none. An example of a trivial node is the zero value for ψ at $r = 0$, exemplified by the 2p function. The function at zero r cannot be a node since the wavefunction does not pass through the origin, and r cannot be negative.

Example 2.10

The position of the node in the $R(r)_{2s}$ function is found by setting the $2s$ radial function to zero. In practice, it is sufficient to set $(2-\rho)$ to zero. Hence, the node lies at $\rho = 2$, or $r = a_0$. Calculate the corresponding result for $R(r)_{3s}$. $[5 \pm \sqrt{7}]$

The form of R_{1s}^2 suggests that the most probable place for the electron, or the maximum of electron density, is at the nucleus ($r = 0$). That it is not so depends upon the balance between the kinetic energy of the radial motion of the electron and its potential energy with respect to the nucleus. The true probability evolves from the radial distribution function $4\pi r^2 R^2$. On the one hand, R^2 examines a volume element $d\tau$ along any radius from the nucleus, and must therefore find a maximum at the nucleus. On the other hand, in the function $4\pi r^2 R_{1s}^2$ the volume element $d\tau$ scans R_{1s}^2 in spherical shells of area $4\pi r^2$ and thickness $d\tau$ at varying distances r from the nucleus.

Plotting exercise 2.1

Run the program RADL by clicking on the program name. Select a file name and then input the function name itself, say $2s$, according to the instructions on the monitor screen, The output consists of a user-named file, which lists the total results numerically, and also to three files named R1, R2 and R3 containing the r, ψ data for plotting the radial functions $R(r)$, $|R(r)|^2$ and $4\pi r^2 |R|^2$, respectively, over an appropriate range of the variable r, which is listed in units of a_0; the wavefunction data are in the form $\psi/(Z_{eff}/a_0)^{3/2}$. Transfer the files R1–R3 to the Python work folder. Refer to Section 7.2 for setting up the Python plotting environment, then plot the data by the following commands in the Python shell for the program GRFN. Each Python command is followed by the ENTER key.

```
import grfn32
grfn32.graf(['R1'])
```

for a single plot, or for all three together

```
grfn32.graf(['R1', 'R2', 'R3'])
```

Add the abscissa line by the command

continued

Plotting exercise 2.1 *continued*

---

   plt.plot([0, p],[0,0])

where p is the maximum value of the abscissa, probably 16 with the files provided.

   The procedure allows a convenient investigation of the wavefunctions for $n = 1$ to 4 ($l = 0$–3).

(a) Plot the three functions on a single graph.

(b) With the mouse pointer, determine from the graph the r/a_0 values where the R_{2s} and $4\pi r^2 |R_{2s}(r)|^2$ functions cross the zero line.

(c) Determine from the graph the r/a_0 values of the turning points for the $4\pi r^2 |R_{2s}|^2$ (assume $Z = 1$).

(d) In (b) and (c), confirm the graph estimates from the radial functions. [Results at the end of the chapter.]

2.11.3　The complete wave equation

The three separate parts into which the Schrödinger equation was divided can now be brought together as a complete solution to the wavefunction for the hydrogen atom and, indeed, other one-electron hydrogen-like species. The one-electron wavefunction, known as an atomic orbital (Section 2.13), may be written as:

$$\psi(r,\theta,\phi) = R(r)\Theta(\theta)\Phi(\phi) = R_{n,l}(r) Y_{l,m_l}(\theta,\phi)$$

As an example, consider the three separate functions for the $2p_z$ electron of hydrogen, for which $n = 2$, $l = 1$ and $m_l = 0$. Then, $R_{2,1}(r) = (Z/a_0)^{3/2}$ $(1/\sqrt{24})\rho \exp(-\rho/2)$, $\Theta_{1,0} = \sqrt{3/4\pi} \cos\theta$ and $\Phi_{1,0}(\phi) = 1/\sqrt{2\pi}$. While these equations are themselves separately normalized, the complete function will not be normalized merely by combining these three equation. The functional parts of the expressions form the equation $N\psi_{2,1,0}$, which is normalized in the usual way. Thus, for the $2p$ wavefunction:

$$N^2\psi_{2,1,0}^2 = N^2 \int_0^\infty (Zr/a_0)^4 \exp(-Zr/a_0)\mathrm{d}r \int_{-\pi}^{\pi} \cos^2\theta \sin\theta\mathrm{d}\theta \int_0^{2\pi} \mathrm{d}\phi = 1$$

$$(2.83)$$

It is straightforward to show that the integrals over θ and ϕ are (2/3) and 2π, respectively. For the integral over r, following Appendix 7, let $t = Zr/a_0$, so that $Z\mathrm{d}r = a_0\mathrm{d}t$. Then the integral over r becomes:

$$(a_0/Z)^3 \int_0^\infty t^4 \exp(-t)\mathrm{d}t = (a_0/Z)^3\Gamma(5) = 24(a_0/Z)^3$$

Table 2.3 *Complete hydrogenic wavefunctions ψ_{n,l,m_l} (including the Euler expansion for $l \geq 1$)* [6]

n	l	m_l	$\psi_{n,l,m_l} = R_{n,l}Y_{l,m_l}$	
1	0	0	$1s$	$\dfrac{1}{\sqrt{\pi}}\left(\dfrac{Z_{\text{eff}}}{a_0}\right)^{3/2}\exp(-Z_{\text{eff}}r/a_0)$
2	0	0	$2s$	$\dfrac{1}{4\sqrt{2\pi}}\left(\dfrac{Z_{\text{eff}}}{a_0}\right)^{3/2}\left(2-\dfrac{Z_{\text{eff}}r}{a_0}\right)\exp(-Z_{\text{eff}}r/2a_0)$
2	1	0	$2p_0$	$\dfrac{1}{4\sqrt{2\pi}}\left(\dfrac{Z_{\text{eff}}}{a_0}\right)^{3/2}\dfrac{Z_{\text{eff}}r}{a_0}\exp(-Z_{\text{eff}}r/2a_0)\cos\theta$
2	1	*	$2p_x$	$\dfrac{1}{4\sqrt{2\pi}}\left(\dfrac{Z_{\text{eff}}}{a_0}\right)^{3/2}\dfrac{Z_{\text{eff}}r}{a_0}\exp(-Z_{\text{eff}}r/2a_0)\sin\theta\cos\phi$
2	1	*	$2p_y$	$\dfrac{1}{4\sqrt{2\pi}}\left(\dfrac{Z_{\text{eff}}}{a_0}\right)^{3/2}\dfrac{Z_{\text{eff}}r}{a_0}\exp(-Z_{\text{eff}}r/2a_0)\sin\theta\sin\phi$
3	0	0	$3s$	$\dfrac{1}{81\sqrt{3\pi}}\left(\dfrac{Z_{\text{eff}}}{a_0}\right)^{3/2}\left(27-18\dfrac{Z_{\text{eff}}r}{a_0}+2\left(\dfrac{Z_{\text{eff}}r}{a_0}\right)^2\right)\exp(-Z_{\text{eff}}r/3a_0)$
3	1	0	$3p_z$	$\dfrac{1}{81}\sqrt{\dfrac{2}{\pi}}\left(\dfrac{Z_{\text{eff}}}{a_0}\right)^{3/2}\left(6-\dfrac{Z_{\text{eff}}r}{a_0}\right)\dfrac{Z_{\text{eff}}r}{a_0}\exp(-Z_{\text{eff}}r/3a_0)\cos\theta$
3	1	*	$3p_x$	$\dfrac{\sqrt{2}}{81\sqrt{\pi}}\left(\dfrac{Z_{\text{eff}}}{a_0}\right)^{3/2}\left(6-\dfrac{Z_{\text{eff}}r}{a_0}\right)\dfrac{Z_{\text{eff}}r}{a_0}\exp(-Z_{\text{eff}}r/3a_0)\sin\theta\cos\phi$
3	1	*	$3p_x$	$\dfrac{\sqrt{2}}{81\sqrt{\pi}}\left(\dfrac{Z_{\text{eff}}}{a_0}\right)^{3/2}\left(6-\dfrac{Z_{\text{eff}}r}{a_0}\right)\dfrac{Z_{\text{eff}}r}{a_0}\exp(-Z_{\text{eff}}r/3a_0)\sin\theta\sin\phi$
3	2	0	$3d_{z^2}$	$\dfrac{1}{81\sqrt{6\pi}}\left(\dfrac{Z_{\text{eff}}}{a_0}\right)^{3/2}\left(\dfrac{Z_{\text{eff}}r}{a_0}\right)^2\exp(-Z_{\text{eff}}r/3a_0)(3\cos^2\theta-1)$
3	2	*	$3d_{xz}$	$\dfrac{1}{81}\sqrt{\dfrac{2}{\pi}}\left(\dfrac{Z_{\text{eff}}}{a_0}\right)^{3/2}\left(\dfrac{Z_{\text{eff}}r}{a_0}\right)^2\exp(-Z_{\text{eff}}r/3a_0)\sin\theta\cos\theta\cos\phi$
3	2	*	$3d_{yz}$	$\dfrac{1}{81}\sqrt{\dfrac{2}{\pi}}\left(\dfrac{Z_{\text{eff}}}{a_0}\right)^{3/2}\left(\dfrac{Z_{\text{eff}}r}{a_0}\right)^2\exp(-Z_{\text{eff}}r/3a_0)\sin\theta\cos\theta\sin\phi$
3	2	*	$3d_{x^2-y^2}$	$\dfrac{1}{81\sqrt{2\pi}}\left(\dfrac{Z_{\text{eff}}}{a_0}\right)^{3/2}\left(\dfrac{Z_{\text{eff}}r}{a_0}\right)^2\exp(-Z_{\text{eff}}r/3a_0)\sin^2\theta\cos 2\phi$
3	2	*	$3d_{xy}$	$\dfrac{1}{81\sqrt{2\pi}}\left(\dfrac{Z_{\text{eff}}}{a_0}\right)^{3/2}\left(\dfrac{Z_{\text{eff}}r}{a_0}\right)^2\exp(-Z_{\text{eff}}r/3a_0)\sin^2\theta\sin 2\phi$

* Real orbitals formed by the standard procedure of linear combinations do not possess true m_l values.

Introducing the other integrals into the normalization equation $N^2 \times 24(a_0/Z)^3 \times 4\pi/3 = 1$, so that $N = (Z/a_0)^{3/2}\sqrt{1/32\pi}$ and the complete equation is:

$$\psi_{2,1,0} = \frac{1}{4\sqrt{2\pi}}\left(\frac{Z}{a_0}\right)^{3/2}\frac{r}{a_0}\exp(-r/2a_0)\cos\theta$$

as listed in Table 2.3, where Z_{eff} is used in place of Z for generality (see Section 2.13.5).

The allowed energies corresponding to acceptable solutions of a wavefunction are governed by n, and are given by the equation:

$$E_n = -\frac{Z_{\text{eff}}^2\mu e^4}{32\pi^2\varepsilon_0^2\hbar^2 n^2} \tag{2.84}$$

where the negative sign indicates bound states for the atom, for which energy is a negative quantity rising to zero as r tends to infinity. Inserting the constants and taking $n = 1$ for the lowest energy state, $E_n = 2.1787 \times 10^{-18}$ J; this value represents the ionization energy of a hydrogen atom. It is interesting to note that the same expression for energy arises from the Bohr theory of the hydrogen atom ($Z_{\text{eff}} = 1$). However, this fact should not be regarded as supporting evidence for that theory: the hydrogen atom is a special case because of the simple form of its potential energy function, and the parameter n evolves in differing ways in the two treatments.

Plotting exercise 2.2

(a) With the aid of data the sets *an2s.txt* and *an3s.txt* (provided through the program PLOT), use program ANFN to describe briefly the structure of the 2*s* and 3*s* atomic orbitals. (b) How many nodes are present in each function? (c) Compare a central section of the plots with the corresponding radial functions (from RADL and GRFN).   [Results at the end of the chapter.]

2.12 Quantum numbers

Three quantum numbers, n, l and m_l, have arisen from the discussion of the Schrödinger equation. Two of them, l and m_l, involve the separable nature of R from Y and θ from ϕ, respectively. The quantum number n arises from the convergent series involved in the radial function. Not all quantum numbers are independent and Table 2.4 indicates the levels of dependency.

Table 2.4 *Interdependence of the quantum numbers n, l and m_l*

n	1	2			3									....
l	0	0	1		0	1			2					$0 ....(n-1)$
m_l	0	0	−1	0	1	0	−1	0	1	−2	−1	0	1	2	$-l ....l$

2.12.1 Angular momentum and spin

An angular momentum quantum number l gives rise to $2l + 1$ orientations of the angular momentum vector, the positions that are specified by m_l. Furthermore, the result of the Stern and Gerlach experiment [4] led to the conclusion that an electron possesses a property equivalent to an intrinsic spin about its own axis. The *spin quantum number* is designated s, and its projection on to the z-axis by m_s. For an electron, the allowed value of s is ½ with a magnitude of $\sqrt{s(s+1)}\hbar$, or $\sqrt{3/4}\hbar$, and its component m_s along the z-axis has $2s + 1$ values. Since s is ½, only two orientations obtain: $m_s = +½$, denoted graphically by ↑ or symbolically α, and $m_s = -½$, denoted by ↓ or β. The spin angular momentum vector lies at an angle to the z-axis equal to $\cos^{-1} 1/\sqrt{3}$, the tetrahedral half-angle, $+\cos^{-1}(1/\sqrt{3})$ for an α electron and $-\cos^{-1}(1/\sqrt{3})$ for a β electron. If an electron has the quantum numbers $n = 5$ and $l = 2$, then m_l could be ±2, ±1 or 0 with $m_s \pm ½$.

Particles with a spin of ½, electrons, protons and neutrons, are known as *fermions*, whereas those with a spin of unity, photons, gluons and mesons, are termed *bosons*. The basic particles constituting matter are fermions, whereas those that transmit electromagnetic bonding forces to form matter are bosons.

The permitted orientations of the angular momentum vector can be represented by a vector model, as illustrated in Fig. 2.14(a) for the case of $l = 2$; the integers indicate the permitted values of m_l for this value of l. The azimuthal orientation of the vector around the z-axis is indeterminate, and any generator of a cone is a possible orientation for the angular momentum vector (Fig. 2.14b).

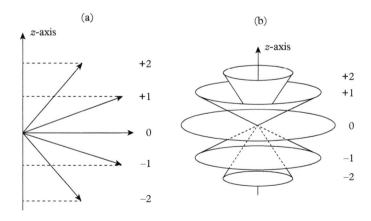

Fig. 2.14 Vector model of angular momentum of an atom. (a) Permitted orientations of the angular momentum vector for $l = 2$. (b) All generators of the cone for a given value of m_l are permitted orientations. The cones have slant lengths of $\sqrt{l(l+1)}$, which represent the magnitudes of the angular momenta.

2.12.2   Pauli's exclusion principle

The *Pauli exclusion principle* states that an atomic orbital can contain up to two electrons, and where there are two electrons in one and the same orbital their spins must be opposed. Paired spins have zero angular momentum because the spin vectors are centrosymmetric with respect to the centre of the atom; their cone generators form the angles $\pm\cos^{-1}(1/\sqrt{3})$ with the z-axis.

2.13   Atomic orbitals

While the wavefunction itself is a purely mathematical entity, it is useful in studies of chemical bonding to have a form of visualizing its meaning in this context; it is achieved through a form of illustration of the atomic orbital.

An *atomic orbital* is a one-electron wavefunction for an electron in an atom. As discussed above, it consists of a radial, or 'size' term and an angular, or 'shape' term; Tables 2.1 and 2.2 list these two parts of the wavefunction for the range 1 to 3 of the principal quantum number n. From Eq. (2.84), it follows that the electron energy level is proportional to $1/n^2$, so that as n increases the energy levels become ever closer, and in the limit as n tends to infinity they form a continuum; at this point, the energy corresponds to the first ionization energy of the species.

2.13.1   Ionization energy

The *ionization energy* of a species is the energy required to just expel an electron from its ground state energy level. In the case of hydrogen, the reaction is:

$$H(g) \rightarrow H^+(g) + e \tag{2.85}$$

and the ionization energy is easily measured from an extrapolation, linear in the case of hydrogen, of spectroscopic frequencies $\overline{\nu}$ against $1/n^2$ to give *ca.* 2.179×10^{-18} J; the value calculated from Eq. (2.84) with $Z_{\mathrm{eff}} = 1$ and $n = 1$ is 2.1787×10^{-18} J (to 5 figures), or 1312.0 kJ mol^{-1}.

The first ionization energies of other species cannot be obtained in quite the same manner. The wavenumber of a line in the absorption spectrum may be given by:

$$\overline{\nu} = \overline{\nu_\infty} - R/(n-\delta)^2 \tag{2.86}$$

where $\overline{\nu_\infty}$ and δ, the *quantum defect*, are both unknown initially. A process for determining the first ionization energy of sodium will be described with reference to the spectral data in Table 2.4, known as the P spectral series for sodium for $n = 10$ to 17.

A plot of $\overline{\nu}$ against $1/n^2$ and analytical extrapolation to $1/n^2 = 0 (\equiv \overline{\nu_\infty})$ gives a starting value of 0.81 for δ. With this value, $\overline{\nu}$ is fitted to the function $R/(n-\delta)^2$, as shown in Fig. 2.15. The new values of both $\overline{\nu_\infty}$ and δ are then used iteratively,

Table 2.5 *Spectral wavenumbers and first ionization energy for atomic sodium*

n	$\bar{v}/\text{cm}^{-1}$	$R/(n-\delta)^2/\text{cm}^{-1}$	$\bar{v}_\infty/\text{cm}^{-1}$	δ	
		$\delta = 0.81$			
10	40137.2	1299.3		0.867	
11	40383.2	1056.8		0.871	
12	40566.0	876.4		0.875	
13	40705.7	738.5		0.880	0.87
14	40814.5	630.8		0.888	
15	40901.1	545.0		0.896	
16	40971.2	475.6		0.905	
17	41028.7	418.7		0.914	
			41452.7		
		$\delta = 0.87$			
		1316.5		0.853	
		1069.4		0.852	
		885.9		0.850	
		745.8		0.848	0.85
		636.5		0.847	
		549.6		0.845	
		479.4		0.842	
		421.8		0.838	
			41448.8		
		$\delta = 0.85$			
		1310.7		0.857	
		1065.2		0.858	
		882.7		0.859	
		743.4		0.859	0.86
		643.6		0.861	
		548.1		0.862	
		478.1		0.863	
		420.7		0.863	
			41450.1		
		$\delta = 0.856^{\text{a}}$			
		1312.4		0.856	
		1066.4		0.857	
		883.6		0.856	
		744.1		0.855	0.856
		635.2		0.857	
		548.5		0.857	
		478.5		0.857	
		421.0	41449.7	0.856	

[a] Estimated from iterations 2 and 3

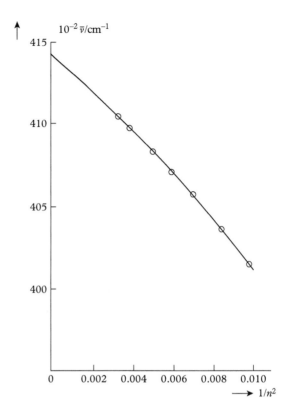

Fig. 2.15 Plot of $\bar{\nu}$ against $1/n^2$ for the *P* spectral series of sodium, extrapolated analytically to $1/n^2 = 0$ (ν_∞); both $1/n^2$ and $1/(n-\delta)^2$ must extrapolate to the same value of ν_∞.

as shown in Table 2.5, until both $\overline{\nu_\infty}$ and δ converge. The convergence limit of 41,449.7 is converted to an ionization energy I by $I = hc\overline{\nu_\infty}$, leading to the value of 5.1390 eV per atom, or 485.65 kJ mol^{-1} for sodium.

2.13.2   Atom shells

Atomic *shells* are determined by the quantum number n, and are given the following notation:

n	1	2	3	4 ...
Shell	K	L	M	N ...

Atomic *sub-shells* are determined according to the value of l:

l		1	2	3	4 ...
Sub-shell		s	p	d	f ...

For $n = 2$, for example, the L shell has sub-shells s (one) and p (three). In general, the number of atomic orbitals in an atom with principal quantum number n is n^2; thus for $n = 2$ (the L shell), there are four subshells, s, p_{-1}, p_0 and p_1, where the subscript is indicative of the value of m_l; for the s sub-shell, $m_l = 0$ is the only permitted value. The alphabetic notation used here for orbitals derives from spectroscopic usage: the terms s (sharp), p (principal), d (diffuse) and f (fundamental) were used to describe spectroscopic transitions with increasing value of l from 0 to 3.

From Table 2.2, the ground state radial wavefunction for hydrogen is $(1/\pi a_0^3)^{1/2} \exp(-r/a_0)$. Thus, the electron lies in the K shell, with the sub-shell as $1s$, since $l = 0$; the wavefunction can be identified conveniently as $\psi(1s)$, $\psi_{1,0,0}$, or similar designation. If electron spin is to be included in an orbital description, the $1s$ one-electron functions are then described as:

$$\psi_{1,0,0,1/2} = (1/\pi a_0^3) \exp(-r/a_0)\alpha$$

and

$$\psi_{1,0,0,-1/2} = (1/\pi a_0^3) \exp(-r/a_0)\beta \tag{2.87}$$

for the two spin states $\pm 1/2$ indicated in the subscripts to ψ.

2.13.2.1 Selection rules for atoms

The spectra of atomic hydrogen can be explained by transitions of electrons between two energy levels (Section 2.6). However, not all possible transitions are allowed. If a photon is expelled from an atom by an electron transition, conservation of angular momentum requires that the electron angular momentum must change to compensate for that carried away by the photon. A photon, being a boson (Section 2.11.1), has an intrinsic spin angular momentum of unity; thus, a p electron ($l = 1$) can fall back to an s orbital ($l = 0$), of lower energy, with emission of radiation. Changes in n are not restricted in this manner because n governs energy rather than angular momentum. The selection rules for atoms are:

$$\Delta n = 1, 2, 3, \ldots$$
$$\Delta l = \pm 1 \tag{2.88}$$
$$\Delta m_l = 0, \pm 1$$

Allowed transitions can be depicted on a Grotrian diagram, of which Fig. 2.16 is a simple example.

2.13.3 Radial functions and size

The radial component of a wavefunction involving the r parameter is governed by the principal quantum number n, which determines the 'size' of the atomic orbital. In the radial functions (Table 2.2), the exponential factor in the wavefunction

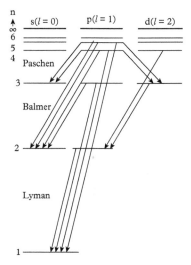

Fig. 2.16 Grotrian diagram showing some of the allowed transitions in atomic hydrogen: changes in n are unrestricted, but those in l and m_l must be ± 1 and 0, ± 1, respectively. [*Introduction to Physical Chemistry*, 3rd ed. 1998; reproduced by courtesy of Cambridge University Press.]

shows that its value decreases exponentially as the distance from the nucleus r is increased. The terms in ρ, where they occur, cause the wavefunctions to fall to zero at the nucleus, and the Laguerre polynomials are responsible for the formation of non-trivial nodes as the wavefunction oscillates through the zero value in the $4\pi r^2 R^2(r)$ function. These features are evident on the illustrations in Fig. 2.13, and it may be helpful for the reader to and plot radial functions, say, the $2p$, $3s$ and $3d$, as indicated in Plotting exercise 2.1.

2.13.4 Angular functions and shape

The angular functions (surface harmonics) for the hydrogenic wavefunction are listed in Table 2.1 for values of l from 0 to 2, which include the sub-shells s, p and d. As the name 'angular' suggests, these components of the wavefunction determine the 'shape' of the atomic orbitals.

2.13.4.1 s Orbitals

Atomic orbitals of the type ns are *spherically symmetrical*, differing mainly in the numbers of nodes. The s orbitals may be pictured as spherical regions of electron density centred on the nucleus and comprising about 95% of the total probability density. The radial probability distribution has been discussed in Section 2.9.

2.13.4.2 p and d Orbitals

The three p orbitals for a given value of $n \geq 2$ are distinguished by their values of m_l, as shown in Table 2.1. For $l = 1$, the values of m_l define the orientations of electron orbital angular momentum about a direction usually chosen as the z reference axis. The $2p$ orbital with m_l equal to zero is a real function, and its shape is governed by $\cos\theta$, so that its probability density is proportional to $\cos^2\theta$. This orbital is named p_z and possesses a nodal plane which is the x, y plane (Fig. 2.17), a region of zero electron density.

The diagram shows both Y and Y^2 with their 95% probability envelopes. Since the density for this orbital is proportional to $\cos^2\theta$, its maxima occur when the vector $\mathbf{v}$ makes angles of 0° and 180° in θ with respect to the z-axis. Electron density contours for a $2p_z$ orbital are illustrated by Fig. 2.18; the 0.1 contour encloses *ca.* 66% of the electron density; 95% is enclosed by a 0.02 contour, not shown on the diagram for reason of space limitation.

This type of representation is reminiscent of an electron density map obtained through x-ray crystallographic structure analysis which, although it does not resolve electron density for the separate orbitals, reveals an average electron density for a given species in a crystal structure. A highly resolved x-ray electron density map of benzene, C_6H_6 [8], is shown in Fig. 2.19; the contours are drawn at number density [9] intervals of 0.25.

The p orbitals with $m_l = \pm 1$ present a new situation. Table 2.1 shows that of the atomic orbitals considered so far, the s and $p_0 (\equiv p_z)$ are real functions, whereas p_1 and p_{-1} and other orbitals for which l is not equal to zero, are complex

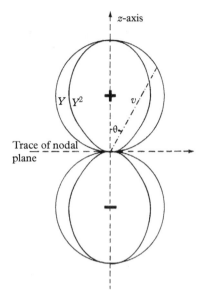

Fig. 2.17 A central section of the boundary surface of the p_z angular function. The length v of any segment line is proportional to $|\cos\theta|$; the positive and negative regions of the Y function are indicated by the $\pm$ signs. The boundary surface labelled Y^2 shows the dumbbell shape of the p_z^2 electron density function.

functions. It is usual to set up real and imaginary functions, using Euler's formula [10], and to designate them p_x and p_y orbital functions, respectively. The three $2p$ angular functions are:

$$\left.\begin{aligned}p_z &= (3/4\pi)^{1/2}R_{n,l}(r)\cos\theta \\ p_1 &= -(3/8\pi)^{1/2}R_{n,l}(r)\sin\theta\exp(i\phi) \\ p_{-1} &= (3/8\pi)^{1/2}R_{n,l}(r)\sin\theta\exp(-i\phi)\end{aligned}\right\} \qquad (2.89)$$

Linear combinations are formed with p_1 and p_{-1} to give the functions p_x and p_y, represented as standing waves:

$$\left.\begin{aligned}p_x &= (1/\sqrt{2})(p_- - p_+) = (3/4\pi)^{1/2}R_{n,l}(r)\sin\theta\cos\phi \\ p_y &= (i/\sqrt{2})(p_- + p_+) = (3/4\pi)^{1/2}R_{n,l}(r)\sin\theta\sin\phi\end{aligned}\right\} \qquad (2.90)$$

In these two equations $1/\sqrt{2}$ is an additional normalization factor arising from the linear combination. Species without defined axes, such as atoms and linear molecules may be examined by the complex functions; with other species, where x, y and z axes have been assigned, the real functions shown on the right-hand sides of Eq. (2.90) are more appropriate. The angular components are of

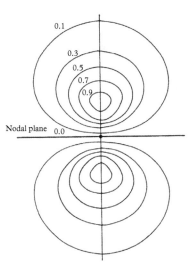

Fig. 2.18 Electron density contours for a $2p_z$ orbital of carbon as fractions of $|\psi_{2p_z}|^2_{\max}$. The 0.1 contour surface encloses *ca.* 66% of the $2p_z$ electron density; 95% is enclosed by a 0.02 contour (not shown).

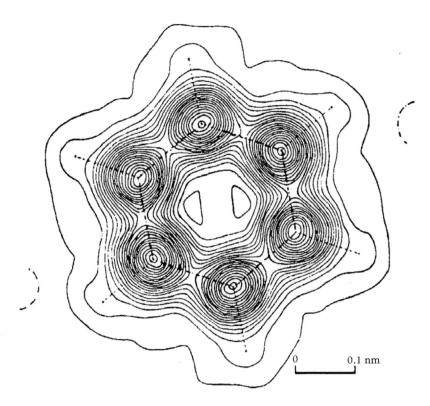

Fig. 2.19 A very highly resolved x-ray crystallographic electron density map of benzene, C_6H_6. Although the hydrogen atoms are not fully resolved, the contour lines indicate clearly the regions of their sites. [Cox EG *et al.* 1958. *Phil. Trans. Roy. Soc.* A247: 1; reproduced by courtesy of the Royal Society, London]

particular interest here, since the associated radial functions do not change parity under symmetry transformations. The linear combinations are standing waves with no net orbital angular momentum around the z-axis, because they represent a superposition of functions with equal and opposite values of m_l.

The p_x, p_y and p_z atomic functions are proportional to their parent wavefunctions p_0, p_1 and p_{-1}, respectively, and Table 2.6 lists the angular atomic orbital functions for $n = 1$ to 3. Fig. 2.20 illustrates the central section of the boundary surface of a $2p_x$ function; the label Y here is not a surface harmonic of Table 2.1, but rather the p_x orbital function listed in Table 2.6; Y^2 is simply the square of that function.

Table 2.6 *Angular components Y' of s, p and d atomic orbitals*

Orbital type	Angular function	$f(x, y, z)$
s	—	—
p_z	$\cos \theta$	z
p_x	$\sin \theta \cos \phi$	x
p_y	$\sin \theta \sin \phi$	y
d_{z^2}	$3 \cos^2 \theta - 1$	$2z^2 - (x^2 + y^2 + z^2) \equiv z^2$
$d_{x^2-y^2}$	$\sin^2 \theta \cos 2\phi$	$x^2 - y^2$
d_{xz}	$\sin \theta \cos \theta \cos \phi$	xz
d_{yz}	$\sin \theta \cos \theta \sin \phi$	yz
d_{xy}	$\sin^2 \theta \sin 2\phi$	xy

Fig. 2.20 A central section of the boundary surface of the p_x orbital function, with Y^2 showing the dumbbell shape of the p_x orbitals. The labels Y and Y^2 serve a function similar to that in Fig. 2.17, but with the proviso given in the text for the p_x and p_y functions; again, the $\pm$ signs relate to the Y function. This figure and that of Fig. 2.17 were obtained by plotting with program ANFN the field figures obtained by program PLOT.

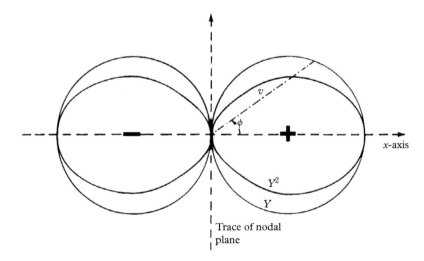

It is normal practice to use these linear combination p_x and p_y orbitals and similar orbital functions in studying covalent bonding. However, it may be noted that they are not true wavefunction: the functions derived by linear combinations such as those in Eq. (2.90) cannot be described by quantum numbers and cannot be used to describe electronic orbital angular momentum or spin [11, 12].

The energies of the three p orbitals are threefold degenerate since they depend upon the principal quantum number n. Similar comments apply to d (and higher) orbitals, where only $d_0 (\equiv d_{z^2})$ is real, and for them the energies are five-fold degenerate. The degenerate p orbitals have the same shape; the d orbitals, although degenerate, have three important shapes. These orbitals are included in the program ANFN and can also be viewed on the 'Orbitron' web site. The s, p and d angular functions corresponding to the orbital types listed in Table 2.6 are frequently illustrated as in Fig. 2.21. It can be seen that the p orbitals differ only in

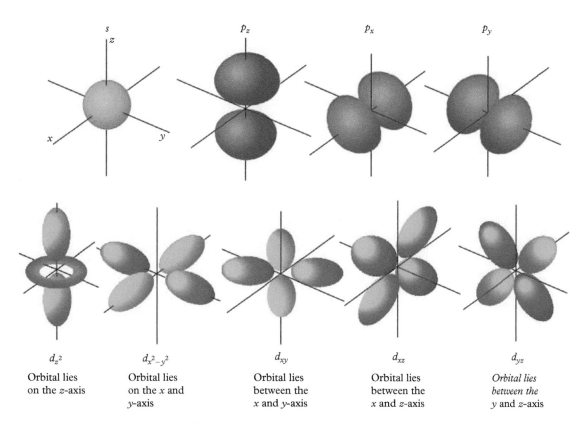

Fig. 2.21 Surfaces of constant $|\psi|^2$ for the s, p and d atomic orbitals, containing 95% of the electron density; the directions of the $+x$, $+y$ and $+z$ axes for all orbitals are shown on the s orbital diagram. The notation for the p orbitals is straightforward, and that for the d orbitals is indicated on the figure. [Reproduced by courtesy of Daniel Damelin, The Concord Consortium.]

their orientation with respect to the reference axes, as do the $3d_{xy}, 3d_{xz}$ and $3d_{yz}$. Fuller tables of surface harmonics, radial functions and complete wavefunctions can be found in the literature [6, 13, 14].

Table 2.7 *Quantum mechanical screening constants and effective nuclear charges [15]; earlier values (in parentheses) are Slater values [16]*

Atom	$\sigma(1s)$	$\sigma(2s)$	$\sigma(2p)$	$\sigma(3s)$	$\sigma(3p)$
H	1				
He	1.6875 (1.70)				
Li	2.6906 (2.70)	1.2792 (1.30)			
C	5.6727 (5.70)	3.2166 (3.25)	3.1358 (3.25)		
N	6.6651	3.8474	3.8340		
O	7.6579	4.4916	4.4532		
Na	10.6359	6.5714	6.8018	2.5074 (2.20)	
Ar	17.5075	12.2304	14.0082	7.7568	6.7641

2.13.5 Screening constant and effective nuclear charge

In multi-electron atoms the outer electrons experience a repulsion by the inner electrons. The result is an effective reduction of the nuclear charge experienced by the outer electrons to an extent σ known as a *screening constant*. The *effective nuclear charge* Z_{eff} is then:

$$Z_{eff} = Z - \sigma \tag{2.91}$$

Values of the screening constant σ have been given by Slater [15], together with a selection from a more recent compilation [16] are given in Table 2.7. Complete wavefunctions using Z_{eff} and based on r rather than ρ are listed in Table 2.3.

2.14 Aufbau principle

The description of atomic orbitals is linked closely with the periodic table of the elements, which is illustrated fully on the inside front cover of the book.

Beginning with hydrogen, electrons may be considered to be fed in, one at a time, into atomic orbitals thus building up electron configurations. This procedure has been termed the *Aufbau* principle (Ger. *Aufbau* = building up).

In Fig. 2.22, atomic orbitals are represented by rectangular cells, and electrons with opposed spins are shown as ↑↓. The elements H, He, Li and Be present no difficulty; in boron the $2p$ electron can occupy any one of three $2p$ orbitals, and it matters not which is used. With carbon, however, the second p electron could be accommodated in two differing ways: either it pairs with a spin opposite to that of a $2p$ electron already present from the build-up, or it can enter an unoccupied atomic orbital. The dilemma is resolved through *Hund's multiplicity rule*, which states that degenerate or near-degenerate orbitals tend to be occupied singly by electrons with parallel spins. The rule follows from the wave mechanical property of *spin correlation*, which describes the tendency of electrons with unpaired spins to repel each other less than do those with paired spins. Spin correlation causes a slight decrease in atomic volume, thereby enhancing the attraction between the electrons and the nucleus [17], and so producing an energetically more stable system.

The stability of the electron configurations of the inert gases was the pillar of Lewis's electron-pair bond theory and the foundation stone of the valence shell electron-pair repulsion theory. It is realized by systems of orbitals, each with paired spins, up to and including those of a given principal quantum number n. Thus, Fig. 2.22 illustrates diagrammatically the build-up of orbitals for

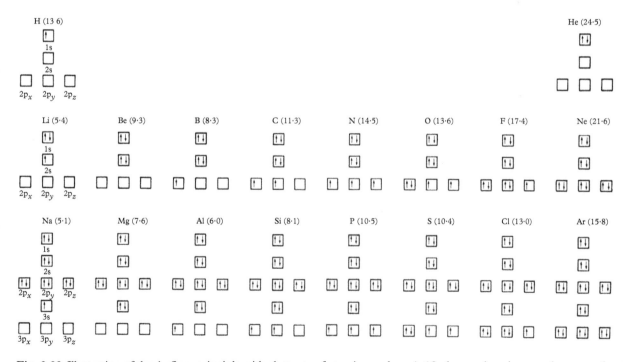

Fig. 2.22 Illustration of the Aufbau principle with elements of atomic numbers 1–18; the numbers in parentheses are the first ionization energies in eV.

$n = 1 - 3$, atomic numbers 1–18, and lists in parentheses the first ionizations energies of the elements in eV. With an obvious interpretation, the following electron configurations can be written:

H	$(1s)^1$
He	$(1s)^2$
Li	$(1s)^2 (2s)^1$
–	
–	
C	$(1s)^2 (2s)^2 (2p)^2$
–	
–	
Ne	$(1s)^2 (2s)^2 (2p)^6$
Na	$(1s)^2 (2s)^2 (2p)^6 (3s)^1$ or (Ne) $(3s)^1$
Na^+	(Ne)
–	
–	
Si	(Ne) $(3s)^2 (3p)^2$
–	
–	
Ar	(Ne) $(3s)^2 (3p)^6$ or (Ar)
K	(Ar) $(4s)^1$
-	
-	

By absorption of energy, an electron may be raised to an energy level higher than that of the lowest energy level, or *ground state*. For example, the excited state

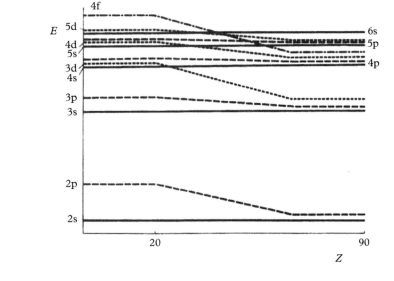

Fig. 2.23 Variation in energy E with atomic number Z for atomic orbitals with principal quantum numbers n of 2–6. In particular, the nd orbitals lie between the $(n + 1)s$ and $(n + 1)p$ for $n \geq 3$ and $Z < 20$, and cross the $(n + 1)s$ line to lower energies for $Z > 20$. [*Introduction to Physical Chemistry*, 3rd ed. 1998; reproduced by courtesy of Cambridge University Press.]

carbon C* may be written as $(1s)^2(2s)^1(2p)^3$ in which a $2s$ electron is promoted to the $2p$ level transiently prior to an imminent bonding operation.

The periodic table is not continued exactly in the manner indicated by the above example. Between calcium $(Ar)(4s)^2$ and zinc $(Ar)(3d)^{10}(4s)^2$ there exists the first transition series of elements, where the $3d$ orbitals are progressively filled. Even this is not as simple as it seems: for example, copper is written as $(Ar)(3d)^{10}(4s)^1$ and not as $(Ar)(3d)^9(4s)^2$ as might be imagined. More detailed calculations indicate that the magnetic properties of copper are best explained by a statistical distribution corresponding to $(Ar)(3d)^{9.5}(4s)^{1.5}$. Energy levels for atoms are indicated qualitatively in Fig. 2.23; it may be noted that for $Z > 20$ the nd levels cross the $(n+1)s$ levels to lower energies. For further discussion on the periodic table the reader is referred to modern textbooks of inorganic chemistry [18, 19].

2.15   Multi-electron species

In species with more than one electron, the analytical solution of the Schrödinger equation is intractable owing to the presence of multiple $1/r$ electron–electron interaction terms which are not separable. In a molecule, each electron comes under the influence of the potential fields of the nucleus and the other electrons. In the simple example of the hydrogen molecule, the Hamiltonian operator becomes:

$$\mathcal{H} = -[\hbar^2/(2M)](\nabla_A^2 + \nabla_B^2) - [\hbar^2/(2\mu)](\nabla_1^2 + \nabla_2^2) -$$
$$\{[e^2/(4\pi\varepsilon_0)][1/r_{A1} + 1/r_{A2} + 1/r_{B1} + 1/r_{B2} + 1/r_{12} + 1/r_e]\} \tag{2.92}$$

where the terms have meanings as before. The six particle interactions for the hydrogen molecule are shown schematically in Fig. 2.24.

In Section 2.11, the *Born–Oppenheimer approximation* was described, and it finds its full force in multi-electron problems, as nuclear interactions are neglected in a first-order approximation. Effectively, it separates the kinetic energy of the nuclei from the electronic energy and permits a calculation of the latter for fixed values of the internuclear distance r. The total energy is then obtained as the sum of the eigenvalues from Eq. (2.29) and the repulsion energy of the two nuclei, at each selected value of r. Using different values of r with the same procedure leads to a potential energy curve (Fig. 2.25). At the equilibrium distance r_e, the energy is solely potential; at this value of r the *total energy* is D_e, whereas the comparable experimental value is the *dissociation energy* D_0, the difference between D_e and D_0 being the zero-point energy of the species. For polyatomic molecules the situation is more complicated, but a position similar to that expressed by Fig. 2.25 obtains.

There are two main methods by which solutions to the Schrödinger equation may be realized within the Born–Oppenheimer approximation, the valence-bond theory and the molecular orbital theory, of which the latter will be considered in the greater detail.

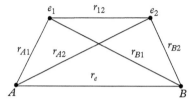

Fig. 2.24 Schematic illustration of the six particle interactions between two nuclei A and B and two electrons e_1 and e_2 in the hydrogen molecule; the distance r_e is that corresponding to the minimum energy configuration of the molecule.

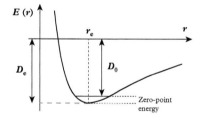

Fig. 2.25 Variation of energy $E(r)$ with internuclear distance r for a diatomic molecule: the minimum theoretical energy D_e differs from the experimental dissociation energy value D_0 by the zero-point energy of vibration.

2.16   Valence-bond theory

Valence-bond theory was introduced by Heitler and London in 1927 and represents the earliest application of wave mechanics to chemical bonding. It leans towards the Lewis electron-pair bonding and introduces the concepts of *hybridization*, as in the sp^3 hybrid orbitals in tetrahedral configurations, and *canonical structures*, as in the structures of benzene shown hereunder where the $\leftrightarrow$ sign represents a wave mechanical combination, and not a simple mixing or equilibrium situation:

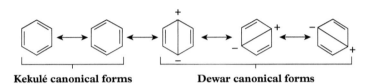

Kekulé canonical forms   **Dewar canonical forms**

2.16.1   Homonuclear diatomic molecules

In the formation of the hydrogen molecule two atoms are brought together from effectively infinite separation, passing through the region of separated atoms until their distance apart r brings them within a *bond-forming region*. A bond is formed by the pairing of opposing electron spins, one spin from each of the atomic orbitals of the two atoms. In the separated atom state electron 1 is on atom A and electron 2 is on atom B, and the eigenfunction may be written as $\Psi = \psi_{A1} \psi_{B2}$, since the total wavefunction of two uncorrelated wavefunction is their product, and the eigenvalue E is given by the sum $E_A + E_B$.

Within the bond-forming region, however, $\Psi = \psi_{A2} \psi_{B1}$ is an equally probable formulation. The wave mechanical construct shows that the true state is best represented as:

$$\Psi = c_1 \psi_{A1} \psi_{B2} \pm c_2 \psi_{A2} \psi_{B1} \tag{2.93}$$

where c_1 and c_2 are the fractions of each function. In the case of a homonuclear diatomic molecule $c_1 = c_2$. Calculations showed that the lower energy system is the *bonding function*:

$$\Psi_+ = \psi_{A1} \psi_{B2} + \psi_{A2} \psi_{B1} \tag{2.94}$$

The formation of an H–H bond implies a high probability that the electrons will occupy the region between nuclei A and B. Fig. 2.26 is a plot of $|\psi|^2$ in arbitrary units against R, the displacement of A and B from r_e, which is marked as 0 on the abscissa. The separate A and B curves correspond to $\psi_{A1} \psi_{B2}$ and $\psi_{A2} \psi_{B1}$, with the AB curve as the combination function $\psi_{A1} \psi_{B2} + \psi_{A2} \psi_{B1}$, where A and B have undergone constructive interference leading to an enhancement of density

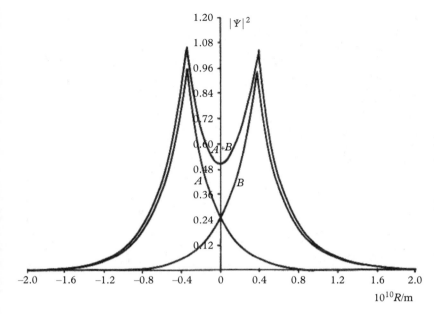

Fig. 2.26 Formation of the hydrogen H_2 molecule. The A and B density curves correspond to $\psi_{A1}\psi_{B2}$ and $\psi_{A2}\psi_{B1}$, plotted against R, the displacement of the nuclei from the equilibrium value r_e ($R = 0$); the AB curve is the sum ($\psi_{A1}\psi_{B2} + \psi_{A2}\psi_{B1}$) and shows an enhancement of density in the AB internuclear region. [*Introduction to Physical Chemistry*, 3rd ed. 1998; reproduced by courtesy of Cambridge University Press.]

in the internuclear region. The H–H bond type has cylindrical symmetry and is termed a σ-*bond*.

This simple application of the valence-bond theory gave r_e = 0.074 nm, whereas the experimental value is 0.087 nm. The calculated energy E^+ = −303 kJ mol^{-1}, and is equivalent to D_e, whereas the corresponding experimental value is −458 kJ mol^{-1}. The agreement is illustrated in the diagram below, which shows also the *antibonding function* $\Psi_- = \psi_{A1}\psi_{B2} - \psi_{A2}\psi_{B1}$ and its corresponding energy is E^-.

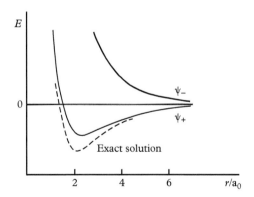

Valence-bond results for the hydrogen molecule: ψ_+ bonding, ψ_- antibonding

By compounding a wavefunction of fifty terms, a value for E^+ of -457.99 kJ mol^{-1} was obtained, the best experimental value being -457.99 ± 0.07 kJ mol^{-1}.

As an example of the application of the valence-bond model to a polyatomic species, the water molecule will be considered. The electron configuration of oxygen is $(1s)^2(2s)^2(2p)^4$ with electron spins paired in one of the three $2p$ orbitals. The other two $2p$ orbitals combine (overlap) the $1s$ orbitals of two hydrogen atoms forming spin-paired σ-bonds. Since the p orbitals are mutually perpendicular, a bond angle of 90° is indicated. The angle is known from experiment to be 105.4°, and it is possible to refine the valence-bond theory by introducing the concept of hybridization.

An O–H hybrid function may be written in the form:

$$\Psi = c_s \psi_s + c_p \psi_p \tag{2.95}$$

where c_s and c_p are fractional contributions of ψ_s and ψ_p functions, respectively. A total of eight valence electrons implies four hybrid orbitals. Two are formed with hydrogen and oxygen $2p_x$ and $2p_y$ and two orbitals are occupied by lone-pair electrons of oxygen. The four hybrid orbitals are directed approximately towards the apices of a regular tetrahedron: two of the orbitals overlap with hydrogen $1s$ orbitals, whereas the other two contain lone-pair electrons. The tetrahedral bond angle is 109.47°, which is nearer to the true value than is 90°. It can be envisaged that the repulsion of the two lone pairs of electrons perturbs the tetrahedral angle such that the smaller, true bond angle of 104.5° obtain.

The two lone pairs of electrons make a major contribution to the dipole moment of 1.8 D.[9] This value may be compared with 0.2 D for the F_2O molecule, where the larger polarizability of the F–O bond, directed away from the lone pairs, reduces their contribution to the dipole moment, thus supporting the picture deduced for the water molecule. Some of this discussion on the water molecule is illustrated through Fig. 2.27a–c.

2.16.2 Heteronuclear diatomic molecules

In a heteronuclear diatomic species, the electron distribution is polarized, that is, there is an accumulation of charge on one species, the more electronegative, and a corresponding depletion on the other; the bond is said to be polar, or possessing a *dipole moment*.

The dipole moment[10] μ of a linear species is given by:

$$\mu = |q|ed \tag{2.96}$$

where $|q|$ is the magnitude of the charge at each end of the representative dipole moment vector, e is the electronic charge and d is the length of the dipole. In hydrogen fluoride, $\mu = 1.80$ D by experiment and the H–F bond length d is 0.0927 nm. Hence, $|q| = 0.404$, so that the nominal charge on the fluorine atoms is -0.40 with $+0.40$ on hydrogen.

[9] 1 D = 3.33564×10^{-30} C m.
[10] The symbol p is used for dipole moment in some literature.

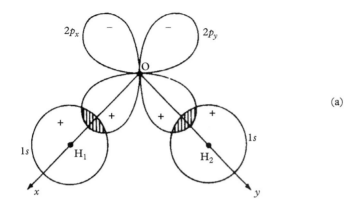

(a)

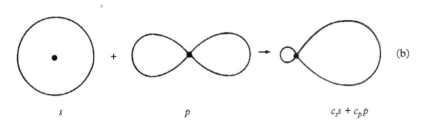

(b)

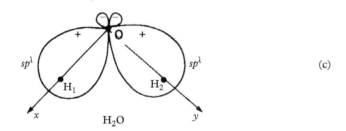

(c)

Fig. 2.27 Diagrammatic valence-bond model for the formation of the water molecule. (a) Bonding by $O(2p)$ and $H(1s)$ overlap would lead to a bond angle of 90°. (b) Wave mechanical mixing forms the hybrid carbon orbital $c_s\psi_s + c_p\psi_p$. (c) Bonding with the hybrid orbital and $H(1s)$ leads to strong overlap and a bond angle of 104.4° in the equilibrium configuration with $c_s^2/c_p^2 \approx 1/4$. [*Introduction to Physical Chemistry*, 3rd ed. 1998; reproduced by courtesy of Cambridge University Press.]

Bond formation can be anticipated between $H(1s)$ and $F(2p)$, and a valence-bond bonding wavefunction would be written as:

$$\Psi = c_H\psi(H_{1s}) + c_F\psi(F_{2p}) \tag{2.97}$$

On account of the lack of symmetry, the proportions of the component orbitals is unequal, and calculations show that $c_H^2/c_F^2 = 0.11/0.89$.

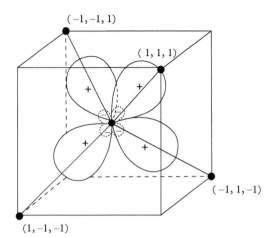

Fig. 2.28 Valence-bond model for hybrid sp^3 carbon orbitals in methane. The carbon atom lies at the centre of the cube, with the C–H bonds along its body diagonals. The coordinates in parentheses specify the corners of the cube with reference to the carbon atom as origin and a unit distance equal to half a cube edge length. In crystallographic terms they are the *directions* <111>.

2.16.3   Other molecular species

The valence-bond method cannot immediately account for four equal bonds from carbon in the methane molecule, bonds that are directed in a regular tetrahedral manner to the four hydrogen atoms. A simple argument based on promotion of a $2s$ electron to a $2p$ level, similar to that used in the discussion of the water molecule, would lead to three bonds of one type and one other. The lack of equivalence is overcome by postulating sp^3 hybrid orbitals, a linear combination of $2s$ and three $2p$ orbitals, which are directed to the corners of a cube, as indicated in Fig. 2.28.

Many other chemical species can be described by the valence-bond method, notably compounds of metals in the *transition series* of elements, and d^2sp^3 is an example configuration found in octahedral coordination compounds such as the hexamminecobalt(III) ion $[Co(NH_3)_6]^{3-}$, chromium hexacarbonyl $Cr[CO_6]$ and uranium hexafluoride UF_6. However, the discussion of the valence-bond method will not be considered further herein, as the molecular orbital method proves to be more exact and more fully developed. Several very comprehensive computer programs are available that are dedicated to large scale calculations in theoretical chemistry, based on molecular orbital theory.

2.17   Molecular orbital method: homonuclear species

The *molecular orbital* theory of the covalent bond was introduced by Mulliken in 1932. In this model, electrons are divorced from their original atoms and pervade the whole molecule, moving under the influence of the nuclei and of one another. The simplest molecular species is the hydrogen molecule-ion, H_2^+. Under the Born–Oppenheimer approximation, the Schrödinger equation for this species can be solved, albeit a somewhat complicated procedure, and it will prove more profitable to adopt a simpler model known as the *linear combination of atomic orbitals*, or concisely, the LCAO molecular orbital method (LCAO_MO). First, however, it is necessary to discuss the *variation principle*, and to use the hydrogen molecule-ion and hydrogen molecule as exemplars.

2.17.1 Variation principle

The lowest energy configuration for hydrogen is $1s$, and it can be assumed that the ground state molecular orbital of the hydrogen molecule-ion is similar. An LCAO_MO wavefunction can be written as:

$$\Psi = c_1\psi_1 + c_2\psi_2 \tag{2.98}$$

Notwithstanding what is known about the species it is not assumed that $c_1 = c_2$, but the atomic wavefunctions are known to be orthonormal, that is, $\int \psi_i\psi_j^* d\tau = \delta_{i,j}(i,j = 1,2)$, where $\delta_{i,j}$ is the *Kronecker delta*.[11] Wave mechanical operators, such as the Hamiltonian and momentum operators are *Hermitian*, that is, they obey the relation:

$$\int \psi_m^* O\psi_n d\tau = \left(\int \psi_n^* O\psi_m d\tau \right)^* \equiv \int (O\psi_m)^* \psi_n d\tau \tag{2.99}$$

where ψ and ψ^* are normalized wavefunctions and O is a wave mechanical operator; the eigenvalues of Hermitian operators are real [6, 7]. Introducing Ψ from Eq. (2.98) into Eq. (2.28) and then multiplying both sides by Ψ^* and integrating, the energy (eigenvalue) E is given as:

$$E = \frac{\int \Psi^* \mathcal{H}_e \Psi d\tau}{\int \Psi^* \Psi d\tau} \tag{2.100}$$

As the individual wavefunctions ψ are also real, the denominator in Eq. (2.100) is, from Eq. (2.98):

$$\int \Psi^2 d\tau = \int (c_1\psi_1 + c_2\psi_2)^2 d\tau = c_1^2 \int \psi_1^2 d\tau + c_2^2 \int \psi_2^2 d\tau + 2c_1c_2 \int \psi_1\psi_2 d\tau$$

$$= c_1^2 + c_2^2 + 2c_1c_2 S \tag{2.101}$$

The parameter S is the *overlap integral*, defined as $\int \psi_1\psi_2^* d\tau$, or here with real functions as:

$$S = \int \psi_1\psi_2 d\tau \tag{2.102}$$

[11] $\delta_{i,j} = 1$ for $i = j$, otherwise $\delta_{i,j} = 0$.

The numerator in Eq. (2.100), taking Ψ as a real function, is:

$$\int \Psi \mathcal{H}_e \Psi \, d\tau = \int (c_1\psi_1 + c_2\psi_2)\mathcal{H}_e(c_1\psi_1 + c_2\psi_2)d\tau$$

$$= c_1^2 \int \psi_1\mathcal{H}_e\psi_1 d\tau + c_2^2 \int \psi_2\mathcal{H}_e\psi_2 d\tau + c_1 c_2 \int \psi_1\mathcal{H}_e\psi_2 d\tau \quad (2.103)$$

$$+ c_1 c_2 \int \psi_2\mathcal{H}_e\psi_1 d\tau$$

Normal practice sets the *Coulomb integral* $\int \psi_1 \mathcal{H}_e \psi_1^* d\tau$ equal to the parameter α, but with real wavefunctions the integrals become:

$$\int \psi_1\mathcal{H}_e\psi_1 d\tau = \alpha_1$$
$$\int \psi_2\mathcal{H}_e\psi_2 d\tau = \alpha_2 \quad (2.104)$$

which equations represent electrons 1 and 2 in their respective atomic orbitals and in the molecular environment. Since the Hamiltonian commutes, the last two integrals of Eq. (2.103) may be written as:

$$2\int \psi_1\mathcal{H}_e\psi_2 d\tau = \beta \quad (2.105)$$

and β is termed the *resonance integral*.

Hence, E may now be formulated from Eq. (2.100) as:

$$E = \frac{c_1^2\alpha_1 + c_2^2\alpha_2 + 2c_1 c_2\beta}{c_1^2 + c_2^2 + 2c_1 c_2 S} \quad (2.106)$$

and represents the interaction between electrons 1 and 2 in the molecular environment.

The minimum energy conformation corresponds to the partial differentials of Eq. (2.106) with respect to c_1 and c_2 set to zero. With a little straightforward manipulation, the results are:

$$\frac{\partial E}{\partial c_1} = \frac{2[c_1(\alpha_1 - E) + c_2(\beta - SE)]}{c_1^2 + c_2^2 + 2c_1 c_2 S} = 0$$

$$\frac{\partial E}{\partial c_2} = \frac{2[c_2(\alpha_2 - E) + c_1(\beta - SE)]}{c_1^2 + c_2^2 + 2c_1 c_2 S} = 0 \quad (2.107)$$

which lead to the *secular equations*:

$$c_1(\alpha_1 - E) + c_2(\beta - SE) = 0$$
$$c_2(\alpha_2 - E) + c_1(\beta - SE) = 0 \quad (2.108)$$

since $\alpha_1 = \alpha_2$, the equations are solvable if the *secular determinant* is equal to zero, that is, if:

$$\begin{vmatrix} (\alpha_1 - E) & (\beta - SE) \\ (\beta - SE) & (\alpha_2 - E) \end{vmatrix} = 0 \quad (2.109)$$

2.17.1.1 *Hydrogen molecule-ion and hydrogen molecule*

In the hydrogen molecule-ion, $c_1 = \pm c_2$ from its symmetry, so that Eq. (2.109) becomes $(\alpha - E)^2 = (\beta - ES)^2$, or:

$$E_\pm = \frac{\alpha \pm \beta}{1 \pm S} \tag{2.110}$$

If these values for E are substituted into Eq. (2.108), then $c_1 = \pm c_2$ as anticipated, and $c_\pm$ is obtained by normalization:

$$1 = c_\pm^2 \int (\psi_1 \pm \psi_2)^2 d\tau = c_\pm^2 \int (\psi_1^2 + \psi_2^2 \pm 2\psi_1\psi_2)d\tau = c_\pm^2(1 + 1 \pm 2S)$$

whence $c_\pm$ becomes $1/(2 \pm 2S)^{1/2}$ for the two wavefunctions $c_\pm(\psi_1 \pm \psi_2)$. The value of S in H_2^+ at its equilibrium internuclear separation of 0.11 nm is 0.56; this value is atypically large, as it is also in the hydrogen molecule. From the normalization condition, the values $c_+ = 0.566$ and $c_- = 1.066$ lead to two wavefunctions for the hydrogen molecule-ion, bonding (+) and antibonding (−):

$$\begin{aligned} \Psi_+ &= 0.566(\psi_1 + \psi_2) \\ \Psi_- &= 1.066\,(\psi_1 - \psi_2) \end{aligned} \tag{2.111}$$

with energies given by Eq. (2.110). Since both α and β are negative quantities, Eq. (2.110) shows that Ψ_+ corresponds to the lower (more negative) energy E_+ (Section 2.16.1). It forms the occupied bonding orbital $1s\sigma$ (or $1\sigma_g$ in an alternative notation), whereas Ψ_- corresponds to the more positive energy E_- and is the unoccupied antibonding orbital $1s\sigma^*$ (or $1\sigma_u$). The energy-level diagram for the hydrogen molecule-ion is shown in Fig. 2.29(a) together with that for the

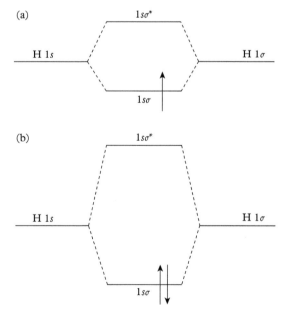

Fig. 2.29 Molecular orbital energy level diagrams, with a commonly used notation for the molecular orbitals. (a) Hydrogen molecule-ion, H_2^+. (b) Hydrogen molecule, H_2. In this and other similar diagrams a single arrow, ↑ or ↓ (or a half-arrow ⇂ or ↿) represents an unpaired electron, and ↑↓ (or two half-arrows) a pair of electrons with opposed spins. The notations $1s\sigma$ and $1s\sigma^*$ refer to bonding and antibonding molecular orbitals respectively. (This notation is not the only one in use).

hydrogen molecule (b); the similarity between these two species is brought out in this diagram. In a more refined calculation on the hydrogen molecule, and in more complex species, other factors such electron–electron repulsion need to be taken into account.

2.17.1.2 Orbital symmetry

Chemical species exhibit symmetry, a property by which they may be usefully classified. With the examples of diatomic molecules, the following classes can be defined:

(a) Orbitals that are symmetric about the direction of an internuclear axis are σ-molecular orbitals (Fig. 2.30a–d);

(b) Orbitals that possess a nodal plane containing the internuclear axis are π-molecular orbitals (Fig. 2.30e–f);

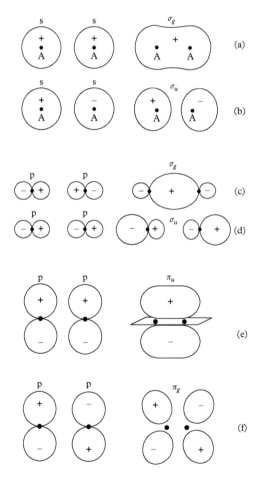

Fig. 2.30 Atomic orbitals and corresponding molecular orbitals for homonuclear diatomic species: σ_g and π_u are bonding, whereas σ_u and π_g are antibonding. The g/u notation relates to the centrosymmetry of the species.

(c) Orbitals may be described as even g (Ger. *gerade* = even) or as odd u (Ger. *ungerade* = uneven, odd) with respect to inversion across the centre of the internuclear axis (Fig. 2.30a–f); the subscripts g and u are termed the *parity* of the orbital.

The figure shows that σ_g- and π_u-molecular orbitals are bonding, whereas σ_u- and π_g-molecular orbitals are antibonding. The notations σ_u and σ^* are used to represent like types of orbitals, similarly with π_g and π^*; σ_g with π_u and σ_u with π_g represent like types of symmetry.

For species with $n = 2$, the outer (valence) electrons are $2s$ and $2p$; the $1s$ electrons form a relatively inert core, the effect of which on the valence electrons may be neglected in a first approximation.

2.17.1.3 Heteronuclear species

As an example of a heteronuclear diatomic molecule the study of hydrogen fluoride (Section 2.16.2) will be continued, now with the molecular orbital method. The electronic configuration of fluorine is $(1s)^2(2s)^2(2p)^5$; ionization energies can provide a guide to the orbitals most likely to be involved in bonding:

	F($2s$)	F($2p$)	H($1s$)
I/eV	40.2	18.6	13.6

These data indicate that bond formation would be expected between H($1s$) and F($2p$), and the expected two LCAO molecular orbitals are:

$$\Psi_{\pm} = c_H H(1s) \pm c_F F(2p) \tag{2.112}$$

Although H($1s$) and F($2s$) have the same symmetry, their energies are too far apart for bond formation when compared with the energies of H($1s$) and F($2p$), which also have correct symmetry for bonding (rotation about the molecular axis leaves these $1s$ and $2p_z$ orbitals unchanged). The ratio of the coefficients determined by the variation method is $c_H^2/c_F^2 = 0.11/0.89$ in the bonding orbital, and $0.89/(0.11)$ for the antibonding orbital (see Section 2.16.2). Again, in this species the core electrons are neglected, and bonding overlap arises only for atomic orbitals of the same symmetry and of similar energy.

A first approximation used often with a heteronuclear diatomic involves setting S in the secular determinant Eq. (2.109) to zero. Then:

$$\begin{vmatrix} (\alpha_H - E) & \beta \\ \beta & (\alpha_F - E) \end{vmatrix} = 0$$

which, after evaluation, becomes:

$$E = \frac{(\alpha_H + \alpha_F) \pm \sqrt{(\alpha_H - \alpha_F)^2 + 4\beta^2}}{2} \tag{2.113}$$

If, one the one hand, the energy difference of the participating atomic orbitals is large, that is, $|\alpha_H - \alpha_F| \gg 2\beta$ then E tends to the two values α_H and α_F; these values represent the energies of the individual orbitals, and form a non-bonding orbital. On the other hand, if the difference is small, so that $(\alpha_H - \alpha_F) \ll 2\beta$, then E tends to the values $(\alpha_H + \alpha_F \pm 2\beta)/2$, corresponding to bonding (+) and antibonding (–) molecular orbitals.

A molecular orbital energy-level diagram for hydrogen fluoride is shown in Fig. 2.31. Hydrogen $1s$ lies above fluorine $2p$, in accordance with the ionization energies listed. Hydrogen $1s$, and fluorine $2p_z$ form a σ-bond, 3σ on this diagram, an alternative notation, implying that 1σ and 2σ molecular orbitals respectively represent the $F(1s)^2$ and $F(2s)^2$ core; the $3\sigma^*$ orbital is antibonding. The $2p_x$ and $2p_y$ orbitals, perpendicular to the internuclear axis, form non-bonding 1π-molecular orbitals, as shown in the diagram. The complete molecular orbital description, using Fig. 2.31, is $[(1s\sigma)^2(2s\sigma)^2](3s\sigma)^2(1p\pi)^4$. The portion of the configuration in brackets contributes to the core, and the *net bonding parameter* κ is 1:

$$\kappa = (n_e - n_e^*)/2 \tag{2.114}$$

where n_e and n_e^* are the numbers of electrons in the bonding and antibonding orbitals, respectively.

2.17.1.4 *Nitrogen and oxygen molecules*

Similar molecular orbital treatments of nitrogen and oxygen lead to the following molecular orbital configurations:

$$N \ [(1s\sigma)^2(1s\sigma^*)^2(2s\sigma)^2(2s\sigma^*)^2](2p\pi)^4(2p\sigma)^2$$

$$O \ [(1s\sigma)^2(1s\sigma^*)^2(2s\sigma)^2(2s\sigma^*)^2](2p\sigma)^2(2p\pi)^4(2p_x\pi^*)(2p_y\pi^*)$$

The energy level diagrams for nitrogen and oxygen are shown in Figs. 2.32 and 2.33, respectively. The $2p\sigma$ and $2p\pi$ energies are *reversed in order* after

Fig. 2.31 Molecular orbital energy level diagram for the hydrogen fluoride HF molecule. The F $(1s)^2$ and $(2s)^2$ electrons form a core; the degenerate 1π orbitals (from F(2) and F($2p_y$) are different in symmetry from H($1s$) and are non-bonding; F($2p_z$) and H($1s$) are correct in symmetry and commensurate in energy for σ-bonding; $\kappa = 1$.

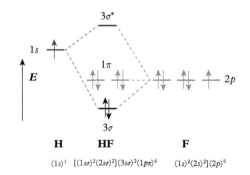

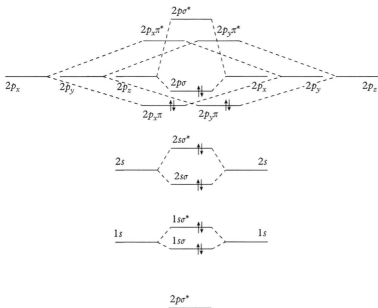

Fig. 2.32 Molecular energy level diagram for nitrogen N_2. Triple bonding is through $(2p\pi)^4$ and $(2p\sigma)^2$; all electrons are paired, and $\kappa = 3$.

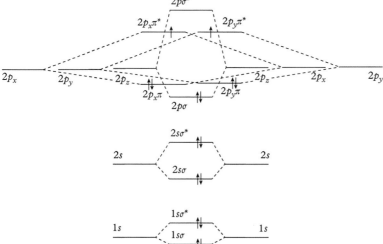

Fig. 2.33 Molecular energy level diagram for oxygen O_2. The positions of $2p\pi$ and $2p\sigma$ energy levels are reversed relative to nitrogen. The electrons are paired except for the $2p_x$ π^* and $2p_y$ π^* molecular orbitals, the single occupancies of which lead to paramagnetism in oxygen; $\kappa = 2$.

nitrogen; for molecules in later horizontal periods, the $p\sigma$ level lies below the $p\pi$ in energy. This change depends upon the separation of the $2s$ and $2p$ energies, which decreases with increasing atomic number, and on the fact that the energy of a half-filled $2p$ orbital is raised because of electron–electron repulsion. The unpaired electrons in $2p_x\pi^*$ and $2p_y\pi^*$ are responsible for the paramagnetic property of oxygen. Time is well spent reviewing the animated σ- and π-bond formations on the 'Orbitron' web site.

2.17.2 Symmetry-adapted molecular orbitals

A further study of methane will enable an introduction to *symmetry-adapted molecular orbitals* with the LCAO_MO procedure. In any set of molecular orbitals the electrons supplied by the atoms are delocalized over the molecule. Each molecular orbital is represented by a wavefunction Ψ that is compounded by the LCAO procedure. Generally, a molecular orbital contains two nuclei and up to two electrons; if two electrons are present in one and the same molecular orbital, their spins are opposed. Bonding molecular orbitals are formed by in-phase interaction between atomic orbitals, antibonding molecular orbitals by out-of-phase interaction: there is no formulation for a non-bonding molecular orbital; in other terminology, it forms a lone pair of electrons on the atom in question. The basic properties of these molecular orbitals may be summarized as follows:

- A *basis set* of orbitals comprises the atomic orbitals that are available for molecular orbital interaction, both bonding and antibonding;

- The *number* of molecular orbitals is always equal to the number of atomic orbitals included in the LCAO basis set;

- If the molecule under consideration has a degree of symmetry greater than identity, the consequent degenerate atomic orbitals are *grouped* by LCAO as symmetry adapted linear combinations of atomic orbitals. They belong to their group theoretical representations of the point group of the molecule, and the wavefunctions that describe them are the symmetry adapted linear combinations, or SALCs;

- The number of molecular orbitals belonging to *one representation* is equal to the number of SALCs belonging to that representation;

- In any representation, the SALCs form more readily if their atomic orbitals are of the *same symmetry* and *similar in energy*.

The applications of SALCs are expounded in detail in most books treating quantum chemistry and group theory [19–25]; here, methane is used as an example to outline of the procedure, but first some related topics must be introduced briefly.

2.17.2.1 *Representations*

In the present context, a *representation* may be considered as a set of matrices that shows how the symmetry operations of a point group act upon a function, such as an atom, a vector or an orbital. For example, a C_2 symmetry operation (twofold rotation) acting along the z-axis of a species and passing through the origin of coordinates has the following action on a function $f(x, y, z)$:

$$x \rightarrow -x, y \rightarrow -y, z \rightarrow z$$

and the symmetry is represented by the matrix operation:

$$\mathbf{C}_2 = \begin{pmatrix} \bar{1} & 0 & 0 \\ 0 & \bar{1} & 0 \\ 0 & 0 & 1 \end{pmatrix} \begin{pmatrix} x \\ y \\ y \end{pmatrix} \begin{pmatrix} \bar{x} \\ \bar{y} \\ z \end{pmatrix}$$

It is often convenient, and neat, to place the negative sign of an element above it. The *trace*, or *character*, of the matrix is, the sum of its diagonal elements; traces of symmetry matrices form the body of a *character table*.

2.17.2.2 *Symmetry elements, symmetry operations and point groups*

A point group may be defined as *a set of symmetry operations all of which act through a single fixed point*; this point is also the origin of the reference axes for the body. It follows that the symmetry operations of a point group must leave at least one point unmoved; in some cases it is a line or plane that is unmoved under the action of the point group.

A *symmetry operation* when applied to a body transforms it to a situation that is indistinguishable from its initial situation and thus reveals the symmetry inherent in the body according to the nature of the operation.

A *symmetry operator* is best regarded as a mathematical function that carries out a symmetry operation in a definite manner and orientation, and is most usefully represented in matrix form. It is that entity by virtue of which its corresponding symmetry operation is executed.

A *symmetry element* is a geometrical entity, a point, line or plane, in a body or assemblage, with which a symmetry operation is associated. A symmetry element is strictly conceptual, but it can be advantageous to endow it with a sense of reality. The symmetry element associates all parts of the assemblage in which it is present into a number of symmetrically related sets.

In the water molecule, the operation σ_v links the two hydrogen atoms by reflection across the $\sigma_v(x, z)$ plane.

All atoms in the water molecule are related spatially by its symmetry operations E, C_2, σ_v and σ_v'; their actions leave at least one point unmoved, namely, the origin at the centre of the oxygen atom.

The symmetry elements are displayed by the following diagram.

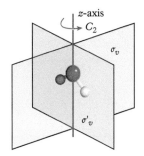

Symmetry elements of point group C_{2v}

They comprise the twofold axis C_2 along z, the symmetry (mirror) plane σ_v in the x, z plane, a second mirror plane σ'_v in the y, z plane (the plane of the molecule) and the identity element E at the origin, the point through which all symmetry elements pass.

The symmetry operators are E, C_2, σ_v and σ'_v and are represented by the matrices:

$$
\begin{pmatrix} 1 & 0 & 0 \\ 0 & 1 & 0 \\ 0 & 0 & 1 \end{pmatrix} \quad
\begin{pmatrix} \bar{1} & 0 & 0 \\ 0 & \bar{1} & 0 \\ 0 & 0 & 1 \end{pmatrix} \quad
\begin{pmatrix} 1 & 0 & 0 \\ 0 & \bar{1} & 0 \\ 0 & 0 & 1 \end{pmatrix} \quad
\begin{pmatrix} \bar{1} & 0 & 0 \\ 0 & 1 & 0 \\ 0 & 0 & 1 \end{pmatrix}
$$
$$\quad\quad E \quad\quad\quad\quad\quad C_2 \quad\quad\quad\quad\quad \sigma_v \quad\quad\quad\quad\quad \sigma'_v$$

which specify their orientation and action in generating the symmetry operations of the point group. The point group symbol for this group is C_{2v}. Point groups and symmetry operations are discussed in detail in the literature [22].

2.17.2.3 Character tables

Consider the point group C_{2v}, the symmetry operators of which have already been described. The actions of these symmetry operations on a point x, y, z lead to the following table, wherein each number indicates how an object or function transforms under a given symmetry; thus, x and xy transform under σ_v as x and $\overline{xy}$. In this point group there is only one symmetry operation in each of the symmetry classes E, C_2, σ_v and σ'_v. In point group T_d, for example, there are eight C_3 operations, one of which is indicated in Fig. 2.34: the others are in the directions ha, gd and fc, together with four C_3^2 operations, where C_3^2 means two successive C_3 operations in each of the same four orientations.

C_{2v}	E	C_2	σ_v	σ'_v
x	1	−1	1	−1
y	1	−1	−1	1
z	1	1	1	1
xy	1	1	−1	−1

For symmetry operations on product functions in C_{2v}, it follows that x^2, y^2 and z^2 transform like z under all operations of the group, yz transforms like y and zx like x; xy transforms differently, therefore it is listed in the above table, and its characters are the *product* of the characters of x and y under each symmetry operation. These four ways in which a function transforms under C_{2v} are tabulated, reordered, labelled A_1 to B_2 and formed into the *character table* for point group C_{2v}:

C_{2v}	E	C_2	σ_v	σ'_v		
A_1	1	1	1	1	z	x^2, y^2, z^2
A_2	1	1	−1	−1	R_z	xy
B_1	1	−1	1	−1	x, R_y	zx
B_2	1	−1	−1	1	y, R_x	yz

(i) Other sources write zx instead of xz.
(ii) x, y and z relate to translational movements and R_x, R_y and R_z to rotational movements.

The four labels A_1, A_2, B_1 and B_2 are the Mulliken notation for the irreducible representations (Section 2.17.2.5) in point group C_{2v}, and the numbers in the body of the table are the *characters* of the irreducible representations (or just *characters*) for the given point group. In terms of the atomic orbitals in this point group, the following transformations apply:

s and p_z	transform as	$A_1(z)$
p_x	transforms as	$B_1(x)$
p_y	transforms as	$B_2(y)$
d_{z^2} and $d_{x^2-y^2}$	transform as	$A_1(z)$
d_{xy}	transforms as	$A_2(xy)$
d_{zx}	transforms as	$B_1(x)$
d_{yz}	transforms as	$B_2(y)$

A short exposé of the properties of character tables is given with Table 2.8 on page 94. Further details on character tables can be gleaned from the literature [22–24, 26, 27].

Example 2.11

From the matrices for the symmetry operations of point group C_{2v} confirm the entries in the character table for C_{2v}. How do x^3y and y^2z transform?

2.17.2.4 Projection operators

In this section and elsewhere, **bold italic** font is used to represent operators, including symmetry operators such as C_2; *italic* for scalar quantities, such as symmetry elements C_2, and for point group symbols, such as C_{2v}; and **bold** for matrices, such as $\mathbf{C_2}$ representing the mathematics of C_2; brief notes on matrices and matrix operations are given in Appendix A10.

Projection operators provide a method of generating SALCs, and will be described briefly. The *projection operator* $\boldsymbol{P_\alpha}$, operating on a function ψ in an irreducible representation α is given by:

$$\boldsymbol{P_\alpha}\psi = \frac{l_\alpha}{h}\sum_R \chi_\alpha \boldsymbol{O_R}\psi \qquad (2.115)$$

where the coefficients χ_α of the operator $\boldsymbol{O_R}$ of symmetry R are the real characters of the symmetry matrix $\mathbf{O_R}$ in the representation α. They are obtained directly from the character table of the appropriate point group. The projection operator $\boldsymbol{P_\alpha}$ acting on a function reproduces that function multiplied by l_α/h. The order h of the point group is the *total number* of its symmetry operators, and l_α is the dimension of the α irreducible representation; any function not belonging to the α representation is annihilated by the application of $\boldsymbol{P_\alpha}\psi$. In the example point group C_{2v}, $h = 4$ and l is unity for each irreducible representation.

As another example the point group T_d, which represents methane, has the following character table.

Table 2.8 *Character table for point group T_d and short exposé of the properties of character tables*

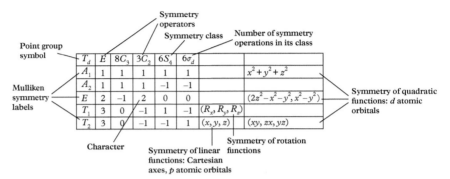

The order of the group is 24, the numbers g of symmetry operators in each *symmetry class* are, in order, 1, 8, 3, 6, 6 (their sum is 24), and the dimensions l of the irreducible representations are 1 for each of A_1 and A_2, 2 for E (doubly degenerate) and 3 for each of T_1 and T_2 (triply degenerate).

In the methane molecule, there are four orbitals linking the atoms, so that four molecular orbitals are expected. A representation of σ-bonding in this species may be obtained by the *unshifted atom* method [22, 25].

The rotational symmetry elements of methane are shown in by Fig. 2.34. The tetrahedron *aceg* has the following symmetry: three *fourfold alternating axes* (S_4) through the mid-points of edges such as *eg* and *ac*; four *threefold rotation axes* (C_3) along lines such as *eb*; six *twofold rotation* axes (C_2) through the mid-points of opposite edges of the circumscribing cube, six *mirror planes* σ_d (S_2) such as *achf*, where σ_d is $\perp$ to S_2.

In this application of the unshifted atom method, the contributions from the symmetry operations are obtained by counting the number of bonds that are invariant under the symmetry operations; only one operation in a class need be counted since all operations in that class follow the same pattern. Thus, in point group T_d only one of the eight C_3 operations need be considered. The following matrix illustrates the operation of C_3 (*gd*) on the four C–H orbitals shown in Fig. 2.35:

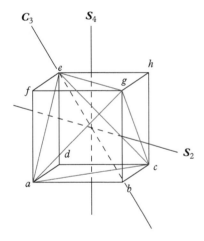

Fig. 2.34 Cube *a, b, c, d, e, f, g, h* and tetrahedron *a, c, g, e*, with corresponding symmetry elements aligned. Symmetry axes of rotation, C_3 (four) and C_2 (six), and of roto-reflection (alternating), S_4 (three), common to both polyhedra are indicated.

$$\begin{pmatrix} 1 & 0 & 0 & 0 \\ 0 & 0 & 1 & 0 \\ 0 & 0 & 0 & 1 \\ 1 & 0 & 0 & 0 \end{pmatrix} \begin{pmatrix} \psi_1 \\ \psi_2 \\ \psi_3 \\ \psi_4 \end{pmatrix} = \begin{pmatrix} \psi_1 \\ \psi_3 \\ \psi_4 \\ \psi_2 \end{pmatrix}$$

The trace of the 4 × 4 matrix is 1; only C–H$_{\psi_1}$ remains unmoved by a C_3 operation through ψ_1, and contributes 1 to the representation. Continuing in this

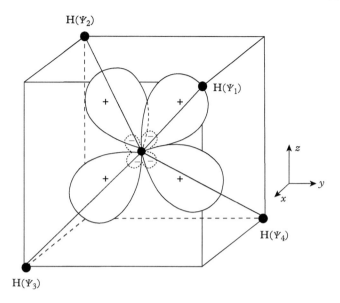

Fig. 2.35 The disposition of C–H molecular orbitals of carbon in methane within a cube; $H(\psi_i)$ $(i = 1\text{–}4)$ indicate the hydrogen atom positions with respect to the carbon atom at the centre of the cube.

manner with all symmetry classes, the following result is obtained for the four orbitals of methane:

E leaves all vectors unmoved	$\chi_E = 4$	
C_3 leaves one vector unmoved	$\chi_{C_3} = 1$	
S_4 moves all vectors	$\chi_{S_4} = 0$	
C_2 moves all vectors	$\chi_{C_2} = 0$	
$S_2 (\equiv \sigma_d)$ leaves two vectors unmoved	$\chi_{\sigma_d} = 2$	

E is the identity (doing nothing) operation, equivalent to no action, and is a member of all point groups Hence, the following reducible representation Γ_σ is derived:

T_d	E	$8C_3$	$3C_2$	$6S_4$	$6\sigma_d$
Γ_σ	4	1	0	0	2

It may be reduced by inspection of the character table for T_d (Table 2.8) to the irreducible representations $A_1 + T_2$ which leads to the orbitals s (A_1-type representations are always spherically symmetrical) and the three degenerate p orbitals x, y, z. The analytical deduction of the number of irreducible representations is shown in the next section.

2.17.2.5 *Reduction of a representation*

If a representation contains more than one of the individual transformations discussed in Section 2.17.2.3, it may be *reducible*. Then, it can be reduced to one or more *irreducible representations* by determining the number of times a given irreducible representation lies within the reducible representation, using the following relation:

$$n_\alpha = \frac{1}{h} \sum_{n_C} g \chi_\alpha \chi_\Gamma \tag{2.116}$$

where n_α is the number of times that the irreducible representation α appears in a given reducible representation Γ, h is the order of the group, g is the number of operations in a symmetry class, χ_α is the character of the α irreducible representation for operations in the given symmetry class (obtained from the appropriate character table) and χ_Γ is the character of the reducible operation in the class; and the sum is over all classes n_C. As an example, consider the irreducible representation T_2 in the reducible representation Γ_σ for point group T_d. Then, from Eq. (2.116):

$$(E) \qquad (C_3) \qquad (C_2) \qquad (S_4) \qquad (\sigma_d)$$

$$n_{A_1} = \frac{1}{24}\{(1 \times 1 \times 4) + (8 \times 1 \times 1) + (3 \times 1 \times 0) + (6 \times 1 \times 0) + (6 \times 1 \times 2)\} = 1$$

$$n_{A_2} = \frac{1}{24}\{(1 \times 1 \times 4) + (8 \times 1 \times 1) + (3 \times 1 \times 0) + (6 \times \bar{1} \times 0) + (6 \times \bar{1} \times 2)\} = 0$$

$$n_E = \frac{1}{24}\{(1 \times 2 \times 4) + (8 \times \bar{1} \times 1) + (3 \times 2 \times 0) + (6 \times 0 \times 0) + (6 \times 0 \times 2)\} = 0$$

$$n_{T_1} = \frac{1}{24}\{(1 \times 3 \times 4) + (8 \times 0 \times 1) + (3 \times \bar{1} \times 0) + (6 \times 1 \times 0) + (6 \times \bar{1} \times 2)\} = 0$$

$$n_{T_2} = \frac{1}{24}\{(1 \times 3 \times 4) + (8 \times 0 \times 1) + (3 \times \bar{1} \times 0) + (6 \times 1 \times 0) + (6 \times 1 \times 2)\} = 1$$

2.17.3 Molecular orbitals for methane

Molecular orbitals, $\Psi_i (i = 1 - 4)$ for methane can be generated from the hydrogen $1s$ atomic orbitals $\psi_i (i = 1 - 4)$, and the $(2s)^2$ and $(2p)^2$ orbitals of carbon. Each of the resulting molecular orbitals consists of a combination of an s orbital *spanning* (corresponding to) the A_1 representation, and three p orbitals spanning T_2. The projection operators P_{A_1} and P_{T_2} are applied to any three of the atomic orbitals; the fourth orbital is not needed at this stage as it is not linearly independent of the other three. In Fig. 2.35, the wavefunctions $\psi_i (i = 1 - 4)$ are indicated, occupying the positions g, e, a and c of Fig. 2.34. The action of *all* symmetry operations of T_d on the orbitals with respect to the carbon atom as origin must be determined. In the result tabulated, the notation $_2C_3^2$, for example, implies two successive anticlockwise C_3 operations acting at ψ_2 (axis direction eb in Fig. 2.34). This operation has the effects $\psi_1 \rightarrow \psi_3, \psi_2 \rightarrow \psi_2$ and $\psi_3 \rightarrow \psi_4$, as entered in rows 1–3 of the table under $_2C_3^2$. The complete table for the operations P_{A_1} on ψ_1, ψ_2 and ψ_3 is listed hereunder. Note that the operation S_2 does not appear explicitly as it is represented by the operation σ_d.

T_d	E	$_1C_3$	$_1C_3^2$	$_2C_3$	$_2C_3^2$	$_3C_3$	$_3C_3^2$	$_4C_3$	$_4C_3^2$	C_{2z}	C_{2y}	C_{2x}
ψ_1	ψ_1	ψ_1	ψ_1	ψ_4	ψ_3	ψ_4	ψ_2	ψ_3	ψ_2	ψ_2	ψ_4	ψ_3
ψ_2	ψ_2	ψ_3	ψ_4	ψ_2	ψ_2	ψ_1	ψ_4	ψ_1	ψ_3	ψ_1	ψ_3	ψ_4
ψ_3	ψ_3	ψ_4	ψ_2	ψ_1	ψ_4	ψ_3	ψ_3	ψ_2	ψ_1	ψ_4	ψ_2	ψ_1

	S_{4z}	S_{4z}^3	S_{4y}	S_{4y}^3	S_{4x}	S_{4x}^3	$_1\sigma_d$	$_2\sigma_d$	$_3\sigma_d$	$_4\sigma_d$	$_5\sigma_d$	$_6\sigma_d$
ψ_1	ψ_4	ψ_3	ψ_3	ψ_2	ψ_2	ψ_4	ψ_1	ψ_2	ψ_1	ψ_4	ψ_1	ψ_3
ψ_2	ψ_3	ψ_4	ψ_1	ψ_4	ψ_3	ψ_1	ψ_2	ψ_1	ψ_2	ψ_3	ψ_2	ψ_4
ψ_3	ψ_1	ψ_2	ψ_4	ψ_1	ψ_4	ψ_2	ψ_3	ψ_4	ψ_3	ψ_2	ψ_3	ψ_1

Applying the projection operators P_{A_1} and P_{T_2} then leads to the SALCs; the operation of $P_{A_1}(\psi_3)$ is straightforward, that of $P_{T_2}(\psi_3)$, for example, is set out in full as an example:

$$P_{A_2}(\psi_3) = \frac{1}{24}(6\psi_1 + 6\psi_1 + 6\psi_3 + 6\psi_4)$$

$$= \frac{1}{4}(\psi_1 + \psi_2 + \psi_3 + \psi_4)$$

$$P_{T_2}(\psi_3) = \frac{3}{24}(3\psi_3 + 0(\psi_4 + \psi_2 + \psi_1 + \psi_4 + \psi_3 + \psi_3 + \psi_2 + \psi_1) - \psi_4 - \psi_2 - \psi_1$$

$$- \psi_1 - \psi_2 - \psi_4 - \psi_1 - \psi_4 - \psi_2 + \psi_3 + \psi_4 + \psi_3 + \psi_2 + \psi_3 + \psi_1)$$

$$= \frac{3}{24}(-2\psi_1 - 2\psi_2 + 6\psi_3 - 2\psi_4)$$

$$= \frac{1}{4}(-\psi_1 - \psi_2 + 3\psi_3 - \psi_4).$$

Discounting the multiplying fractions, the complete, normalized set follows:

$$\Psi_1 = P_{A_1}\psi_1 = \frac{1}{2}(\psi_1 + \psi_2 + \psi_3 + \psi_4) \tag{2.117}$$

$$\Psi_2 = P_{T_2}\psi_1 = \frac{1}{\sqrt{12}}(3\psi_1 - \psi_2 - \psi_3 - \psi_4) \tag{2.118}$$

$$\Psi_3 = P_{T_2}\psi_2 = \frac{1}{\sqrt{12}}(-\psi_1 + 3\psi_2 - \psi_3 - \psi_4) \tag{2.119}$$

$$\Psi_4 = P_{T_2}\psi_3 = \frac{1}{\sqrt{12}}(-\psi_1 - \psi_2 + 3\psi_3 - \psi_4) \tag{2.120}$$

These SALCs are next established in terms of s, p_x, p_y and p_z functions and then renormalized. Equation (2.117) is the fully symmetric A_1 and refers to an s orbital; it is already normalized since $(1/2)^2(\psi_1^2 + \psi_2^2 + \psi_3^2 + \psi_4^2) = 1$. So the first SALC is:

$$\frac{1}{2}(\psi_1 + \psi_2 + \psi_3 + \psi_4) \tag{2.121}$$

Writing Eq. (2.118) as a normalized, linear combination of p functions:

$$\frac{1}{\sqrt{12}}(3\psi_1 - \psi_2 - \psi_3 - \psi_4) = (c_x^2 + c_y^2 + c_z^2)^{1/2}(c_x p_x + c_y p_y + c_z p_z) \tag{2.122}$$

where $1/\sqrt{12}$ is the normalization factor in Eq. (2.118). Using Fig. 2.34 to clarify the relationship of the symmetry elements of the tetrahedron to the reference axes, the effects of the C_3 operation in the direction of ψ_1 are seen to be:

$$p_x \rightarrow p_y \quad p_y \rightarrow p_z \quad p_z \rightarrow p_x$$

so that $(c_x p_x + c_y p_y + c_z p_z)$ in Eq. (2.122) transforms to $(c_x p_y + c_y p_z + c_z p_x)$. By comparing the coefficients of $p_i (i = x, y, z)$, it follows that $c_x = c_z$, $c_y = c_x$ and $c_z = c_y$; then

$$\frac{1}{\sqrt{12}}(3\psi_1 - \psi_2 - \psi_3 - \psi_4) = \frac{1}{\sqrt{3}}(p_x + p_y + p_z) \tag{2.123}$$

since resolving $\psi_i(i = 1 - 4)$ into its components shows that the positive square root is required, and the normalization factor is now $1/\sqrt{3}$.

Consider next the operation of C_3 in the direction of ψ_2, then:

$$p_x \rightarrow -p_y \quad p_y \rightarrow p_z \quad p_z \rightarrow -p_x$$

so that with Eq. (2.119), $(c_x p_x + c_y p_y + c_z p_z)$ transforms to $(-c_z p_x - c_x p_y + c_y p_z)$ and again the positive square root applies:

$$\frac{1}{\sqrt{12}}(-\psi_1 + 3\psi_2 - \psi_3 - \psi_4) = \frac{1}{\sqrt{3}}(-p_x - p_y + p_z) \tag{2.124}$$

By similar reasoning with ψ_3 ψ_3 and Eq. (2.120):

$$\frac{1}{\sqrt{12}}(-\psi_1 - \psi_2 + 3\psi_3 - \psi_4) = \frac{1}{\sqrt{3}}(p_x - p_y - p_z) \tag{2.125}$$

Solving Eqs. (2.122) to (2.125) for p_x, p_y and p_z, and including Eq. (2.121), leads to the result, expressed in matrix form as:

$$\begin{pmatrix} 1/2 & 1/2 & 1/2 & 1/2 \\ 1/2 & \overline{1/2} & 1/2 & \overline{1/2} \\ 1/2 & \overline{1/2} & \overline{1/2} & 1/2 \\ 1/2 & 1/2 & \overline{1/2} & \overline{1/2} \end{pmatrix} \begin{pmatrix} \psi_1 \\ \psi_2 \\ \psi_3 \\ \psi_4 \end{pmatrix} = \begin{pmatrix} s \\ p_x \\ p_y \\ p_z \end{pmatrix}$$

The 4 × 4 matrix is orthogonal, as is readily confirmed (Appendix A10); hence, multiplication by its inverse, which here is the same as its transpose, gives

$$
\begin{pmatrix}
1/2 & 1/2 & 1/2 & 1/2 \\
1/2 & \overline{1/2} & \overline{1/2} & 1/2 \\
1/2 & 1/2 & \overline{1/2} & \overline{1/2} \\
1/2 & \overline{1/2} & 1/2 & \overline{1/2}
\end{pmatrix}
\begin{pmatrix}
s \\ p_x \\ p_y \\ p_z
\end{pmatrix}
=
\begin{pmatrix}
\psi_1 \\ \psi_2 \\ \psi_3 \\ \psi_4
\end{pmatrix}
\tag{2.126}
$$

and the final form of the molecular orbitals may be written as:

$$
\left.
\begin{aligned}
\Psi_1 &= \tfrac{1}{2}(s_1 + p_x + p_y + p_z) \\
\Psi_2 &= \tfrac{1}{2}(s_1 - p_x - p_y + p_z) \\
\Psi_3 &= \tfrac{1}{2}(s_1 + p_x - p_y - p_z) \\
\Psi_4 &= \tfrac{1}{2}(s_1 - p_x + p_y - p_z)
\end{aligned}
\right\}
\tag{2.127}
$$

Another way of looking at the methane bonding is to note that Eq. (2.118) cannot be an individual function as written, since the three T_2 SALCs need to combine with the $2p$ orbitals of carbon in sign, as well as in symmetry. The lobes of the p orbitals are directed from the carbon atom at the centre of the cube towards each of the cube faces. In the diagram, shading represents positive signs of the wavefunctions:

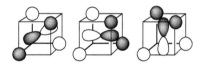

p-Orbitals in methane

Hence, for the signs of the orbitals on carbon to match the hydrogen wavefunction, $2p_x$ must use the ψ_1 and ψ_3 orbitals positively and ψ_2 and ψ_4 negatively; a similar argument applies for the for $2p_y$ and $2p_z$ orbitals. Thus, Eqs. (2.117) to (2.120) can be written as:

$$
\begin{array}{ll}
\tfrac{1}{2}(s_1 + s_2 + s_3 + s_4) & s \\
\tfrac{1}{2}(s_1 - s_2 + s_3 - s_4) & p_x \\
\tfrac{1}{2}(s_1 - s_2 - s_3 + s_4) & p_y \\
\tfrac{1}{2}(s_1 + s_2 - s_3 - s_4) & p_z
\end{array}
\tag{2.128}
$$

In matrix notation:

$$
\underbrace{
\begin{pmatrix}
1/2 & 1/2 & 1/2 & 1/2 \\
1/2 & \overline{1/2} & 1/2 & \overline{1/2} \\
1/2 & \overline{1/2} & \overline{1/2} & 1/2 \\
1/2 & 1/2 & \overline{1/2} & \overline{1/2}
\end{pmatrix}
}_{\mathbf{C}}
\underbrace{
\begin{pmatrix}
\Psi_1 \\ \Psi_2 \\ \Psi_3 \\ \Psi_4
\end{pmatrix}
}_{\mathbf{M}}
=
\underbrace{
\begin{pmatrix}
s \\ p_x \\ p_y \\ p_z
\end{pmatrix}
}_{\mathbf{A}}
\tag{2.129}
$$

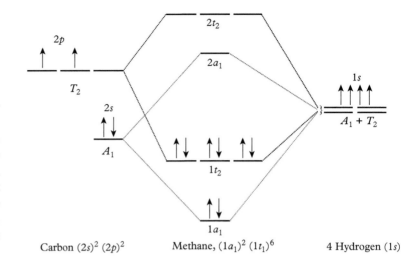

Fig. 2.36 Molecular orbital energy level diagram for methane: A_1 and T_2 are the Mulliken labels for the irreducible representations in point group T_d. The two $1s$ electrons of from carbon remain as a relatively inert core in the methane molecule. Bonding orbitals are $1a_1$ and $1t_2$; $\kappa = 4$. (Compare the notation with that in Fig. 2.29)

Carbon $(2s)^2 (2p)^2$ Methane, $(1a_1)^2 (1t_1)^6$ 4 Hydrogen $(1s)$

Since matrix $\mathbf{C}$ is orthogonal, its inverse $\mathbf{C}^{-1}$ is equal to its transpose; thus, $\mathbf{M} = \mathbf{C}^{-1}\mathbf{A} = \mathbf{C}^{\mathrm{T}}\mathbf{A}$. Hence, the matrix form:

$$
\underbrace{\begin{pmatrix} \Psi_1 \\ \Psi_2 \\ \Psi_3 \\ \Psi_4 \end{pmatrix}}_{\mathbf{M}} = \underbrace{\begin{pmatrix} 1/2 & 1/2 & 1/2 & 1/2 \\ 1/2 & \overline{1/2} & \overline{1/2} & 1/2 \\ 1/2 & 1/2 & \overline{1/2} & \overline{1/2} \\ 1/2 & \overline{1/2} & 1/2 & \overline{1/2} \end{pmatrix}}_{\mathbf{C}^{-1}} \underbrace{\begin{pmatrix} s \\ p_x \\ p_y \\ p_z \end{pmatrix}}_{\mathbf{A}}
\tag{2.130}
$$

which is in agreement with Eq. (2.127).

The character table for T_d shows that T_2 is spanned by the three product terms d_{xy}, d_{yz} and d_{zx} as well as by the three p functions. It follows that symmetry arguments alone would predict molecular orbitals formed from either the combination of $2s$ and three $2p$, or $2s$ and three $3d$ orbitals. However, chemical knowledge of energy levels leads to the adoption of the former combination as it corresponds to a lower energy state for the molecule. A molecular orbital energy level diagram is shown in Fig. 2.36, using a notation that reflects the relationship of the molecular orbital energy level to the representations of the point group. The orbital occupancy may be written as $1a_1^2, 1t_2^6$, indicating the threefold degeneracy of T_2 symmetry; the $(1s)^2$ electrons remain as the relatively inert core, $1a_1$.

Other polyatomic molecules can be treated in a similar manner, the complexity of the calculations increasing, not surprisingly, with increasing size of the molecule. Another approach, applicable to π electron systems is afforded by Hückel theory, which will be considered in its applications *inter alia* to ethene and benzene.

2.18   Hückel molecular orbital theory

Molecular orbitals for *conjugated* molecules can be studied under the *Hückel molecular orbital* (HMO) method, an approximation applicable to π-electron systems. It assumes that the σ-bonds of the molecular species form a rigid core governing the general molecular shape, like the bonds in the plane of the six-membered ring of benzene and its derivatives, and that the π electrons can be treated separately from the σ electrons.

2.18.1   Ethene

The two carbon atoms in ethene may be identified by the subscripts 1 and 2; then wavefunctions can be written as:

$$\Psi_\pi = c_1 \psi_1 \pm c_2 \psi_2 \tag{2.131}$$

where $\psi_i (i = 1, 2)$ represent the two $2p_z$ orbitals of the carbon atoms. Following the variation method as discussed in Section 2.17.1, and on account of the symmetry of the species, the secular determinant evolves as:

$$\begin{vmatrix} \alpha - E & \beta - ES \\ \beta - ES & \alpha - E \end{vmatrix} \tag{2.132}$$

The roots of this equation can be computed readily as before.

In the Hückel approximation, three assumptions are made:

1. All overlap integrals S can be equated to zero.
2. All resonance integrals β other than those between adjacent carbon atoms can be set to zero.
3. All other resonance integrals take one and the same value β in hydrocarbon species.

Then, the Hückel determinant of Eq. (2.132) assumes the form:

$$\begin{vmatrix} \alpha - E & \beta \\ \beta & \alpha - E \end{vmatrix} \tag{2.133}$$

which is solved readily for the roots

$$E_\pm = \alpha \pm \beta \tag{2.134}$$

where E_+ corresponds to the fully occupied, bonding π molecular orbital of Eq. (2.131), and E_- relates to its corresponding, unoccupied antibonding π^*

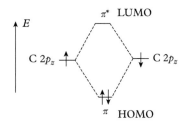

Fig. 2.37 Molecular orbital π energy level diagram for ethene. The electrons in the two $2p_z$ orbitals of carbon have opposed spins, and unite to form the bonding π-molecular orbital (the highest occupied molecular orbital, or HOMO), of energy E_+ and the antibonding π^* molecular orbital (the lowest unoccupied molecular orbital, or LUMO), E_-. Fig. 2.30(e) and (f) illustrate the $2p$ atomic orbitals and the π bonding (u) and antibonding (g) orbitals.

molecular orbital. The two molecular orbitals in Eq. (2.131) form the *frontier molecular orbitals* (FMO): they comprise the *highest (energy) occupied molecular orbital* (HOMO), and the *lowest (energy) unoccupied molecular orbital* (LUMO). The π molecular orbital energy level diagram for ethene is shown in Fig. 2.37. The total energy is $2(\alpha + \beta)$; both α and β are negative quantities. Since α represents the energy level for the atoms, the stabilization energy is β per atom. The energy amount $(\alpha + \beta)$ can be used a reference level for estimating delocalization energies in other conjugated systems.

As the zero of this energy is α, the amount 2β is the energy required for the promotion transition $\pi \rightarrow \pi'$ which occurs, for example, in the catalytic hydrogenation of ethene:

$$H_2C' = C'H_2 + H' \longrightarrow H_3C = C'H_2 + H' \dashrightarrow H_3C - CH_3$$

2.18.2   Allyl radical

A slightly more complicated example involves the allyl radical $^\cdot C_3H_5$,

H-C $\overset{H}{\underset{H}{\mid}}$ C $\overset{\cdot}{C}^+$ $\overset{H}{\mid}$ $\longleftrightarrow$ H-C $\overset{H}{\underset{H}{\mid}}$ C $= C$ $\overset{H}{\mid}$ $\equiv$ H-C $\overset{H}{\underset{H}{\mid}}$ C $\cdots$ C $\overset{H}{\mid}$

(a)   (b)   (c)

Allyl radical: (a) and (b) are canonical forms

where (a) and (b) are canonical structures, (c) indicates a more realistic distribution of electron density and the sign $\leftrightarrow$ has the meaning as before.

Three $2p$ atomic orbitals of carbon combine to form the expected three molecular orbitals of the form:

$$\Psi = c_1\psi_1 + c_2\psi_2 + c_3\psi_3 \tag{2.135}$$

where the coefficients c_1, c_2 and c_3 are to be determined. From the previous arguments, the secular equations may be written as:

$$c_1(\alpha - E) + c_2\beta \qquad\qquad = 0$$
$$c_1\beta + c_2(\alpha - E) + c_3\beta \qquad = 0 \tag{2.136}$$
$$c_2\beta + \qquad c_3(\alpha - E) = 0$$

with the secular determinant:

$$\begin{vmatrix} \alpha - E & \beta & 0 \\ \beta & \alpha - E & \beta \\ 0 & \beta & \alpha - E \end{vmatrix} = 0 \tag{2.137}$$

Writing $(\alpha - E)/\beta = y$ leads to:

$$\begin{vmatrix} y & 1 & 0 \\ 1 & y & 1 \\ 0 & 1 & y \end{vmatrix} = 0 = y^3 - 2y \tag{2.138}$$

which is readily factorized to $y(y^2-2)$, giving the roots $y = 0, \pm\sqrt{2}$, with energies α and $\alpha \pm \sqrt{2}\,\beta$. Next the coefficients in Eq. (2.136) are calculated. Taking first the value of y as $-\sqrt{2}$, c_2 becomes $\sqrt{2}c_1$. Also, $c_1 + \sqrt{2}c_1(-\sqrt{2}) + c_3 = 0$, whence $c_1 = c_3$. Then, by normalization:

$$c_1^2 + c_2^2 + c_1^2 = (c_2/\sqrt{2})^2 + c_2^2 + (c_2/\sqrt{2})^2 = 1$$

so that $c_2 = +1/\sqrt{2}$, whereupon $c_1 \, (= c_3) = 1/\sqrt{2}$.

The first molecular orbital is:

$$\Psi_1 = \frac{1}{2}\psi_1 + \frac{1}{\sqrt{2}}\psi_2 + \frac{1}{2}\psi_3 \tag{2.139}$$

Using next the result, $y = 0$, it follows that $c_1 = c_3$ and $c_2 = 0$. By normalization, $c_1 = \pm 1/\sqrt{2}$. Then, if the positive square root be applied, $c_1 = 1/\sqrt{2}$, so that $-1/\sqrt{2} = c_3$. Thus, a second molecular orbital is:

$$\Psi_2 = \frac{1}{\sqrt{2}}\psi_1 - \frac{1}{\sqrt{2}}\psi_3 \tag{2.140}$$

By symmetry, or by using the result $y = +\sqrt{2}$, the third orbital is

$$\Psi_3 = \frac{1}{2}\psi_1 - \frac{1}{\sqrt{2}}\psi_2 + \frac{1}{2}\psi_3 \tag{2.141}$$

The correctness of the last result, for Ψ_3, may be confirmed as follows. Recalling that atoms 1 and 3 are not adjacent:

$$E = \int \Psi_3 \mathcal{H}_e \Psi_3 d\tau = \int \tfrac{1}{2}(\psi_1 - \sqrt{2}\psi_2 + \psi_3)\mathcal{H}_e\tfrac{1}{2}(\psi_1 - \sqrt{2}\psi_2 + \psi_3)d\tau$$

$$= \tfrac{1}{4}\left\{ \begin{array}{l} \int \psi_1 \mathcal{H}_e\psi_1 d\tau - \sqrt{2}\int \psi_1 \mathcal{H}_e\psi_2 d\tau + \int \psi_1 \mathcal{H}_e\psi_3 d\tau - \\[4pt] \sqrt{2}\int \psi_2 \mathcal{H}_e\psi_1 d\tau + 2\int \psi_2 \mathcal{H}_e\psi_2 d\tau - \sqrt{2}\int \psi_2 \mathcal{H}_e\psi_3 d\tau + \\[4pt] \int \psi_3 \mathcal{H}_e\psi_1 d\tau - \sqrt{2}\int \psi_3 \mathcal{H}_e\psi_2 d\tau + \int \psi_3 \mathcal{H}_e\psi_3 d\tau \end{array} \right\}$$

$$= \tfrac{1}{4}(\alpha - \sqrt{2}\beta + 0 - \sqrt{2}\beta + 2\alpha - \sqrt{2}\beta + 0 - \sqrt{2}\beta + \alpha\}$$

$$= \tfrac{1}{4}(4\alpha + 4\sqrt{2}\beta) = \alpha - \sqrt{2}\beta$$

which is the energy of Ψ_3.

2.18.3 Benzene

In this molecule, the six carbon atoms will generate six molecular orbitals. The 6×6 secular determinant can be written without difficulty as follows:

$$
\begin{vmatrix}
\alpha - E & \beta & 0 & 0 & 0 & \beta \\
\beta & \alpha - E & \beta & 0 & 0 & 0 \\
0 & \beta & \alpha - E & \beta & 0 & 0 \\
0 & 0 & \beta & \alpha - E & \beta & 0 \\
0 & 0 & 0 & \beta & \alpha - E & \beta \\
\beta & 0 & 0 & 0 & \beta & \alpha - E
\end{vmatrix} = 0 \qquad (2.142)
$$

If the substitution $y = (\alpha - E)/\beta$ is made, the determinant becomes:

$$
\begin{vmatrix}
y & 1 & 0 & 0 & 0 & 1 \\
1 & y & 1 & 0 & 0 & 0 \\
0 & 1 & y & 1 & 0 & 0 \\
0 & 0 & 1 & y & 1 & 0 \\
0 & 0 & 0 & 1 & y & 1 \\
1 & 0 & 0 & 0 & 1 & y
\end{vmatrix} = 0 \qquad (2.143)
$$

which evaluates by the procedure discussed in Appendix A10 as the equation:

$$ y^6 + 6x^4 + 9x^2 - 4 = 0 \qquad (2.144) $$

which factorizes readily as

$$ (y + 2)(y - 2)(x^2 - 1)^2 = 0 \qquad (2.145) $$

leading to the following results:

Solution	Energy	Molecular orbitals
$y_1 = 2$	$\alpha - 2\beta$	Highest antibonding
$y_2 = 1$	$\alpha - \beta$	Doubly degenerate, antibonding ; LUMO
$y_3 = 1$	$\alpha - \beta$	
$y_4 = -1$	$\alpha + \beta$	Doubly degenerate, bonding; HOMO
$y_5 = -1$	$\alpha + \beta$	
$y_6 = -2$	$\alpha + 2\beta$	Lowest bonding

The molecular orbitals for benzene are illustrated in Fig. 2.38, and Fig. 2.39 shows a π energy-level diagram for this species in an equivalent notation.

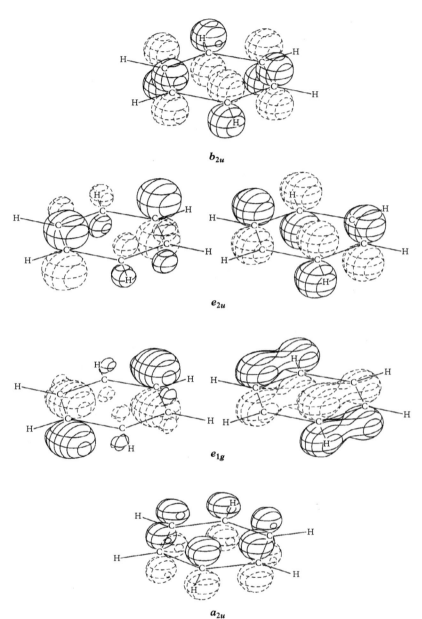

b_{2u}

e_{2u}

e_{1g}

a_{2u}

Fig. 2.38 π-Molecular orbitals for benzene. The order from top to bottom corresponds to an increase in bonding energy, and is $\alpha - 2\beta, (\alpha - \beta, \alpha - \beta), (\alpha + \beta, \alpha + \beta)$ and $\alpha + 2\beta$, respectively; the $\alpha \pm \beta$ molecular orbitals are degenerate pairs. [Jorgensen WL and Salem L. *The Organic Chemist's Book of Orbitals*, 1973; reproduced by courtesy of Elsevier.]

From the data above, the energy for π bonding is clearly $2(\alpha + 2\beta) + 4(\alpha + \beta) = 6\alpha + 8\beta$. The corresponding energy for one ethenic double bond, as in ethene, is $(\alpha + \beta)$ so that the expectation for three such bonds would be $6\alpha + 6\beta$. Thus, benzene is more stable than expected for three ethenic double bonds by the

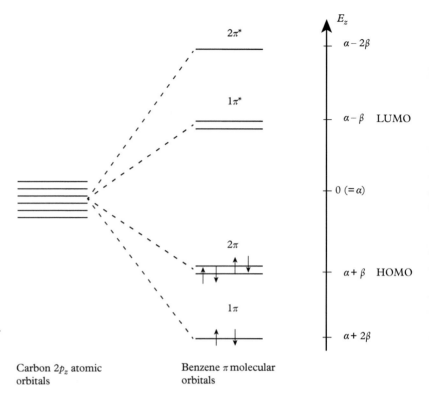

E_z

$2\pi^*$ $\alpha - 2\beta$

$1\pi^*$ $\alpha - \beta$ LUMO

$0 \, (= \alpha)$

2π $\alpha + \beta$ HOMO

1π $\alpha + 2\beta$

Fig. 2.39 π-Molecular orbital energy level diagram for benzene; the delocalization energy is 2β. [*Symmetry of Crystals and Molecules*, 2014; reproduced by courtesy of Oxford University Press.]

Carbon $2p_z$ atomic orbitals

Benzene π molecular orbitals

amount 2β; this is the delocalization energy for the molecule. The magnitude of β is not well defined, and would not be expected to be a constant for all conjugated molecules. The value for benzene is approximately $-85 \, \text{kJ mol}^{-1}$, so that the delocalization energy (2β) is *ca.* $-170 \, \text{kJ mol}^{-1}$. This value is determined by comparing the enthalpy of hydrogenation of benzene ($\Delta H_h = 206.3 \, \text{kJ mol}^{-1}$) with that of a hypothetical cyclohexatriene ($\Delta H_h = 376.5 \, \text{kJ mol}^{-1}$) treated as containing three ethenic double bonds.

The canonical (aka resonance) structures of benzene illustrated in Section 2.16 evolve mainly from the valence-bond treatment of the covalent bond, and the term resonance energy is sometime used for the quantity 2β.

As final example of the application of the projection operator, benzene will be considered further. The point group of the benzene molecule is D_{6h}. Again, only rotations are required to specify the symmetry of the LCAOs [25] so that point group C_6 may be used; thus, the six symmetry operations of C_6 are involved instead of the twenty-four in D_{6h}. In point group C_6, point group C_2 is a subgroup,[12] and the operation C_2 acts in the same direction as C_6. The representation generated by the unshifted atom procedure (Section 2.17.2.4) on the $2p_z$ orbitals is seen readily to be:

[12] A subgroup is a set of symmetry operators of a group that form a group in its own right; thus, C_2 is a subgroup, *inter alia*, of S_4.

C_6	E	C_6	C_3	C_2	C_3^2	C_6^5
Γ_π	6	0	0	0	0	0

since all atoms are shifted by rotational symmetry.

The representation Γ_π reduces to $A + B + E_1 + E_2$. The projection operator is used as before to form LCAO_MOs of the form $\Psi_i = \sum\limits_{i=1}^{6} \psi_i$ for each of these symmetry types. The character table for point group C_6 is given as Table 2.9; as will be seen, it contains complex characters but that does not really make for difficulty.

Following the procedure used in Section 2.17.3, the following SALCs are obtained:

$$P_A(\Psi) = \Psi_1 = (\psi_1 + \psi_2 + \psi_3 + \psi_4 + \psi_5 + \psi_6)$$

$$P_B(\Psi) = \Psi_2 = (\psi_1 - \psi_2 + \psi_3 - \psi_4 + \psi_5 - \psi_6)$$

$$P_{E_1}(\Psi) = \left\{ \begin{array}{l} \psi_1 + \varepsilon\psi_2 - \varepsilon^*\psi_3 - \psi_4 - \varepsilon\psi_5 + \varepsilon^*\psi_6 \\ \psi_1 + \varepsilon^*\psi_2 - \varepsilon\psi_3 - \psi_4 - \varepsilon^*\psi_5 + \varepsilon\psi_6 \end{array} \right\} \qquad (2.146)$$

$$P_{E_2}(\Psi) = \left\{ \begin{array}{l} \psi_1 - \varepsilon^*\psi_2 - \varepsilon\psi_3 + \psi_4 - \varepsilon^*\psi_5 - \varepsilon\psi_6 \\ \psi_1 - \varepsilon\psi_2 - \varepsilon^*\psi_3 + \psi_4 - \varepsilon\psi_5 - \varepsilon^*\psi_6 \end{array} \right\}$$

The first two wavefunctions Ψ_1 and Ψ_2 are as given in Eq. (2.146). The P_E equations can be converted to real forms by taking positive and negative linear combinations thereof, but first noting certain properties of ε and ε^*. From Euler's theorem,[13] the expression $\exp(i2\pi/6)$ in the character table has the following values for $\theta = 60°$ $(\pi/6)$: $\varepsilon + \varepsilon^* = 1$ and $\varepsilon - \varepsilon^* = i\sqrt{3}$. Proceeding:

$$P_{E_1+} = 2\psi_1 + (\varepsilon + \varepsilon^*)\psi_2 - (\varepsilon^* + \varepsilon)\psi_3 - 2\psi_4 - (\varepsilon + \varepsilon^*)\psi_5 + (\varepsilon + \varepsilon^*)\psi_6$$

so that

$$\Psi_3 = 2\psi_1 + \psi_2 - \psi_3 - 2\psi_4 - \psi_5 + \psi_6 \qquad (2.147)$$

[13] $\exp(i2\pi\theta) = \cos 2\pi\theta + i\sin 2\pi\theta$.

Table 2.9 *Character table for point group C_6*

C_6	E	C_6	C_3	C_2	C_3^2	C_6^5		$\varepsilon = \exp(i2\pi/6)$
A	1	1	1	1	1	1	z, R_z	$x^2 + y^2, z^2$
B	1	-1	1	-1	1	-1		
E_1	$\left\{ \begin{array}{c} 1 \\ 1 \end{array} \right.$	$\begin{array}{c} \varepsilon \\ \varepsilon^* \end{array}$	$-\varepsilon^*$	$\begin{array}{c} -1 \\ -1 \end{array}$	$\begin{array}{c} -\varepsilon \\ -\varepsilon^* \end{array}$	$\left. \begin{array}{c} \varepsilon^* \\ \varepsilon \end{array} \right\}$	$(x, y), (R_x, R_y)$	(xz, yz)
E_2	$\left\{ \begin{array}{c} 1 \\ 1 \end{array} \right.$	$\begin{array}{c} -\varepsilon^* \\ -\varepsilon \end{array}$	$-\varepsilon$	$\begin{array}{c} 1 \\ 1 \end{array}$	$\begin{array}{c} -\varepsilon^* \\ -\varepsilon \end{array}$	$\left. \begin{array}{c} -\varepsilon \\ -\varepsilon^* \end{array} \right\}$		$(x^2 - y^2, xy)$

and by similar manipulations with P_{E_1-}, P_{E_2+} and P_{E_1-} :

$$\Psi_4 = i\sqrt{3}(\psi_2 + \psi_3 - \psi_5 - \psi_6)$$
$$\Psi_5 = (2\psi_1 - \psi_2 - \psi_3 + 2\psi_4 - \psi_5 - \psi_6) \tag{2.148}$$
$$\Psi_6 = i\sqrt{3}(\psi_2 - \psi_3 + \psi_5 - \psi_6)$$

The set of six functions Ψ_1 to Ψ_6 are now normalized, discounting numerical multipliers as in Section 2.17.3, leading to six orthonormal molecular orbitals:

$$\Psi_1 = (1/\sqrt{6})(\psi_1 + \psi_2 + \psi_3 + \psi_4 + \psi_5 + \psi_6)$$
$$\Psi_2 = (1/\sqrt{6})(\psi_1 - \psi_2 + \psi_3 - \psi_4 + \psi_5 - \psi_6)$$
$$\Psi_3 = (1/\sqrt{12})(2\psi_1 + \psi_2 - \psi_3 - 2\psi_4 + \psi_5 + \psi_6)$$
$$\Psi_4 = (1/2)(\psi_2 + \psi_3 - \psi_5 - \psi_6) \tag{2.149}$$
$$\Psi_5 = (1/\sqrt{12})(2\psi_1 - \psi_2 - \psi_3 + 2\psi_4 + \psi_5 - \psi_6)$$
$$\Psi_6 = (1/2)(\psi_2 - \psi_3 + \psi_5 - \psi_6)$$

The E_1 functions have one *nodal plane*, the E_2 have two and B has three (Fig. 2.40); this diagram may be compared with Fig. 2.38, wherein the nodal planes are also clear. The greater the number of nodal planes, the higher (more positive) is the associated energy.

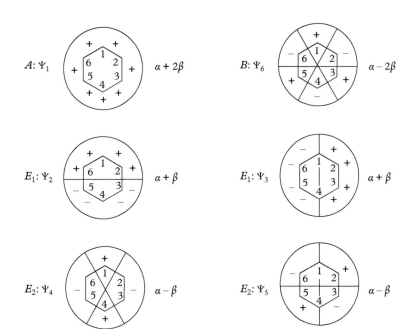

Fig. 2.40 Nodal planes, symmetries and energies for the six molecular orbitals of benzene: A; no nodal planes; E_1, one nodal plane; E_2, two nodal planes; B, three nodal planes. (Compare with Fig. 2.38.)

The Hückel procedure outlined above becomes more involved for larger molecules, but the same principles apply. However, as will be described shortly, a program is available for implementing the Hückel molecular orbital method with conjugated species, and including heteroatoms.

2.18.4 π-Bond order

The π electrons in conjugated hydrocarbons are not localized in pairs, so that classical double bonds do not exist. Each electron contributes to the bonding in the molecule, and the relative π bonding between pairs of adjacent atoms i, j is related to the coefficients c_n in the wavefunction by the π-*bond order* $p_{i,j}$:

$$p_{i,j} = \sum_k \eta_k c_{i,k} c_{j,k} \qquad (2.150)$$

where η_k is the number of electrons in the k occupied π molecular orbitals, subscript i in $c_{i,k}$ refers to atom i in molecular orbital k, and j in $c_{j,k}$ refers to atom j in molecular orbital k, the sum being taken over all k *occupied* π molecular orbitals. The Hückel treatment of benzene leads to the molecular orbitals in Eq. (2.149). Consider the 1, 2 C–C bond; then from Eq. (2.150):

$$p_{1,2} = 2 \left(\frac{1}{\sqrt{6}} \frac{1}{\sqrt{6}} + \frac{2}{\sqrt{12}} \frac{1}{\sqrt{12}} \right) = 0.667$$

and the same result obtains for each C–C bond because they are equal. The total bond order is obtained by adding 1 for the σ-bond, giving 1.667 as the total bond order $P_{i,j}$, more than that implied by a Kekulé structure for benzene (Section 2.16), which suggests a π-bond order of 0.5; the extra one-sixth arises through the delocalization energy based on that of three ethenic double bonds. Bond order correlates well with bond length, as shown by Fig. 2.41, which presents the total $(\pi + \sigma)$ bond order.

2.18.5 Free valence index

The *free valence index* $\mathfrak{F}$ measures the extent to which an atom is bonded to its neighbours. A large value corresponds to weak bonding, and such results have a bearing on the reactivity of a species. The index is defined by

$$\mathfrak{F} = M_i - \sum_k P_{i,k} \qquad (2.151)$$

where M_i is the *maximum bonding power* of the i atom, and the sum is taken over all k occupied orbitals involving that atom. It can be shown that M_i for carbon is 4.732, from the known but unstable trimethylenemethane molecule $C(CH_2)_3$: the central carbon atom has three σ-bonds, and a Hückel calculation

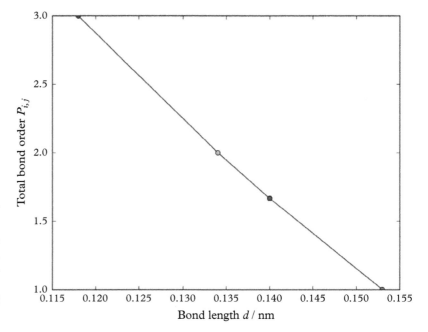

Fig. 2.41 Total bond order $P_{i,j}$ against bond length d for orders 1 (C_2H_6, 0.153 nm), 1.667 (C_6H_6, 0.140 nm), 2 (C_2H_4, 0.134 nm), and 3 (C_2H_2, 0.118 nm). The curve is fitted well by the quadratic $P = 2.749 \times 10^{-4}d^2 - 0.1316d + 14.70$, with d in pm.

on the molecule gives $1/\sqrt{3}$ for each π-bond order; hence the total bond order is $3 + 3/\sqrt{3} = 4.732$.

Trimethylenemethane molecule

Thus, for atom 1 in benzene, there are two bonds involved, with of total order 5/3 and one bond of order 1 (σ C–H bond); hence

$$\mathscr{F} = 4.732 - [(1 + 2(5/3)] = 0.399 \qquad (2.152)$$

and will be the same for all six atoms of benzene. Discounting all but the π bonds, $\mathscr{F}_A$ may be written simply as $\mathscr{F}_A = \sqrt{3} - \sum_A p_{A,B}$.

2.18.6 Charge distribution

If a carbon atom forms three σ-bonds and is also π bonded, it will remain neutral if there is an average of one electron in each π orbital. The *charge distribution* q_i may be taken as the deviation from neutrality of the i atom, and is defined by:

$$q_i = 1 - \sum_k \eta_k c_{i,k}^2 \qquad (2.153)$$

where the symbols have meanings as before and the sum is over all k occupied molecular orbitals. In a full analysis of benzene, the coefficients c_{11}, c_{12} and c_{13} are 0.40825, 0.49735 and 0.29322, and there are two electrons per orbital. Hence, $q_1 = 1 - 2(0.40825^2 + 0.49735^2 + 0.29322^2) = 6 \times 10^{-6}$, which is effectively zero, the difference from zero arising from rounding errors; benzene is a neutral species.

2.19 Extended Hückel molecular orbital theory

The Hückel theory can be extended to encompass the treatment of conjugated species containing heteroatoms by a modification of the secular determinant. In principle, the α and β terms are adjusted according to the nature of the heteroatom and its bonding. The model frequently applied adjusts the values of α and β to $\alpha_X = \alpha + h_X$ and $\beta_X = k_X \beta$, for the species X. In the examples considered here for nitrogen, N·, as in pyridine, $h = 0.5$ and $k = 1.0$, whereas for N:, as in pyrrole, $h = 1.5$ and $k = 0.8$; other values have been put forward from time to time. The procedure will be illustrated with pyridine (**I**), pyrrole (**II**) an aniline (**III**):

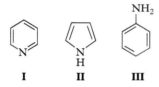

I   **II**   **III**

In pyridine (**I**), the lone pair electrons on nitrogen form no part of the ring system; they are localized in a $2p_z$ orbital and do not enter into π-bonding, although they open the molecule to electrophilic attack. In pyrrole (**II**), the lone pair electrons on nitrogen are part of the ring system and contribute to π-bonding; both of these species have six valence π-electrons. Pyridine and pyrrole can be analysed by the program HUCK in the Web Program Suite. Guidance on the use of the program is given in Chapter 7, but additional notes are given here.

The crucial stage in using HUCK is in forming the correct secular determinant, which means setting the correct atom connectivity, and choosing the correct values for α and β. In the example of pyridine, the input is as follows:

Atom	Connectivity	k	
1	2	1	
2	3	1	
3	4	1	
4	5	1	
5	6	1	
6	1	1	
1			(Number of heteroatoms)
1	0.5		(Heteroatom number and its h-value)

In using the program it is helpful to make a numbered molecular drawing, and essential to determine the exact number of electrons involved in π-bonding.

The Hückel method leads to interesting results relating to chemical reactivity; data below on aniline and pyridine show charge distributions on the carbon atoms of these species:

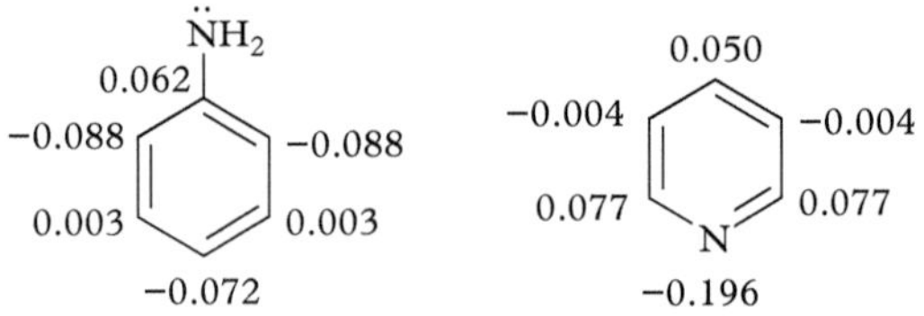

Aniline   **Pyridine**

Aniline reacts with bromine water to give 2,4,6-tribromoaniline: the active brominating agent in this reaction is the Br^+ species, which seeks preferentially the relatively negative 2-, 4- and 6-centres. With a nucleophilic reagent, such as acetyl chloride, the partially positive carbonyl carbon of the reagent seeks the most negative centre. The lone-pair electrons on the amine nitrogen atom are the most negative centre and acetylation leads to acetanilide (*N*-phenylacetamide), as in a Schotten–Baumann reaction. The carbonyl carbon atom is relatively positive owing to the higher relative electronegativities of oxygen and chlorine, thus preparing the carbon atom for nucleophilic attack on the most negative centre of aniline, as the following scheme shows:

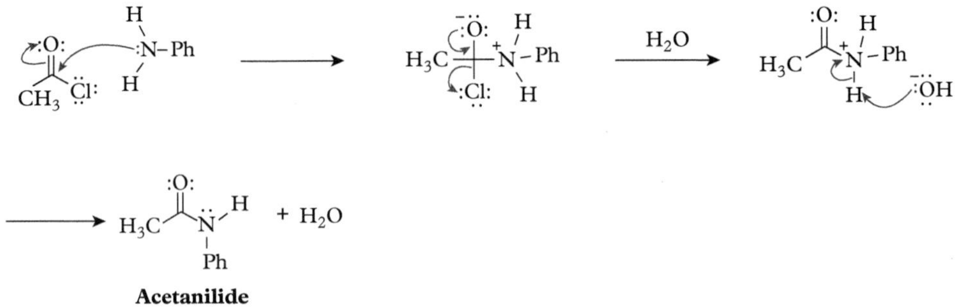

Acetanilide

In a similar way, the nitration of pyridine by the active nitronium ion $(NO_2)^+$, leads to 3-nitropyridine. A nucleophilic reagent, however, seeks the 2-position; in the Chichibarin reaction, sodamide at *ca.* 120 °C followed by treatment with water produces 2-aminopyridine:

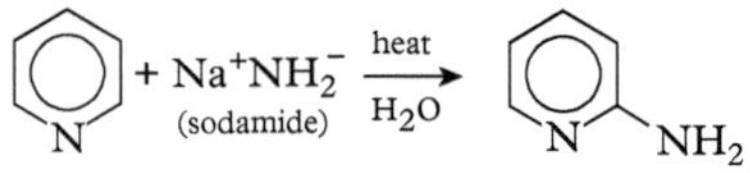

2-Aminopyridine

Practice with the HMO procedure arises through the problems that are included at the end of this chapter.

2.20   Valence shell electron pair repulsion theory

A useful, qualitative approach to molecular geometry can be achieved sometimes by the *valence shell electron pair repulsion* theory, generally written as VSEPR, of which detailed descriptions are available in most modern textbooks on chemistry. The theory attempts to predict the shapes of polyatomic species in terms of the repulsions between pairs of electrons, particularly lone pair electrons, on the component atoms. It begins with a Lewis bonding-pair/lone-pair model, and assumes that a molecular species adopts that shape which minimizes the repulsions between pairs of electrons; in other words, it seeks to place electron pairs as far apart as possible.

Consider a molecular species AB_n. It will have bonding pairs of electrons defining a region or *domain* between the A and B atoms where the electrons are most probably to be found. A non-bonding (lone) pair of electrons defines a domain located on the central atom A of a molecule; this domain is larger than that of a bonded pair. Domains are negative entities and tend to repel one from the other. The preferred molecular arrangement is that which minimizes these repulsions, leading to the minimum energy conformation for the molecule. For the example species AB_n there are five basic b/n domain geometries: linear 2/0, trigonal planar 3/0, tetrahedral 4/0, trigonal bi-pyramidal 5/0 and octahedral 6/0, where b/n represents the numbers of bonding and non-bonding electron pairs, respectively. Variations from these geometries are occasioned by the number and type of domains present.

The theory is applied by determining the numbers of bonding and non-bonding electrons from a Lewis model of the molecule, and then applying the results to the standard conformations of Table 2.10, or similar compilation, which have been confirmed experimentally. VSEPR is a point charge model in which the following order for electron pair repulsion exists:

lone pair/lone pair > lone pair/bonding pair > bonding pair/bonding pair

In Table 2.10, 'electron distribution' refers to the electron geometry on the central atom, the $b/0$ configuration, whereas 'molecular geometry' is the shape of the molecule after the repulsion effects of substituent atoms have acted on the central atom electron distribution, b/n, $n > 0$ configuration. Thus, in the water molecule, the electron distribution is nominally tetrahedral. However, only two of the four tetrahedral positions are occupied, by hydrogen atoms. The remaining two lone pairs exert a mutual repulsion which travels to and repels the bonded hydrogen atoms, leading to a decrease in the H–O–H angle from the tetrahedral value to the known value of 104.5°.

Table 2.10 *Molecular shapes according to VSEPR theory*

Number of bonding electron pairs	Number of non-bonding electron pairs	Electron distribution	Molecular geometry	Angle/deg	Examples
2	0	linear	linear	180	BeF_2, CO_2
1	1	linear	linear	180	CO, N_2
3	0	trigonal planar	trigonal planar	120	BF_3
2	1	trigonal planar	bent	<120	SO_2
1	2	trigonal planar	linear	180	O_2
4	0	tetrahedral	tetrahedral	109.5	CH_4
3	1	tetrahedral	trigonal pyramidal	<109.5	NH_3
2	2	tetrahedral	bent	<109.5	H_2O
1	3	tetrahedral	linear	180	HF
5	0	trigonal bi-pyramidal	trigonal bi-pyramidal	90, 120 and 180	PF_5
4	1	trigonal bi-pyramidal	trigonal bi-pyramidal	90, 120 and 180	SF_4
3	2	trigonal bi-pyramidal	T-shape	90, 180	ClF_3
2	3	trigonal bi-pyramidal	linear	180	XeF_2
6	0	octahedral	octahedral	90, 180	SF_6
5	1	octahedral	square pyramidal	90, 180	$[SbCl_5]^{2-}$
4	2	octahedral	square planar	90, 180	$[ICl_4]^-$

The molecules SF_2 and SCl_2 resemble the water molecule in structure. Since fluorine is more electronegative than chlorine, it will draw more electron charge from sulphur than will chlorine. This effect will, in turn lead to a smaller bond angle in SF_2 than in SCl_2. From experiment, the angles are 98° in SF_2 and 102° in SCl_2. In these three examples, tetrahedral configurations are perturbed by the lone pairs thus decreasing the bond angle below the ideal value.

Sulphur dioxide with multiple bonds has the formal structure:

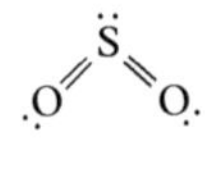

Sulphur dioxide

Its electron distribution is trigonal planar, but larger *electron domains* for the multiple bonds produce only a small repulsion effect, and the bond angle is just less than that for trigonal-planar at 119.1°.

The VSEPR theory is a simple and straightforward method for predicting the shapes of small molecular species. It is particularly successful where M is a main-group element, but is less satisfactory for transition metal compounds because the central atom does not always have a spherical or near-spherical shape. However, the particular configurations d^0, d^5 and d^{10} respond fairly well to VSEPR treatment: $TiCl_4 (d^0)$ tetrahedral, $[CoF_6]^{2-} (d^5)$ octahedral and $[Ag(NH_3)_2]^+ (d^{10})$ linear are all predicted correctly. Table 2.10 is a guide to the application of VSEPR theory, and it is discussed in more detail in books on inorganic chemistry [17, 18, 28].

Example 2.12

How would the VSEPR theory determine the probable structure of the ammonia molecule NH_3?

The Lewis structure for ammonia is:

$$H : \overset{\cdot\cdot}{\underset{H}{H}} : H$$

Number of bonding electron pairs 3; number of non-bonding (lone) pairs 1. Thus, the indicated electron distribution is tetrahedral. Repulsion of the bonding pairs by the lone pair acts so as to decrease the H–N–H angles below the value for tetrahedral geometry. The resulting molecular geometry is trigonal pyramidal (see also Table 2.10):

$$H^{\prime\prime\prime} \overset{\overset{\cdot\cdot}{N}}{\underset{\underset{H}{107°}}{\diagup}} H$$

Pyramidal structure for ammonia

(By experiment, each H–N–H bond angle is *ca.* 107° whereas the tetrahedral geometry has bond angles of 109.47°.)

2.21   Crystal-field and ligand-field theories

The bonding of chemical groups to transition metals was explained initially by the point charge *crystal-field model* which considered that the interactions between a transition metal and the bonding groups arise through a purely electrostatic attraction between the positive central metal cation and the surrounding anions, or other species that are negative on account of their electron distribution.

The geometry of coordination compounds involves mainly MX_6 octahedral, and MX_4 tetrahedral or square-planar complexes. The negative species X coordinating a metal M are donors such as NH_3 and H_2O which form σ-bonds or acceptors like CO and $(CN)^-$, or donors Cl^- and $:NH_2$ that form π-bonds; a combination of donor groups can also form the coordinating species in complex compounds.

In a complex of the type MX_6 for example, the focus is on the energy changes of the d levels of the metal cation as the negative species, originally effectively at infinity, approach it. If the charges of six negative ions were distributed uniformly over a spherical surface around the metal ion, the energy of the d orbitals would increase (become more positive) because of electron–electron repulsions while the fivefold degeneracy of the d orbitals remains. If instead the negative species occupy the vertices of an octahedron with the metal atom at its centre then the degeneracy is split into two groups.

By reference to Fig. 2.21 it is evident that as anions approach the metal, the group consisting of d_{z^2} and $d_{x^2-y^2}$ increase in energy whereas the group d_{xy}, d_{xz} and d_{yz} decrease in energy. The average energy of the d electrons remains unchanged from that of the *barycentre*, the energy of the degenerate d *orbitals* in a field of spherical symmetry:

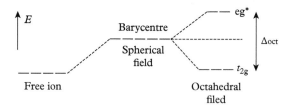

d Orbitals of the metal ion, free and under spherical and octahedral fields

The extent of splitting depends of several factors:

- the metal ion and, particularly, the number of its d electrons;
- the charge on the metal ion: the larger the charge, the greater the splitting;
- the coordination pattern of the negative species around the cation;
- the electric field strength of the ligands: the stronger the field the greater the splitting.

In the MX_6 octahedral compounds, the group of three t_{2g} orbitals are directed between the negatively charged ligands whereas the group of two e_g^* orbitals are directed towards the ligands. The energy difference between the two groups is Δ_{oct}, the *crystal-field stabilization energy* (CFSE).

The next most common coordination compounds are of type MX_4, in which four ligands form a tetrahedral coordination pattern around the metal ion. In a tetrahedral crystal field, the *d* orbitals are again split into two groups, with an energy difference of Δ_{tet} but the lower energy levels are d_z^2 and $d_{x^2-y^2}$ with d_{xy}, d_{zx} and d_{yz} in the higher energy state. Furthermore, since the donor electrons in tetrahedral symmetry are not oriented directly towards the *d* orbitals, the energy splitting is lower in magnitude than in octahedral complexes. Square planar and other complex geometries can also be described by this theory.

However, crystal field theory fails to take account of overlap between the orbitals of the metal and the attached groups, an implicit covalency. This bonding is embraced in the *ligand-field model*, in which the electrons on the central metal ion are delocalized, a sort of expansion of the *d* orbitals. Repulsion is decreased by overlap, and this effect, which depends principally on the nature of the ligand, gives rise to the *nephelauxetic series* of ligands (Gk. *nephēle* = cloud; *auxēsis* = growth) which is related to the size of Δ and thus to electron spin:

$$I^- < Br^- < SCN^- < Cl^- < NO_3^- < F^- < OH^- < H_2O < NCS^- < NH_3 < NO_2^- < CN^- < CO$$

weak field (high spin)   strong field (low spin)

A typical transition metal coordination compound is hexamminecobalt(III) chloride $[Co(NH_3)_6]Cl_3$ (Fig. 2.42); this compound has octahedral symmetry, point group O_h. It still remains that the degeneracy of the metal *d* orbitals is split into a group of two, the e_g^* molecular orbitals, and a group of three, the t_{2g} molecular orbitals of lower energy. The energy difference between the e_g^* and t_{2g} levels is Δ_{oct}, the *ligand-field stabilization energy* (LFSE) for octahedral coordination, and its magnitude is governed by the factors discussed earlier.

As another example, consider chromium(II) as a d^4 species. In the presence of *weak field (high spin)* ligands, such as $(SCN)^-$ or Cl^- the *d* electrons are split in the manner of Fig. 2.43a and Δ_{oct} is small in magnitude. However, if the ligand is *strong field (low spin)*, then the electron distribution of Fig. 2.43b is adopted and Δ_{oct} is large in magnitude. In both cases, the symmetry remains as O_h.

2.21.1 The hexacyanoferrate(II) ion

In this example σ-bonding will be considered in the octahedral hexacyanoferrate(II) ion $[Fe(CN)_6]^{4-}$. The significant atomic orbitals of this metal from the first transition series are again subscribed from the $3d$, $4s$ and $4p$ energy levels of the atom.

The point group symmetry of the hexacyanoferrate(II) ion is O_h, and the six Fe–(CN) bond vectors can be used to generate a representation for the Fe–(CN) σ-bonding. Since a centre of symmetry is present in this group, the g/u notation (Section 2.17.1.2) is given with the labels of the irreducible representations, thus indicating the orbital symmetry with respect to i, the centrosymmetric operator. A representation can be deduced readily by first setting the octahedral ion within the framework of a cube, with the apices of the octahedron occupying the centres

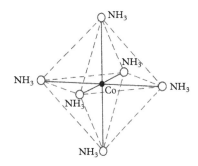

Fig. 2.42 Octahedral configuration of the complex ion $[Co(NH_3)_6]^{3+}$, a typical transition metal coordination compound with symmetry O_h. Compare this figure with the *d* orbitals shown in Fig. 2.21.

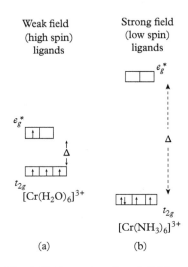

Fig. 2.43 Examples of weak field and strong field electron configurations in chromium coordination compounds; the quantity Δ between the t_{2g} and e_g^* levels is the ligand-field stabilization energy. [*Symmetry of Crystals and Molecules*, 2014; reproduced by courtesy of Oxford University Press.]

of the faces of the cube, whereupon the corresponding symmetry directions can be visualized readily; hence the following reducible representation may be derived as in earlier examples:

O_h	E	$8C_3$	$6C_2$	$6C_4$	$3C_2(\equiv 3C_4^2)$	i	$6S_4$	$8S_6$	$3\sigma_h$	$6\sigma_d$
Γ	6	0	0	2	2	0	0	0	4	2

It reduces to $A_{1g} + E_g + T_{1u}$, and the character table for point group O_h is given as Table 2.11; the ligand σ-bonding orbitals are illustrated by Fig. 2.44.

Table 2.11 *Character table for point group O_h*

O_h	E	$8C_3$	$6C_2'$	$6C_4$	$3C_2(\equiv 3C_4^2)$	i	$6S_4$	$8S_6$	$3\sigma_h$	$6\sigma_d$		
A_{1g}	1	1	1	1	1	1	1	1	1	1		$x^2 + y^2 + z^2$
A_{2g}	1	1	-1	-1	1	1	-1	1	1	-1		
E_g	2	-1	0	0	2	2	0	-1	2	0		$(2z^2 - x^2 - y^2, x^2 - y^2)$
T_{1g}	3	0	-1	1	-1	3	-1	0	-1	1	(R_x, R_y, R_z)	
T_{2g}	3	0	1	-1	-1	3	-1	0	-1	1		(xz, yz, xy)
A_{1u}	1	1	1	1	1	-1	-1	-1	-1	-1		
A_{2u}	1	1	-1	-1	1	-1	1	-1	-1	1		
E_u	2	-1	0	0	2	-2	0	1	-2	0		
T_{1u}	3	0	-1	1	-1	-3	-1	0	1	1	(x, y, z)	
T_{2u}	3	0	1	-1	-1	-3	1	0	1	-1		

Fig. 2.44 Ligand molecular orbitals for the hexacyanoferrate(II) ion formed from the $4s$ and $4p$ atomic orbitals of iron with the labels of the irreducible representations to which they correspond in symmetry, and arranged in order $A_{1g} \rightarrow E_g$ of decreasing (less negative) energy.

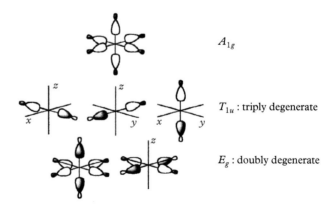

A_{1g}

T_{1u} : triply degenerate

E_g : doubly degenerate

Symmetry labels can now be assigned:

	AO	$f(x, y, z)$	Label
The first ligand orbital is totally symmetric	$4s$	$x^2 + y^2 + z^2$	A_{1g}
The next three orbitals (triply degenerate)	$4p$	x, y, z	T_{1u}
The last two orbitals (doubly degenerate)	$3d$	z^2, x^2-y^2	E_g

A molecular orbital energy-level diagram to illustrate the bonding of the Fe^{2+} cation with six coordinating $(CN)^-$ anion ligands is shown in Fig. 2.45. The six cyanide ligand wavefunctions ψ_i ($i = 1$ to 6) for $x, \bar{x}, y, \bar{y}, z, \bar{z}$ in order, as determined through projection operators, are listed hereunder together with their orbital types and symmetry labels:

s	$1/\sqrt{6}(\psi_1 + \psi_2 + \psi_3 + \psi_4 + \psi_5 + \psi_6)$	A_{1g}
p_x	$1\sqrt{2}(\psi_1 - \psi_2)$	T_{1u}
p_y	$1\sqrt{2}(\psi_3 - \psi_4)$	T_{1u}
p_z	$1\sqrt{2}(\psi_5 - \psi_6)$	T_{1u}
d_{z^2}	$1/\sqrt{12}(-\psi_1 - \psi_2 - \psi_3 - \psi_4 + 2\psi_5 + 2\psi_6)$	E_g
$d_{x^2-y^2}$	$1/2\,(\psi_1 + \psi_2 - \psi_3 - \psi_4)$	E_g

The d_{z^2} orbital has twice the probability along the z-axis compared with the x and y directions. Three d functions on the metal atom, d_{xy}, d_{zx} and d_{yz}, of representation T_{2g} have no matching representations among the functions of the cyanide ligands, so that these d functions remain as the non-bonding t_{2g}.

There is a total of eighteen electrons in the molecular orbitals of the complex ion; each of the molecular orbitals a_{1g}, t_{1u} and e_g contains a pair of bonding electrons with opposed spins. Where symmetry functions of differing energy are combined, the resulting molecular orbitals are different in energy. The lower energy bonding molecular orbitals assume the character of the ligand, and the twelve electrons involved are effectively donated to the metal atom. The six electrons from the metal occupy the t_{2g} non-bonding orbitals.

The *energy gap* between the t_{2g} (HOMO) and e_g^* (LUMO) levels represents the ligand-field stabilization energy. The d_z^2 orbital can be resolved into a linear combination of $d_{z^2-x^2}$ and $d_{z^2-y^2}$ ($2z^2-x^2-y^2$) each of which is spatially equivalent to $d_{x^2-y^2}$ but such that d_z^2 has a probability of twice that of $d_{x^2-y^2}$; the three d_{xy}, d_{zx} and d_{yz} orbital functions are also equivalent.

Valence electrons from the metal atom engage with the t_{2g} and e_g^* molecular orbitals. The first three electrons occupy the t_{2g} levels singly. The fourth to sixth electrons can either pair with t_{2g} or enter the higher e_g^* level. If the d orbital splitting energy is less than the pairing energy these electrons occupy the e_g^* level, whereas conversely, as in the case of the $[Fe(CN)_6]^{4-}$ ion, they pair in t_{2g}. The choice

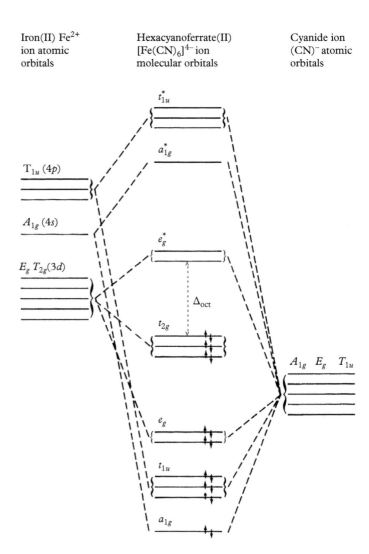

Iron(II) Fe^{2+} ion atomic orbitals

Hexacyanoferrate(II) $[Fe(CN)_6]^{4-}$ ion molecular orbitals

Cyanide ion $(CN)^-$ atomic orbitals

Fig. 2.45 Molecular orbital energy level diagram for the hexacyanoferrate(II) ion $[Fe(CN)_6]^{4-}$. In the ion, molecular orbitals a_{1g}, t_{1u} and e_g are bonding, t_{2g} and e_g^* are non-bonding, and a_{1g}^* and t_{1u}^* are antibonding; the distance between the non-bonding t_{2g} and e_g^* is the ligand-field splitting energy Δ_{oct}; the CN species is a strong field ligand, so the e_g^* levels lie above the t_{2g}. [Reproduced by courtesy of Woodhead Publishing, UK.]

depends on the field strength of the ligand; the $(CN)^-$ ion lies at the strong field (low spin) end of the nephelauxetic series of ligands, a position that favours pairing. The value and position of Δ_{oct} relative to the t_{2g} and e_g^* energy levels varies with the strength of the ligand field. The e_g^* levels in this example lie at $\frac{3}{5}\Delta$ above the atomic degenerate d levels with t_{2g} at $\frac{2}{5}\Delta$ below it. An example of the variation of Δ with field strength is shown in Fig. 2.46, which shows typical low spin (a) and high spin (b) configurations.

The three non-bonding d functions on iron that do not match in symmetry with respect to σ-bonding are involved in π-bonding with the cyanide ligands, with the d_{xy}, d_{zx} and d_{yz} from the metal held in the non-bonding t_{2g} now

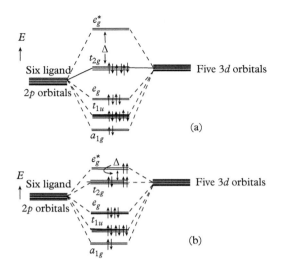

Fig. 2.46 Molecular orbital energy level diagrams for octahedral coordination. (a) Strong field (low spin), as with the $[Fe(CN)_6]^{4-}$ ion, $t_{2g}^6(e_g^{*0})$; as this species has all spins coupled it is diamagnetic. (b) Weak field (high spin), as in the $[Co(SCN)_6]^{3-}$ ion, $t_{2g}^4 e_g^{*2}$; four unpaired electrons lead to significant paramagnetic properties. [*Introduction to Physical Chemistry*, 3rd ed. 1998; reproduced by courtesy of Cambridge University Press.]

participating with the six T_{2g} (p_z) ligand atomic orbitals to form π-bonding molecular orbitals, and the corresponding π^*-antibonding orbitals. The donation of electrons in this way is often called back-bonding, which stabilizes the relatively more electron-rich low oxidation state. It is more likely to occur with metals in high oxidation states, and particularly with d^0–d^3 species.

The hexacyanoferrate(III) ion is similar to the hexacyanoferrate(II) ion but with one less electron, which is taken from the t_{1u} energy level; thus the Fe(III) complex is paramagnetic, whereas the hexacyanoferrate(II) complex is diamagnetic. The Fe(III) is the less stable of the two and is readily oxidized:

$$[Fe^{III}(CN)_6]^{3-} + e^- \rightarrow [Fe^{II}(CN)_6]^{4-}$$

The ligand field stabilization energy of transition metals has been illustrated through graphs of the enthalpy of hydration of the cations of the first or second series of transition metals plotted against atomic number; a typical curve shows a 'double hump' along a series such as calcium to zinc or strontium to cadmium, which have d^0 to d^{10} configurations. Fig. 2.47 shows the plots of the enthalpy of hydration $\Delta H_h(M^{2+}, aq)$ from the ideal ion-gas level, the enthalpy of formation of the gaseous cation $\Delta H_f(M^{2+}, g)$ from its elements, and the difference of these quantities $\Delta H_h(M^{2+}, aq) - \Delta H_f(M^{2+}, g)$, each in relation to a straight line joining the positions of the end members of the first transition series of metals.

It is interesting that a plot of the terms $\Delta H_f(M^{2+}, g)$, which is the sum of only the enthalpies of sublimation and the first and second ionizations of the elemental metals, shows a similar double hump curve. The 'subtraction curve' $\Delta H_h(M^{2+}, aq) - \Delta H_f(M^{2+}, g)$ indicates stabilization energy present in the metal which from the reversed position would seem to indicate a highly significant situation in the metal itself [29]; each atom in a metal is surrounded by a

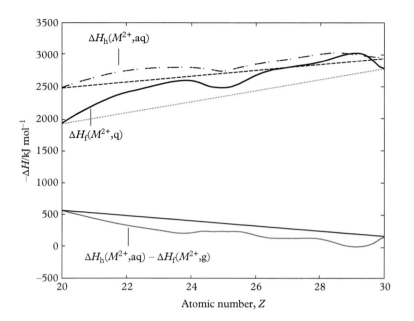

Fig. 2.47 Graphs of the enthalpy of hydration of the divalent cations Ca^{2+} to Zn^{2+}, $\Delta H_h(M^{2+}, aq)$, the enthalpy of formation of the corresponding gaseous cations $\Delta H_f(M^{2+}, g)$ and the difference of these quantities, each referred to a straight line joining its end members. A stabilization effect is clearly present in the elemental metals themselves, varying with atomic number, which enhances that arising from the hydration of M to $[M(H_2O)_6]^{2+}$.

field generated by its neighbours. The other terms in the cycle illustrated do not exhibit double hump curves:

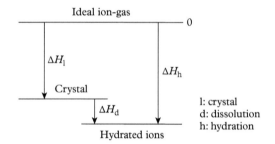

Formation ΔH_l, dissolution ΔH_d and hydration ΔH_h enthalpies.

The fascinating field of coordination chemistry has a vast literature, of which [17, 18, 25, 28] are significant examples.

2.22 Paramagnetism

A *magnetic moment* may arise in an atom from electron spin or from a coupling of electron spin and orbital motion. The relationship between the number of unpaired electrons on a species and the total magnetic moment is indicated by Table 2.12 for a selection of transition metal species.

Table 2.12 *Theoretical and experimental magnetic moments in Bohr magneton units*

Species	Unpaired electrons	μ_B (spin)	μ_B (spin + orbital)	μ_B (experimental)
V(IV)	1	1.73	3.00	1.7–1.8
Cu(II)	1	1.73	3.00	1.7–2.2
V(III)	2	2.83	4.47	2.6–2.8
Ni(II)	2	2.83	4.47	2.8–4.0
Cr(III)	3	3.87	5.20	≈ 3.8
Co(II)	3	3.87	5.29	4.1–5.2
Fe(II)	4	4.90	5.48	5.1–5.5
Co(III)	4	4.90	5.48	≈ 5.4
Mn(II)[b]	5	5.92	5.92	≈ 5.9
Fe(III)[b]	5	5.92	5.92	≈ 5.9

[a] $\mu_B = e\hbar/(2m_e)$.
[b] Mn(II) and Fe(III) have no orbital component of magnetic moment; both species are d^5.

A common type of complex has the tetrahedral configuration, as in the $[Zn(NH_3)_4]^{2+}$ ion; its point group is T_d and it will be examined in Problem 2.24 of this chapter.

The molar *magnetic susceptibility*,[14] χ_m, of a species is related to the magnetic moment μ by:

$$\mu = \sqrt{3kT\chi_m/L\mu_B^2} \qquad (2.154)$$

where μ_B is the *Bohr magneton*.[15]

In practice, magnetic susceptibility is measured with a Gouy balance that has been calibrated with a standard substance of known susceptibility, such as $MnSO_4.4H_2O$ or $(NH_4)_2Fe(SO_4)_2.6H_2O$. A sample of the compound under examination is suspended between the poles of the magnet of the balance; a species that is paramagnetic is oriented with the field, whereupon its apparent mass is increased. The magnetic susceptibility is related to the apparent change in mass through the calibration. If values of χ_m are measured at different temperatures, then μ can be deduced from a plot of χ_m against $1/T$. Equation (2.154) is embodied in the *Curie–Weiss law*: $\chi_m = C/(T - T_c)$, where T_c is the temperature at the *Curie point* and C is a constant. In an experiment with a cobalt compound μ was deduced as 4.17×10^{-23} J T^{-1}, or 4.5 μ_B; from Table 2.12, three unpaired electrons are indicated, which implies that cobalt is in the oxidation state Co(II).

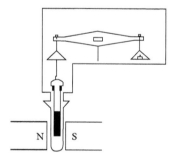

Schematic diagram of a Gouy balance

[14] Volume susceptibility = mass susceptibility/density.
[15] μ_B (Bohr magneton) $= e\hbar/(2m_e) = 9.2740 \times 10^{-24}$ JT^{-1}.

At the *Curie point*, for example, 630 K, 1043 K and 1400 K for Ni, Fe and Co, respectively, a paramagnetic compound exhibits a singularity, with a phase transition to a state that exhibits large domains with parallel spins. This alignment is known as *ferromagnetism* and gives rise to strong magnetism. Conversely, the spins may alternate, giving rise to *antiferromagnetism*, or zero magnetization; the temperature at which this change occurs is known as the *Néel temperature*.

A compound with no unpaired electrons is *diamagnetic*, and in the Gouy balance experiment a diamagnetic substance moves out of the magnetic field and appears of decreased mass. All species have a diamagnetic component of their susceptibility, but in the presence of unpaired electrons this component is swamped by paramagnetism (see also Section 6.4ff).

2.23 Apparently abnormal valence and three-centre bonding

Certain molecular species appear to exhibit abnormal valence, or *hypervalency*; compounds such as PCl_5, SF_6 and ClF_3 have more valence electrons available than are required for an outer octet. In other *electron-deficient* species, there are too few electrons than are required by the octet theory. Both valence-bond and molecular orbital methods have been applied to these apparently anomalous molecules.

2.23.1 Sulphur hexafluoride

The electron configuration of sulphur is $(Ar)(3s)^2(3p)^4$, and in the SF_6 molecule, an example of hypervalency, the sulphur atom has been stated to have an 'expanded octet' in order to explain the existence of six bonds to the atom. Calculations indicate the use of $3d$ orbitals, which are not much higher in energy than the $2p$ orbitals on the fluorine atoms, so as to allow more than eight electrons in the valence shell; in valence-bond terminology d^2sp^3 hybrid orbitals are formed to account for the bonding pattern.

The molecular orbital approach takes the $3s$ and $3p$ valence shells of sulphur and the $2p$ shell of fluorine and twelve molecular orbitals are constructed, four bonding, four antibonding and two non-bonding. The first two electrons are fed into $1a_1$ (Fig. 2.48), the next six into the triply degenerate $1t_1$ and the remaining four electrons into the doubly degenerate, non-bonding $1e$ molecular orbitals. Hence, the bonding the SF_6 molecule is explained without invoking either the d orbitals or an expanded octet construct. The $3d$ electrons of sulphur could participate, being similar in energy to fluorine $2p$, but such participation is not demanded by the theory. Whereas the valence-bond treatment requires each atomic orbital on the central sulphur atom to participate in bonding, the molecular orbital approach deals readily with hypervalency because of the number of molecular

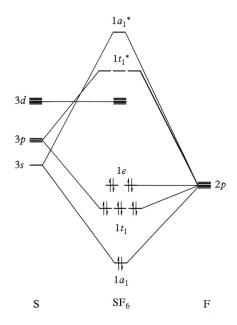

Fig. 2.48 Molecular orbital energy-level diagram for bonding in the SF_6 molecule. The $3d$ energy levels are available for bonding but are not needed, since the $1a_1, 1t_1$ and $1e$ molecular orbitals accommodate the twelve electrons in the valence shell.

orbitals available. Factors other than d orbital participation, such as atomic size, may be responsible for hypervalency.

2.23.2 Allyl anion

An example of electron deficient, *three-centre* bonding occurs in the allyl anion, a species nominally C_3H_5 but which has one electron more than occur in the allyl radical studied in Section 2.18.2: the anion may be formulated as:

Canonical forms of the allyl anion

where the $\longleftrightarrow$ symbol indicates that a resonance hybrid of feasible canonical forms is a representation of the true molecular configuration.

The point group of the allyl anion is C_{2v}: the C_2 axis passes through the central carbon atom and lies in the plane of the molecule; the σ_v plane passes through the central carbon atom and lies normal to the plane of the molecule; and the σ_v' plane is that of the molecule itself. Now, sketch the molecule and insert the symmetry elements. Consider the three $2p$ orbitals coinciding with the orthogonal

x, y and z axes. Then, the following unshifted atom contributions arise for the carbon atoms: for E none, contribution 3; for C_2 one, contribution -1; for σ_v one, contribution 1; for σ_v' three (inversion of p orbital signs), contribution -3. So the representation is:

C_{2v}	E	C_2	σ_v	σ_v'
Γ_π	3	-1	1	-3

which reduces to $A_2 + 2B_1$. The character table for C_{2v} is given in Table 2.13.

Applying projection operators to $\psi_1(C_1)$ and normalizing:

$$P_{A_2}\psi_1 = (1/\sqrt{2})(\psi_1 - \psi_2)$$
$$P_{B_1}\psi_1 = (1/\sqrt{2})(\psi_1 + \psi_2)$$

(2.155)

Atoms C(1) and C(3) are related by symmetry and atom C(2) must use the second B_1 irreducible representation:

$$P_{B_1'}\psi_2 = \psi_2$$

(2.156)

For the A_2 SALC, the energy is clearly zero, and for the B_1 SALCs, the secular determinant is:

$$\begin{vmatrix} H_{B_1 B_1} - E & H_{B_1 B_1'} \\ H_{B_1 B_1'} & H_{B_1' B_1'} - E \end{vmatrix}$$

(2.157)

where the terms are evaluated as follows:

$$H_{B_1 B_1} = 1/2 \int (\psi_1 + \psi_2)\mathcal{H}_e(\psi_1 + \psi_2)d\tau = \alpha$$
$$H_{B_1 B_1'} = (1/\sqrt{2}) \int \psi_2 \mathcal{H}_e(\psi_1 + \psi_2)d\tau = \sqrt{2}\beta$$
$$H_{B_1' B_1'} = \int \psi_2 \mathcal{H}_e \psi_2 d\tau = \alpha$$

Table 2.13 *Character table for point group C_{2v}*

$C2v$	E	C_2	$\sigma(xz)$	$\sigma(yz)$		
A_1	1	1	1	1	z	x^2, y^2, z^2
A_2	1	1	-1	-1	R_z	xy
B_1	1	-1	1	-1	x, R_y	zx
B_2	1	-1	-1	1	y, R_x	yz

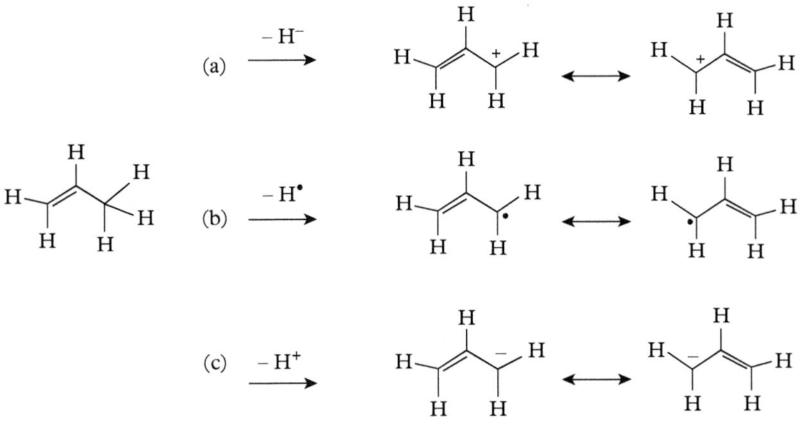

Fig. 2.49 Molecular structures of the allyl species: (a) cation, (b) radical, (c) cation.

——————— π_3-antibonding

—— $+$ $+\!\!+$ π_2-nonbonding

$+\!\!+$ $+\!\!+$ $+\!\!+$ π_1-bonding

(a) Allyl cation (b) Allyl radical (c) Allyl anion

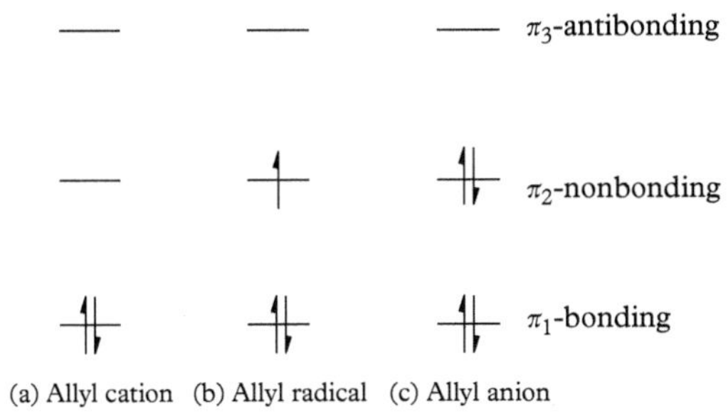

Fig. 2.50 π-Molecular orbital relative energy-level diagrams for the allyl species: (a) cation, (b) radical, (c) cation.

Substituting these values into Eq. (2.157) gives the equation:

$$(\alpha - E)^2 = 2\beta^2$$

or

$$E = \alpha \pm \sqrt{2}\beta$$

leading to a total energy $4\alpha + 2\sqrt{2}\beta$. If the allyl anion were not delocalized, the energy would be $4\alpha + 2\beta$. Hence, the delocalization energy is $(2\sqrt{2}-2)\beta$, or 0.828β. The molecular structures of the three allyl species, cation, radical and anion, and the relative energy levels are illustrated in Figs. 2.49 and 2.50, respectively.

2.23.3 λ^5-Borane

Boron hydrides appear to have insufficient valence electrons for the number of bonds present. The explanation of their stability hinges of the formation of *three-centre* bonds for these electron-deficient species that have less than eight electrons in the valence shell. A three-centre, two-electron bond involves three atoms sharing two electrons. The combination of three atomic orbitals produces three molecular orbitals, one bonding, one non-bonding, and one antibonding. The two electrons occupy the bonding orbital, resulting in a net bonding effect.

In the λ^5-borane B_2H_6 molecule (aka diborane), a total of twelve molecular orbitals can be formed, three from each boron atom $[(1s)^2](2s)^2(2p)^1$ excluding the use of the $(1s)^2$ core, and one from each hydrogen atom $(1s)$, leading to six bonding and six antibonding molecular orbitals. According to the Aufbau principle, the twelve valence electrons fill the six lower, bonding orbitals. In the actual structure, two B–H–B molecular orbitals exist, which give rise to two, three-centre, two-electron, bridging bonds. The structure of λ^5-borane may be written as:

Diborane molecule

The molecule is isoelectronic with $H_3C^+–C^+H_3$, which suggests a way of looking at the bonding beginning with an ethane-like structure:

$$H_2B^- = B^-H_2$$

which is then stabilized by the approach of two protons

$$\downarrow H^+$$
$$H_2B^- = B^-H_2$$
$$H^+ \uparrow$$

to give the doubly-bridged structure. Many other examples of three-centre bonds are now recognized.

The molecular orbital theory provides neat explanations for this and similar compounds that were insuperable obstacles for the Lewis electron-pair hypothesis, and also not easily resolvable by the valence-bond model.

2.23.4 Shifting bonds

While the Lewis electron-pair mechanism remains a basic feature in most chemical covalent bonding situations, relatively recent work has introduced the idea of *charge-shift bonding*. It has been suggested that in some species normally considered to be covalently bonded, such as $H_2N–NH_2$, F–F and HO–OH, the covalent component of the bonding is either weak or repulsive. Instead, the atoms are held together in the molecule by a resonance between ionic and covalent canonical structures [30, 31], with resonance energies as large as 300 kJ mol^{-1}. It could perhaps be understood if, in the process of bonding of the two atoms, the kinetic energy of the system were increased to such an extent that it overcame the potential energy of attraction of the species, thus separating them into positive and negative entities. They would then interact to restore stability, and a dynamic process of separation and reunion could be envisaged. Further work on this topic will undoubtedly emerge.

2.23.5 σ-Hole bonds

It was discovered many years ago that when iodine is added to a saturated solution of ammonium nitrate the species H_3N-I_2 is formed, but it decomposes rapidly into its components when exposed to air. This type of linkage in which a donor species, such as the lone-pair electrons on a nitrogen or an oxygen atom, induces polarity and subsequently bonds with a positive region on a halogen atom has been termed a halogen bond. The positive region on the acceptor halogen atom or 'hole' in its electron density distribution, has led to the term σ-*hole bond* for this class of interaction; the hole lies on an area of the acceptor species opposite to that of the covalent bond [32].

An x-ray crystallographic study of the 1:1′-bromine-1,4-dioxane complex: revealed an

1:1′-Bromine-1,4-dioxane

intermolecular O–Br distance of 0.271 nm [34]. Since the sum of the van der Waals radii of the oxygen and bromine atoms is 0.347 nm, a strong interaction between these atoms is indicated; the O-Br–Br linkage is very closely linear. From the geometry of the structure, it appears that the halogen acceptor atoms are linked to the oxygen donor atoms in the direction of the axes of the orbitals of the lone pairs.

2.23.6 Diamondoid compounds

Investigations on diamondoid compounds has revealed C–C single bond distances as large as 0.17 nm. *Diamondoids* are cage-like molecules, the simplest of which is adamantane $C_{10}H_{16}$, illustrated here together with its numbering system:

Adamantane

Condensed adamantane structures include one or more cages, and these diamondoids occur naturally in petroleum deposits, from which they have been extracted and purified to form large, pure crystals of polymantane molecules having more than a dozen adamantane cages per molecule [34]. These species simulate the cubic diamond framework, but with terminal C–H bonds.

An interesting molecule in this class of compounds is 2-(1-diamantyl) [121]tetramantane:

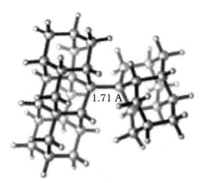

2-(1-Diamantyl)[121]tetramantane

It has been synthesized by reacting 1-bromodiamantane and 2-bromo-[121]tetramantane in xylene with metallic sodium under argon. Pure stable crystals were obtained, and the structure determined by x-ray diffraction was refined to R = 4.5%. Thus, the unusually long C–C single bond length of 0.1704 nm is confirmed [35]. One reason for the stability of long bonds in these compounds is said to arise from the fact that dispersion interactions (Section 3.4.3) play an important role in binding the two fragments and so enhance the stability of the long bond. In fact, calculations have indicated that a juxtaposition of sufficient surface area for large dispersion interactions could stabilize C–C bonds of lengths up to 0.4 nm [36]. Diamondoid compounds have found applications in nanotechnology, biomedicine and materials and petroleum sciences [37].

Notwithstanding these relatively new forms of interaction, the four bond types on which the classification discussed in the first chapter is based remain a valid basis for the discussion of a very large area of chemical combination. However, it is evident that the distinction between a bond and a non-bond is not clear-cut: any form of chemical interaction might be termed a bond.

2.24   Theoretical techniques

Theoretical chemistry is the application of mathematical methods to study the structure and reactions of chemical species. While theoretical methods may, in principle, be computational or non-computational, most activity in the theoretical field centres on computational studies.

Computational chemistry involves the numerical calculation of electronic structures of molecules and their interactions. It can be a useful way to investigate difficultly obtainable or costly materials, and is useful for making predictions preparatory to experimental procedures. It involves quantum and classical mechanics and statistical procedures; highly developed computer software exists that permits calculations in many areas of chemistry:

- electronic structure determination and optimization;
- determination of reaction paths and potential energy surfaces;
- charge distributions on atoms in molecular environments;
- prediction of crystal structures;
- protein docking;
- thermodynamic parameters;
- statistical mechanics.

The most used theoretical techniques are *ab initio* (L. 'from the beginning') and *semi-empirical* techniques, together with *molecular mechanics* based on classical physical force fields or approximations to them. The main applications of these techniques may be summarized as follows:

Ab initio methods apply quantum mechanics to small chemical systems (10 + atoms) without introducing any empirical parameters. Good accuracy can be achieved, but it decreases as the system size increases.

Semi-empirical techniques involve quantum mechanics but include some empirical parameters, such as two-electron integrals. They can be applied to larger systems (100+ atoms), and are useful in applications to organic chemistry, where only a small number of elements is involved.

Molecular mechanics uses the force fields of classical mechanics parameterized against experimental data; the procedures of *Monte Carlo* for equilibrium thermodynamic parameters and radial distribution functions, and *molecular dynamics* for time-dependent processes such as liquid transport are included in this group. These procedures are computationally faster than the *ab initio* and semi-empirical methods and can be used with large systems (1000 + atoms), but they cannot be applied to processes that involve making or breaking of chemical bonds.

A discussion of theoretical calculations lies outside the scope of this book. The number of programs now developed and supported for theoretical work is very large, and the reader who is interested in this aspect of chemistry should study the literature of the subject available in text and on web sites [38–40].

2.25 Structural and physical characteristics of covalent solids

From the discussions in this chapter, it is evident that the number of compounds with atoms bonded by covalency is vast. Covalent *solids*, in which covalent bonding is responsible for cohesion in the solid state, are relatively few in number. The atoms in these covalent compounds tend to have similar electronegativities, and bonding relies mainly upon an electron sharing process that leads to molecular orbital formation. In organic compounds, such as those studied herein, the bonding in the species includes terminal hydrogen, oxygen, fluorine or other atoms in order to satisfy valence requirements. Consequently, there is no mechanism for extending covalent bonding between molecules; intermolecular bonds rely on other forces that will be the subject of the next chapter.

The best example of a covalent solid is the diamond allotrope of carbon. It can be imagined to be based on a methane-like structure, but with the hydrogen atoms replaced by tetrahedral carbon atoms, thus providing a mechanism for building up a three-dimensional solid solely by covalent forces. The bonding is strong, and one result is that the solid has a very high the melting point, approximately 3550°C, but only under very high pressure or in the presence of an inert gas, because of oxidation to carbon dioxide.

The crystal structure of diamond is shown in Fig. 2.51; remember that these diagrams of structures are representative portions of infinitely large structures, obtained by stacking the unit cells regularly in three-dimensional space, and that the covalent forces in diamond are continued throughout the whole structure. The diamond structure is also adopted by several well-known semi-conducting materials, such as silicon and germanium, to name but a few. A summarizing comparison of the classification of compounds is given in Section 1.8ff.

Graphite (Fig. 2.52) is not a covalent solid, yet its melting point is slightly higher than that of diamond. Although it is relatively easy to break the interlayer

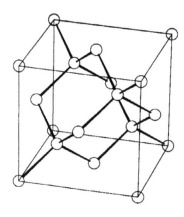 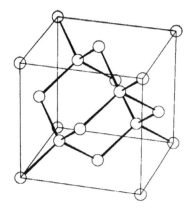

Fig. 2.51 Stereoscopic view of the unit cell and environs of the diamond structure of carbon. Each atom is bonded covalently to its neighbours in a four-coordinated, tetrahedral pattern; the C–C bond length is 0.1544 nm and the C–C–C bond angle 109.47°.

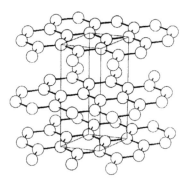
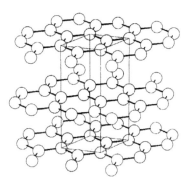

Fig. 2.52 Stereoscopic view of the unit cell and environs of the graphite structure of carbon. Each atom is bonded covalently to its neighbours in layers, in a three-coordinated, trigonal-planar pattern; the C–C bond length is 0.142 nm and the C–C–C bond angle is 120°. Between the layers, the distance is 0.335 nm, a little less than twice the van der Waals radii for carbon and indicating an enhanced interlayer attraction from π overlap. The weakness of this interlayer bonding imbues graphite with its lubricant properties.

van der Waals links in graphite, the layers contain an effectively infinite number of fused six-membered rings that benefit in stability from delocalization energy, as in benzene; the free energy of formation of diamond is only 2.90 kJ mol^{-1} greater than that of graphite, the thermodynamically stable form of carbon. A development of the bonding situation of graphite is found in *graphene*, which will be discussed in a later chapter

The covalent elements in periodic group 14 (IV A) are carbon–α-tin; other covalent solids are quartz (Fig 2.53), boron nitride and carbides, particularly titanium hafnium carbide.

The change in bond type through the group 14 (IVA) elements is paralleled by significant changes in the *electrical resistivity*:

	C(diamond)	Si	Ge	α-Sn	β-Sn	Pb
ρ/ohm m	5×10^{12}	2×10^3	5×10^{-1}	1×10^{-5}	1×10^{-7}	2×10^{-7}

Covalent solids form strong, hard crystals, with low compressibility and high melting points. In these properties they are very similar to ionic solids. However,

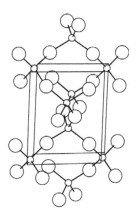
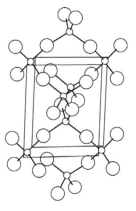

Fig. 2.53 Stereoview of the unit cell and environs of the crystal structure of β-quartz, SiO$_2$: circles in decreasing order of size represent O and Si, respectively.

in strong contrast to ionic solids, they are insulators both in solid and in melt. Covalent solids are chemically rather inert substances, insoluble in all usual solvents. Partial covalent bonding, as in zinc sulphide and the silver halides, will be studied in the chapter on ionic solids, and a discussion of semiconducting materials follows the discussion on metals.

....

REFERENCES 2

[1] Newton I. *Philosophiæ Naturalis Principia Mathematica*, Books 1–3; 1687.
[2] Planck M. *Ann. Phys.* 1901; 4: 553.
[3] Bracewell R. *The Fourier Transformation and its Applications*. McGraw-Hill, 1999.
[4] Atkins P and de Paula J. *Atkins' Physical Chemistry*, 9th ed. Oxford University Press, 2009.
[5] Whitman LJ *et al. Phys. Rev. Lett.* 1991; 66: 1338.
[6] Pauling L and Wilson EB. *Introduction to Quantum Mechanics with Applications to Chemistry*. Dove Books, 1986. (First published, 1935).
[7] Margenau H and Murphy GM. *The Mathematics of Physics and Chemistry*. Van Nostrand, 1943.
[8] Ladd M and Palmer RA. *Structure Determination by X-ray Crystallography*, 5th ed. Springer Science+Business Media, 2013.
[9] Cox EG *et al. Phil. Trans. Roy. Soc. Lond.* 1958; A247: 1.
[10] Euler L. *Introductio in Analysin Infinitorum*. Bosquet, Lausanne, 1748. (Eng. trans. Blanton J. *Introduction to Analysis of the Infinite, Book I*. Springer Verlag, 1988.)
[11] Boeyens JCA. *Chemistry from First Principles*. Springer, 2008.
[12] Shpenkov GP. *Hadronic J.* 2006; 29: 455; http://shpenkov.janmax.com/HybridizationShpenkov.pdf
[13] House JE. *Fundamentals of Quantum Chemistry*. Academic Press, 2003.
[14] McQuarrie DA. *Quantum Chemistry*. University Science Books, 2007.
[15] Slater JC. *Phys. Rev.* 1930; 36: 57.
[16] Clementi E and Raimondi DL. *J. Chem. Phys.* 1963; 38: 2686.
[17] Atkins P *et al. Shriver and Atkins' Inorganic Chemistry*, 5th ed. Oxford University Press, 2009.
[18] Kettle SFA. *Physical Inorganic Chemistry*. Oxford University Press, 1998.
[19] Pilar FL. *Elementary Quantum Chemistry*, 2nd ed. McGraw-Hill, 1990.
[20] Levine IN. *Quantum Chemistry*, 3rd ed. Prentice Hall, 2000.
[21] Lowe JP and Peterson K. *Quantum Chemistry*. Academic Press, 2005.
[22] Ladd M. *Symmetry of Molecules and Crystals*. Oxford University Press, 2014.
[23] Kettle SFA. *Readable Group Theory for Chemists*. Wiley-Blackwell, 2007.
[24] Atkins PW. *Molecular Quantum Mechanics*. Oxford University Press, 2010.

[25] Cotton FA. *Chemical Applications of Group Theory*, 3rd ed. John Wiley & Sons Inc, 1990.

[26] Salthouse JA and Ware MJ. *Point Group Character Tables and Related Data.* Cambridge University Press, 1972.

[27] Atkins PW *et al. Tables for Group Theory*. Oxford University Press, 2006. Online at http://www.cinam.univ-mrs.fr/klein/teach/inorganic/tables_for_group_theory.pdf

[28] Figgis BN and Hitchman MA. *Ligand Field Theory and its Applications.* Wiley-VCH, 2000.

[29] Ladd MFC and Lee WH, *J. Chem. Soc.* 1962; p. 2837.

[30] Shaik S *et al. Nat. Chem.* 2009; 1: 449.

[31] *idem. Chem. Eur. J.* 2005; 11: 6358.

[32] Politzer P *et al. Phys. Chem. Chem. Phys.* 2013; 15: 11178.

[33] Hassel O *et al. Acta Chem. Scand.* 1954; 8: 873.

[34] Dahl JE *et al. Science* 2003; 299: 96.

[35] Fokin AA *et al. J. Am. Chem. Soc.* 2012; 134: 13641.

[36] Schreiner PR *et al. Science* 2011; 477: 308.

[37] Mansoori GA *et al. Diamondoid Molecules.* World Scientific, 2012.

[38] Jensen K. *Introduction to Computational Chemistry.* John Wiley and Sons, 1996.

[39] Young D. *Computational Chemistry.* John Wiley and Sons, 2001.

[40] Gaussian Inc. 2014. http://www.gaussian.com/g_prod/g09b.htm

Problems 2

2.1. A projectile moves off from rest at a velocity of 40.00 m s^{-1} and at an angle to the ground of 35°. What are (a) the time taken for the projectile to reach the maximum height, (b) the maximum height attained by the projectile, and (c) its kinetic energy at maximum height? The gravitational acceleration is 9.812 m s^{-2}, and an absence of air resistance may be assumed.

2.2. Calculate the energy and momentum of an electron travelling at one tenth the speed of light *in vacuo* if the energy is wholly kinetic.

2.3. At what wavelength is the energy density for a black-body radiator at a maximum at 500 K?

2.4. Compare the Rayleigh–Jeans equation with the Planck equation for black-body radiation by calculating the energy density of radiation in the wavelength range 590 nm to 600 nm in a black-body cavity of volume 10 cm^3 at a temperature of 500 °C. The mean wavelength may be taken as 595 nm and the wavelength range 10 nm.

2.5. A clean surface of pure, metallic tin *in vacuo* was exposed to monochromatic radiation from a quartz-mercury arc source. The emitted photoelectrons were collected in an oxidized copper Faraday cylinder. The current obtained for different values of the applied voltage V was measured with an electrometer.

In different experiments differing wavelengths λ were used. The deflection θ recorded by the electrometer was proportional to the photoelectric current, and θ increased as V was made more negative at the electrode metal. A field was set up in the space between the dissimilar tin and oxidized copper metals, and a contact potential difference arose, acting from tin to copper, that is, tin was positive with respect to the Faraday cylinder. The following results were obtained for three wavelengths:

$\lambda = 546.1$ nm		$\lambda = 365.0$ nm		$\lambda = 312.6$ nm	
$-V/V$	$\theta/\deg$	$-V/V$	$\theta/\deg$	$-V/V$	$\theta/\deg$
2.257	28	1.157	67.5	0.5812	52
2.205	14	1.105	36	0.5288	29
2.152	7	1.0525	19	0.4765	12
2.100	3	1.0002	11	0.4242	5.7
		0.9478	4	0.3718	2.5

(a) Plot V_0 against θ for each wavelength, and estimate the minimum applied voltage that prevents the fastest moving photoelectrons from reaching the Faraday cylinder. Program LSLI in the Web Program Suite may be helpful in extrapolating for V_0 and GRFN for clear, interpretable plots.

(b) An electron moving through a potential difference V acquires energy eV eV. This energy is equivalent to the kinetic energy $\frac{1}{2}m_e v^2$, where m_e and v are, respectively, the mass and speed of the electron. (Distinguish between v velocity and ν frequency.) Show that an equation $V_0 e = \eta\nu - \zeta$ represents the variation of V_0 with frequency ν. Find the values of the parameters η and ζ. What do they represent?

2.6. In an electron diffraction experiment, electrons of wavelength 0.5 nm were employed. What would be the velocity of such electrons?

2.7. The Balmer spectral series for atomic hydrogen was analysed originally in terms of wavelength λ through the equation $\lambda = K[n^2/(n^2 - 4)]$, where K is a constant and n is an integer greater than 4. Show that this equation is equivalent to Eq. (2.15), and find K in terms of R_H. What is the energy associated with the spectral line nearest to the red end of the Balmer series?

2.8. What is the minimum uncertainty in the speed of an electron travelling in a one-dimensional box of length $2a_0$?

2.9. An electron is confined to a one-dimensional box. Assuming that $\int_x^{x+\Delta x} |\psi(x)|^2 dx \approx |\psi(x)|^2 \Delta x$, find the fraction m of the length a of the box at which the probability of the electron position is a maximum for the ground state and first two excited states.

2.10. What is the smallest value of the kinetic energy of an electron in a cubical box of side 10^{-15} nm?

2.11. Calculate the probability of finding a hydrogen $1s$ electron in the volume bounded by $r = 1.10a_0$ to $1.11a_0$, $\theta = 0.20\pi$ to 0.21π and $\phi = 0.60\pi$ to 0.61π. The wavefunction is given in Table 2.3, and may be assumed constant over the small volume considered.

2.12. Write the electron configurations for (a) N, (b) Al, (c) Cl$^-$ and (d) Y.

2.13. Calculate the energy of the species He$^+$ in the ground state.

2.14. Set up fully the Schrödinger wavefunction for the helium atom—but do not attempt its solution.

2.15. By considering the orthogonality criterion determine whether or not, and under what restraints if any, the linear combination of hydrogen-like atomic orbitals (a) $1s + 2s$, and (b) $1s + 2p$, will lead to bonding molecular orbitals for the heteronuclear diatomic species H–F.

2.16. Obtain data with the program RADL for plotting the hydrogenic radial distribution functions with the program GRFN for the $1s$, $2s$ and $2p$ atomic orbitals. (a) With the mouse over the plots, determine the value of r/a_0 (abscissa) at which maxima occur. (b) Confirm your findings from the first derivative with respect to r of the $1s$, $2s$ and $2p$ radial functions (Table 2.2). Note that terms in the radial functions not involving r can be omitted from the differentiation procedure.

2.17. The overlap integral for two hydrogenic s atomic orbitals may be given by $S_{s,s} = \exp(-\rho)(1 + \rho + \rho^2/3)$, and for s and p by $S_{s,p} = \rho \exp(-\rho)(1 + \rho + \rho^2/3)$, where $\rho = r/a_0$. (a) Plot both functions from $r = 0$ to $r = 3a_0$; program GRFN will produce a clear plot for a set of ρ, S data. (b) Determine the value of the overlap integral $S_{s,s}$ for (i) H$_2^+$ ($r_e = 0.106$ nm) and (ii) H$_2$ ($r_e = 0.0741$ nm). (c) At what value of ρ is the $S_{s,p}$ overlap integral a maximum. Confirm the graphical result by differentiation of the appropriate function.

2.18. The dipole moment of the water molecule is 1.8 D. Given that the O–H length is 0.096 nm and that the H–O–H bond angle is 104.4°, calculate the partial charges on the oxygen and hydrogen atoms.

2.19. Show than an sp^x hybrid from orthonormal atomic orbitals has a normalized wavefunction given by

$$\psi_{sp^x} = (1 + x)^{-1/2}(\psi_s + x^{1/2}\psi_p)$$

2.20. Determine and solve the secular determinant for 1,3-butadiene, C_4H_6, in the Hückel approximation. Obtain the energies E_π of the molecular orbitals, the delocalization energy, the coefficients of the wavefunction, the bond orders, and the free valence and charge for each atom.

2.21. Apply Hückel molecular orbital theory to 1,3-cyclobutadiene (remember that atoms 1 and 4 are adjacent), and determine the energies E_π for the first four molecular orbitals, and the delocalization energy D_π. Which π orbitals constitute the frontier molecular orbitals for this molecule? What can be said of the ground state of this molecule? [Note that 1,3-cyclobutadiene is not aromatic: aromaticity is governed by the *Hückel rule*, namely, that the number of π electrons must be representable as $4n + 2$, $n = 1, 2, \ldots$]

2.22. The 1,1'-diethyl-4,4'-cyanine cation may be treated as a 'box' of length from N: to N^+, including the N: atom.

1,1'-Diethyl-4,4'-cyanine cation

The mean bond length along the conjugated chain (which includes the lone pair on the N: atom) is 0.140 nm. The box contains $(2n + 2)$ π electrons, that is, two from the neutral nitrogen atom N: and two from each double bond; n is the number of double bonds, and the electrons occupy the first $n + 1$ molecular orbitals. The colour of the dye arises from the transition of an electron from uppermost occupied level to the next highest level. Show that the wavelength λ of the first excited transition is

$$\lambda = 3.2972 \times 10^{-6} l^2 / (2n + 1)$$

where l is the bond length in nm. Calculate λ and determine the colour of the dye.

2.23. For the transition metal ions $(3d)^1$ to $(3d)^9$ write the d electron configurations and the number N of unpaired electrons in (a) weak ligand field and (b) strong ligand field configurations.

2.24. The $[Zn(NH_3)_4]^{2+}$ ion has regular tetrahedral symmetry, point group T_d with a small magnitude of d orbital splitting. By considering how a regular tetrahedron is related in symmetry to an octahedron and a cube (Fig. 2.34), and with regard to Figs. 2.21 and 2.42, show what splitting is to be expected for the given tetrahedral complex. Would this zinc compound be paramagnetic or diamagnetic?

2.25. A coordination compound containing iron has an experimental molar susceptibility χ_m of 1.13×10^{-1} m^3 mol^{-1} at 25°C. Given that the diamagnetic susceptibility is negligible, determine the probable number of unpaired electrons in the species and the oxidation state of iron in the coordination compound.

2.26. Use VSEPR theory to predict the probable shapes of the fluorine containing compounds, (a) F_2O, (b) NF_3, (c) SiF_4, (d) SF_4, (e) SF_6 and (f) XeF_4. (Table 2.10 may be helpful).

2.27. Use the method of projection operators to obtain SALC_MOs for ammonia, NH_3, point group C_{3v}. Character tables for this point group and for C_3 are given below. Obtain orthogonal SALCs for E symmetry by linear combination of those derived. It is possible to work in the simpler symmetry of point group C_3. Check the final SALCs for orthogonality properties and draw a molecular orbital energy level diagram.

C_{3v}	E	$2C_3$	$3\sigma_v$		
A_1	1	1	1	z	$x^2 + y^2, z^2$
A_1	1	1	−1	R_z	
E	2	−1	0	$(x, y), (R_x, R_y)$	$(x^2 - y^2, xy), (zx, xy)$

C_3	E	C_3	C_3^2		$\varepsilon = \exp(i2\pi/3)$
A	1	1	1	z, R_z	$x^2 + y^2, z^2$
E	$\begin{Bmatrix} 1 \\ 1 \end{Bmatrix}$	$\begin{matrix} \varepsilon \\ \varepsilon^* \end{matrix}$	$\begin{matrix} \varepsilon^* \\ \varepsilon \end{matrix}$	$(x, y), (R_x, R_y)$	$(x^2 - y^2, xy), (yz, zx)$

2.28. Set up a representation for the water molecule based on the movements of the O–H vectors under the point group symmetry C_{2v}. Find by inspection of the character table for C_{2v}, or otherwise, the irreducible representations. Determine wavefunctions by the projection operator method and normalize them. Check for orthogonality of the functions determined. The character table for C_{2v} is given below.

C_{2v}	E	C_2	$\sigma(xz)$	$\sigma(yz)$		
A_1	1	1	1	1	z	x^2, y^2, z^2
A_2	1	1	−1	−1	R_z	xy
B_1	1	−1	1	−1	x, R_y	zx
B_2	1	−1	−1	1	y, R_x	yz

2.29. Phosphorus pentafluoride PF_5 has a trigonal bi-pyramidal structure:

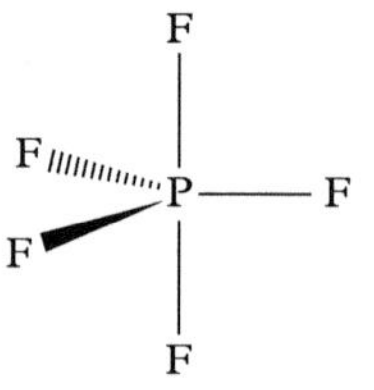

Phosphorus pentafluoride

Its point group is D_{3h}, the character table for which is listed as Table 2.14. Set up a representation based on the five P–F vectors, and determine what combinations of orbitals would be most likely to satisfy the geometry of the molecule.

2.30. *In vivo* oxidation of β-carotene (Example 2.6) produces two molecules of (11*cis*)-retinal (aka retinaldehyde or vitamin A):

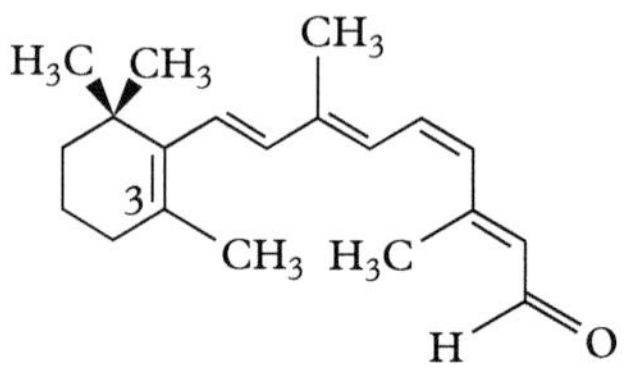

(11*cis*)-retinal

The process of optical vision begins with its isomerization to (11*trans*)-retinal by absorption of a γ photon. The retinal molecule portion from C(3) to the terminal =O may be likened to a 'box', with an average bond distance of 0.140 nm. Determine (a) the energy separation between the

Table 2.14 *Character table for point group* D_{3h}

D_{3h}	E	$2C_3$	$3C_2$	σ_h	$2S_3$	$3\sigma_v$		
A_1'	1	1	1	1	1	1		$x^2 + y^2, z^2$
A_2'	1	1	−1	1	1	−1	R_z	
E'	2	−1	0	2	−1	0	(x, y)	$(x^2 − y^2, xy)$
A_1''	1	1	1	−1	−1	−1		
A_2''	1	1	−1	−1	−1	1	z	
E''	2	−1	0	−2	1	0	(R_x, R_y)	(zx, yz)

ground state and the first excited state of retinal; (b) the wavelength of the radiation produced when the excited state returns to the ground state; (c) the change in shift of the polyene emission spectrum contingent upon a change in the length of the conjugated chain.

2.31. Calculate by the Hückel MO method the relative stabilities of tri-hydrogen species: (a) $(H_1–H_2–H_3)^+$, $(H_1–H_2–H_3)$ and $(H_1–H_2–H_3)^-$, and (b) the same species in the cyclic forms below:

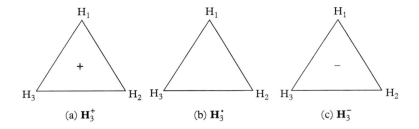

(a) $\mathbf{H_3^+}$　　　　(b) $\mathbf{H_3^{\cdot}}$　　　　(c) $\mathbf{H_3^-}$

(b) By inspection, or otherwise, write the numerical form of the secular determinant for each of the species (a)–(c); draw energy level diagrams, and list the energy levels in terms of α and β.

2.32. The Hückel determinant for 1,3,5-hexatriene:

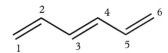

Hexatriene

is

$$
\begin{vmatrix}
y & 1 & 0 & 0 & 0 & 0 \\
1 & y & 1 & 0 & 0 & 0 \\
0 & 1 & y & 1 & 0 & 0 \\
0 & 0 & 1 & y & 1 & 0 \\
0 & 0 & 0 & 1 & y & 1 \\
0 & 0 & 0 & 0 & 1 & y
\end{vmatrix} = 0
$$

Solve the determinant (Appendix A10) and list the values of y, and the energies in terms of α and β; Bairstow's method (program ROOT) may be needed in order to solve the derived equation. Write the total energy E and the delocalization energy D_π. Confirm your results with the program HUCK. Check by hand some of the results from the program for bond orders, charge distributions and free valence indexes.

2.33. Apply the program HUCK to the molecules of pyridine and pyrrole; for pyrrole, assume $h = 1.5$ and $k = 0.8$. The crucial factor in using HUCK is the setting up of the correct secular determinant; follow the example of pyridine (Section 2.19). From the results for pyridine, predict the likely products of its reaction with (i) sodium amide (sodamide) in liquid ammonia, (ii) gaseous bromine and (iii) nitration acids (HNO_3/H_2SO_4).

2.34. What is the important difference between the lone-pair electrons on benzophenone ($C_6H_5(CO)C_6H_5$ and phenol C_6H_5OH?

2.35. How is pyridine a stronger base than pyrrole and, therefore, forms hydrochloride salts more readily than does pyrrole?

2.36. (a) Use HUCK to determine the bond orders of anthracene:

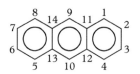

Anthracene

(b) Coulson's formula for deriving a C–C bond length d (in Å) from a π-bond order p is:

$$d = 1.515 - \frac{0.180}{1 + 1.050\left(\frac{1-p}{p}\right)}$$

Investigate this formula for $p_{1,2}, p_{2,3}, p_{4,12}, p_{9,11}, p_{11,12}$ from HUCK, and compare the results with data from the literature below. (c) What are the most chemically reactive sites of anthracene?

(a) Bragg WH. *Proc. Phys. Soc. Lond.* 1922; 36: 167.

(b) Robertson JM. *Proc. Roy. Soc. Lond.* 1933; 140: 79.

(c) Mathieson AMcL *et al. Acta Crystallogr.* 1950; 3: 245.

(d) Sutton LE. (Ed.) *Tables of Interatomic Distances and Configuration in Molecules and Ions*, Special Publication No.18, The Chemical Society (now the Royal Society of Chemistry), 1965. (*Supplement* 1956–1959). Online at https://archive.org/details/TablesOfInteratomicDistancesAndConfigurationInMoleculesIons

(e) Cruickshank DWJ. *Acta Crystallogr.* 1956; 9: 915.

2.37. (a) Sketch the geometrical isomers of the neutral octahedral cobalt coordination compound [$Co(NH_3)_2H_2O(NH_2)_2Cl$]. (b) List the point group of each isomer. (c) Which isomers, if any, exhibit optical isomerism?

2.38. The wavelength λ and frequency v in a wave guide are related by the equation

$$\lambda = \frac{c}{\sqrt{v^2 - v_0^2}}$$

Express the group velocity v_g in terms of c and the phase velocity $v_p (= \lambda v)$.

2.39. How many electrons can an atom possess if (a) $n = 4$ and $l = 1$, and (b) $n = 4, l = 0$?

2.40. (a) What are the possible quantum numbers for a $4p$ electron with $m_s = -\frac{1}{2}$? (b) What are the spin directions of the valence electrons of magnesium?

Solution for Plotting exercise 2.1

(a)

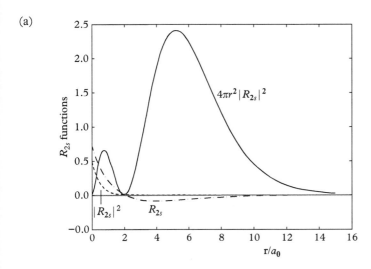

$R_{2s}, |R|^2_{2s}$ and $4\pi r^2 |R|^2_{2s}$ functions

(b) $R_{2s} = 0$ at $r/a_0 = 2$.

(c) Turning points for the $4\pi r^2 |R_{2s}|^2$ function, measured from the graph, are 2, 0.74 and 5.2.

(d) The appropriate radial distribution function, using Table 2.2 and neglecting all multiplying constants, is $r^2 (2 - r/a_0)^2 \exp(-r/a_0)$. The turning points occur for the derivative of this function set to zero. Straightforward manipulation leads to the equation $\sigma^3 - 8\sigma^2 + 16\sigma - 8 = 0$, where $\sigma = r/a_0$. This equation solves by program ROOT to a precision of 10^{-6}, giving the values of r/a_0 as 2, 0.7639 and 5.236.

Solution for Plotting exercise 2.2

(a)   The graphs below show the functions ψ_{2s} and ψ_{2s}, and the following points emerge:

ψ_{2s}	ψ_{3s}
Positive at the centre	Positive at the centre and larger than ψ_{2s}
Amplitude decreases from the centre	Amplitude decreases more rapidly than does ψ_{2s}
Passes through zero once ($r/a_0 = 2$)	Passes through zero twice ($r/a_0 = 1.273$ and 4.732)

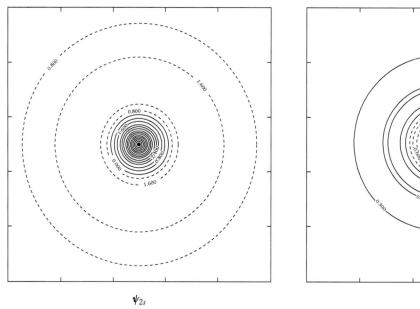 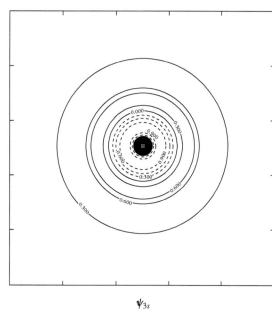

ψ_{2s} ψ_{3s}

(b) ψ_{2s} One node; ψ_{3s} two nodes. [In general the number of nodes is $n - l - 1$.]

(c)   The plot of the radial functions below at (i) confirms the above findings. The link with the transverse central sections of the two radial functions shown at (ii) is evident.

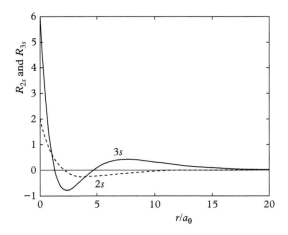

(i) R_{2s} and R_{3s} functions

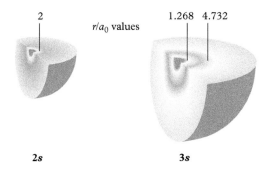

(ii) Exploded R_{2s} and R_{3s} functions

Molecular Compounds

3

Opposition brings concord. Out of discord comes the fairest harmony.

Heraclitus

3.1 Introduction

All forms of attraction between adjacent entities of any chemical substance or between substances themselves depend ultimately on the interaction of positive and negative constituents of matter. The interactions are most notable in the solid or liquid state, and involve coulombic-type electrical forces. When two atomic species are very close together, their potential energy is large and they tend to repel each other. Conversely, when they are very far apart their potential energy is effectively zero, but they can be attracted one to the other if brought sufficiently close to feel a force of attraction. In this intermediate, bond-forming region their attractive potential energy is operative. A position exists at an equilibrium distance of separation r_e at which the system has a minimum potential energy and thus adopts a stable state. Then, an expenditure of work on the system is required in order to separate the atoms or even just to change the distance between them, whether or no they be of the same or of different species.

In chemistry, bonding involves two types of forces which might be called qualitatively *strong* or *weak* in terms of the cohesive energy U, with shades in between. The following values are a guide to the strength of the differing interactions:[1]

	Ionic (S)	Covalent (S)	Metallic (M-S)	Ion–dipole (M-W)	'van der Waals' (W)
U/kJ mol^{-1}	400–4000	150–1100	75–1000	50–500	<1–50

In the previous chapter, bonding through covalent forces, was considered and shown to involve essentially the sharing of pairs of electrons between atoms, leading to a stable balance of the attractive and repulsive forces acting on them. For many such species, the sharing of electrons allows each atom to attain the equivalent of a full outer shell of eight electrons, a particularly stable configuration. Additionally, it is known that electrons do not remain necessarily attached to the atom from which they are derived; they can become delocalized through chemical combination, and so pervade the molecule as a whole.

[1] S, strong; M, medium; W, weak.

Bonding, Structure and Solid-State Chemistry. First Edition. Mark Ladd.
© Mark Ladd 2016. Published in 2016 by Oxford University Press.

Weak and medium-weak interactions involve species carrying partial charges: they are *dipolar interactions*, of either a permanent or a transient nature. They are often denoted collectively as *van der Waals interactions*, and form the main topic of this chapter. They relate to interactions between closed-shell species, by means of forces of attraction termed ion (monopole)–dipole, dipole–dipole, dipole–induced dipole and induced dipole–induced dipole that are proportional to varying powers n in $1/r^n$, rather than to the $1/r$ of the coulombic interactions discussed in the previous chapter, r being the distance apart of the two species under consideration. There are also interactions involving quadrupoles, albeit of much smaller magnitude. This chapter will study these dipolar interactions and compounds derived from polar and non-polar entities, together with topics such as π-electron overlap, hydrogen-bonded compounds, and clathrate structures.

3.2 Electric moments and partial atomic charges

The *first electric moment* of a charge distribution on a species is its *dipole moment*, which has been encountered briefly in Section 2.16.2. It may be a permanent feature of a molecule, as in hydrogen fluoride, for example, or it may be transient, dependent on some method of induction. A dipole moment is a vector property, but a linear species presents no difficulty in calculating its dipole moment.

Example 3.1

Determine the dipole moment of 1,2-difluorobenzene, given that the C–F bond moment is 1.51 D.[2] The dipole moment is the vector sum of the two bond moments which lie at an angle of 60° to each other. Thus, $\mu = 2 \times 1.51 \text{ D} \times \cos 30° = 2.615 \text{ D}$ [by experiment, 2.59 ± 0.02 D].

A more accurate calculation of a dipole moment involves the *vector sum* of the products of partial charges and their position in space with reference to a given origin:

$$\mu = \sum_i q_i e \mathbf{r}_i \tag{3.1}$$

where q_i is the partial charge on atom i, e the electronic charge and $\mathbf{r}_i$ is the vector distance of the charge on atom i from a convenient origin. Both μ and $\mathbf{r}_i$ can be resolved into Cartesian components, which are then compounded vectorially:

$$|\mu| = \left(\mu_x^2 + \mu_y^2 + \mu_z^2\right)^{1/2} = \left\{ \sum_{j=1}^{3} \left[\left(\sum_{i=1}^{N} q_i x_i \right)^2 \right] \right\}^{1/2} \tag{3.2}$$

[2] Debye unit; $1 \text{ D} = 3.33564 \times 10^{-30}$ C m.

N is the total number of atoms in the species and j and i runs over the three coordinates x, y and z; in practice, the partial charge on an atom is found to vary with its environment.

Example 3.2

Four charged species are distributed in a plane as follows:

q	x/nm	y/nm
0.45	0	0
−0.36	0.132	0.000
−0.38	−0.0620	0.107
0.18	0.182	−0.087

Determine the magnitude and direction of the dipole moment of the array of charges. The first atom may be taken as a convenient origin. Then from Eq. (3.2):

$$\mu_x/e = (-0.36 \times 0.132 \text{ nm}) + (0.38 \times 0.062 \text{ nm}) + (0.18 \times 0.182 \text{ nm})$$
$$= 8.800 \times 10^{-3} \text{ nm}$$
$$\mu_y/e = (-0.38 \times 0.107 \text{ nm}) + (-0.18 \times 0.087 \text{ nm}) = -5.632 \times 10^{-2} \text{ nm}$$

Hence, $\mu = [(8.800 \times 10^{-3} \text{ nm})^2 + (-5.632 \times 10^{-2} \text{ nm})^2]^{1/2} \times 1.6022 \times 10^{-19}$ C $= 9.133 \times 10^{-30}$ C m, which evaluates further to 9.133×10^{-30} C m $/ 3.3356 \times 10^{-30} = 2.74$ D. The direction of μ is that of a vector $x = 8.800$ nm, $y = -5.632$ nm from the origin, and lying in the x, y plane.

Attempts have been made to determine partial charges from *relative electronegativities*; a fairly satisfactory formulation computes fractional ionic character q as $0.16|\Delta\chi| + 0.035\Delta|\chi^2|$, where $\Delta\chi$ is the difference in the relative electronegativities of two atomic species [1]. The relative electronegativity χ of a species expresses its tendency to attract electron density to itself in combination with other species. Several calculations of electronegativities have been devised, the best known being that of Pauling and a selection of his results is given here [2, 3]:

Atom	H	Li	B	C	N	O	F	Cl	Na
χ	2.20	0.98	2.04	2.55	3.04	3.44	3.98	3.16	0.931

Relative electronegativities are used as though they are dimensionless quantities, which is not strictly true.

Partial charges are obtained most reliably by *ab initio* theoretical calculation (Section 2.24) on the species in question, or on a moiety containing them. The programs for doing so generally calculate the dipole moment as well, with results in good agreement with those obtained experimentally.

The *second moment* of a charge distribution is the *quadrupole moment*, which consists of four finite charges but with an overall charge of zero and, thus, a zero dipole moment. Carbon dioxide, an example of a linear quadrupole, has a zero dipole moment, but a quadrupole moment Θ of -14.3×10^{-40} C m^2 has been reported.

Example 3.3

(a) Calculate the partial ionic character of hydrogen fluoride (i) from dipole moment data ($\mu = 6.00 \times 10^{-30}$ C m and $d = 0.927$ nm) and (ii) from relative electronegativities.

   (i) $\mu = 6.00 \times 10^{-30}$ C m $= q \times 1.6022 \times 10^{-19}$ C $\times 0.927 \times 10^{-9}$ m, whence $q = 0.404$.

   (ii) $|\Delta\chi| = 1.78$, whence $q = (0.16 \times 1.78) + (0.035 \times 1.78^2) = 0.396$. Thus, 40% is a good estimate for the percentage ionic character in hydrogen fluoride.

(b) The quadrupole moment of the linear system of charges in the CO_2 molecule is $2(qe)d^2$, with the C–O distance d of 0.116 nm. Approximating the value for q_O in oxygen to -0.32, as found for the water molecule, the quadrupole moment for CO_2 is $2 \times (-0.32) \times 1.6022 \times 10^{-19}$ C $\times (0.116 \times 10^{-9}$ m$)^2 = -13.8 \times 10^{-40}$ C m^2, in good agreement with the experimental result.

3.3 Polarization and polar molecules

Polarization of molecules is present in all states of matter, and may be defined as the *total dipole moment per unit volume*. In some species polarization is so small as to be negligible in a first approximation, whereas in other species it is the sole means of cohesion in the condensed state. In between, not surprisingly, there are shades of variation. Polar interactions have in common a mechanism dependent upon electron displacement in atoms or molecules, but across a range of substances there are significant differences.

Any substance that is a non-conductor of electric current but capable of being polarized by an electric field, intrinsic or applied, may be termed a *dielectric*.

Hydrogen fluoride is a polar molecule with $\mu = 1.80$ D; (Z)-1,2-dichloroethene is polar with $\mu = 1.89$ D but (E)-1,2-dichloroethene is non-polar on account of its symmetry:

(Z)-1,2-dichloroethene   (E)-1,2-dichloroethene

In passing, Z implies here that the two Cl substituents lie on the same side of the double bond (Ger. *zusammen* = together) whereas E implies the Cl atoms on opposite sides (Ger. *entgegen* = opposite).

All these species are examples of dielectrics; their component atoms are displaced from their equilibrium positions in the presence of an applied electric field.

The total polarization P of a molecule is compounded of three parameters

$$P = P_e + P_a + P_o \tag{3.3}$$

where the subscripts refer respectively to electron, atom and orientation polarizations. The following processes can occur to varying degrees in the polarization of a species:

(a) The electron density around the nucleus of each atom is displaced (*electron polarization*); this effect results in the centres of gravity of positive and negative charge ceasing to coincide;

(b) Bonded atoms or groups of atoms are displaced relative to one another (*atom polarization*); this effect is particularly important for polar molecules, as it alters the bending and stretching modes of vibration;

(c) Permanent molecular dipoles tend to become aligned by the field (*orientation polarization*), although this effect is opposed by thermal motion of the molecules;

(d) At high field strengths, species which can be polarized to differing extents in different molecular directions will tend to be aligned such that their direction of highest intrinsic polarization turns into the field; this effect may be assumed to be incorporated into the orientation polarization.

In these displacement mechanisms (a) and (b) are concerned with translational movements and (c) and (d) with rotational movements of the electron density of the atoms and molecules comprising the dielectric species.

Electron polarization is a distortion of the electron density of a species by the presence of an applied electric field. The electron density is displaced in one direction and the nucleus in the opposite direction, with the gradient.

Atom polarization P_a is small in magnitude, often less than one tenth of electron polarization, and is estimated by subtraction from the total polarization of the species the sum of the electron and orientation polarization contributions.

Orientation polarization, where present, is of the order of one hundred times the electron polarization, and must be determined separately.

In a homonuclear molecule, for example, there will be no permanent relative displacement of the effective centres of positive and negative charge, and the molecule is non-polar. In a heteronuclear species, differing electronegativities cause an electron displacement towards the more negative centre, and a dipole is created.

Example 3.4

The dipole moment for gaseous hydrogen fluoride is 6.00×10^{-30} C m. If there are N molecules per cubic metre, then $N = L/V_m$. At ambient conditions, the molar volume V_m is 0.02447 m^3 mol^{-1}, so that $N = 2.461 \times 10^{25}$ m^{-3}. The polarization (dipole moment per unit volume) is $N\mu$, so that $P = 1.477 \times 10^{-4}$ C m^{-2}.

The polarization field, to be discussed shortly, is P/ε_0; thus, from the hydrogen fluoride example the polarization field would be *ca.* 1.67×10^7 V m^{-1}, or 167 kV cm^{-1}, which would be sufficient to cause electrical breakdown of the dielectric. That this does not occur implies that not all hydrogen fluoride dipoles are aligned simultaneously.

The tendency for molecular orientation by an applied field is always opposed by the thermal motion of the molecules at any temperature above absolute zero. It is necessary, therefore, to determine the effective fraction of polar molecules that are aligned with a given applied field.

3.3.1 The Langevin function

Consider one of an assembly of molecules of dipole moment μ, all of which have freedom of movement. The potential energy V of the dipole may be equated to the work done on the dipole by rotating it from a position of zero potential energy to any other position in the field.

If the dipole axis be assumed initially perpendicular to the direction of the field E, then its potential energy would be zero because there is then no component of μ in the direction of the field. Let the dipole be turned so that it makes an angle θ with E (Fig. 3.1). The work done on a charge $+q$, resolved *in the direction of E*, is $-qE\cos\theta$, or $-qEx$; the negative sign indicates a spontaneous movement of charge down the field gradient. In a similar manner, the work done on the

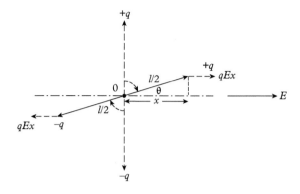

Fig. 3.1 The potential energy V of a dipole of moment μ having length l and charges $\pm qe$ and set at an angle θ to the field E is given by $V = -\mu E \cos \theta$, where $\mu = qel$.

charge $-q$ is also $-qEx$. Hence, V is $-2qEx$, or $-2qE(l/2) \cos \theta$. But ql is the dipole moment μ. Thus:

$$V = -\mu E \cos \theta \tag{3.4}$$

The minimum (most negative) energy condition corresponds with $\theta = 0$. However, such an alignment is countered by the thermal motion of the molecules. The fractional number of dipoles lying at an angle θ to the field may be represented by the number of dipoles lying within a solid angle $\delta \omega$. In other words, it is the fraction of dipoles the axes of which lie at angles to the field between θ and $\theta + \delta \theta$ (Fig. 3.2); θ is any angle lying in the space between the two coaxial cones of semi-vertical angles θ and $\theta + \delta \theta$.

A sphere of arbitrary radius r is constructed with its centre O being at the mid-point of the dipole (Fig. 3.3). The generators of the two cones described in Fig. 3.2 define an annular region of width $r\delta \theta$ on the sphere. The solid angle is *defined* as the ratio of the area of the annulus to the square of the distance from the annulus to the solid angle point O. Thus, since $\delta \theta$ is small:

$$\delta \omega = 2\pi r \sin \theta \, r \delta \theta / r^2 = 2\pi \sin \theta \, \delta \theta \tag{3.5}$$

Thus, the solid angle is independent of r; a sphere is chosen because it has a convenient geometrical shape. For zero field, the distribution of dipoles is

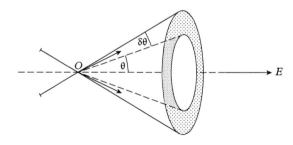

Fig. 3.2 The solid angle $\delta \omega$ (see text) contains those dipoles included by the space within the two concentric cones of semi-angles θ and $\theta + \delta \theta$.

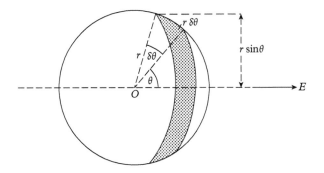

Fig. 3.3 Annulus of width $r\,\delta\theta$ on a sphere of arbitrary radius r, and defined by the generators of the two concentric cones shown in Fig. 3.2.

random, and the number of them per unit volume δN with their centres at O in the solid angle $\delta\omega$ is proportional to the value $\delta\omega$; thus:

$$\delta N = N_0 2\pi \sin\theta\,\delta\theta \qquad (3.6)$$

where N_0 is a constant dependent upon the total number of molecules under consideration. When E is non-zero, the equation for δN is weighted by a factor representing the energy V needed to turn the dipoles into the angle θ. Not all dipoles will require the same energy; in other words, different initial orientations have differing probabilities. A Boltzmann distribution (Appendix A4) will be assumed applicable, with V given by Eq. (3.4). Hence:

$$\delta N = N_0 2\pi \sin\theta \exp(\mu E \cos\theta / kT)\,\delta\theta \qquad (3.7)$$

If $<\mu>$ is the average moment in the direction of the field, then $<\mu> = \mu <\cos\theta>$, since μ is constant with θ being the variable parameter. Thus:

$$P_o = N<\mu> = N\mu<\cos\theta> \qquad (3.8)$$

As a general result, the average value $<X>$ in a distribution of function $\phi(X)$ is given by:

$$<X> = \frac{\displaystyle\int X\phi(X)\,\mathrm{d}X}{\displaystyle\int \phi(X)\,\mathrm{d}X} \qquad (3.9)$$

Applying this relationship to the problem:

$$<\cos\theta> = \frac{\displaystyle\int_0^\pi \cos\theta N_0 2\pi \sin\theta \exp(\mu E \cos\theta / kT)\,\mathrm{d}\theta}{\displaystyle\int_0^\pi N_0 2\pi \sin\theta \exp(\mu E \cos\theta / kT)\,\mathrm{d}\theta} \qquad (3.10)$$

Because of the construction used, integration from 0 to π gives the average value of $\cos\theta$ for all dipoles in the sphere.

To evaluate the integrals, let $\cos\theta = y$, so that $d\theta = -dy/\sin\theta$, and let $\mu V/kT = a$. Then:

$$<\cos\theta> = \frac{\int_1^{-1} -y\exp(ay)\,dy}{\int_1^{-1} -\exp(ay)\,dy} = \frac{\int_1^{-1} y\exp(ay)\,dy}{\int_1^{-1} \exp(ay)\,dy} \tag{3.11}$$

Equation (3.11) may be written as:

$$<\cos\theta> = \frac{d(\ln\xi)}{da}$$

where $\xi = \int_{-1}^{1}\exp(ay)dy$, a result verifiable readily by differentiation. Hence

$$\xi = \frac{1}{a}[\exp(a) - \exp(-a)]$$

so that

$$\ln\xi = \ln(1/a) + \ln(\sinh a) + \ln(2)$$

Then

$$\frac{d(\ln\xi)}{da} = \frac{-1}{a} + \frac{\cosh a}{\sinh a}$$

whence

$$<\cos\theta> = \coth a - \frac{1}{a} \tag{3.12}$$

and the *Langevin function* $\mathcal{L}(a)$ is written as:

$$\mathcal{L}(a) = \coth a - \frac{1}{a} \tag{3.13}$$

Expanding $\coth a$:

$$\coth a = \frac{1}{a} + \frac{a}{3} - \frac{a^3}{45} + \frac{2a^5}{945}\cdots$$

and neglecting terms in a^3 and higher:

$$\mathcal{L}(a) = a/3 \tag{3.14}$$

Hence, replacing a by $\mu E/kT$:

$$<\cos\theta> = \mu E/(3kT) \tag{3.15}$$

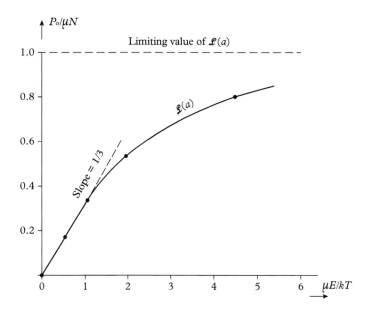

Fig. 3.4 Graph of the variation of the orientation polarization P_o, normalized to a maximum of unity by $1/\mu N$, as a function of $\mu E/kT$; the limiting slope at small values of a is 1/3.

and from Eq. (3.8):

$$P_o = N\mu^2 E/(3kT) \qquad (3.16)$$

Approximating $\exp(ay)$ to $(1 + ay)$ in Eq. (3.11) leads to the same result, but without knowledge of the magnitudes of the terms neglected. Consider again the problem in Example 3.3. If dielectric breakdown occurs at a potential of 1.67×10^7 V m^{-1}, then since $\mu E/kT$ at 298 K is 0.024, the error in neglecting terms in coth a of powers three and higher is only about 5×10^{-6}, so that the approximation is very satisfactory.

Experimental results for gaseous ammonia show that a plot of $P_o/\mu N$ against $\mu E/kT$ produces the curve in Fig. 3.4. At small values of $\mu E/kT$, the slope is 1/3, in accord with Eq. (3.16). The maximum value of $P_o/\mu N$ is 1, since the maximum $<\cos\theta>$ is unity from Eq. (3.10).

Example 3.5

The dipole moment for ammonia is 5.0×10^{-30} C m at 298.15 K. From Eq. (A11.34) and Eq. (3.16), and taking N from Example 3.4, $(P/E) = N\mu^2/3kT = 4.985 \times 10^{-14}$ F^{-1} m. Since $P/E = \varepsilon_0(\varepsilon_r - 1)$, as will appear shortly, $(\varepsilon_r - 1) = 4.985 \times 10^{-14}$ F^{-1} m$/8.8542 \times 10^{-12}$ F m$^{-1} = 56.3 \times 10^{-4}$. Comparison with Table 3.1 confirms that the orientation polarization is the major contributor to the relative permittivity of ammonia.

3.3.2   Electron polarization

Consider a molecule, initially non-polarized, introduced into an electric field of strength E. There results an electron displacement and the molecule acquires an induced dipole moment μ which is proportional to the field strength:

$$\mu = \alpha E \tag{3.17}$$

where the constant of proportionality α is the *polarizability* of the species. If there are N identical molecules per unit volume then, since polarization has been defined as the total dipole moment per unit volume, the electron polarization P_e is

$$P_e = N\mu = N\alpha_e E \tag{3.18}$$

and α_e is the polarizability of the species with respect to electron displacement.

3.3.2.1   *Digression on electrical units*

A dipole moment μ as defined above is measured in C m, and field strength has the units V m^{-1}. From Eq. (3.17), the units of α are C m^2 V^{-1}. This unit is slightly cumbersome, and it is conventional to define α', a *volume polarizability* (aka polarizability volume), in the usual units of cm^3 by:

$$\alpha' = 10^6 \alpha / (4\pi\varepsilon_0) = \mathbb{R}\alpha \tag{3.19}$$

where $\mathbb{R}$/cm^3 = $(10^6) \times 8.9875 \times 10^9$ F^{-1} m. The units of polarizability volume are, therefore $\frac{\text{C V}^{-1} \text{ m}^2 \,(\equiv\text{F m}^2)}{\text{F m}^{-1}}$ = m^3, and the factor of 10^6 ensures the conversion of m^3 to cm^3; the inclusion of 4π links ε_0 with the baseline physical constant μ_0, the *magnetic constant* (aka the *permeability of a vacuum*).

Example 3.6

The hydrogen chloride molecule has a polarizability α of 2.93×10^{-40} F m^2. From Eq. (3.19), it follows that the volume polarizability α' is 2.63×10^{-24} cm^3. Volume polarizabilities are mostly of the order of 10^{-24} cm^3.

3.3.2.2   *Polarization in gaseous species*

The discussion on electron polarization can be amplified by reference to a monatomic species, such as gaseous argon, which contains a central nucleus of charge $+q$ surrounded by an electron density of total charge $-q$; the electron density is assumed to be uniform within a sphere of radius r (Fig. 3.5a). When an electric field E is applied to the atom, the electron density is displaced relative to the nucleus. The centres of gravity of positive and negative charge no longer coincide, and a dipole of length x is created within the atom, Fig. 3.5b.

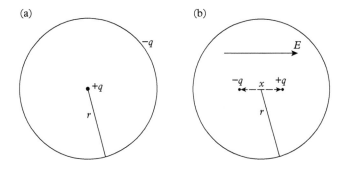

The electron density experiences a distorting force of magnitude qE owing to the applied field, and an attractive force $-q|q/(4\pi\varepsilon_0 x^2)$ between the nucleus and that part of the electron density lying within a sphere of radius x. Since the electron density is uniform, at equilibrium:

$$qE = (x/r)^3 q^2/(4\pi\varepsilon_0 x^2) \tag{3.20}$$

or

$$E = qx/(4\pi\varepsilon_0 r^3) \tag{3.21}$$

Since qx represents a first moment μ created by the charge displacement within the atom, it follows from Eq. (3.17) that

$$\alpha_e = 4\pi\varepsilon_0 r^3 \tag{3.22}$$

Hence, the electron polarization P_e for N species is given by:

$$P_e = N4\pi\varepsilon_0 r^3 E \tag{3.23}$$

If the gaseous material under discussion is held in a medium of relative permittivity ε_r, then from Eq. (A11.34):

$$P_e = \varepsilon_0(\varepsilon_r - 1)E \tag{3.24}$$

whence

$$(\varepsilon_r - 1) = N4\pi r^3 = \frac{N\alpha_e}{\varepsilon_0} \tag{3.25}$$

If the Avogadro constant is L and the molar volume of the substance V_m, then:

$$\frac{N}{L} = \frac{1}{V_m} = \frac{D}{M_m} \tag{3.26}$$

where D is the density and M_m the molar mass. Hence:

$$(\varepsilon_r - 1)\frac{M_m}{D} = \frac{L\alpha_e}{\varepsilon_0} \tag{3.27}$$

Example 3.7

The molar mass of argon is 0.03995 kg mol^{-1}, its density is 1.78 kg m^{-3} and relative permittivity 1.000517.

$$\alpha_e = \frac{(1.000517 - 1) \times 0.03995 \text{ kg mol}^{-1} \times 8.8542 \times 10^{-12} \text{ F m}^{-1}}{1.78 \text{ kg m}^{-3} \times 6.0221 \times 10^{23} \text{ mol}^{-1}}$$

$= 1.706 \times 10^{-40}$ F m^2, and from Eq. (3.19):

$\alpha' = 1.706 \times 10^{-40}$ F m$^2 \times (10^6) \times 8.9875 \times 10^9$ F^{-1} m $= 1.53 \times 10^{-24}$ cm^3.

3.3.3 Refractive index

According to the Maxwell electromagnetic theory, the relative permittivity ε_r is related to the refractive index n, at comparable frequencies, by the equation:

$$\varepsilon_r = n^2 \tag{3.28}$$

Since $\varepsilon_r \approx 1$ for a gas:

$$\varepsilon_r^{1/2} = [1 + (\varepsilon_r - 1)]^{1/2}$$

Expanding by the binomial theorem to two terms and rearranging gives:

$$(\varepsilon_r - 1) = 2(n - 1) \tag{3.29}$$

The error in the approximation leading to Eq. (3.29) is of the order of the value of the next term of the binomial expansion, namely, $\frac{1}{2}(\frac{1}{2} - 1)(\varepsilon_r - 1)^2/2 = -3.3 \times 10^{-8}$ in the case of argon ($\varepsilon_r = 1.000517$), a negligible amount in a good first approximation.

3.3.4 Van der Waals equation of state

The van der Waals equation of state for n moles of a gas may be written as:

$$p = \frac{nRT}{(V - nb)} - n^2a/V^2 \tag{3.30}$$

where a is determined by van der Waals attractive forces between the gas molecules and b is the *excluded volume* of the gas (Appendix A11.8). For a gas in the critical state, both $(\partial p/\partial V)_T$ and $(\partial^2 p/\partial V^2)_T$ are zero. It can be shown [4] that for one mole of gas, the critical volume V_c is equal to $4b$, which, from Appendix A11.8, gives:

$$V_c = 12v \qquad (3.31)$$

where v is the volume of *one molecule*; thus, the term $4\pi Nr^3$ represents three times the volume of spherical atoms of radius r per unit volume of gas. Hence:

$$(\varepsilon_r - 1) = 3v = V_c/4 \qquad (3.32)$$

Since density is inversely proportional to volume, then, for one mole of gas, $V_c/1 = D/D_c$, and:

$$(\varepsilon_r - 1) = D/(4D_c) \qquad (3.33)$$

where D_c is the density of the gas in the critical state, a measurable property. Table 3.1 compares $(\varepsilon_r - 1)$ with $2(n - 1)$ and $D/(4D_c)$; the refractive index n is quoted for sodium D radiation of average wavelength 589.3 nm.

The non-polar molecules show fair agreement between the three measures $(\varepsilon_r - 1)$, $2(n - 1)$ and $D/(4D_c)$, which implies a similar level of agreement for α_e, *vide* Eq. (3.25), and it may be concluded that each of these three parameters gives a medium–good representation of the electron polarizability. In polar molecules, however, the results based on $(\varepsilon_r - 1)$ are of a different order of magnitude. Relative permittivity is determined usually in static or low frequency fields. In such situations, the orientation of permanent dipoles (Section 3.3.1) makes a major contribution to the total polarization.

The refractive index is a measure of the interaction of the high-frequency electric vector of incident radiation with the electron density of the atom; in other words, it is concerned only with P_e. It is necessary to test theory with appropriate data, since polarization is dependent upon the frequency of an applied electric field.

Table 3.1 *Comparison of properties of monatomic and polyatomic gases[a]*

	D/kg m^{-3}	$10^{-3}D_c$ / kg m^{-3}	$10^4 D/(4D_c)$	$10^4 2(n - 1)$	$10^4 (\varepsilon_r - 1)$
Monatomic, Ar	1.78	0.53	8.4	5.6	5.2
non-polar Xe	5.90	1.15	12.8	14.0	14.7
Polyatomic, O_2	1.43	0.43	8.3	5.4	4.9
non-polar CO_2	1.98	0.46	10.8	9.0	9.2
Polyatomic, NH_3	0.77	0.23	8.4	7.5	72
polar H_2O	0.60	0.32	4.7	5.1	79

[a] Subscript 'c' indicates the critical state of a gas.

3.4 Polarization in condensed states

In gases, the interaction between individual molecules is negligible under ambient conditions. This situation is not normally obtained with gases under high pressure, and certainly not in the liquid and solid states, and a discussion of the various dipolar interatomic forces is now appropriate. Of the several interactions that are often grouped together under the heading of 'van der Waals forces', three will be discussed here in detail and others indicated diagrammatically.

In practical calculations involving polar interactions in the condensed states, where permittivities ε are significantly greater than those of gases, ε is often replaced by the dimensionless relative permittivity ε_r, where $\varepsilon_r = \varepsilon/\varepsilon_0$; thus, water at 20 °C has a permittivity of 7.092×10^{-10} F m^{-1}, so that ε_r at that temperature is 80.1 K.

3.4.1 Ion–dipole interaction

In a polar solid, consider the interaction of a charge $+Q_2$ with a dipole of charge Q_1, length l and moment μ_1, where Q is a numerical charge q multiplied by the electron charge e (Fig. 3.6a). The ion–dipole potential energy of interaction $V_{i,d}$ is based on the coulombic formula $-Q_1 Q_2/(4\pi\varepsilon_0 r)$, and is given by two pairwise additive terms:

$$V_{i,d} = \frac{1}{4\pi\varepsilon_0}(Q_1 Q_2/AC - Q_1 Q_2/BC) \tag{3.34}$$

For AC:

$$1/AC = (r^2 + rl\cos\theta + l^2/4)^{-\frac{1}{2}} = r\{1 + [(l/r)\cos\theta + l^2/4r^2]\}^{-\frac{1}{2}}$$

and similarly for BC:

$$1/BC = (r^2 - rl\cos\theta + l^2/4)^{-\frac{1}{2}} = r\{1 - [(l/r)\cos\theta - l^2/4r^2]\}^{-\frac{1}{2}}$$

whence

$$1/AC = \left\{\left[1 + \left(\frac{l}{r}\cos\theta + l^2/4r^2\right)\right]^{-\frac{1}{2}} - \left[1 - \left(\frac{l}{r}\cos\theta - l^2/4r^2\right)\right]^{-\frac{1}{2}}\right\}$$

and for BC

$$1/BC = \left\{\left[1 + \left(\frac{l}{r}\cos\theta + l^2/4r^2\right)\right]^{-\frac{1}{2}} - \left[1 - \left(\frac{l}{r}\cos\theta - l^2/4r^2\right)\right]^{-\frac{1}{2}}\right\}$$

Making the reasonable assumption that r is significantly larger than l, a valid binomial expansion of $(1 \pm x)^{-\frac{1}{2}}$, ignoring powers of x greater than 2, is $1 \mp x/2 + (3/8)x^2$; thus:

$$\frac{1}{r}\left(\frac{1}{AC} - \frac{1}{BC}\right) = \left[1 - \left(\frac{l\cos\theta}{2r} + \frac{l^2}{8r^2}\right) + \frac{3}{8}\left(\frac{l^2\cos^2\theta}{r^2} + \frac{l^4}{16r^4} + \frac{l^3\cos\theta}{2r^2}\right)\right]$$

$$- \left[1 + \left(\frac{l\cos\theta}{2r} - \frac{l^2}{8r^2}\right) + \frac{3}{8}\left(\frac{l^2\cos^2\theta}{r^2} + \frac{l^4}{16r^4} - \frac{l^3\cos\theta}{2r^2}\right)\right]$$

$$= -\frac{l\cos\theta}{r} + \frac{3l^3\cos\theta}{8r^2} = -(l/r)\cos\theta\left(1 - \frac{3l^2}{8r}\right)$$

In most problems, the dipolar length l is very small compared with the distance r between the ion and the centre of the dipole, in which case $\left(1 - \frac{3l^2}{8r}\right)$ is approximated to unity, whereupon

$$V_{i,d} = \frac{-\mu_1 Q_2 \cos\theta}{4\pi\varepsilon_0 r^2} \tag{3.35}$$

Maximum interaction arises for $\cos\theta = \pm 1$, that is, when Q_2 is *collinear* with the dipole axis.

Solutes consisting of polar molecules yield stable, solvated, ionic species on dissolution in a polar solvent on account of the interaction between the ionic component and the polar, solvent molecule. Solvents such as water, the lower alcohols and ketones and some nitro-compounds are ionizing solvents; these solutes are not ionized to a measurable extent in non-polar solvents such as carbon tetrachloride or toluene.

3.4.2 Dipole–dipole interactions

The dipoles of two molecules attract each other and a force is set up between them. In a fluid, the molecules are free to rotate, and the field of one dipole tends to orientate the dipole of a neighbouring species. Furthermore, attractive forces tend to dominate as they act at a longer range than do repulsion forces, and a net attractive potential exists.

3.4.2.1 Dipole–dipole interaction in a solid

Consider two molecules with permanent dipole moments μ_1 and μ_2, separated by a distance r and fixed in orientation, and lying in one plane (not the most general case), as shown in Fig. 3.6b. For the purposes of calculation, the dipoles

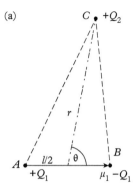

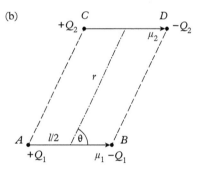

Fig. 3.6 Dipolar interaction. (a) Ion–dipole interaction: AB represents the length l of the dipole, with charges $\pm Q_1$ and dipole moment μ_1; Q_2 is a positive point charge at C distant r from the centre of the dipole, such that r makes an angle θ with the dipole axis. (b) Dipole–dipole interaction: Dipoles AB and CD, each of length l, parallel and coplanar, with r the distance between their centres. In both diagrams, Q represents a numerical charge q multiplied by the electron charge e.

are given the same length l by considering Q_2 modified as necessary with respect to Q_1. Proceeding with an argument similar to that of the ion–dipole interaction:

$$V_{d,d} = \frac{1}{4\pi\varepsilon_0}\left(\frac{2Q_1Q_2}{r} - \frac{Q_1Q_2}{AD} - \frac{Q_1Q_2}{BC}\right)$$

$$= \frac{Q_1Q_2}{4\pi\varepsilon_0}\left(\frac{2}{r} - \frac{1}{AD} - \frac{1}{BC}\right)$$

Expanding:

$$V_{d,d} = \frac{Q_1Q_2}{4\pi\varepsilon_0 r}\left[2 - \frac{1}{(r^2 + l^2 + 2lr\cos\theta)^{\frac{1}{2}}} - \frac{1}{(r^2 + l^2 - 2lr\cos\theta)^{\frac{1}{2}}}\right]$$

$$= \frac{Q_1Q_2}{4\pi\varepsilon_0 r}\left[2 - \frac{1}{r}\left(1 + \frac{l^2}{r^2} + \frac{2l\cos\theta}{r}\right)^{-\frac{1}{2}} - \frac{1}{r}\left(1 + \frac{l^2}{r^2} - \frac{2l\cos\theta}{r}\right)^{-\frac{1}{2}}\right]$$

Using the binomial expansion as before and making assumptions similar to those for the ion–dipole interaction, the *dipole–dipole* potential energy evaluates as:

$$V_{d,d} = \frac{-\mu_1\mu_2(3\cos^2\theta - 1)}{4\pi\varepsilon_0 r^3} \tag{3.36}$$

The *general form* of Eq. (3.36) is shown in Fig. 3.7f; in the argument leading to Eq. (3.36), $\theta_1 = \theta_2$ and $\phi = 0$. Maximum interaction corresponds to either a *collinear head-to-tail* orientation ($\theta_1 = \theta_2 = 0$) or an *antiparallel* situation ($\theta_1 = 0$, $\theta_2 = \pi$); for these cases, in order:

$$V_{d,d} = \mp\left(\frac{2\mu_1\mu_2}{4\pi\varepsilon_0 r^3}\right) \tag{3.37}$$

3.4.2.2 *Dipole–dipole interaction in a fluid*

In a fluid, where the dipoles are mobile, an average potential energy may be written in the form:

$$<V_{d,d}> = \frac{-\mu_1\mu_2}{4\pi\varepsilon_0 r^3}<f(\theta,\phi)P(\theta,\phi)> \tag{3.38}$$

If ϕ is given the constant value of zero, in accord with Fig. 3.7f, then $f(\theta,\phi)$ is just the expression $(3\cos^2\theta - 1)$ in Eq. (3.36). For the polarization $P(\theta,\phi)$, a Boltzmann distribution (Appendix A4) is assumed, dependent on θ, namely, $\exp(-V_\theta/kT)$, where V is given by Eq. (3.36). Since V_θ is very much smaller than kT, $P(\theta,\phi)$ may be approximated by $(1 - V_\theta/kT)$. Hence:

$$<V_{d,d}> = \left(\frac{-\mu_1\mu_2}{4\pi\varepsilon_0 r^3}\right)(<3\cos^2\theta - 1>)(1 - <V_\theta>/kT)$$

$$= \left(\frac{-\mu_1\mu_2}{4\pi\varepsilon_0 r^3 kT}\right)<(3\cos^2\theta - 1)> - \left(\frac{-\mu_1\mu_2}{4\pi\varepsilon_0 r^3}\right)^2<[3\cos^2\theta - 1]^2>$$

$$\tag{3.39}$$

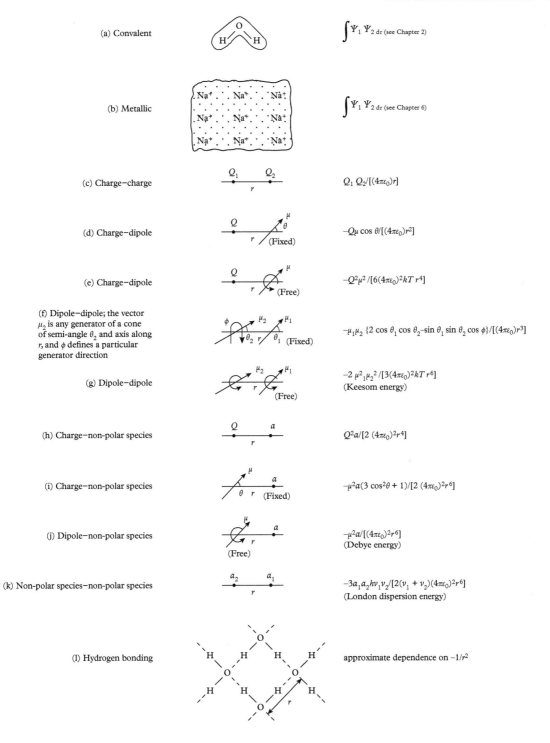

(a) Convalent

$\int \Psi_1 \Psi_2 \, d_\tau \text{ (see Chapter 2)}$

(b) Metallic

$\int \Psi_1 \Psi_2 \, d_\tau \text{ (see Chapter 6)}$

(c) Charge–charge

$Q_1 Q_2 / [(4\pi\varepsilon_0)r]$

(d) Charge–dipole

$-Q\mu \cos\theta / [(4\pi\varepsilon_0)r^2]$

(e) Charge–dipole

$-Q^2\mu^2 / [6(4\pi\varepsilon_0)^2 kT \, r^4]$

(f) Dipole–dipole; the vector μ_2 is any generator of a cone of semi-angle θ_2 and axis along r, and ϕ defines a particular generator direction

$-\mu_1\mu_2 \{2\cos\theta_1 \cos\theta_2 - \sin\theta_1 \sin\theta_2 \cos\phi\} / [(4\pi\varepsilon_0)r^3]$

(g) Dipole–dipole

$-2\,\mu^2_1\mu_2^2 / [3(4\pi\varepsilon_0)^2 kT \, r^6]$
(Keesom energy)

(h) Charge–non-polar species

$Q^2a / [2\,(4\pi\varepsilon_0)^2 r^4]$

(i) Charge–non-polar species

$-\mu^2a(3\cos^2\theta + 1) / [2\,(4\pi\varepsilon_0)^2 r^6]$

(j) Dipole–non-polar species

$-\mu^2a / [(4\pi\varepsilon_0)^2 r^6]$
(Debye energy)

(k) Non-polar species–non-polar species

$-3a_1a_2 h\nu_1\nu_2 / [2(\nu_1 + \nu_2)(4\pi\varepsilon_0)^2 r^6]$
(London dispersion energy)

(l) Hydrogen bonding

approximate dependence on $-1/r^2$

Fig. 3.7 Schematic representation of types of intermolecular forces. (a) Covalent. (b) Metallic. (c) Charge–charge; r is the distance between the centres of the charges. (d) Charge–fixed dipole; r is the distance between the charge and the centre of the dipole. (e) Charge–free dipole; an inverse dependence on temperature T shows that the tendency towards an orientation of the dipoles decreases as T is increased. (f) Fixed dipole–fixed dipole; ϕ is the angle of rotation of moment μ_2 relative to moment μ_1 with respect to the dipole axis r. (g) Free dipole–free dipole (aka Keesom energy), and temperature dependent. (The fraction 2/3 is sometimes reported in lieu of 4/5.) (h) Charge–non-polar species; α is the polarizability of the non-polar species. (i) Fixed dipole–non-polar species; α is the polarizability of the non-polar species. (j) Free dipole–non-polar species (aka Debye energy). (k) Non-polar species–non-polar species (aka London energy, or dispersion energy). (l) Hydrogen bonding; there is an approximate $-1/r^2$ dependence, where r is the closest non-bonded distance between two non-hydrogen atoms, for example O$\cdots$O in O–H$\cdots$O. It is common to refer to the 'strength' of a hydrogen bond by the O$\cdots$O length; thus, 0.25 nm for this length would be termed 'strong', while 0.29 nm for the same type of liaison is 'weak'. [*Introduction to Physical Chemistry*, 3rd ed. 1998; reproduced by courtesy of Cambridge University Press.]

It is straightforward to show, in polar coordinates, that $<3\cos^2\theta - 1> = 0$, and that $<(3\cos^2\theta - 1)^2> = 4/5$ (see Example 3.9). Hence, the dipole–dipole energy is

$$<V_{d,d}> = -\frac{4}{5}\left(\frac{\mu_1^2\mu_2^2}{(4\pi\varepsilon_0)^2 kTr^6}\right) \tag{3.40}$$

This expression is the *Keesom energy* [5, 6] (Fig. 3.7g); it is not strictly a potential energy as it is temperature dependent. The potential energy of two dipoles is proportional to $1/r^3$, and the Keesom equation is an orientational average, assuming that polarization is isotropic.

Example 3.8

Water at 298.15 K has a dipole moment of 1.84 D, relative permittivity of 78.4 and a distance of closest approach of molecules 0.2 nm. Since ε_r is a significant quantity, it multiplies ε_0 in Eq. (3.40):

$$<V_{d,d}> = -\frac{4}{5 \times 1.3807 \times 10^{-23}\ \text{J K}^{-1} \times 298.15\,\text{K}}$$

$$\left(\frac{(1.84 \times 3.3356 \times 10^{-30}\ \text{C m})^2}{(4 \times \pi \times 78.4 \times 8.8542 \times 10^{-12}\ \text{F m}^{-1}}\right)^2 \frac{1}{(0.2\ \text{nm} \times 10^{-9})^6\ \text{m}^6}$$

$$= -5.67 \times 10^{-23}\,\text{J}$$

Example 3.9

In the above context, the average $< 3\cos^2\theta - 1 >$ over a sphere is obtained as $\int_0^{\pi/2} (3\cos^2\theta - 1)\sin\theta\,d\theta == -\cos^3\theta + \cos\theta|_0^{\pi/2} = 0$; the upper limit of $\pi/2$ applies since $\cos^2\theta$ repeats at this value of θ. In a similar manner, show that $< (3\cos^2\theta - 1)^2 > = \int_0^{\pi/2} (3\cos^2\theta - 1)^2 \sin\theta\,d\theta = 4/5$.

Thus, the Keesom energy V_K for water at 298 K is $L < V_{d,d} >$, which evaluates to -34.1 kJ mol^{-1}. The numerical result depends strongly on the value of r; here, a distance of twice d_{O-H} (≈ 0.1 nm) in the water molecule has been chosen.

A refinement of this equation was given by Debye, to take account of the interaction between a permanent and an induced dipole. He showed that and additional term, the *Debye energy* V_D, should be added to the Keesom energy term $< V_{d,d} >$:

$$V_D = \frac{-\mu^2\alpha}{(4\pi\varepsilon_0\varepsilon_r)^2 r^6} \tag{3.41}$$

where μ is the permanent dipole moment, α the polarizability of a neighbouring molecule and r their distance of approach. Generally, V_D is a relatively small addition to the total interactive energy.

3.4.3 Induced dipole–induced dipole interaction

The induced dipole–induced dipole dispersion energy was deduced first by London in 1930 [7, 8]. The energy is quantum mechanical in origin and the derivation of its expression is lengthy; a simplified discussion will suffice here to show the form of the interaction.

Consider two atoms of a non-polar system, such as argon. The atom has no permanent dipole moment because the electron density distribution about its nucleus has spherical symmetry, when averaged over a time that is large compared with the period of fluctuation of electron density. However, a transient moment μ will exist for any given argon atom Ar(1) because of the fluctuation of its electron density. Thus, an induced dipole arises in that atom on account of the asymmetry contingent upon the fluctuation (Fig. 3.8a). Its moment μ induces a field E at a distance r in the direction of the dipole given by:

$$E = \mu/(4\pi\varepsilon_0 r^3) \tag{3.42}$$

The field E then polarizes a neighbouring argon atom Ar(2), producing in it a dipole of moment μ' (Fig. 3.8b). These instantaneous dipoles can interact one with the other or by induction with a neutral argon atom. The interactional energy is proportional to $\mu\mu'$, which is proportional to $1/r^6$ (Fig. 3.7), where r is the distance between any two interacting species. As one transient dipole changes

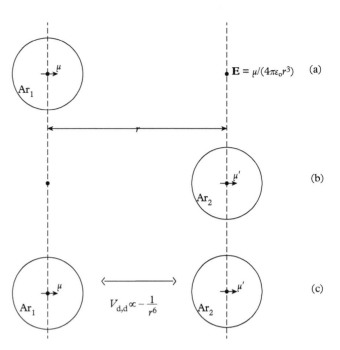

Fig. 3.8 London (dispersion) energy. (a) Instantaneous dipole of moment μ in the first argon atom Ar_1 caused by a fluctuating electron density in the atom produces a field E equal to $\mu/(4\pi\varepsilon_0 r^3)$ at a distance r from it. (b) A second argon atom Ar_2 at the distance r in the direction of the field is polarized by the field E, producing a dipole of moment μ', commensurate with μ. (c) Potential energy $V_{d,d}$ between the two argon atoms, proportional to $-1/r^6$.

its value and orientation, so the induced dipole will follow it. Because of correlation, the multitude of transients do not average to zero, but provide an overall attraction which leads to an attractive induced energy (Fig. 3.8c). Whatever the direction of the instantaneous variation in electron density within the atoms, the overall effect is always an attraction. The result from the quantum mechanical calculation shows the important dependence on r^{-6}, and the formulations of the *induced dipole–induced dipole* energy given by London are

$$V_{\mathrm{id,\,id}} = -\frac{3}{4}\frac{\alpha^2 h\nu_0}{(4\pi\varepsilon_0)^2 r^6}$$

and in the more general case of dissimilar species: (3.43)

$$V_{\mathrm{id,\,id}} = -\frac{3}{2}\frac{\alpha_1\alpha_2 h\nu_{0,1}\,\nu_{0,2}}{(\nu_{0,1}+\nu_{0,2})(4\pi\varepsilon_0)^2 r_{12}^6}$$

where ν refers to a characteristic frequency of fluctuation of the electron density, and r_{12} is the distance between species 1 and 2.

The equations may be expressed in terms of the volume polarizability (in m^{-3}) and with the first ionization energy I replacing $h\nu_0$, but they are best applied to the lighter atomic species:

$$V_{\mathrm{id,\,id}} = -\frac{3}{4}\frac{(4\pi\varepsilon_0\alpha')^2 I}{(4\pi\varepsilon_0)^2 r^6} = -\frac{3}{4}\frac{(\alpha')^2 I}{r^6}$$

and for dissimilar species: (3.44)

$$V_{id,id} = -\frac{3}{2} \frac{\alpha_1' \alpha_2' I_1 I_2}{(I_1 + I_2) r_{12}^6}$$

In general, equations (3.43) are satisfactory for distances r less than *ca.* 30 nm. For greater distances, say 300 nm, the field generated at one atom takes approximately $(3 \times 10^{-7}$ m$) / (3 \times 10^8$ m s$^{-1})$ to reach a neighbour. This time of *ca.* 10^{-14} s is commensurate with the period of fluctuation of the electron density and consequently interference occurs. A detailed treatment shows that in these circumstances the energy of attraction is proportional to $1/r^7$; this interaction and other dipolar interactions discussed herein are treated in detail in the literature [9].

Example 3.10

The volume polarizability of hydrogen is 0.911×10^{-30} m^3 and the ionization energy is 1312 kJ mol^{-1}. The distance of separation of two hydrogen atoms may be taken as twice the van der Waals radius of hydrogen, 0.12 nm. The London dispersion energy is then:

$$V_{id,id} = -\frac{3}{4} \frac{(0.911 \times 10^{-30} \text{ m}^3)^2 \times 1312 \text{ kJ mol}^{-1}}{(0.24 \times 10^{-9} \text{ m})^6}$$

$$\approx -\frac{10^{-57} \text{ m}^6 \text{ kJ mol}^{-1}}{(2 \times 10^{-4} \times 10^{-54}) \text{ m}^6} \quad \text{or} \quad -5 \text{ kJ mol}^{-1}; \text{ specifically, } -4.27 \text{ kJ mol}^{-1}$$

The significant interaction energies, Keesom, Debye and London, although small individually, all contribute to cohesion. The energy needed to disperse the particles in a molecular solid into the gas phase may be represented approximately by the enthalpy of sublimation ΔH_{sub}; hence, this quantity provides a useful experimental quantity against which the energy models may be tested. In Table 3.2 the results are tested against the theory discussed, and the agreement is supportive for both non-polar and polar species.

'Van der Waals forces', a term often used to embrace dipolar interaction energies, exist in all liquids and solids, but less conspicuously in gases under ambient conditions. As London forces they are responsible for cohesion in the liquid and solid states of the so-called inert gases. As would be expected, an increase in atomic size leads to a greater degree of electron displacement, or polarizability, and thus a stronger interaction, as the following boiling point data indicate:

	He	Ne	Ar	Kr	Xe	Rn
BP/K	4	27	87	121	165	211

Table 3.2 *Dipolar interaction energies/kJ mol^{-1} for selected molecular compounds*

	V_K	V_D	V_L	Total	ΔH_{sub}
Ar	0	0	−8.5	−8.5	−8.4
H$_2$	0	0	−0.8	−0.8	−0.8
CO	0	0	−8.4	−8.4	−7.9
HCl	−3.3	−1.0	−16.8	−21.1	−20.0
H$_2$O	−34.1	−1.9	−9.0	−45.0	−47.3

K = Keesom, D = Debye, L = London

Van der Waals forces play a significant role in the binding processes of biological materials, and have small but significant effects even in mainly ionically bonded solids. They occur in many guises in nature: a somewhat unusual example is the locomotion of *Uroplatus fimbriatus*, a small lizard species found in hot countries and known as a gecko (Indonesian-Malay: *gēkoq*, imitative of the call of the animal). On account of its specialized toe-pads it can climb smooth, vertical surfaces, such as glass, and cross indoor ceilings with ease. It is believed that this ability arises from London forces acting between the toe-pads of the gecko and the material of the smooth surface (Fig. 3.9).

Fig. 3.9 *Uroplatus fimbriatus*, or gecko; its locomotive ability on glass walls and on ceilings is believed to arise through van der Waals forces between the pads of its feet and the smooth surface. [Public domain.]

3.4.3.1 *Vibrational bonding*

The idea of vibrational bonding was introduced to explain a possible stability in the heavy atom–light atom–heavy atom system of iodine–hydrogen–iodine. Semi-empirical theoretical calculations predicted bound states in the absence of a minimum in the potential energy, but it was not substantiated at that time [10]. Recently, theoretical experiments on the reaction HBr with Br to form BrBBr, where a bridging species B was one of the isotopes ^{1}H, ^{2}H, ^{3}H or the very light muonium species Mu. The potential energies and zero-point energies were mapped, but no minimum potential energy was found. However, in the case of Br Mu Br stabilization arose through a decrease in zero-point energy in excess of the increase of potential energy of the system. For the triatomic systems with $B =\,^1$H, ^{2}H, ^{3}H bonding is through van der Waals interactions, but with the very light muonium species stabilization arises by vibrational bonding [11]. The triparticle systems can be envisaged as:

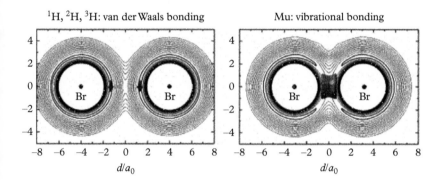

^{1}H, ^{2}H, ^{3}H: van der Waals bonding   Mu: vibrational bonding

and the BrBBr distances are 0.410 nm ($B =\,^1$H, ^{2}H, ^{3}H) and 0.326 nm ($B=$ Mu). The extra stability in the BrMuBr system, as indicated by the decrease in $d_{Br...Br)}$, represents a lowering of the zero-point energy of vibration in this species (see also Section 3.11.1).

3.4.4 Surface tension

A liquid forms normally on condensation from its gaseous state and it then takes up the shape of its containing vessel. The shape of a pure liquid without restraint corresponds to a minimum energy condition; thus, small liquid drops are spherical because the ratio of surface area to volume is least for this shape.

The work $\delta\omega$ needed to change a surface area A by an amount δA may be written as:

$$\delta\omega = \gamma\,\delta A \qquad (3.45)$$

where γ is the *surface tension* of the liquid. The minimization of surface area leads to a curved surface of the liquid, and the pressure on its concave side is proportional to γ/r, where r is the radius of curvature of the surface. The pressure difference between the inner and outer surfaces of a curved liquid interface tends to zero as r tends to infinity, that is, when the surface is flat.

3.4.4.1 Capillarity

The rise of liquid in a tube, or *capillary action*, arises on account of the surface tension force. A concave curvature means that the pressure just below a meniscus is less than atmospheric pressure. The pressure outside the tube at a level parallel to the tangent to the meniscus is atmospheric, whereas inside the tube at that level, the pressure is less than atmospheric by the amount $2\gamma/r$, where r is the radius of the tube. Thus, the liquid rises in the tube to a height h, such that the hydrostatic pressure is balanced by the pressure arising from surface tension, that is,

$$hDg = 2\gamma/r \tag{3.46}$$

where D is the density of the liquid and g is the gravitational acceleration.

The *contact angle* ζ between the inner wall of the tube and the meniscus is inversely proportional to γ: a liquid wets a surface for $0° < \zeta < 90°$, but fails to do so for $90° < \zeta < 180°$. In water $\zeta \approx 27°$, whereas for mercury it is $140°$ and so does not wet the glass tube wall, indicating the stronger cohesive forces in elemental mercury than between the molecules in liquid water. Surface tension and capillary action are discussed in detail in the literature [9].

Mercury is interesting as the only metal that is liquid under ambient conditions. The explanation lies in an *inert pair* effect, well documented for thallium, tin and lead compounds. It arises from a relativistic contraction of the $6s$ electrons, which constrains them closer to the nucleus than would be expected. Bonding such as Hg–Hg, a type known for gold in $Au_2(g)$, would mean that the third and fourth valence electrons would enter an antibonding, destabilizing molecular orbital. Thus, mercury is monomeric: the bonding arises most probably through the London dispersion interaction, producing a bonding interaction that is readily overcome by thermal motion at room temperature.

The cohesive forces in water are greatly enhanced by hydrogen bonding. The melting point of ice in the absence of hydrogen bonding would be less than about $-90°C$; thus, the London dispersion forces are greater in mercury, compared to water, than the melting point of $-38.8°C$ would appear to indicate.

Example 3.11

A column of water in a capillary tube of radius 0.25 mm rises to a height of 6.76 cm at a temperature of 20°C; the density of water at this temperature is 998.2 kg m^{-3}. The surface tension at the given temperature is expressed by

$\gamma = \frac{1}{2}Dgrh$, where D is the density of water at $20°C$, g is the gravitational acceleration, r is the inner radius of the tube and h is height of the water meniscus in the tube; hence, the surface tension is:

$$\gamma = \frac{1}{2}(998.2 \text{ kg m}^{-3} \times 9.812 \text{ m s}^{-2} \times 6.76 \times 10^{-2} \text{ m} \times 0.25 \times 10^{-3} \text{ m})$$

$$= 0.828 \text{ kg s}^{-2} \equiv 0.828 \text{ N m}^{-1}$$

3.5 Effect of frequency on polarization

The total polarization is expressed as the sum of the three terms that represent electronic, atomic and orientational effects:

$$P = N\alpha_e E + P_a + N\mu^2 E/(3kT) \tag{3.47}$$

and Fig. 3.10 shows a diagrammatic expression of these three components.

In a low frequency or static applied field all three processes represented on the right-hand side of Eq. (3.47) contribute to the total polarization in a gaseous polar species. As the frequency of the field is increased, the orientational component ceases to follow the alternations of the field, and P_o falls to zero. The atom polarization falls to zero at a higher frequency still, and above 10^{15} Hz, which is the optical frequency range, only P_e remains operative.

At the frequency at which a polarization component ceases to follow the alternations of the field, there is a discontinuity in the curve of polarization against frequency. In addition, both P_a and P_e may exhibit characteristic resonance absorption effects, a so-called *anomalous dispersion*; Fig. 3.11 illustrates the form of frequency dependence of polarization for a polar molecule with single atomic and electronic anomalous dispersion frequencies.

The total polarization P and its electronic component P_e may be investigated experimentally through measurement of the static relative permittivity and refractive index parameters respectively. The orientation component P_o is determined from dipole moment measurements; then, the atomic contribution P_a is obtained by difference; an example will be considered shortly. It may be noted in passing that the characteristic frequency marked ν_0 on the dispersion curve corresponds to that which occurs in a detailed derivation of the London energy, a simple treatment of which was given in Section 3.4.3.

The dispersion of polarization in condensed states follows the pattern already discussed for gases. However, owing to their much higher viscosity, liquids show a more rapid decrease in polarization with increase in frequency than do gases. In the solid state, orientation effects may be absent altogether, since molecular rotation occasioned by an applied field is hindered by both strong intermolecular attractions and steric factors; water provides an illustration (Fig. 3.12).

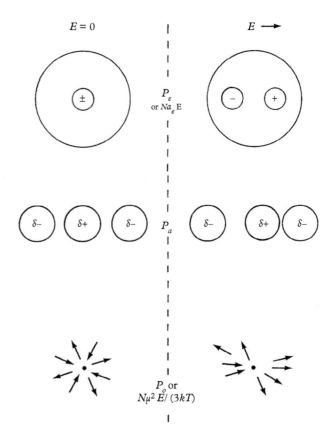

Fig. 3.10 Schematic representation of the components of total polarization P in an electrostatic field of strength E: electron polarization P_e, atom polarization P_a and orientation polarization $P = P_o + P_a + P_e$; thus, $P_a - (P_e + P_o)$.

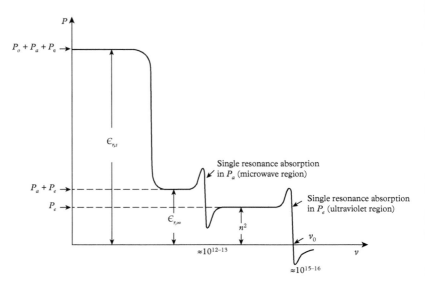

Fig. 3.11 Variation of total polarization P with frequency ν of an applied alternating field. As the frequency of the field is increased, components P_o, P_a and P_e, in that order, fail to follow the alternation of the field and fall to zero value. Discontinuities arise as first P_o and then P_a cease to follow the field; resonance absorptions may arise at these discontinuities.

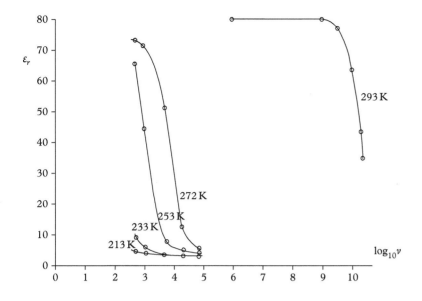

Fig. 3.12 Variation of relative permittivity ε_r for ice and water, as a function of frequency ν at different temperatures.

The water molecule is strongly polar, with a dipole moment μ of 6.0×10^{-30} C m, and the attractive forces between molecules in both the liquid and solid states are enhanced by hydrogen bonding. The dispersion of ε_r between 213 K and 293 K is shown by Fig. 3.12. The orientation effect falls off very rapidly at a frequency of 10^9–10^{10} Hz at 293 K; at 253 K the fall-off occurs at *ca.* 10^3 Hz, and P_o is almost non-existent at 213 K.

3.6 Molar polarization

Molar polarization is chemical term rather than a true polarization, and is represented by the left-hand term of the Clausius–Mosotti equation (A11.45):

$$\frac{(\varepsilon_r - 1)}{(\varepsilon_r + 2)} \frac{M}{D} = \frac{L\alpha}{3\varepsilon_0}$$

This equation is applicable to chemical systems where there is no contribution to the total polarization from permanent dipole moments, or when the applied field is sufficiently high in frequency that permanent dipoles in the material cannot follow its directional fluctuations.

The molar polarization of a mixture is an approximately *additive property* of its components. In a solution of n_A mole of A and n_B mole of B, the molar polarization $P_m(AB)$ of the solution is given by:

$$P_m(AB) = x_A P_m(A) + x_B P_m(B) \tag{3.48}$$

where $x_A = n_A(n_A + n_b)$ is the *mole fraction* of the A component in the two-component solution (and x_B *mutatis mutandis*). In a similar manner, the left-hand side of Eq. (A11.46) represents the *molar refraction*:

$$\frac{(n^2 - 1)}{(n^2 + 2)} \frac{M}{D} = \frac{L\alpha}{3\varepsilon_0}$$

it, too, is an additive property.

Example 3.12

Table 3.3 lists data on a dilute solution of chlorobenzene (polar) in benzene (non-polar). Preliminary calculations to obtain P_m for chlorobenzene, extrapolated to infinite dilution, from the total molar polarization Eq. (3.3) have been carried out.

Table 3.3 *Molar polarization of chlorobenzene in benzene*

T/K	$(10^3/T)/K^{-1}$	$10^4 P_m(C_6H_5Cl)/m^3\ mol^{-1}$
193	5.18	1.065
213	4.69	1.000
233	4.29	0.940
253	3.95	0.893
273	3.66	0.855
293	3.41	0.815
313	3.19	0.778
333	3.00	0.755

By analogy with Eq. (3.3) and in molar terms:

$$P_m = P_{m,e} + P_{m,a} + P_{m,o} \tag{3.49}$$

From Eq. (3.16), $\alpha_o = N\mu^2/(3kT)$, and using Eq. (3.49) with Eq. (A11.45):

$$P_m = P_{m,e} + P_{m,a} + L\mu^2/(9\varepsilon_0 kT) \tag{3.50}$$

Hence, a graph of P_m against $1/T$ should be linear, with a slope $L\mu^2/(9\varepsilon_0 k)$. The plot is shown in Fig. 3.13, and the slope, from a least squares calculation with program LSLI, is 1.4337×10^{-2} m^3 K^{-1} mol^{-1}; the Pearson coefficient r is 0.9996 for the least-squares fit. The dipole moment μ is 5.118×10^{-30} C m, or 1.53 D. The intercept, which is the sum $P_{m,e} + P_{m,a}$, is 3.2542×10^{-5} m^3 mol^{-1}.

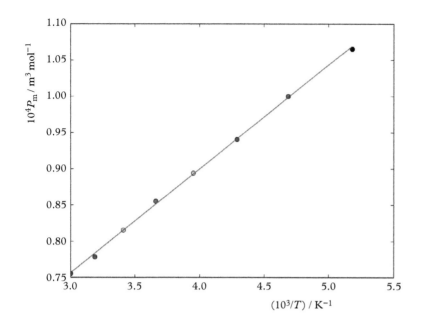

Fig. 3.13 Variation of the molar polarization P_m with temperature T for a solution of chlorobenzene in benzene; the slope is $L\mu^2/(9\varepsilon_0 k)$, and the intercept is the sum $P_e + P_a$.

The refractive index for chlorobenzene is 1.524 and from the Lorentz–Lorenz equation (A.11.46), the molar refraction $R_{m,e}$, which here is the same as $P_{m,e}$, is:

$$\frac{(1.524^2 - 1)}{(1.524^2 + 2)} \times \frac{112.57 \times 10^{-3}\ \text{kg mol}^{-1}}{1106\ \text{kg m}^{-3}} = 3.114 \times 10^{-5}\ \text{m}^3\ \text{mol}^{-1}$$

Hence, $P_{m,a}$ is 1.40×10^{-6} m^3 mol^{-1}, the expected small fraction *ca.* (1/20) of P_e.

3.7 Static permittivity

From Eq. (3.47) and the Clausius–Mosotti equation (A11.45), but ignoring the strong interaction term $3/(\varepsilon_r + 2)$ on its left-hand side, the total molar polarization of a gas may be written as:

$$P_m = (\varepsilon_{r,s} - 1)\frac{M}{D} = \frac{L}{\varepsilon_0}[\alpha_e + \alpha_a + \mu^2/(3kT)] \tag{3.51}$$

where $\varepsilon_{r,s}$ is the *static relative permittivity*. The orientation contribution is strongly dependent on temperature, a fact well illustrated by three chloro-derivatives of methane (Fig. 3.14); the relevant dipole moments are listed:

	CH$_4$	CH$_3$Cl	CH$_2$Cl$_2$	CHCl$_3$	CCl$_4$
$10^{30}\mu$/C m	0	3.4	5.2	6.2	0

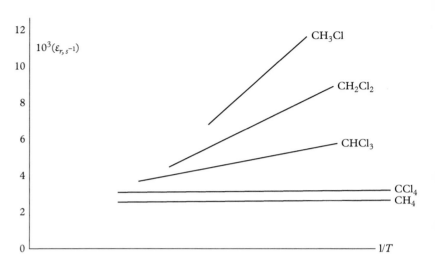

Fig. 3.14 Variation of the static relative permittivity $\varepsilon_{r,s}$ with temperature T for methane and its chloro-substituted compounds: CH_4, CH_3Cl, CH_2Cl_2, $CHCl_3$, CCl_4. In the first and last members of this series the individual bond moments cancel one with the other, leading to an invariance of $\varepsilon_{r,s}$ with temperature.

The relative permittivities of the species with zero dipole moment should not vary with temperature. Of the others, the higher the dipole moment the greater should be the temperature dependence, which is borne out by experiment, as shown by Fig. 3.14.

Example 3.13

The data below refer to measurements on the polar gas, hydrogen bromide. A linear relationship between P_m and $1/T$ is expected:

T/K	$\varepsilon_{r,s}$	$(10^3/T)/K^{-1}$	$10^6(\varepsilon_r - 1)M_r/D)/m^3\ mol^{-1}$
268	1.00323	3.7313	70.88
294	1.00279	3.4014	67.26
339	1.00223	2.9499	62.21

From a least squares fit (Fig. 3.15) the slope, $L\mu^2/(3\varepsilon_0 k)$, is 1.1381 m³ K⁻¹ mol⁻¹ so that $\mu = 2.633 \times 10^{-30}$ C m $\equiv 0.789$ D. The strong interaction term $3/(\varepsilon_{r,s} + 2)$ is *ca.* 0.9991, and its neglect is of only slight consequence for the value of μ. The refractive index n for hydrogen bromide at $25°C$ is 1.00051 and the density is 3309 kg m⁻³ and neglecting the term $3/(\varepsilon_{r,s} + 2)$ leads to

2.495×10^{-5} m^3 mol^{-1} for $P_{\mathrm{m,e}}$. Since the least squares intercept is 2.846×10^{-5} m^3 mol^{-1}, $P_{\mathrm{m,a}} = 0.351$ m^3 mol^{-1}.

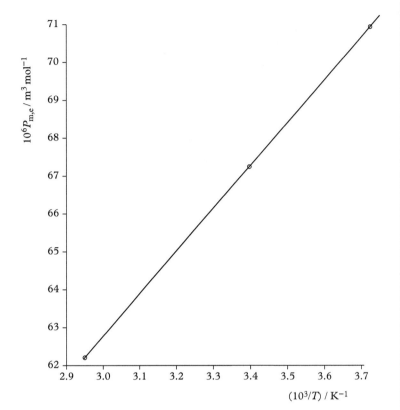

Fig. 3.15 Variation of the static relative permittivity $\varepsilon_{\mathrm{r,s}}$ with temperature for hydrogen bromide; the slope of the line is $L\mu^2/(3\varepsilon_0 k)$.

3.7.1   Isotropic solids

Isotropic solids can be considered in three groups:

1. $P_{\mathrm{a}} = P_{\mathrm{o}} = 0$ with $P_{\mathrm{e}} \neq 0$: this group condition applies to elements in their solid state. Equation (A11.45) is relevant, but it is difficult to verify experimentally. In order to change the value of M/D the temperature must be varied, with a concomitant restriction on the range of applicability of the procedure. However, in this group, P_{e} may be measured from both $\varepsilon_{\mathrm{r,s}}$ and n. For example, the relative permittivity $\varepsilon_{\mathrm{r,s}}$ of the diamond

allotrope of carbon is 5.7, and the refractive index n is 2.42, so that n^2 is 5.86, thus following Eq. (3.28) closely.

2. $P_o = 0$ with P_a and $P_e \neq 0$: this group condition applies to solids with more than one atomic species, but possessing no permanent dipole moment. Ionic crystals are examples of materials with high P_a values:

	KF	KCl	KBr	KI
n^2	1.90	2.22	2.43	2.81
$\varepsilon_{r,s}$	5.5	4.7	4.9	5.1

The vibrations of the positive and negative ions in an ionic solid are perturbed by an applied field, producing appreciable atom polarization effects because of elastic ion displacements; they decrease with a lowering in temperature because the thermal vibrations of the ions decrease in amplitude. Equation (3.28) is not applicable to this group of substances. The topic of electron and atom polarization will be encountered again in studying ionic crystals.

3. P_a, P_o and $P_e \neq 0$: solids to which this group condition applies, although possessing dipolar character, may not exhibit large values for $\varepsilon_{r,s}$ because the molecular dipoles cannot rotate easily, if at all, in the solid state. Nitrobenzene is a striking example of this group. As the crystal is brought towards its melting point temperature, a partial alignment of dipoles is attained. A discontinuity occurs and $\varepsilon_{r,s}$ rises sharply, but subsequently the

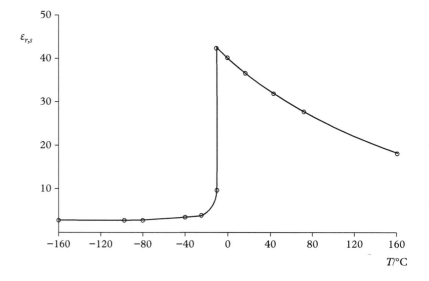

Fig. 3.16 Variation of the static relative permittivity $\varepsilon_{r,s}$ with temperature T for nitrobenzene. The alignment of dipoles as the melting point temperature is approached is reflected in the large increase in orientation polarization; subsequently the polarization decreases owing to the increased thermal agitation of the molecules as the temperature rises.

value of the permittivity decreases (Fig. 3.16). The fall-off occurs because of an increase in thermal motion, as expressed by a decrease in the term $\mu^2 E/(3kT)$ in the Langevin function.

3.8 Intermolecular potentials

At large molecular separations r there is little or no interaction, and the intermolecular potential energy of the system is effectively zero. Attractive forces come into play as r decreases into a bonding region, but at distances equal to or less than the collision diameter strong repulsion between the electrons of the species causes a steep rise in potential energy. The superposition of these two potential energies leads to a curve of the form of Fig. 3.17a, with a minimum energy occurring at the equilibrium separation r_e. Simpler, but less effective potential functions are the hard-sphere and square-well models, Fig. 3.17b and c, respectively.

The curve Fig. 3.17a is the Lennard-Jones 12-6 potential model [12], a particular case of the Mie potential $(-A/r^n + B/r^m)$, and has been applied successfully over a wide range of problems. It has the form:

$$V(r) = 4\varepsilon_{LJ}\left[\left(\frac{\sigma}{r}\right)^{12} - \left(\frac{\sigma}{r}\right)^6\right] \tag{3.52}$$

where ε_{LJ} is an energy parameter for the system under consideration; σ is the distance at which $V(r) = 0$, and for monatomic species is similar in magnitude to twice the van der Waals radius of the species.

At small separations, the r^{-12} term is dominant, and the potential is strongly positive. Hence this term describes the short range repulsive potential arising from the distortion of the electron density at small interatomic distances. In contrast the r^{-6} term is dominant when the separation r is of increased magnitude. Hence this term describes the long range tail of the attractive potential function between two species. In working with the Lennard-Jones potential, the mathematical advantage of m being twice n becomes clear.

If Eq. (3.52) is differentiated with respect to r and the derivative equated to zero, then the equilibrium distance of separation r_e at the minimum of the curve is given by $r_e = 2^{1/6}\sigma$. Applying this value to Eq. (3.52) shows that $V(r)_{r=r_e} = -\varepsilon_{LJ}$; some values for the parameters σ and ε_{LJ} are listed in Table 3.4; ε_{LJ} is usually listed as the parameter ε_{LJ}/k.

Fig. 3.17 Intermolecular potential energy functions, $V(r)$. (a) Lennard-Jones 12–6 potential function: $V(r) = 0$ when r is equal to the Lennard-Jones collision diameter σ; $V(r) = -\varepsilon_{LJ}$ at $r = r_e$, the equilibrium intermolecular distance. This distance r_e is equal to $2^{1/6}\sigma$; and if $\sigma = 0.34$ for argon, then the corresponding value of r_e is 0.38 nm. The Lennard-Jones parameter ε is itself a positive quantity. (b) Hard-sphere model showing a discontinuity at $r = \sigma$, the hard-sphere diameter. This model cannot be used for studying equilibrium properties because it has no minimum in $V(r)$. (c) Square-well model; a minimum in $V(r)$ exists but discontinuities are present.

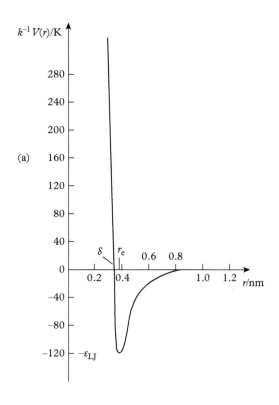

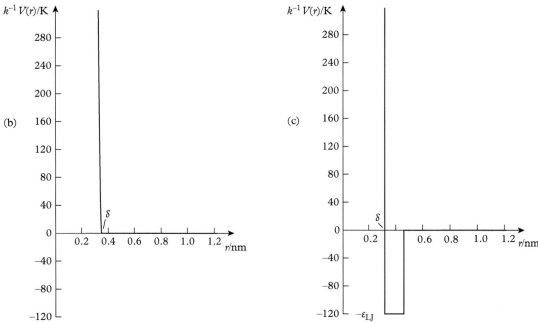

Approximate values for these parameters are indicated in parentheses in the table. They were determined as $\varepsilon_{LJ}/k = 0.775T_c$ and $\sigma = 0.1(V_{m,c}/3)^{1/3}$, where T_c and $V_{m,c}$ are the temperature and molar volume of a gas at the critical state; the values for σ are very close to the distances of closest approach for these gases. This similarity is reasonable: the essential difference is that whereas the values of collision diameters are obtained through experimental measurements on viscosity, those of σ and ε_{LJ} are chosen as best-fit parameters for Eq. (3.52) to reproduce molecular properties. For argon and nitrogen, ε_{LJ} has the approximate values 1.3 kJ mol^{-1} and 0.76 kJ mol^{-1}, respectively. The corresponding classical thermal energies at 298.15 K are 3.7 kJ mol^{-1} or $(3/2)RT$, and 6.2 kJ mol^{-1} or $(5/2)RT$. The larger value for nitrogen arises from a rotational contribution to the energy that is not present in argon.

Table 3.4 *Parameters for the Lennard-Jones 12-6 potential function (values in parentheses derive from critical constants)*

	(ε_{LJ}/k)/K	σ/nm
Ar	124 (117)	0.34 (0.29)
Kr	155 (162)	0.39 (0.31)
N_2	92 (98)	0.37 (0.31)
CO_2	190 (235)	0.40 (0.32)

Plotting exercise 3.1

Run program LJON to obtain files lj12.txt and lj6.txt. which contain data for curves proportional to $1/r^6$ and $1/r^{12}$ respectively (Section 7.11) Transfer the files to the Python work folder, and follow Section 7.2.1 to obtain the program GRFN in the Python editor window. Then carry out the following commands in the Python shell; Enter means depress the 'Enter' key (Note: Lower case is necessary for commands, and each command is followed by the ENTER key):

import grfn (Enter)

Plot the r^{-6} and r^{-12} graphs:

grfn.graf(['lj6', 'lj12'])

Note the greater steepness of the r^{-12} graph, which is clarified by adding the zero line:

plt.plot([0.10, 0.20], [0.0, 0.0])

Now plot the sum of the two graphs. Make sure that the cursor is in the shell. Then, up-arrow on the shell to convert the **grfn.graf** instruction to;

grfn.graf(['lj6', 'lj12'], True)

'True' indicates that a sum is required; the default option is False, for no summation. The curve now has the form of a Lennard-Jones 12-6 potential.

continued

Plotting exercise 3.1 *continued*

Add the zero line to this graph. Up-arrow in the shell to the previous plot will suffice:

plt.plot([0.1,0.20], [0,0])

The line $y = 0$ shows again the greater steepness of the r^{-12} plot compared to the r^{-6}.

Add the two commands

xlabel(r'$ r/ \rm nm$')
ylabel(r'$ V(r)/ \rm Energy\,(relative\,units)$')

The plot should now look like this:

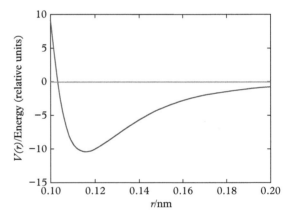

Lennard-Jones 12-6 potential energy function

What are the values of r_e and the energy at the minimum of the curve? Use the mouse pointer and read the values from the bottom of the graph. [*ca.* 0.115, 10.5]

3.9 Structure of liquids

In this section the term liquid implies a pure substance, which may be monatomic like argon or krypton, or polyatomic as in nitrogen or water. These species are often termed classical liquids, because their behaviour can be studied by classical physics.

3.9.1 Liquid–gas equilibrium

If the difference between a liquid and its vapour in the critical region is indistinct, the properties of a simple liquid might be expected to follow initially from an equation such as the van der Waals equation of state, Eq. (3.30).

In the case of an ideal gas, the molar internal energy U_m at a temperature T is $C_{V,m}T$, but with a real gas interparticle interactions must be considered. Thermodynamic arguments [13] show that:

$$p = T(\partial p/\partial T)_V - (\partial U/\partial V)_T \qquad (3.53)$$

The van der Waals equation may be written as

$$p = RT/(V_m - b) - a/V_m^2 \qquad (3.54)$$

whence from Eqs. (3.53) and (3.54):

$$(\partial U/\partial V_m)_T = a/V_m^2 \qquad (3.55)$$

Integrating Eq. (3.55) with respect to V_m leads to:

$$U_m = -a/V_m + f(T) \qquad (3.56)$$

The limiting value of $f(T)$ as a tends to zero is the same as that for an ideal gas, and is $3RT/2$ for a monatomic species. The term $-a/V_m$ in Eq. (3.56) represents the contribution of the attractive forces to the internal energy of a gas that obeys the van der Waals equation.

The molar heat capacity at constant volume is given by

$$C_{V,m} = (\partial U_m/\partial T)_V \qquad (3.57)$$

and if the van der Waals equation were obeyed by a monatomic liquid, its molar heat capacity would be $3RT/2$ (Appendix 5); in fact, for argon it is closer to $5RT/2$. Evidently a better equation of state is needed for a liquid.

3.9.2 Radial distribution function

The intrinsic energy of a monatomic liquid comprises kinetic energy of $3kT/2$ per molecule, and a potential energy[3] term $V(r)$ represented by the Lennard-Jones 12-6 function modified by a radial distribution function. The difference from a gas arises because of stronger intermolecular forces in a liquid compared to those in the corresponding gas under ambient conditions.

A *radial distribution*, or *pair correlation function*, symbolized here by $g(r)$, characterizes the average structure of a liquid. More precisely, it measures the average distribution of the atoms or molecules of the liquid relative to one another.

[3] Distinguish between the commonly used notation V (volume) and V (potential energy).

In a spherical shell of radius r and thickness dr the number $n(r)$ of atoms distributed uniformly in the shell distant r from a given central atom is:

$$n(r) = 4\pi r^2 \rho \, dr \tag{3.58}$$

where ρ is the bulk (number) density[4] of the species, and may be termed the local density $\rho(r)$ at that value of r; the bulk density ρ of the liquid is N/V, the number N of atoms in the volume V.

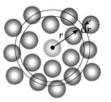

$$n(r)_{\text{shell}} = 4\pi r^2 \rho dr$$

A determination is carried out for all atoms through shells defining the whole sphere and the results averaged. The radial distribution function $g(r)$ is then defined as:

$$g(r) = <\rho>/\rho = n(r)/(4\pi r^2 N)dr \tag{3.59}$$

where the angular brackets mean the average value of $\rho(r)$ taken over all atoms and times, and N is the number density of the species. This value of $g(r)$ does not depend on any specific atom being chosen as the centre or any particular time; it is an equilibrium property of the system.

Atomic sizes range from 0.1 nm to 0.3 nm, and at distances up to 0.05 nm from the central atom $g(r)$ is effectively zero. At r values between 0.1 nm and 0.5 nm a first set of adjacent molecules exists (see Fig. 3.19b, discussed shortly); the local density exceeds the bulk density and $g(r)$ is large. Between this set and a second set, the average density is small and the value of $g(r)$ falls. At large values of r the distribution tends to a uniform state, so that the average density $<\rho>$ then approaches the bulk value ρ and the limiting value of $g(r)$ becomes unity, as it is for a gas. The influence of the central atom is now insignificant.

The importance of the radial distribution function $g(r)$ is that the macroscopic properties of gases and liquids can be obtained in terms of $g(r)$, thus providing a link between the microscopic and bulk (thermodynamic) parameters of fluids.

The average intrinsic energy $<U_m>$ per molecule[5] of a monatomic liquid consists of kinetic energy term and a potential energy function (Appendix A12), and may be written in the form:

$$< U_{\text{m}}>/L = (3/2)kT + 2\pi \left(\frac{N}{V}\right) \int_0^\infty r^2 V(r)g(r)\, dr \tag{3.60}$$

---

[4] In this context, ρ is used for density in place of D.

[5] U is commonly used for potential energy in this work context.

where (N/V) is a number density and $V(r)$ is a pair potential function, such as Eq. (3.52). Each pair of interactions in $V(r)$ is included once only. On multiplication of both sides by L, Eq. (3.60) evaluates as a molar energy. The experimental determination of $g(r)$ for a liquid can be achieved by diffraction studies with x-rays or neutrons, or by computer simulation techniques. For a more specific treatment of this subject, the reader is referred to the literature [14–16].

3.9.3 Diffraction studies on liquids

Diffraction studies using either x-rays or neutrons provide evidence of order in liquids. A gas such as argon does not give a spot diffraction pattern because there are no characteristic, regular spacings between the atoms. Some polyatomic gases give diffraction patterns, but they arise from intramolecular rather than from intermolecular spacings.

Diffraction patterns may be recorded on a photographic film, as in Fig. 3.18, or with far higher resolution by diffractometric techniques. The intensity of scattering $I(S)$ by the liquid is the product of the scattering function for the radiation employed and the radial distribution function, integrated over all space. Since the integrals over θ and φ, in spherical polar coordinates (Appendix A6), evaluate to 4π, the intensity distribution is

$$I(S) = 4\pi \, \Phi \int_0^\infty r^2 g(r) \frac{\sin Sr}{Sr} \, dr \qquad (3.61)$$

where Φ is a function of the x-ray atomic scattering process and S is $\frac{\sin\theta}{\lambda}$. The Fourier transform of Eq. (3.61) is

$$g(r) = \frac{1}{2\pi^2 \Phi \, r^2} \int_0^\infty I(S) \frac{\sin Sr}{Sr} \, dS \qquad (3.62)$$

which is an alternative definition of the radial distribution function and shows how it may be deduced from x-ray data.

An x-ray diffractogram for argon in the gaseous and liquid states is shown in Fig 3.19. The result for the gas (a) shows no order, that is, no particular value of r predominates in $g(r)$. Therefore, from the foregoing discussion, $g(r)$ is unity and the number of atoms in spherical shells of radius r will be a constant, proportional to r^2. Thus, the form of the curve is parabolic (Fig. 3.19a), and the density parameter ρ in molar terms is L/V_m. If the temperature of the gas be increased, then the parabolic form of the curve would remain but the value of $g(r)$ would decrease because a given spherical shell would contain fewer molecules, that is, ρ would be smaller. Realistically, the curve (a) should follow the dashed line at $r = \sigma$, the hard-sphere, or collision, diameter (*ca.* 0.34 nm for argon).

An x-ray diffractogram for liquid argon is shown in Fig. 3.19b. Here, the radial distribution function is modified by the correlation between pairs of atoms

Fig. 3.18 X-ray diffraction pattern from a liquid: the typical, diffuse, dark diffraction rings on the photographic film correspond to the more highly correlated intermolecular distances in the liquid.

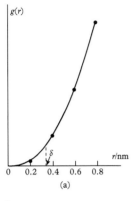

(a)

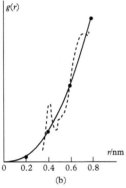

(b)

Fig. 3.19 (a) Radial distribution function $g(r)$ for argon gas; realistically, the curve falls to zero at r equal to the collision diameter δ, because of the strong forces of repulsion acting at $r < \delta$. (b) X-ray diffractogram of liquid argon (dashed line) near the triple point (83 K) superimposed on to the random distribution in (a). The area under the peak at $r \approx$ 0.37 nm ($r_{\text{van der Waals}}(\text{Ar}) = 0.188$ nm) corresponds to an average coordination number of 10.4; a second enhanced distribution occurs at *ca.* 0.70 nm.

that form short-range order in the liquid. The first peak at $r = 0.37$ nm represents nearest neighbours and the area A under that peak is the number of such neighbours:

$$A = 4\pi n \int_0^{r_{\text{max}}} r^2 g(r)\, dr \tag{3.63}$$

The upper limit r_{max} is best determined as the value of r_{peak} and the calculated value of A doubled to represent the whole area. As r is increased, the diffractogram tends to a parabolic form since the distribution again becomes uniform at large values of r.

Near the triple-point temperature, the first coordination number for argon is 10.4. As the temperature is increased, this number decreases to 4.2. The range and non-integral nature of the coordination numbers of a liquid distinguishes it sharply from a solid, in which the coordination number does not vary with temperature unless a phase transition occurs.

3.9.4 Internal energy of fluid

The internal energy of a liquid is not easily evaluated because $g(r)$ is not a simple function. The following approximate procedure examines argon as both gas and liquid at the critical temperature.

Example 3.14

At the critical state of argon, the temperature is 150.7 K, and the molar volume is $L/(0.0966 \times 10^{-3})$ m³ mol⁻¹. Since no correlation is assumed for the gas, $g(r) = 1$. The Lennard-Jones potential energy function of Eq. (3.52) is employed with $\sigma = 0.34$ nm and $\varepsilon_{\text{LJ}}/k = 120$ K; thus, the molar internal energy is, from Eq. (3.60):

$$U_{\text{m}} = (3/2)RT + 2\pi L^2 \varepsilon_{\text{LJ}} k/V_{\text{m}} \int_{r_0}^{\infty} r^2 [(\sigma/r)^{12} - (\sigma/r)^6]\, dr$$

Evaluating: $(3/2)RT = 1.88$ kJ mol⁻¹, and $2\pi L^2 \varepsilon_{\text{LJ}} k/V_{\text{m}} = 4.0384 \times 10^{31}$ J m⁻³ mol⁻¹. The integral is $-\frac{\sigma^{12}}{9r^9} + \frac{\sigma^6}{3r^3}\Big|_{r_0}^{\infty}$ and if the lower limit r_0 is chosen as σ, the result is $-2\sigma^3/9$, which is -8.73×10^{-30} m³. Hence, the *configurational energy* is -0.36 kJ mol⁻¹. Adding the kinetic energy term of 1.88 kJ mol⁻¹ gives the value of 1.52 kJ mol⁻¹ for the total internal energy U_m for the gas. This value is of the correct order, but a better result is obtained by simulation (see Table 3.5).

The calculation for liquid argon is complicated by the form of the integrand in Eq. (3.60), and is best carried out by a simulation procedure. However, as an approximation, using the average value of $g(r)$ over the range 0 to 0.8 nm from Fig. 3.19b, results in the value of -2.72 kJ mol^{-1} for the configurational energy and, hence, -0.84 kJ mol^{-1} for the total energy U_m, which may be compared with both experimental values and those from the simulation techniques that are described next.

3.9.5 Simulation techniques

Two theoretical methods for investigating the structure of a liquid are *Monte Carlo*, which leads to equilibrium thermodynamic parameters and radial distribution functions $g(r)$, given a form for the potential $V(r)$, and *molecular dynamics*, which is a time-dependent procedure and provides measures of transport properties.

3.9.5.1 Monte Carlo

In this technique, an atomic model is set up in a cubic unit cell, and an initial configuration of 200 to 300 atoms is repeated by three-dimensional face-to-face stacking, so generating a macroscopic model by the translation group of a lattice. Fig. 3.20 shows a projection of the model after an elapsed simulation time. The cubic cell side a is chosen such that the desired number density N at a temperature T is achieved ($N = L/a^3$). The initial configuration of atoms for a monatomic liquid is based on a face-centred cubic array.

The model of Fig. 3.20 contains a degree of translational symmetry that is not present in the liquid. It is assumed that, provided the range of the potential

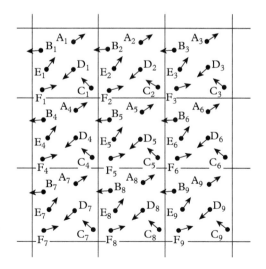

Fig. 3.20 Projection of a configuration of atoms in a Monte Carlo simulation of liquid argon. The positions of the atoms in any one cubic cell are random; the other cells are formed by the three-dimensional lattice translation group. After sufficient simulation time, an equilibrium arrangement is attained; as one molecule moves out of a given cell so another moves in from an adjacent cell so as to maintain a constant number density. [Reproduced by courtesy of Professor A. J. C. Ladd, University of Florida.]

Table 3.5 *Experimental and theoretical thermodynamic data for liquid argon*

T/K	$10^6 V/m^3 mol^{-1}$	p/atm[a]		−U/kJ mol⁻¹	
		Calc	Expt	Calc	Expt
100.0	29.7	116	115	5.52	5.54
140.0	41.8	18	37	3.81	3.86
150.7	75.2	49	49	2.48	2.47

[a] 1 atm = 1.0132501 bar = 101,325.01 Pa (N m⁻²).

Fig. 3.21 Models of argon in the solid and liquid states from computer simulation of the Newtonian equations of motion. (a) Equilibrium state of a crystal, showing small vibrations around the cubic, face centred unit cell lattice sites. (b) Equilibrium liquid state, showing random configurations with only localized regions of order. [Reproduced by courtesy of Professor A. J. C. Ladd, University of Florida.]

(a) 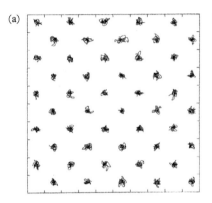   (b)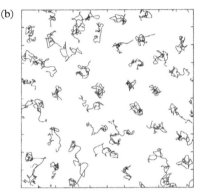

energy of interaction between atoms is less than $a/2$, the potential experienced by any given atom is not affected by the symmetry of the model. It is clear that motion of the species will take atoms out of a unit cell. However, as an atom B, say, moves out from its cell, a translation image moves in to take its place. In this way the atom density is conserved in each cell. A selection of results for liquid argon is listed in Table 3.5; the agreement with experimental data is of a high quality [17–19].

3.9.5.2 Molecular dynamics

Molecular dynamics treats the evolution with time of systems of particles that interact through conservative forces[6] operating under the laws of classical mechanics. In effect it tracks the motions of species in condensed phases by solving the Newtonian equations of motion, and has been particularly useful in elucidating transport parameters.

In applying molecular dynamics to a simulated liquid system, a set of initial coordinates is generated, usually in the form of a face-centred unit cell of a cubic Bravais lattice, at a required density. Initial momenta configurations are assigned randomly, such that the system has the desired total energy, and boundary conditions are imposed in the manner of the Monte Carlo method.

[6] Forces that are derivable from the potential energy function through $F = -dV/dr$.

Table 3.6 *Thermodynamic properties of water by computer simulation*

System	$-E_m$/kJ mol^{-1}	pV_m/RT	$C_{v,m}$/J K^{-1} mol^{-1}
216[a]	43.1	0.05	100
256[b]	39.9 ± 0.3	0.6 ± 0.3	70
Experiment	41.1	0.05	75

The superscripts *a* and *b* link with results in Table 3.7.

Many molecular dynamics calculations have been carried out with a hard sphere potential function. It is computationally straightforward, and shows that the structure of simple liquids is almost independent of their chemical nature, and may be approximated as an interaction of rigid spheres. This idea was present in the early work of Bernal with physical models [20]; a good résumé on the structure of water has been given by Finney [21].

The results of molecular dynamics calculations on the crystal–liquid interface of argon, using 1500 atoms, are illustrated in Fig. 3.21a and b. In (a), the atoms are vibrating about their mean positions in the solid state, the sites of a face-centred unit cell of a cubic lattice, whereas (b) shows the traces of the atoms, now in a typically liquid phase.

Many attempts have been made to simulate the properties of water, both by Monte Carlo and molecular dynamics techniques. In the case of water molecules and other dipolar molecules, care is needed in specifying the pair potential between species, because of the relatively long range effect of dipolar interactions. Table 3.6 lists the configurational energy, an equation of state function and the constant volume heat capacity of water, and compares the results from molecular dynamics with those from experiment.

Table 3.7 *Structural properties of water by computer simulation: positions r and heights M of maxima in radial distribution functions g_{I-J}*

		r_1/nm	M_1	r_2/nm	M_2
g_{O-O}	{216[a]	0.285	3.09	0.470	1.13
	{256[b]	0.285	3.11	0.530	1.06
Experiment		0.283	2.31	0.425	1.08
g_{O-H}	{216[a]	0.190	1.38	0.340	1.60
	{256[b]	0.191	1.24'	0.332	1.53
Experiment		0.190	0.80	0.335	1.70
g_{H-H}	{216[a]	0.250	1.50	0.390	1.20
	{256[b]	0.250	1.15	0.375	1.07
Experiment		0.235	1.04	0.400	1.08

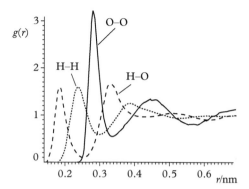

Fig. 3.22 Radial distribution for water as a function of distance, calculated by averaging over a molecular dynamics simulation experiment; the curves are labelled according to the interactions that they represent.

The structural properties of water were addressed by computing the radial distribution functions for O–O, O–H and H–H interactions, by sampling pair distributions after every 250 configurations. Table 3.7 lists the results for the positions and heights of the maxima for a 256 molecule system, and compares them with experimental results. The simulation results are very satisfactory in representing thermodynamic and structural properties of the liquids studied [17–19]. A recent study [22] includes a radial distribution function for water (Fig. 3.22), with results in very close agreement with those in Table 3.7.

3.10 Structural and physical characteristics of molecular solids

The dipolar forces described in this chapter can link an atom to a small, indefinite number of neighbours, and are undirected spatially. In the solid state of an inert gas, London dispersion forces provide the sole means of cohesion between the atoms. Simple structures based on the face-centred unit cell of a cubic lattice are formed, as exemplified by Fig. 1.21. In other molecular compounds, relatively short bonds exist between atoms, such as 0.20 nm in molecular chlorine Cl–Cl, 0.140 nm for the C–C bonds and 0.108 nm for C–H in benzene, but characteristically longer contact distances, 0.35–0.38 nm between nearest non-bonded neighbours in the solid state of these compounds.

In polar substances, the strong dependence of cohesive energy on polarizability is revealed, for example, by the trend in the melting-point temperatures of the silicon tetrahalides:

	SiF_4	$SiCl_4$	$SiBr_4$	SiI_4
r_{Si-X}/nm	0.155	0.202	0.220	0.243
$10^{-40}\alpha$/C m	1.0	3.4	4.8	7.3
MP/K	183	203	278	394

In these compounds, the polarizability of the halogen increases more rapidly from fluorine to iodine than does the intermolecular distance, with consequent enhancement of the melting-point temperature. Over a corresponding range of molar mass among the alkanes, the increase in melting-point temperature is approximately 140 K.

Molecular solids generally form soft, brittle crystals with anisotropic physical properties, such as low melting-point temperatures and high thermal expansivities. Their properties are similar in the melt and in solution. The electrical and optical properties of molecular solids are approximately the aggregate of those properties of their components since the electron systems do not interact strongly in the solid state: for example, boron, fluorine and boron trifluoride have electrical conductivities in the range 10^{-6}–10^{-8} Ω cm.

Most molecular solids are usually soluble in common solvents. The adage 'like dissolves like' applies generally. For example, naphthalene dissolves in benzene but not in water, whereas glucose dissolves in water but not in benzene to any significant extent.

Solubility is a chemical reaction between solute and solvent and involves both enthalpic and entropic changes. While these parameters can be compounded into a free energy of dissolution that explains the observed solubility, there is merit in considering the interplay of these parameters in the dissolution process. Molecular solids that dissolve in a solvent do so usually with small changes in enthalpy, and the entropic component is frequently a significant factor.

Water is a strongly hydrogen-bonded liquid, and dissolution in water involves the breaking of hydrogen bonds in order to free water molecules to solvate the solute molecules. If the enthalpy decrease on solvation is numerically very small and the entropy gain relatively negligible, then the solvent–solvent forces of attraction will not be broken by any form of interaction with the solute Then, dissolution will not take place, or is very small in magnitude, as with naphthalene and water. However, if a solvent for naphthalene is chosen in which its forces of attraction are weak, then the entropy gain of its solution over the (solvent + solute) system is dominant and dissolution will occur, as with benzene solvent. In the case of glucose and water, the hydrogen bonds in water are broken by interaction with the polar –OH groups of glucose and dissolution takes place. Benzene, however, has negligible interaction with glucose and this compound is insoluble in benzene. The interesting topic of solubility will arise again in the study of ionic solids.

3.11   Hydrogen-bonded solids

A wide range of compounds exhibits hydrogen bonding which, in some cases, is responsible for cohesion in the solid state. A *hydrogen bond* is an attractive interaction between a hydrogen atom from a donor molecule or a molecular fragment X–H, in which X is a species that is more electronegative than hydrogen, and an acceptor atom or a group of atoms. It may exist in either the same molecule (*intramolecular hydrogen bonding*) or a different but neighbouring

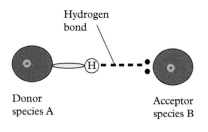

Hydrogen bond

Donor species A

Acceptor species B

Hydrogen bond A—H----B

molecule (*intermolecular hydrogen bonding*). In both situations interaction arises and hydrogen bonds are formed.

Hydrogen is thought of normally as a monovalent species, and is placed at the head of group 1 of the periodic table [23], and for good reason [24]. However, numerous compounds exist in which hydrogen atoms link with two or more other atoms or groups, mainly those of high electronegativity such as fluorine, oxygen and nitrogen, $-NH_2$ and $>C=O$ groups, but to a lesser extent with carbon, sulphur and chlorine. The bonds vary in strength, the variation linking with the bonding distance, as the following compilation shows. Here, atom *A* is the normally bonded atom and H–*A* provides the electron donor, while *B* is the acceptor species. A hydrogen-bonded system may be two-centred, either linear or slightly bent, or three-centred, bifurcated, or even multi-centred in some biological molecules:

	Strong	**Medium**	**Weak**
$U/\text{kJ mol}^{-1}$	15–50	5–15	0–5
Bonding	Charge transfer	Electrostatic	Electrostatic
r_{A-B}/nm	0.22–0.25	0.25–0.30	0.30–0.35
$r_{H\cdots B}/\text{nm}$	0.12–0.15	0.15–0.22	0.22–0.35
$A-H\cdots B/\text{deg}$	175–180	135–180	90–150
Examples	HF, H_2O hydrates	Alcohols, carboxylic acids	$HC\cdots O$, $HC\cdots N$

The bonding is mostly electrostatic in nature, between partial charges on the donor and receptor systems, or of a *charge-transfer* type of covalency [25, 26]. The current view on the hydrogen bond is actually very similar to that given earlier by Pauling [3].

Intramolecular hydrogen bonds arise in species where polar groups in one and the same molecule are sufficiently close for dipolar attraction to lead to bonding. Examples are carboxylic acid dimers where, typically, the O–H bond length is 0.097 nm in the simple acid, but 0.107 nm in the dimer; and *o*-nitrophenol.

Carboxylic acid dimer **o-Nitrophenol**

Most hydrogen bonding examples are intermolecular, linking polar groups in adjacent chemical species, as in many of the examples that will be encountered

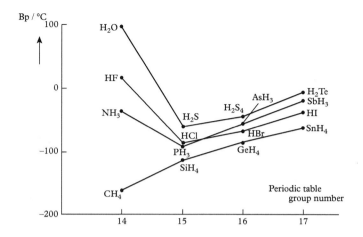

Fig. 3.23 Boiling-point temperatures for the hydrides of elements in groups 14, 15, 16 and 17 (aka IVA to VIIA) of the periodic table; the effect of hydrogen bonding in the highly electronegative species breaks the decreasing sequences of temperature variation of boiling points of these hydrides.

in the following discussion on the classification of molecular compounds. It is especially important in proteins and other molecules of biological importance.

Evidence of hydrogen bonding arises in the physical properties of many chemical substances: the boiling point data for the hydrides of groups 14–17 of the periodic table are well known, and Fig. 3.23 shows clearly the effect of strong hydrogen bonding in the hydrides of nitrogen, oxygen and fluorine. In the absence of hydrogen-bonding the freezing point of water would be *ca.* $-90\,°C$, with unwelcome consequences.

3.11.1 Atomic force microscopy and hydrogen bonds

As well as the evidence of hydrogen bonds inferred from experimental studies involving x-ray and neutron crystal structure analysis, further evidence has arisen from experiments with atomic force microscopy [27].

Atomic force microscopy (AFM was discussed briefly in Section 2.10.1.1. It is a type of scanning probe microscopy having a very high resolution, of the order of fractions of a nanometre, which is significantly greater than that of the best optical diffraction. The atomic force microscope explores the physical properties of a sample by passing a fine pointed probe, supported by a cantilever or other mounting, over the surface of the sample under investigation (see also Section 6.6). The surface exerts London dispersion forces on the probe and deflections in the probe are detected by a linked laser-optical system (Fig. 3.24a). The force on the sample can be estimated by assuming a Lennard-Jones potential $V(r)$ acting between the molecules of the surface. The force F is given by $-dV(r)/dr$, and differentiating Eq. (3.52) with respect to r leads to:

$$\frac{d\left\{4\varepsilon_{LJ}\left[\left(\frac{\sigma}{r}\right)^{12} - \left(\frac{\sigma}{r}\right)^{6}\right]\right\}}{dr} = 24\varepsilon\left(\frac{2\sigma^{12}}{r^{13}} - \frac{\sigma^{6}}{r^{7}}\right) \qquad (3.64)$$

The force is a maximum at the minimum of the curve in Fig. 3.17a, that is, for $-\mathrm{d}V(r)/\mathrm{d}r = 0$, and under this condition $r = 2^{1/6}\sigma$ (see also Section 3.8). Taking this value of r together with $\sigma = 0.3$ nm and $\varepsilon/k = 120$ K, the force evaluates to -8.6 pN.

A recent example of the power of atomic force microscopy is revealed by Fig. 3.24b, which is an AFM photograph of a cluster of hydrogen-bonded 8-hydroxyquinoline molecules (Fig. 3.24d). The photograph shows hydrogen bond electron density as 'lines' between hydrogen atoms and polar groups on the molecules. The photograph of Fig. 3.24c is Fig. 3.24b reversed; now, the hydrogen bonds show up as dark lines against a white background. The molecular structure of 8-hydroxyquinoline is shown by Fig. 3.24e.

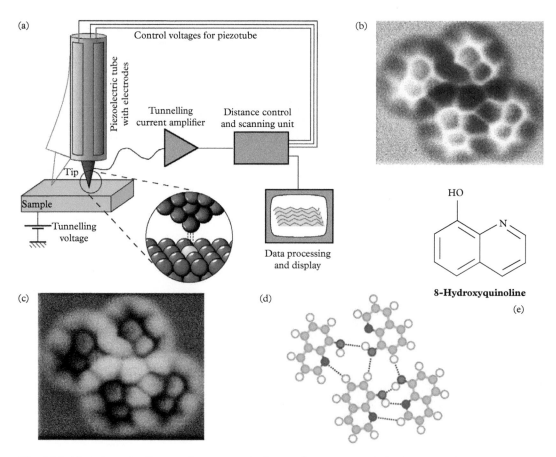

Fig. 3.24 (a) Schematic diagram for an atomic force microscopy analysis. The tip of the piezoelectric tube probe experiences a force when in close proximity to the surface of the material under examination; it produces a tunnel current that is amplified, scanned and passed to a display unit. The STM equipment enables the sample to be moved in *x*, *y* and *z* directions, thus ensuring good vibrational isolation. Feedback circuitry adjusts the tip–sample distance for constancy of the current. [Reproduced by courtesy of Professor M Schmid, TU Wien.] (b) Atomic force microscopy (AFM) picture of the electron density in the crystal structure of 8-hydroxyquinoline:

The interatomic and intermolecular bonds in a four-molecule group of the structure are clearly visible; the 'lines' between the non-bonded oxygen and nitrogen species are believed to represent hydrogen-bond electron density. Some distances involving hydrogen are N···H–C. 0.26 nm; O···H–C, 0.23 nm; O···H–O, 0.20 nm. (See also Section 3.11.1 for a challenge to this interpretation.) (c) The AFM picture in (b) reversed, highlighting the electron density of the molecular group. (d) Structure of the four-molecule group: the colours green, red, blue and white represent respectively the carbon, oxygen, nitrogen and hydrogen atoms. (e) The 8-hydroxyquinoline molecular structure. [Reproduced by courtesy of Professor Xiaohui Qiu.]

Recently, however, this interpretation has been called into question. Molecules of bis(4-pyridyl)acetylene,

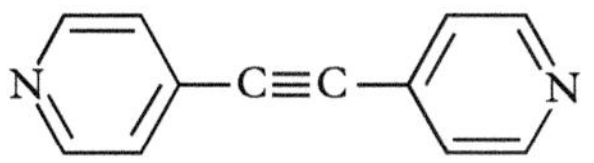

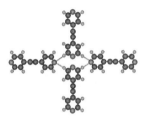

Bis(4-pyridyl)acetylene

condense to form the tetramer

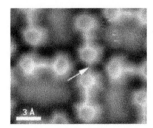

Bis(4-pyridyl)acetylene tetramer

which, when examined by AFM showed faint lines between non-bonded nitrogen atoms at the centre of the tetramer:

AFM photograph of bis(4-pyridyl) acetylene tetramer

They cannot be hydrogen bonds, and they have been interpreted as 'ridges' in the force field between the AFM tip and the tetramer [28, 29]; there is some evidence for N$\cdots$H–C bonding.

At the time of writing, ambiguity exists over the interpretation of the lines in the AFM photograph of 8-hydroxyquinoline, adduced for hydrogen-bond density. The distance between the non-bonded nitrogen atoms in the bis(4-pyridyl)acetylene tetramer is *ca.* 0.32 nm, approximately the same as twice the van der Waals radius of nitrogen (0.310 nm). In the AFM results for 8-hydroxyquinoline, the apparent hydrogen bond lengths (Fig. 3.24b) show significant divergence from the sum of the van der Waals radii. There would seem to be no *a priori* reason why hydrogen bond density would not appear in AFM photograph of 8-hydroxyquinoline; it would not discount the existence of a possible vibrational contribution to the bonding similar that responsible for the faint lines on the AFM picture of the tetramer (see also Section 3.4.3.1).

Closely related work is the more recent report on imaging the exact positions of single atoms in a crystal structure by means of STM measurements, corrected for aberration arising from expansion of the sample mount by changes in the ambient temperature. The substance under examination was the perovskite structure type $(La_{0.18}Sr_{0.82})(Al_{0.59}Ta_{0.41})O_3$ with the cubic space group $Pm\bar{3}m$. The oxygen atoms occupy the centres of the faces of the cubic unit cells, with statistical occupations of the body-centring positions by (Al, Ta) and of the corners of the unit cells by (La, Sr) [30].

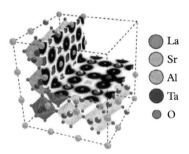

Stacked unit cells of the $(La_{0.18}Sr_{0.82})(Al_{0.59}Ta_{0.41})O_3$ structure [30]

In a highly significant development of AFM, it has been found that magnetic resonance can utilize the spins of electrons in materials as precise probes for imaging. Spin resonance frequency has a high sensitivity to the magnetic fields generated by the proximity of surrounding atoms and molecules. Thus, magnetic resonance spectroscopy provides a unique detector for a particular molecular arrangement and its interaction with the immediate environment. Magnetic resonance imaging is important in medical diagnosis because it provides non-invasive, three-dimensional images of objects. The work has extended the limits of magnetic resonance imaging into the nanoscale region. Sensitive optical detection techniques and microwave manipulation of spins measure the magnetic resonance from a nanoscale group of nuclear spins, achieving a resolution of 10–20 nm.

High sensitivity is needed in an understanding of processes at the molecular level in biological function. Experimental work has used properties of a nitrogen–vacancy (NV) defect in diamond. This defect is a vacancy in a crystal structure adjacent to a nitrogen atom. Electrons are thus trapped in the vacancies and can be manipulated with high precision. Nanosize diamonds appear to be amenable to very precise manipulation and are also biocompatible.

Nanoscale magnetic resonance imaging was first achieved using a sensitive cantilever the behaviour of which could be modified by nuclear spin forces, leading to a resolution of a few nanometres albeit at temperatures less than 1 K. Recent work has used negatively charged nitrogen-vacant (NV) centres, or flaws, in diamond fragments. Nanoscale thermal and magnetic field detectors that can be inserted into living cells will enhance the understanding of chemical and physical *in vivo* reactions [31–34].

3.12 Classification of molecular solids

Molecular solids are diverse in character and extensive in number, but they may be classified and discussed under several headings:

1. The inert gases
2. Elements in groups 15, 16 and 17 (formerly VB–VIIB) of the periodic table
3. Small inorganic molecules
4. Clathrate structures
5. Charge-transfer complexes
6. π-Electron-overlap compounds
7. Organic compounds, including hydrogen-bonded species

3.12.1 Inert gases

The inert gases are monatomic, non-polar species, and remain so in the solid state. Cohesion in the solid arises through London dispersion forces (Section 3.4.3) and, not surprisingly, their cohesive energies are very low, as the following physical data show:

	He	Ne	Ar	Kr	Xe	Rn
Melting-point temperature/K	0.95	24.7	83.6	115.8	161.7	202.2
Boiling-point temperature/K	4.4	27.3	87.4	121.5	166.6	211.5
Enthalpy of vaporization/kJ mol^{-1}	0.08	1.74	6.53	9.05	12.7	18.1

In the solid state, the structures are cubic face-centred, that of Fig. 1.21 being representative of them all. Of the elements in the rows of the periodic table in which they occur they have the highest ionization energies, a reflection of the high stability and low reactivity conferred on them by the configuration of a complete outer electron octet.

3.12.2   Groups 15, 16 and 17 elements

The elements in these groups form a divers set of solid-state structures. Thus, nitrogen has a close-packed hexagonal structure, white phosphorus consists of discrete, tetrahedral P_4 molecules and sulphur exhibits an eight-membered crown-like ring structure:

Crown structure of sulphur S_8

The structures in any one periodic group show a gradual increase towards metallic character as the group is descended:

	N	P	As	Sb	Bi
Electrical resistivity, $10^8 \rho$/ohm m	<1	10	33	40	130

Good sources for the structures of this varied group of chemical species are the American Mineralogist Crystal Structure Database Interface [35] and the Cambridge Structural Database [36].

3.12.3   Small inorganic molecules

A large number of inorganic substances form individual molecules, and their structures and properties have been reviewed in the literature [35]. Elemental mercury (periodic group 12, or IIB in earlier notation) is unique as a liquid metal at room temperature, *ca.* 20 °C. It combines to form Hg_2 molecules of Hg–Hg bond length 0.301 nm that are then linked by relatively weak London dispersion forces. The outermost *s* electrons are significantly inert to oxidative reaction and are retained strongly retained by the nucleus, an *inert pair* effect, and it forms compounds with reluctance (see also Section 3.4.4.1).

The inert pair effect is responsible for the abnormal electrical conductivity of elements in this group; cadmium exhibits an inert pair effect that is weaker than that in mercury, and zinc has the properties of a metal:

	Zn	Cd	Hg
$10^{-7}/ S\ m^{-1}$	1.66	1.47	1.04

Typical compounds of mercury are exemplified by its chlorides. There are no true Hg^I species; the monovalent species exists in the chloride, for example, as Hg_2Cl_2. Linear molecules Cl–Hg–Hg–Cl have an Hg–Hg bond length of 0.253 nm and Hg–Cl 0.243 nm. The coordination of any one mercury atom is a distorted octahedron; in addition to the two Hg–Cl bonds at 0.243 nm; there exist another four Cl species at the longer distance of 0.321 nm.

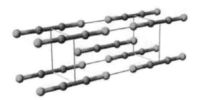

Mercury(I) chloride[7]

In the higher oxidation state, $Hg^{II}Cl_2$ forms two-coordinate linear molecules with two Hg–Cl bond lengths of 0.238 nm and 0.300 nm, thus showing a similarity to the bonding in Hg_2Cl_2. Four other chlorine atoms lie at a distance of 0.338 nm. Similar structures are found for $HgBr_2$ and HgBrCl, and the structures form ribbons of the type X–Hg–X (X= Cl, Br, I). The Cl–Hg–Cl system is almost linear, with a bond angle of 179°. The weak intermolecular links also are typical of the London dispersion interactions.

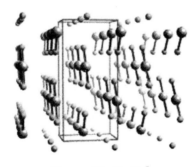

Mercury(II) chloride[7]

[7] Reproduced by courtesy of Professor Mark Winter, University of Sheffield, UK.

3.12.4 Clathrate compounds

A clathrate compound (L. *clatratus* = with bars) consists of a structure containing included, or trapped, molecules of another species. They exist only in the solid state, and may exhibit variable composition. The host can be ice and the trapped material a non-polar gas, such as oxygen, nitrogen or methane, forming the so-called clathrate hydrates that were first documented by Humphry Davy in 1810. A methane clathrate of composition $CH_4 \cdot 5.75H_2O$ is a cubic crystalline solid, in which methane is trapped in the ice structure, and is found under the sediment of the ocean floors (Fig. 3.25).

Not only gases, but other chemical substances are trapped as clathrate compounds. The tetragonal structures of urea $CO(NH_2)_2$ and thiourea $CS(NH_2)_2$, for example, form cage-like structures that are able to occlude small molecules such as benzene and xylene, which are released only on decomposition or dissolution of the host structure.

Cage-like clathrate structure showing cavities

Another well-known host structure is that of 1,4-dihydroxybenzene (quinol); this compound can form clathrate structures with small molecules, such as sulphur dioxide, methanol and acetonitrile CH_3CN. The quinol–acetonitrile complex has a low value of relative permittivity, whereas that of quinol–methanol is large. It may be inferred that acetonitrile molecules are in orientational or dynamic disorder, not linked strongly to the host structure. In the quinol–methanol clathrate, however, the molecules of methanol are locked in position, aided by hydrogen bonding between the polar groups of the two species.

Ammonia, nickel cyanide and benzene form the clathrate $Ni(NH_3)_2Ni(CN)_4 \cdot 2C_6D_6$ (Fig. 3.26),[8] also known as Hofmann's clathrate [37]. The Ni^{2+} cations are coordinated octahedrally to two ammonia molecules and four cyanide ions, and also to four cyanide ions in a square-planar array. This cage-like structure traps two benzene molecules in the host unit cell.

[8] Deuterated benzene; H^2 atoms are more easily detected that H^1 by neutron diffraction.

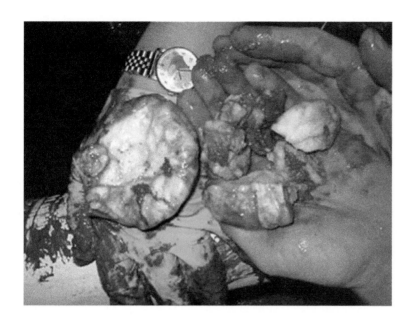

Fig. 3.25 Fragments of methane gas hydrate recovered from the Gulf of Mexico, in 2002. [Reproduced by courtesy of Dr Bill Winters, U. S. Geological Survey.]

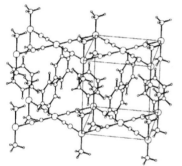 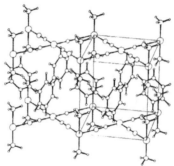

Fig. 3.26 Stereoview of the unit cell and environs of the nickel cyanide–ammonia–benzene clathrate compound; circles in decreasing order of size represent the Ni, N, C and H(D) atoms; the shortest contact distance of 0.36 nm is between C (benzene) and C (CN) group.

3.12.5 Charge-transfer structures

A charge-transfer structure involves a form of covalent bonding in which two or more molecules combine in a donor-acceptor mode. Typical compounds are the benzene-halogen structures, and Fig. 3.27 illustrates the benzene-chlorine charge-transfer complex. The chlorine molecule is orientated normally to the plane of the benzene ring, and the shortest Cl–C(ring centre) is 0.33 nm. The charge-transfer molecules are readily raised to an excited state in the visible region of the electromagnetic spectrum, which accounts for the *charge-transfer spectral* colour observed for these complexes.

Charge-transfer compounds occur for both inorganic and organic materials. A well-known example is the blue-coloured compound formed between iodine and starch, which is used as an indicator in quantitative iodometric analysis. Soluble

Fig. 3.27 Stereoview of the unit cell and environs of the benzene–chlorine charge-transfer complex; circles in decreasing order of size are represent the Cl, C and H atoms; the shortest distance, Cl to the benzene ring centre, is 0.33 nm; the stability of this compound has been attributed to Cl–π interaction.

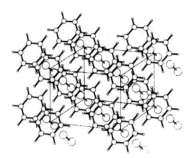 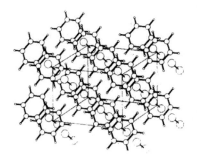

starch is β-amylose, a helical structure; iodine in the anionic form $[I_3]^-$ couples with β-amylose and a charge transfer takes place between the two species. The starch forces the iodine atoms into a linear arrangement in the amylose central region, and a transfer of electron charge between starch and iodine occurs with concomitant energy level changes that give rise to the characteristic blue colour.

In inorganic compounds, charge-transfer complexes involve mainly electron transfer between transition-metal atoms and ligands. The charge-transfer spectral bands of transition-metal coordination compounds result from movement of electron density between molecular orbitals. This can occur either from the metal to the ligand, which is an oxidation of the metal, or from the ligand to the metal, which corresponds to its reduction.

3.12.6 π-Electron overlap compounds

π-Electron overlap is important in the solid state of many aromatic compounds, particularly in aromatic hydrocarbons, where other stronger interactions are absent. In anthracene, for example (see Fig. 3.34), the molecular planes are orientated such that the delocalized electrons in π orbitals of adjacent molecules overlap, thus leading to enhanced stability. Table 3.8 shows an interesting relationship between melting-point temperature and relative molar mass. Normally, an increase in molar mass leads to an increase in the melting-point temperature. Many aromatic compounds, however, are stabilized relative to their alicyclic analogues by π-electron overlap.

Biphenyl $(C_6H_5)_2$ is another molecule of interest in this context. In the gaseous, free molecule state the phenyl rings lie at an angle of about 45° to each other, as shown by both spectroscopic experiment and theoretical calculation. In the crystal, however, the rings are coplanar with the molecules aligned almost parallel one to the other (Fig. 3.28). The mid-points of the C–C′ bonds lie on centres of symmetry on the monoclinic space group[9] $P2_1/c$. Again, π-electron overlap is playing a significant role in the cohesion of the solid state of this species.

For fuller information on molecular compounds, attention is directed to very thorough analyses of molecular complexes and compounds the literature [38, 39].

[9] Reference [23] of Chapter 2.

Table 3.8 *Melting-point temperatures T and relative molar masses M_r for some aromatic hydrocarbons and their fully saturated, alicyclic counterparts*

		T/K	M_r
Benzene	C_6H_6	279	78.1
Cyclohexane	C_6H_{12}	280	84.2
Naphthalene	$C_{10}H_8$	353	128.2
Decahydronaphthalene	$C_{10}H_{18}$	230, 241[a]	138.3
Anthracene	$C_{14}H_{10}$	490	178.2
Tetradecahydroanthracene	$C_{14}H_{24}$	335, 366[a]	192.4

[a] Polymorphs.

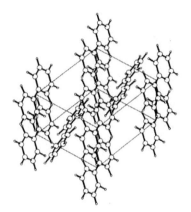 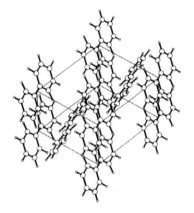

Fig. 3.28 Stereoview of the unit cell and environs of biphenyl $C_{12}H_{10}$; circles in decreasing order of size represent the C and H atoms; the shortest intermolecular contact distance is 0.37 nm. The molecules lie on centres of symmetry in the crystal structure, thus showing that the molecule is centrosymmetric and planar. The planarity implies conjugation throughout the molecule, which is supported by two different C–C distances, C–C' (linking rings) 0.149 nm and C–C (within ring) 0.140 nm.

3.12.7 Organic compounds

As with the classification of solids *in toto*, so there are different bases over which organic compounds may be discussed. One such scheme, as used herein, is by division in terms firstly of molecular shape wherein three main types and are identified qualitatively as *equant*, *flat* and *long*, and then within each of these types both *polar* and *non-polar* species can be recognized. In this way, a useful classification can be devised that encompasses a wide range of organic compounds. Examples of the structural classes will be illustrated by stereoscopic diagrams, but first some characteristic structural data will be reviewed.

Among organic compounds, structural units can be identified which preserve their shape, size and symmetry over a range of compounds: bond lengths, bond angles, and larger structural moieties, such as the six-membered phenyl ring. Tables 3.9–3.11 list typical values for these quantities and also the van der Waals radii of a selection of atoms. The value of a linear or angular property of a bond between atoms depends firstly upon its electronic structure, and these tables show this property by means of the notation C4, N3, O2, and so on, where C4–C4

Table 3.9 *Characteristic bond lengths/nm*

Single bond lengths			
C4–H	0.109	C3–C2	0.145
C3–H	0.108	C3–N3	0.140
C2–H	0.106	C3–N2	0.140
N3–H	0.101	C3–O2	0.136
N2–H	0.099	C2–C2	0.138
O2–H	0.096	C2–N3	0.133
C4–C4	0.154	C2–N2	0.133
C4–C3	0.152	C2–O2	0.136
C4–C2	0.146	N3–N3	0.145
C4–N3	0.147	N3–N2	0.145
C4–N2	0.147	N3–O2	0.136
C4–O2	0.143	N2–N2	0.145
C3–C3	0.146	N2–O2	0.141

Double bond lengths			
C3–C3	0.134	C2–O1	0.116
C3–C2	0.131	N3–O1	0.124
C3–N2	0.132	N2–N2	0.125
C3–O1	0.122	N2–O1	0.122
C2–C2	0.128	O1–O1	0.121
C2–N2	0.132		

Triple bond lengths		Aromatic bond lengths	
C2–C2	0.120	C2–C3	0.140
C2–N1	0.116	C2–N2	0.134
N1–N1	0.110	N2–N2	0.135

implies a single bond between tetrahedral carbon atoms, C3–O2 the geometry >C=O and so on.

Van der Waals radii are related approximately to the size of the outer orbitals of atoms. Thus, a carbon $2p$ orbital containing 95% of the total $2p$ electron density extends from the nucleus to a distance of *ca.* 0.185 nm, and approximately twice this value is an intermolecular distance in the absence of other interactions.

Two non-bonded species in neighbouring molecules in a structure may lie further apart than the sum of their van der Waals radii because of steric effects involving the whole molecule. In other compounds the distance between such non-bonded atoms may be significantly less than the van der Waals radii sum. Thus, the intermolecular moiety O $\cdots$ H–O may result in an intermolecular O $\cdots$ O distance significantly less than twice the sum of the van der Waals radius

Table 3.10 *Characteristic bond angles/°*

Atom[a]	Geometry	Angle	Example
C4	Tetrahedral	109.5	CH_4
C3	Planar	120	C_2H_4
C2	Bent	109.5	–CHO
C2	Linear	180	HCN
N4	Tetrahedral	109.5	NH_4^+
N3	Pyramidal	109.5	NH_3
N3	Planar	120	$H_2N–CHO$
N2	Bent	109.5	H_2CHN
N2	Linear	180	HNC
O3	Pyramidal	109.5	H_3O^+
O2	Bent	109.5	H_2O

[a] The number following the element symbol in column 1 refers to linkages.

of oxygen in the presence of hydrogen bonding. The *o*- and *p*-nitrophenols are interesting examples of hydrogen bonding. In *o*-nitrophenol the hydrogen bonding is intramolecular, whereas it is intermolecular in *p*-nitrophenol, which explains how *o*-nitrophenol can be separated from a mixture of *o*- and *p*-nitrophenols by steam-distillation: BP 215 °C (*ortho*-nitrophenol), 279 °C (*para*-nitrophenol):

Table 3.11 *Van der Waals radii/nm*

Atom	Radius	Atom	Radius
H	0.120	F	0.135
C[a]	0.185	Cl	0.180
N	0.150	Br	0.195
P	0.190	I	0.215
O	0.152	$-CH_3$	0.200
S	0.183	$>CH_2$	0.200
Se	0.200	$-C_6H_5$	0.185[b]

[a] Reported as 0.168 nm in graphite.
[b] Half-thickness of a phenyl ring.

o-Nitrophenol   **p-Nitrophenol**

It is evident that –OH is too remote from $-NO_2$ in *p*-nitrophenol for intramolecular interaction, but it can link to neighbouring molecules in the solid state (see Fig. 3.36).

Those atoms that can exist as well-defined ions exhibit ionic radii that are very similar in magnitude to the van der Waals radii. This result is in agreement with the fact that repulsion forces build up rapidly at small distances of separation, because of the $1/r^{12}$ function in the Lennard-Jones intermolecular potential. Thus, the effective size of a bromine species, for example, is closely similar when the repulsion is balanced against either the strong interionic attraction in potassium bromide, K^+Br^-, or the relatively weaker intermolecular attraction in bromobenzene, C_6H_5Br.

3.12.7.1 *Classification system for organic compounds*

The chosen classification for organic compounds is described in terms of Table 3.12, and enhanced by stereoviews (Figs. 3.29–3.40) of the *unit cell and environs* of each of the example species. Significant features are on the one hand, a similarity of the nearest non-bonded distances in non-polar molecules to the sum of the corresponding van der Waals radii, and on the other hand a negative departure from the van der Waals radii sum of the nearest neighbour distances in the presence of hydrogen bonding in polar species (see also Section 2.23.6).

Table 3.12 *Classification of organic molecular solids*

Molecular shape

	Non-polar	Polar
Equant	Small, isometrically shaped molecules, forming approximately close-packed structures. Examples are methane, ethane, hexachloroethane (Fig. 3.29), adamantane, and cubane (Fig. 3.30). In some structures, such as methane near its melting point, rotation can exist in the solid state	Small, isometrically shaped molecules, with dipolar or hydrogen-bonded (or both) interactions rather than close-packing dominating the structural configuration. Examples include urea, methanol, pentaerythritol (Fig. 3.31), methylamine, and oxalic acid dihydrate (Fig. 3.32).
Flat	Molecules lying with their planes nearly parallel, Staggered configurations may be adopted where π-electron overlap is possible, Examples are benzene (Fig. 3.33), anthracene (Fig. 3.34), and phthalocyanine.	Molecular packing dominated by dipolar or intermolecular hydrogen-bonded interactions Examples are 1,4-dinitrobenzene (Fig. 3.35) and 4-nitrophenol (Fig. 3.36).
Long	Molecules lying parallel or in staggered configurations. Examples are octane (Fig. 3.37), hexane (Fig. 3.38) and dicyanoethyne (Fig. 3.39). With increasing temperature, some paraffins, such as hexane, rotate in the solid-state imparting cylindrical symmetry to the molecule.	Long, polar molecules, tending to associate in pairs through dipolar and or hydrogen-bonded interactions. Alkanammonium salts are ionic, often with the carbon chains in free rotation. Examples are adipic acid, decanamide (Fig. 3.40), potassium caprate and propan-1-ammonium chloride.

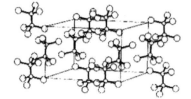 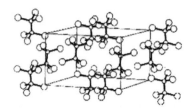

Fig. 3.29 Stereoview of the unit cell and environs of hexachloroethane C_2Cl_6; circles in decreasing order of size represent the Cl and C atoms. Equant; non-polar; shortest intermolecular contact distance 0.37 nm.

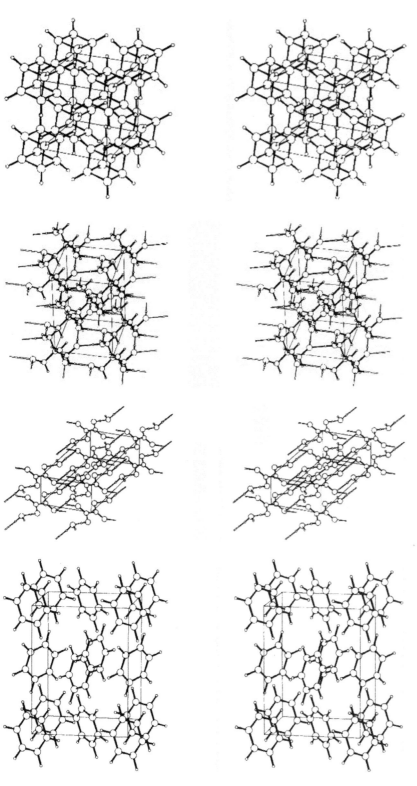

Fig. 3.30 Stereoview of the unit cell and environs of cubane C_8H_8; circles in decreasing order of size represent the C and H atoms. Equant; non-polar; shortest intermolecular contact distance 0.38 nm.

Fig. 3.31 Stereoview of the unit cell and environs of pentaerythritol $C(CH_2OH)_4$; circles in decreasing order of size represent the O and C atoms; hydrogen bonds are shown by double lines. Equant; polar; shortest $O-H \cdots O$ intermolecular contact distance 0.27 nm, indicating medium strength intermolecular hydrogen bonding.

Fig. 3.32 Stereoview of the unit cell and environs of oxalic acid dihydrate $(CO_2H)_2.2H_2O$: circles in decreasing order of size represent the O, C and H atoms; hydrogen bonds shown by double lines. Equant; polar; shortest $O-H \cdots O$ intermolecular contact distance 0.257 nm indicates strong intermolecular hydrogen bonding.

Fig. 3.33 Stereoview of the unit cell and environs of benzene C_6H_6; circles in decreasing order of size represent the C and H atoms. Flat; non-polar; shortest intermolecular contact distance 0.36 nm.

Fig. 3.34 Stereoview of the unit cell and environs of anthracene $C_{14}H_{10}$; circles in decreasing order of size represent the C and H atoms. Flat; nonpolar; shortest intermolecular contact distance 0.36 nm. In this structure and that of benzene, the small reduction in the intermolecular contact distance from twice the van der Waals radius of carbon (0.37 nm) may indicate a degree of stabilization from intermolecular π-bonding.

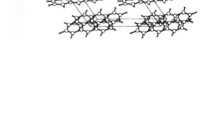
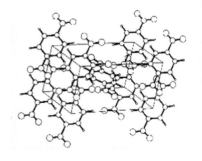

Fig. 3.35 Stereoview of the unit cell and environs of 1,4-dinitrobenzene $C_6H_4(NO_2)_2$; circles in decreasing order of size represent the O, N, C and H atoms. Flat; polar; shortest intermolecular contact distance 0.32 nm, indicating energy-enhancing dipole–dipole bonds.

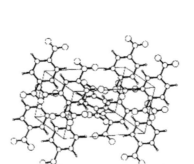
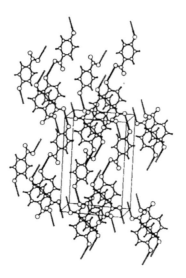

Fig. 3.36 Stereoview of the unit cell and environs of 4-nitrophenol $C_6H_4(NO_2)OH$; circles in decreasing order of size represent O, N, C and H; hydrogen bonds shown by double lines. Flat; polar; shortest O–H · · · ·O intermolecular contact distance 0.29 nm, indicating weak intermolecular hydrogen-bonding.

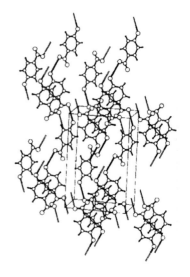

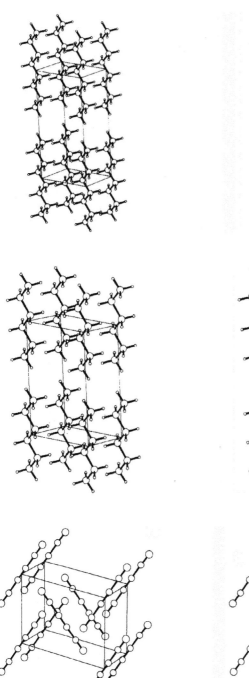

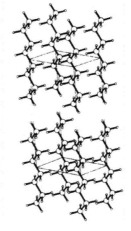

Fig. 3.37 Stereoview of the unit cell and environs of octane C_8H_{18}; circles in decreasing order of size represent the C and H atoms. Long; non-polar; shortest intermolecular contact distance 0.37 nm (twice the van der Waals radius of carbon) indicating only London energy as the source of interaction.

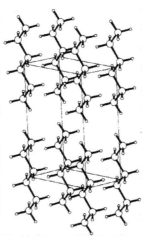

Fig. 3.38 Stereoview of the unit cell and environs of hexane C_6H_{14}; circles in decreasing order of size represent the C and H atoms. Long; non-polar; shortest intermolecular contact distance 0.36 nm.

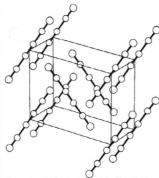

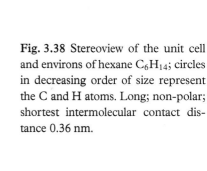

Fig. 3.39 Stereoview of the unit cell and environs of dicyanoethyne C_4N_2; circles in decreasing order of size represent the N and C atoms. Long; non-polar; shortest intermolecular contact distance 0.33 nm.

Fig. 3.40 Stereoview of the unit cell and environs of decanamide $CH_3(CH_2)_8CONH_2$; circles in decreasing order of size represent the O, C and H atoms. Long; polar; shortest intermolecular O–H $\cdots$ N contact distance 0.29 nm, indicating medium–weak hydrogen bonding.

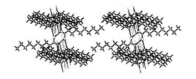 

REFERENCES 3

[1] Hannay NB and Smith CP. *J. Am. Chem. Soc.* 1946; 68: 71.

[2] Pauling L. *J. Am. Chem. Soc.* 1932; 54: 3570.

[3] Pauling L. *The Nature of the Chemical Bond.* Cornell University Press, 1932.

[4] Salzman WR. 2004; http://www.chem.arizona.edu/~salzmanr/480a/480ants/vdwcrit/vdwcrit.html

[5] Keesom WH. *Phys. Z.* 1921; 22: 129, 643.

[6] Magnasco V. *Models for Bonding in Chemistry.* Willey-Blackwell, 2010.

[7] London FW. *Z. Phys.* 1930; 63: 249.

[8] London FW. *Z. Phys. Chem.* 1937; 33: 8.

[9] Israelachvili J. *Intermolecular and Surface Forces.* Academic Press, 1992.

[10] Atabek O and Lefebvre R. *Chem. Phys. Lett.* 1983; 98: 559.

[11] Fleming DG *et al. Angew. Chem. Int. Ed.* 2014; 53: 13706.

[12] Lennard-Jones JE. *Proc. Roy. Soc. Lond.* 1924; A106: 463.

[13] Atkins P and de Paula J. *Atkins' Physical Chemistry*, 9th ed. Oxford University Press, 2009.

[14] Marcus Y. *Introduction to Liqui8d State Chemistry*, Wiley, 1977.

[15] Hansen J-P and McDonald IR. *Theory of Simple Liquids*, 3rd ed. Academic Press, 2006.

[16] Allen MP and Tildesley DJ. *Computer Simulation of Liquids.* Clarendon Press, 1989.

[17] Ladd AJC. *Computer Simulation of Liquids.* PhD Thesis, Cambridge University, 1977.

[18] Ladd AJC and Woodcock LV. *J. Phys. Chem.* 1978; 11: 3565.

[19] Ladd AJC. *Mol. Phys.* 1977; 33: 1039

[20] Bernal JD and Fowler RH. *J. Chem. Phys.* 1933; 1: 515.

[21] Finney JL. *Phil. Trans. Roy. Soc. Lond.* 2004; B359: 1145

[22] Furmanchuk A *et al. Phys. Chem. Chem. Phys.* 2010; 12: 3363.

[23] http://www.nist.gov/pml/data/upload/periodic-table.pdf

[24] Cousins DM *et al. Chem. Commun.* 2013; 49: 11809.

[25] Wang Y. *Phys. Chem. Chem. Phys.* 2014; 16: 3153

[26] Cressy D. *Nat. News Blog* 2010; 3 November, and references therein.

[27] Zhang J *et al. Science* 2013: 342: 611; Supplementary material online at http://www.sciencemag.org/content/suppl/2013/09/25/science.1242603. DC1.html

[28] Hämäläinen S *et al. Phys. Rev. Lett.* 2014; 113: 186102

[29] Hapala P *et al. Phys. Rev. B.* 2014 ; 90 : 085421.

[30] Sang X *et al. Appl. Phys. Lett.* 2015; 106: 61913.

[31] Bhallamudi VP and Hammel CP. *Nat. Nanotechnol.* 2015; 10: 104.

[32] Rugar D *et al. ibid.* 10: 120

[33] Häberle T. *ibid.* 10: 125.

[34] DeVievce SJ *et al. ibid.* 10: 129.

[35] American Mineralogist Crystal Structure Database, 2014. http://rruff. geo.arizona.edu/AMS/amcsd.php

[36] Cambridge Crystallographic Data Centre, 2015. http ://www.ccdc.cam. ac.uk/Solutions/CSDSystem/Pages/CSD.aspx

[37] Büttner HG *et al. Acta Crystallogr.* 1994; B50: 431

[38] Herbstein FH. *Crystalline Molecular Complexes and Compounds*, Vols. 1 and 2. OUP/IUCr, 2005.

[39] Kitaigorodskii AI. *Organic Chemical Crystallography*. New York: Consultants Bureau, 1961.

Problems 3

3.1. The planar molecule methanal (formaldehyde) HCHO in the gas phase has the following atomic data: C–H, 0.110 nm; C=O, 0.122 nm (x-axis); H–C–H $=116°$. The atom charges are C $+0.45$, H $+0.11$ and O -0.38. Determine the magnitude of the dipole moment of the molecule. Compare your result with a literature value, such as in Nelson RD *et al. Selected Values of Electric Dipole Moments in the Gas Phase*, National Bureau of Standards Circular No. 537, 1957 (online at http://www.nist.gov/data/nsrds/NSRDS-NBS-10.pdf).

3.2. Three charges $+q_1$, $-q_2$ and $-q_3$ lie in a line such that $r_{12} = r$ and $r_{23} = x-r$. Charges q_1 and q_2 are fixed and q_3 movable. Show that, at equilibrium, (a) $q_1 r_{23}^2 = -q_2 r_{13}^2$, and (b) $x = \frac{r}{(1-\sqrt{|q_2|/q_1})}$. (c) What can be said of the equilibrium between positions the 0 and x?

3.3. The C=O bond moment is 2.50 D and the bond length is 0.121 nm. Calculate the effective charges on the two atoms and hence estimate the percentage ionic character of the bond. Compare the result with that derived from electronegativities (Section 3.2).

3.4. (a) The relative permittivity and density of α-Se at 298.15 K are 6.701 and 4290 kg m^{-3}, respectively. Calculate the polarizability and the volume polarizability of α-Se. (b) If the polarizability of gaseous argon is 1.706×10^{-40} C m, estimate a radius for the argon atom.

3.5. The following data were obtained for β-camphor at temperatures between 273K and 373 K:

T/K	273	293	313	333	353	373
ε_r	12.5	11.4	10.8	10.0	9.50	8.90

β-**Camphor**

Determine (a) the dipole moment and (b) the volume polarizability for this species. The variation in density from the measured value of 992 kg m^{-3} at 0°C is negligible over the above temperature range, and atom polarization may be considered negligible.

3.6. The dipole moment of the hydrogen fluoride molecule at a given temperature is 6.07×10^{-30} C m. What are (a) the electrostatic potential and (b) the magnitude of the electric field strength, at a point P situated 50 nm from the centre of the H–F bond and making an angle of 30° with it.

3.7. If the C–Cl bond moment from chlorobenzene is 1.53 D, calculate the dipole moments for all tetra- and penta-substituted chlorobenzenes.

3.8. Compare the structure of hexane H_3C–CH_2–$(CH_2)_2$–CH_2–CH_3 (Fig. 3.38), shortest intermolecular contact 0.36 nm for C $\cdots$ C, with that of dicyanoethyne $N\equiv C$–$C\equiv C$–$C\equiv N$ (Fig. 3.39), shortest intermolecular contact 0.33 nm for C $\cdots$ N. Suggest a reason why the molecules of dicyanoethyne do not pack in the simple manner shown by hexane.

3.9. In *Polar Molecules* (1929) by PJW Debye (online at http://babel.hathitrust.org/cgi/pt?id=mdp.39015016069091;view=1up;seq=26), page 20, the following data on the molar polarization of ammonia, obtained by M. Jona, are reported:

T/K	292.2	309.0	333.0	387.0	413.0	446.0
$10^6 P_m/m^3$ mol^{-1}	57.57	55.01	51.22	44.99	42.51	39.59

(a) Determine the dipole moment and volume polarizability of ammonia.

(b) What is the value of the molar polarization of ammonia at 0 °C?

(c) Evaluate the volume polarizability at this temperature.

3.10. The Lennard-Jones potential $V(r)$ is given by Eq. (3.52) in Section 3.8. (a) Determine the equilibrium value r_e from the minimum of $V(r)$; take $\sigma = 0.380 \times 10^{-9}$ nm. (b) Find an expression for $V(r)$ at $r = r_e$. (c) For xenon, the value of ε_{LJ}/k is 214 K. Calculate the Lennard-Jones energy of two xenon atoms at $r = r_e$ in both J and J mol^{-1}.

3.11. If a monovalent ion interacts with the electron density of a non-polar atom, its field $|\mathbf{E}|$ shifts the electron density of the atom an amount Δx from its centre, so that $\mu_{ind} = xe$. If a sodium ion Na$^+$ lies 0.4 nm from the centre of a methane molecule of diameter *ca.* 0.4 nm, how far is the electron density of methane shifted? What is the fractional change in the molecular diameter of the methane molecule? ($\alpha_{CH_4} = 2.89 \times 10^{-40}$ C m^2 V^{-1})

3.12. The first ionization energy of neon is 2080.6 kJ mol^{-1}. Obtain an estimate of the London dispersion energy for a pair of neon atoms.

3.13. (a) Calculate the interactional energy of a pair of krypton atoms at their equilibrium interatomic distance by the Lennard-Jones equation; $\varepsilon = 155$ K, $\sigma = 0.390$ nm; first obtain a value for the equilibrium distance r_e in terms of σ. (b) From the Lennard-Jones equation obtain an expression representing the attractive force F between two atoms. (c) What can be said of the attractive force as a function of the variable r?

3.14. The Lennard-Jones equation is a special form of the Mie pair potential, $V(r) = -A/r^n + B/r^m$, where $n = 6$ and $m = 12$. Values for A and B are 10^{-77} J m^6 and 10^{-134} J m^{12}, respectively.

(a) What is the value of σ, *vide* Eq. (3.52)?

(b) Determine r_m, the value of r at the maximum force F_{max}, and find the value of the F_{max}.

(c) If $V(r) = 0$ at $r = r_0$, what is the ratio r_e/r_0?

(d) Set up, preferably by writing a program, values of $V(r)$ *v.* r for the curve $V(r) = -\frac{A}{r^6} + \frac{B}{r^{12}}$, with $A = 10^2$ J m^6 and $B = 10^2$ J m^{12}. Set an integer variable $i = 95$ to 250 in steps of 1, so that r becomes float(i)/100.0 in the units 10^{-10} m. Then, plot a Lennard-Jones potential curve with GRFN. Use the mouse pointer over the curve and read the values of r_e and r_0, and compare the results with $(2B/A)^{1/6}$ and $(B/A)^{1/6}$, respectively. Is it practicable to read r_{max} (r at F_{max})?

3.15. A molecule of dipole moment p of magnitude 1 D at 25 °C interacts with a unit charge Q at a distance of 1 nm.

(a) If the molecule is assumed to be freely rotating, calculate the interactional energy.

(b) Is the assumption justified at the given temperature?

(c) If the assumption holds, at what distance would free rotation begin to be hindered? For (c), calculate the interactional energy at distances d of $1.0 - 0.4$ in steps of 0.1, and of $0.4 - 0.1$ in steps of 0.05; then plot the results with GRFN.

3.16. The molar polarization of water above its boiling point under a pressure of 100 kPa and has been determined as follows:

$T/°C$	111	147	172	211	249
$10^5 P_m/m^3 \ mol^{-1}$	5.74	5.35	5.01	4.68	4.31

Determine the dipole moment and polarizability of water under the given conditions.

Ionic Compounds

4

> *I have no satisfaction in formulas unless I feel their arithmetical magnitude*
>
> Lord Kelvin

4.1 Introduction

Among the four classes of solids set down in Chapter 1 ionic compounds are, to a certain extent, less complicated than those discussed so far. While the energetics of many types of ionic solid can be discussed on the basis of coulomb interaction, it is necessary also to draw upon some of the material of the previous two chapters for precise results. Ionic structures illustrate clearly the general electrostatic rule that 'unlike charges attract, like charges repel'. This simple principle is often useful in forming a mental picture of bonding.

4.2 Ionic bond

The ionic, or *electrovalent*, bond is formed between atoms of widely different electronegativities. One of the species becomes ionized, typically a metallic element, forming a positive ion, or cation:

$$M \longrightarrow M^+ + e^-$$

and a second stage would be

$$M^+ \longrightarrow M^{2+} + e^-$$

A second species, typically a non-metallic element, acquires an electron to form a negative ion, or anion:

$$X + e^- \longrightarrow X^-$$

The ionized atoms then undergo coulombic interaction, the force F of which is inversely proportional to the square of the distance r between the ions and directly proportional to the magnitudes of the ionic charges:

$$F \propto 1/r^2$$

and

$$\text{(4.1)}$$

$$F \propto z_1 e\, z_2 e$$

Bonding, Structure and Solid-State Chemistry. First Edition. Mark Ladd.
© Mark Ladd 2016. Published in 2016 by Oxford University Press.

where z_1 and z_2 are the magnitudes of the two charged species. The energy $\int F \, dr$ leads to the energy of the system, and may be expressed as

$$U(r) = -z_1 z_2 e^2 / r \tag{4.2}$$

the negative sign implying an attractive energy. It will be noted that as ions become more highly charged the cations have an increased polarizing effect and the anions become more polarizable. Together these changes increase the polarization in the species which ultimately shades into covalent bonding and molecular compounds (see Fig. 4.18).

In general, the atom of an element is a stable entity; it does not ionize spontaneously. Work must be done on the atom system in order to remove an electron, and this work is the *ionization energy* of the atom. Two or more stages of ionization lead to the first, second and higher ionization energies through the processes:

$$M(s) \xrightarrow{\Delta H_{sub}} M(g) \xrightarrow{I_1} M^+(g) + e^-$$

and

$$\tag{4.3}$$

$$M^+(g) \xrightarrow{I_2} M^{2+}(g) + e^-$$

where ΔH_{sub} (aka S_M) is the *enthalpy of sublimation* of the element M, and I_1 and I_2 are its first and second ionization energies. Further ionization energies may be possible for the M species, depending on its atomic number. An element in group n (earlier style numbering) of the periodic table will have a significantly higher $(n + 1)$ ionization energy than that for the group number n:

I_n/kJ mol^{-1}	I (1) Na ($n = 1$)	II (2) Mg ($n = 2$)	III (13) Al ($n = 3$)
I_1	495.8	737.7	577.6
I_2	4562.4	1450.6	1816.6
I_3		7732.6	2744.7
I_4			11577.0

The numbers in parentheses are current periodic group numbering.

The reason for the large increase in the $(n + 1)$ ionization energy of a group n element lies in the attraction by the inner shells, but more particularly because of the high stability of the inert gas configuration of an M^{n+} species: $Na^+(Ne)$, $Al^{3+}(Ne)$, $Ca^{2+}(Ar)$. The normal unit of ionization energy is the joule J, but an alternative unit in common use is the electron-volt eV; 1 eV is the energy acquired by an electron moving through a potential difference of 1 volt and is equal to 1.60218×10^{-19} J, or 96,485.3 J mol^{-1}.

4.2.1   Crystal (cohesive) energy

The *crystal (cohesive) energy* of an ionic crystal, frequently termed (incorrectly) lattice energy, is the *energy liberated when one mole of its component ions in the ideal gas phase condense to form the crystalline solid*; there is no energy of interaction in the ideal gas state. The crystal energy value is expressed generally at either 0 K or 298.15 K, and it can be discussed on both a thermodynamic and an electrostatic basis; at 0 K the crystal energy and crystal enthalpy are of one and the same value.

4.3   Thermodynamic approach to crystal energy

The thermodynamic model for crystal energy of a crystalline ionic compound will be approached first in terms of the corresponding enthalpy change ΔH_c and subsequently a relationship between the quantities ΔH_c and U_c will be established.

Sodium chloride is a typical ionic compound, and its formation may be represented as:

$$Na(s) + \tfrac{1}{2}\, Cl_2(g) \rightarrow NaCl(s) \qquad (4.4)$$

NaCl may be written more precisely as $Na^+Cl^-(s)$, and the letters in parentheses refer to the thermodynamic state, solid (s), liquid (l) or gas (g). The above apparently simple equation contains implicitly several processes which must be examined.

Solid sodium is sublimed:

$$Na(s) \rightarrow Na(g) \qquad (4.5)$$

and the enthalpy of sublimation ΔH_{sub} (aka S_{Na}) is 110.2 kJ mol^{-1} at 298.15 K, the temperature at which these processes here described apply. The gas is next ionized:

$$Na(g) \rightarrow Na^+(g) + e^- \qquad (4.6)$$

and the enthalpy change ΔH_i (aka I_{Na}) is 495.8 kJ mol^{-1}; an amount $(5/2)RT$ for one mole of electrons at 298.15 K and constant pressure is 6.2 kJ mol^{-1} and should be added to ΔH_i but it will be ignored as a cancellation occurs, as will be seen shortly. The sum of the enthalpy changes for the two processes is 606.0 kJ mol^{-1}, and it is clear that there is no tendency for the reactions in Eqs. (4.5) and (4.6) to occur spontaneously.

The reactions of chlorine must be considered next. The gas is dissociated into atoms:

$$\tfrac{1}{2}\, Cl_2(g) \rightarrow Cl(g) \qquad (4.7)$$

and the enthalpy change is one half the *dissociation enthalpy* ΔH_{dis} (aka D_0) of the Cl_2 molecule, namely, $\frac{1}{2} \times 243.4$ kJ mol^{-1} or 121.7 kJ mol^{-1}. The Cl atom then receives an electron released by the ionization of sodium:

$$Cl(g) + e^- \rightarrow Cl^-(g) \tag{4.8}$$

and the enthalpy change or *electron affinity*, ΔH_{ea} (aka E_{Cl}) is -348.6 kJ mol^{-1} $- (5/2)RT$. In some sources the electron affinity is listed as a positive quantity defined by the reverse process of Eq. (4.8). It seems to this author that since Eq. (4.8) is a spontaneous process, as the term 'affinity' suggests, the negative sign and Eq. (4.8) as written are the more appropriate.

The enthalpy sum for processes Eqs. (4.7) and (4.8), neglecting the cancelling terms for 1 mole of electrons, is -226.9 kJ mol^{-1}. The four reactions now total an enthalpy change of 379.1 kJ mol^{-1} for the production of one mole of each of $Na^+(g)$ and $Cl^-(g)$, which still does not indicate a spontaneous process. However, the driving force for Eq. (4.4) lies in the stability conferred on the system when the gaseous ions condense to form crystalline sodium chloride, and this enthalpy change is characterized by the *crystal enthalpy* ΔH_c.

The crystal *energy* and *enthalpy* are negative quantities, determined by differences between two states, those of the crystal and of the gaseous component ions, both at the same temperature and pressure:

$$\Delta U_c(MX, \text{ s}) = U_c(MX, \text{ s}) - [U(M^+, \text{ g}) + U(X^-, \text{ g})] \tag{4.9}$$

Since the reference state is zero in energy (and enthalpy) for the gaseous ions, it follows that at 0 K:

$$\Delta U_c(MX, \text{ s}) = \Delta H_c(MX, \text{ s}) = U_c(MX, \text{ s}) \tag{4.10}$$

Thus, although the energy is a process between two states, the crystal energy is normally given the symbol U_c.

It is important to note that the thermodynamic *reference state of an element* is defined as zero for the pure element in its normal state, for example, hydrogen $H_2(g)$, bromine $Br_2(l)$, carbon C(s, graphite) under the conditions of 1 bar and a temperature of 298.15 K. The often used terms 'room temperature' or 'ambient conditions' generally imply similar situations. The change

$$\text{Elements(reference state)} \xrightarrow{\Delta H_f^{\ominus}} \text{Pure substance} \tag{4.11}$$

under these conditions is represented by the enthalpy of formation $\Delta H_f^{\ominus}$ of the substance and applies to the process of Eq. (4.4). It is normally a negative quantity for thermodynamically stable compounds, and although it is not always possible to form a compound by direct combination, the process of Eq. (4.11) is realized through an appropriate cyclic procedure.

4.3.1 Born–Fajans–Haber cycle

The processes considered above may be represented diagrammatically by a thermodynamic cycle that was put forward by Born, Fajans and Haber. It is interesting to note that these three authors published this cyclic procedure, independently, in the same issue of the same journal in 1919 [1–3]. For some reason, the thermodynamic cycle is almost always referred to as the *Born–Haber cycle*, which is hard luck on Fajans. The cycle in Fig. 4.1 uses the symbols frequently encountered in this context, and have been noted above.

Although it is convenient to be able to equate enthalpy and energy at 0 K, indicated by the primed symbols in Fig. 4.1, when it comes to considering the crystal energy by an electrostatic process, the parameters, such as interionic distances, refer to an ambient temperature, 20°C or 25°C. Then, from equilibrium thermodynamics for the process ion-gas → crystal:

$$\Delta U_c = \Delta H_c - p\Delta V \tag{4.12}$$

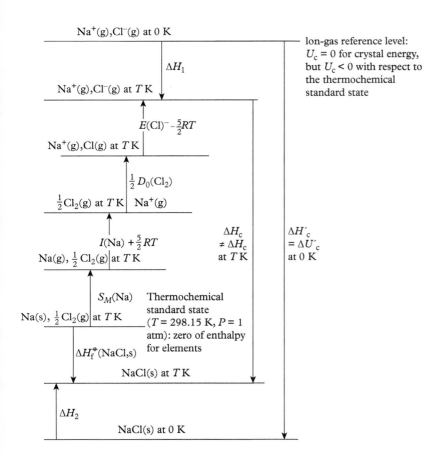

Fig. 4.1 A Born–Fajans–Haber thermodynamic cycle illustrated for the crystal energy of sodium chloride; distinguish carefully between the two reference levels in this cycle; the thermochemical reference state pressure is, strictly, 1 bar (0.9869 atm).

and since $V_{\text{ion-gas}}$ is very much greater than V_c under the given conditions, $\Delta V = -V_{\text{ion-gas}}$ and Eq. (4.12) may be reformulated as:

$$\Delta U_c = \Delta H_c + p V_{\text{ion-gas}} \qquad (4.13)$$

and because the ion-gas is ideal, by definition, $p V_{\text{ion-gas}}$ may be replaced by nRT, where n is the number of moles of gaseous species in a given reaction. Hence:

$$\Delta U_c = \Delta H_c + nRT \qquad (4.14)$$

In the cycle of Fig. 4.1,

$$\Delta H_1 = \int_0^T C_{p,\text{m}}(\text{g}) \mathrm{d}T$$

and

$$\Delta H_2 = \int_0^T C_{p,\text{m}}(\text{s}) \mathrm{d}T \qquad (4.15)$$

where $C_{p,\text{m}}$ represents the molar heat capacity at constant pressure. The term ΔH_1 for 2 mol of ideal gas in the cycle of Fig. 4.1 is equal to $(2 \times 5R/2 \text{ J K}^{-1} \text{ mol}^{-1} \times 0.298 \text{ K} = +12.4 \text{ kJ mol}^{-1})$, ΔH_2 evaluates from published molar heat capacity data for sodium chloride to 10.1 kJ mol^{-1}, and $\Delta H_f^{\ominus}$ (-412.5 kJ mol^{-1}, Table 2.1) hence:

$$\Delta H_c = \Delta H_f - S_M - I_M - \tfrac{1}{2} D_0 - E_X + \Delta H_1 - \Delta H_2 \qquad (4.16)$$

where the symbols are defined above. Inserting data for sodium chloride, ΔH_c (NaCl, s) $= -789.3 \text{ kJ mol}^{-1}$, which is the crystal enthalpy of sodium chloride at 298.15 K. From Eq. (4.14), the crystal energy at the same temperature is

$$U_c = \Delta H_c + 2RT = -789.3 \text{ kJ mol}^{-1} + 4.96 \text{ kJ mol}^{-1} = -784.3 \text{ kJ mol}^{-1}$$

On account of the measures of uncertainty in the thermodynamic parameters in Eq. (4.16), the precision in $U_c(\text{NaCl})$ is *ca.* $\pm 1 \text{ kJ mol}^{-1}$, a value not often obtainable with other compounds.

The exact nature of the binding energy is not explicit in the thermodynamic description of crystal energy. The bonding interactions are effectively locked together in the practical quantity ΔH_f and, of course, in U_c. It is important to consider the individual terms of the Born–Haber cycle, so as to understand the methods of their measurement and the probable precisions of the results. In the ensuing discussion, molar properties will be assumed unless otherwise stated.

4.3.2   Sublimation enthalpy

The enthalpy change appertaining to the process of Eq. (4.5) may be obtained by applying the Clausius–Clapeyron equation to measurements of the vapour pressure of the metal over a range of temperatures. Vapour pressures of metals are small, but they can be measured satisfactorily by an effusion technique.

Vapour is streamed through a hole, the diameter of which is small in comparison to the *mean free path* of the atoms in the vapour. The mean free path λ is the distance travelled by a molecule between collisions and equals $1/(\pi N d^2)$, where N is the number of molecules for a unit volume V and d is the diameter of the molecule, assumed to be spherical. An atom reaching a hole will pass through it, and the number passing through the hole in unit time is that number which would strike the same surface area as that of the hole in the same time. In other words, the rate of effusion is proportional to the vapour pressure. The pressure of the streaming vapour is measured, over the desired range of temperature, by a previously calibrated torsion-fibre apparatus.

The Clausius–Clapeyron equation for the sublimation change of state at a temperature T is:

$$\mathrm{d}p/\mathrm{d}T = \Delta H / \{T[V(\mathrm{g}) - V(\mathrm{s})]\} \qquad (4.17)$$

Sufficiently accurate assumptions are that $V(\mathrm{g})$ is very much greater than $V(\mathrm{s})$, and that the vapour behaves ideally. Then:

$$\mathrm{d}p/\mathrm{d}T = \Delta H / T[V(\mathrm{g})] = \Delta H p / (RT^2)$$

or

$$(1/p)\mathrm{d}p/\mathrm{d}T = \Delta H / (RT^2)$$

and may be written as:

$$\mathrm{d}(\ln p)/\mathrm{d}T = \Delta H / (RT^2)$$

and since $\mathrm{d}(1/T) = -(1/T^2)\mathrm{d}T$:

$$\mathrm{d}(\ln p)/\mathrm{d}(1/T) = -\Delta H / R \qquad (4.18)$$

Thus, a plot of $\ln p$ against the reciprocal of the independent variable T should be linear, with a slope $-\Delta H/R$. A best fit may be obtained by least squares, and the result applies at the average temperature of the relatively small experimental temperature range.

In the Born–Haber cycle, the value ΔH_1 at $(T_1)_{298\,\mathrm{K}}$ is required, and experiment has determined the value ΔH_2 at the average temperature T_2 of the vapour. The cyclic process in Fig. 4.2 illustrates the process to be carried out in order to obtain S_M at 298.15 K. Since $\sum\limits_{\mathrm{cycle}} \Delta H = 0$ from Hess's law

$$\Delta H_{\mathrm{sub}} = \Delta H_1 + \Delta H_2 - \Delta H_3 \qquad (4.19)$$

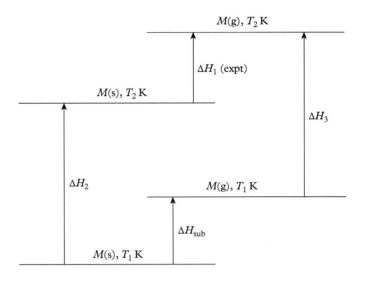

Fig. 4.2 Thermodynamic cycle for extrapolation of the experimental enthalpy of vaporization ΔH_1 from T_2 to T_1; for those metals that are liquid at the upper temperature T_2, the enthalpy of fusion must be added in to the cycle.

which may be recast as

$$\Delta H_{\text{sub}} = \Delta H_1(\text{exprt}) + \int_{T_1}^{T_2} C_p(\text{s})\mathrm{d}T - \int_{T_1}^{T_2} C_p(\text{g})\mathrm{d}T \qquad (4.20)$$

The first integral on the right-hand side of Eq. (4.20) can be evaluated by numerical integration of C_p v. T data (Appendix A13). The second integral is $5R(T_2 - T_1)/2$, since the vapour was assumed to behave like a monatomic, ideal gas. Standard enthalpies of sublimation for the alkali metals are listed in Table 4.1.

Table 4.1 *Standard enthalpies of sublimation of the alkali metals*

	$S_M/\text{kJ mol}^{-1}$
Li	161.5
Na	110.2
K	90.0
Rb	85.8
Cs	78.7

Example 4.1

The following vapour pressure data apply to metallic silver between 900 K and 1000 K; MP(Ag) = 1234 K. Determine the enthalpy of sublimation over the given temperature range.

T/K	900	920	940	960	980	1000
$10^7 p/\text{mbar}$	1.41	3.19	6.98	14.8	30.3	60.3
$10^3(1/T)/\text{K}^{-1}$	1.111	1.087	1.064	1.042	1.020	1.000
$-\ln(p/p_0)$	15.77	14.96	14.18	13.42	12.71	12.02

(p_0 is a reference pressure of 1 bar.) By least squares of $\ln p$ against $10^3(1/T)$, the slope is $-3.376 \times 10^4\text{K}$; hence, $\Delta H_{\text{sub}} = -(-3.376 \times 10^4 \text{ K} \times 8.3145 \text{ J K}^{-1} \text{ mol}^{-1} \times 10^{-3}) = 280.7 \text{ kJ mol}^{-1}$ at the average temperature of 950 K.

The procedure for extrapolation of a value of ΔH_{sub} to 298.15 K from a higher temperature will be considered through Problems 4.2 and 4.3.

4.3.3 Ionization energy

A typical determination of the ionization energy for a metal, Eq. (4.6), was discussed in detail for sodium in Section 2.13.1, and the value of 485.65 kJ mol^{-1} determined. A more accurate value is listed in Table 4.2 together with values for other alkali metals.

4.3.4 Dissociation enthalpy

The process of Eq. (4.7) represents the enthalpy of dissociation, often written as D_0, and may be obtained from an analysis of the electronic absorption spectra of molecular species, halogen molecules in the present context. When a gas is heated, electronic transitions occur in the molecule as radiation is absorbed. In diatomic molecules, a banded spectrum gives way to a continuum that marks the dissociation of the molecule into its component atoms:

$$\tfrac{1}{2}Cl_2(g) \rightarrow Cl(g) \tag{4.21}$$

A *convergence limit* is established, and the frequency at the low-energy end of the continuum represents the minimum energy $D_0(X_2)$ needed to dissociate the molecule; D_e (see Fig. 2.8(b)) is more negative than D_0 by the zero-point energy. For chlorine gas, the experimental convergence limit $\bar{\nu}$ is 21,189 cm^{-1}, which corresponds to an energy of 2.627 eV per molecule. However, the dissociated atoms will not both be in the ground state (Fig. 4.3). The excitation energy ε_X

Table 4.2 *Standard enthalpies of ionization of the alkali metals*

	I_M/kJ mol^{-1}
Li	520.22
Na	495.85
K	418.81
Rb	403.04
Cs	375.70

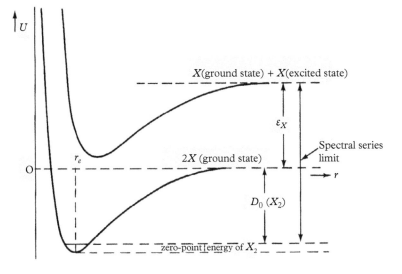

Fig. 4.3 Variation of potential energy U with interatomic distance r for a diatomic molecule X_2; r_e is the equilibrium interatomic distance in the ground state of the species X_2. For clarity, the value depicted for ε_X has been exaggerated relative to D_0. The zero-point energy of vibration at 0 K added to the practical quantity D_0 is the parameter D_e (see also Fig. 1.4).

is obtainable from observations on the spectrum of the gas, and for chlorine it is 0.109 eV. Hence, $D_0(Cl_2)$ is 2.518 eV, or 242.9 kJ mol^{-1}.

4.3.5 Electron affinity

Table 4.3 *Standard dissociation energies and electron affinities of the halogens*

	D_0/kJ mol^{-1}	E/kJ mol^{-1}
F	157.0	−328.0
Cl	242.9	−348.6
Br	223.8	−324.5
I	213.7	−295.5

The electron affinity E_X refers to the process of Eq. (4.8). Single electron affinities have been measured experimentally for the halogens and a few other species, such as OH.

Alkali-metal halides, such as RbI and CsCl, when heated by shock waves produce a vapour containing I$^-$ or Cl$^-$ species in abundance. Their ultraviolet absorption spectra are continua with sharp low-energy thresholds that correspond to the photochemical process:

$$Cl^-(g) \xrightarrow{h\nu} Cl(g) + e^- \tag{4.22}$$

The threshold wavelengths for the halogens have been ascribed precisely, and their electron affinities are listed in Table 4.3, together with the data on the dissociation enthalpies discussed in the previous section. Electrons affinities of many other species may be deduced from crystal energies in a manner to be described shortly.

4.3.6 Enthalpy of formation

The standard enthalpy of formation of a halogen-containing compound MX is the enthalpy change for the process:

$$M(s) + \tfrac{1}{2}X_2(g) \rightarrow MX(s) \tag{4.23}$$

in the reference state of pure substances, 298.15 K and 1 bar. Thus:

$$\Delta H = H(MX, \text{ s}) - [H(M, \text{ s}) + H(\tfrac{1}{2}X_2, \text{g})] \tag{4.24}$$

Since the enthalpy (and energy) of elements in the reference state are defined as zero, this enthalpy change represents the standard enthalpy of formation, $\Delta H_f^\ominus(MX, \text{ s})$. This quantity may be measured by direct combination of the elements in the case of lithium and iodine, but for cesium and chlorine it would be wise to use a more indirect approach. The following sequence of reactions may be considered, and the reader is invited to construct a thermodynamic cycle to show that $\Delta H_f^\ominus$ (CsCl, s) is −433.0 kJ mol^{-1}:

	$\Delta H^\ominus$/kJ mol^{-1}
$H_2(g) + \tfrac{1}{2} O_2(g) \rightarrow H_2O(l)$	−285.9
$Cs(s) + H_2O(l) \rightarrow Cs^+(aq) + OH^-(aq) + \tfrac{1}{2} H_2(g)$	−191.8
$\tfrac{1}{2} H_2(g) + \tfrac{1}{2} Cl_2(g) \rightarrow HCl(g)$	−92.3

	$\Delta H^{\ominus}/\text{kJ mol}^{-1}$
$HCl(g) + aq \rightarrow H_3O^+(aq) + Cl^-(aq)$	-75.1
$Cs^+(aq) + OH^-(aq) + H_3O^+(aq) + Cl^-(aq) \rightarrow Cs^+(aq)$ $+ Cl^-(aq) + H_2O(l)$	-55.8
$CsCl(s) + aq \rightarrow Cs^+(aq) + Cl^-(aq)$	$+18.0$
$Cs(s) + \tfrac{1}{2} Cl_2(g) \rightarrow CsCl(s)$	$\Delta H_f^{\ominus}(CsCl,\ s)$

The complete data on the enthalpies of formation for the alkali-metal halides are listed in Table 4.4.

4.3.7   Precision of crystal energies

The terms discussed in Sections 4.3.2–4.3.5 are known with good precision; enthalpies of formation vary in precision from less than ± 1 kJ mol^{-1} (NaCl) to ± 15 kJ mol^{-1} (CsF), and the precision in crystal energy varies from ± 2 kJ mol^{-1} (*ca.* 0.2%) to ± 20 kJ mol^{-1}. Among the various quantities in Born–Haber cycle calculations for crystal energies, enthalpies of formation have the highest probable errors.

4.4   Electrostatic model for ionic crystals

A crystal energy derived from thermodynamic parameters has a well-defined value, with a precision dependent upon those of its several components. However, it does not reveal, or need, any knowledge of the nature of the ionic bond. It is necessary, therefore, to construct a model based on the interaction of charged species, the simplest of such being a point charge electrostatic model. The structures of the alkali-metal halides were known from x-ray crystallographic analysis

Table 4.4 *Standard enthalpies of formation of alkali-metal halides,* $-\Delta H_f^{\ominus}(MX, s)$ */kJ mol^{-1}*

	Li	Na	K	Rb	Cs
F	612.1	571.1	562.7	549.4	530.9
Cl	405.4	412.5	436.0	430.5	433.0
Br	348.9	361.7	392.0	389.1	394.6
I	271.1	290.0	327.6	328.4	336.8

as early as 1913, and Appendix A14 provides a short summary of some of the very early x-ray structure analyses.

Consider two ions of point charges $z_1 e$ and $z_2 e$ of opposite sign distant r apart. Their electrostatic, or coulombic, energy u_E is:

$$u_E = -z_1 z_2 e^2 / (4\pi \varepsilon_0 r) \tag{4.25}$$

where e is the magnitude of the electronic charge; the negative sign implies an attractive energy.

In a crystal of the sodium chloride structure type (Fig. 1.5), for example, it is necessary to consider the effect on the energy of a pair of ions Eq. (4.25) of all other ions in the structure. The crystal energy for this structure type may be written as

$$U_E = -A z_1 z_2 e^2 / (4\pi \varepsilon_0 r) \tag{4.26}$$

where A is the Madelung constant for the given structure type [4].

4.4.1 Madelung constant

The meaning of *Madelung constant* may be illustrated first in one dimension, by an infinite row of regularly spaced, alternating positive and negative charges, each of magnitude e:

$$-\infty \ldots + \; - \; + \; - \; + \; - \ldots \infty$$
$$\overset{r}{\longleftrightarrow}$$

Any charge may be taken as an origin: its immediate neighbours produce an attractive potential energy of $-2e^2 / (4\pi \varepsilon_0 r)$; its next nearest neighbours give rise to a repulsion of $+2e^2 / [2(4\pi \varepsilon_0 r)]$, and the next nearest again, an attraction of $-2e^2 / [3(4\pi \varepsilon_0 r)]$, and so on. Thus, a sequence arises:

$$-\frac{e^2}{4\pi \varepsilon_0 r} 2 \left(1 - \frac{1}{2} + \frac{1}{3} - \frac{1}{4} + \frac{1}{5} - \frac{1}{6} \cdots \right) \tag{4.27}$$

which sums to $-\frac{e^2}{4\pi \varepsilon_0 r} 2 \, (\ln 2)$. Hence, the effect of the infinite row of charges is to modify the interactive energy of a pair of charges given in Eq. (4.25) by the factor $2 \ln 2$, or *ca.* 1.3863; this number is the Madelung constant for the one-dimensional structure under consideration.

In three dimensions the problem is a little more difficult. Consider the Cl$^-$ ion at the centre of the unit cell of sodium chloride (Fig. 1.5) as an origin. The six nearest neighbours, distant r from this origin, produce an attraction of $-6e^2 / (4\pi \varepsilon_0 r)$; the next nearest neighbours lies at the centres of the unit cell edges, at a distance $r\sqrt{2}$, and produce a repulsion of $+12e^2 / (4\pi \varepsilon_0 r\sqrt{2})$; next again, an attraction of $-8e^2 / (4\pi \varepsilon_0 r\sqrt{3})$, and so on. Writing these terms as a series gives:

$$U_E = -\frac{e^2}{4\pi \varepsilon_0 r} \left(\frac{6}{\sqrt{1}} - \frac{12}{\sqrt{2}} + \frac{8}{\sqrt{3}} - \frac{6}{\sqrt{4}} + \frac{24}{\sqrt{5}} - \frac{24}{\sqrt{6}} \cdots \right) \tag{4.28}$$

which is very slowly converging. If this series were continued, no term divided by $\sqrt{7}$ would occur because there is no lattice point of integers P, Q, R such that $P^2 + Q^2 + R^2 = 7$; in general, the expression $N = m^2(8n - 1)$ where m and n are integers indicates 'forbidden' numbers for N, the sum $P^2 + Q^2 + R^2$, in cubic crystals.

It was shown by Evjen [5] that the rate of convergence could be improved by working with nearly neutral blocks, since potential energy falls off more rapidly with distance in a neutral group than with one dependent upon $1/r$ ion-ion interaction.

In the unit cell of the sodium chloride structure type, the sharing principle in a unit cell of an infinite lattice results in the charge contributions of $\frac{1}{8}$ for corners, $\frac{1}{4}$ for edges and $\frac{1}{2}$ for faces. Thus, the sum of the first three terms of the series in Eq. (4.28) is modified to $\left(\frac{6}{2} - \frac{12}{4\sqrt{2}} + \frac{8}{8\sqrt{3}}\right)$, or 1.456. By taking a cube of side $2a$ the result is 1.75; a more accurate value of the Madelung constant for this structure type is 1.74756. It is interesting to note in relation to Evjen's method that one of Pauling's empirical rules for crystal structures, namely that 'the charge on an ion tends to be neutralized by its nearest neighbours' has here a clear expression.

The Madelung constant can be computed on another model. Imagine an ion of charge e existing in free space. The effect of incorporating it into a crystal structure can be simulated by surrounding it by an earthed, spherical, conducting shell, since in both situations the electric field of the ion is neutralized by its surroundings.

The electrostatic potential (Appendix A11) arising from a charge e at a distance r is $e/(4\pi\varepsilon_0 r)$. The work done in decreasing the charge by an amount δe is $e\delta e/(4\pi\varepsilon_0 r)$. In neutralizing the whole charge, the work done per ion W is

$$W = 1/(4\pi\varepsilon_0) \int_0^e (e/r)\,\mathrm{d}e = 0.5e^2/(4\pi\varepsilon_0 r) \tag{4.29}$$

The volume of the spherical shell of radius r is $4\pi r^3/3$, so that $r = (3V/4\pi)^{1/3}$; hence, Eq. (4.29) becomes

$$W = 0.5e^2/\left[(4\pi\varepsilon_0)(3V/4\pi)^{1/3}\right] = 0.8060e^2/\left(4\pi\varepsilon_0 V^{1/3}\right) \tag{4.30}$$

In the cesium chloride structure type (Fig. 1.11), for example, the volume is a^3, where $a\sqrt{3} = 2r_e$, r_e being the equilibrium Cs^+–Cl^- distance. The unit cell volume is $8r_e^3/(3\sqrt{3})$, or $4r_e^3/(3\sqrt{3})$ per ion, each ion having an identical environment. Identifying r_e with r, the radius of the spherical shell, $V^{1/3}$ becomes $4^{1/3}r/\sqrt{3}$ and W is now $0.8794e^2/(4\pi\varepsilon_0 r)$. The work per ion pair is thus $2W$, or $1.759e^2/(4\pi\varepsilon_0 r)$. Thus, the Madelung constant for the cesium chloride structure type is 1.759; the accepted value is 1.76267.

Madelung constants can be calculated (Appendix A15) with the program MADC in the Web Program Suite.

4.4.1.1 Series convergence

It is interesting to consider briefly the convergence of series at this stage. The series of terms $\sum_0^\infty r_n$ is *absolutely convergent* if $\sum_0^\infty |r_n| = \phi$, where ϕ is a finite number. However, the same series is *conditionally convergent* if $\lim_{n\to\infty} \sum_0^n r_n = \phi'$, where ϕ' is also a finite number, but additionally where $\sum_0^\infty |r_n| = \infty$. As an example, the series in parentheses in Eq. (4.27), which may be expressed as $\sum_1^\infty \frac{(-1)^{n+1}}{n}$, is conditionally convergent to $\ln(2)$, but it matters how the order of the terms of the series are taken. The same series could be written, for example, from the general form $\left[\left(\frac{1}{2n-1} - \frac{1}{4n-2}\right) - \frac{1}{4n}\right]$; thus, the terms for $n = 1, 2, 3$, etc. are

$$\left(1 - \frac{1}{2}\right) - \frac{1}{4} + \left(\frac{1}{3} - \frac{1}{6}\right) - \frac{1}{8} + \left(\frac{1}{5} - \frac{1}{10}\right) - \frac{1}{12} + \ldots = \frac{1}{2} - \frac{1}{4} + \frac{1}{6} - \frac{1}{8} + \frac{1}{10} - \frac{1}{12} + \ldots$$

which from Eq. (4.27) is clearly $\frac{1}{2}\ln(2)$. An example of an absolutely convergent series is $\sum_{n=1}^\infty \frac{(-1)^{n+1} 2^n}{(2n+1)!}$.

The lattice sums that lead to Madelung constants are conditionally convergent, and are summed over a regularly increasing radial distance from an origin point in the crystal lattice; this procedure is implicit in Eq. (4.27), and in Evjen's method described above.

In connecting the Madelung constant calculation with a physical system, a macroscopic boundary condition for the sum must be set. Imagine the finite crystal wrapped in a material of given relative permittivity ε_r. Summing the terms in real space corresponds to a vacuum boundary condition $\varepsilon_r = 1$, whereas summing in Fourier space corresponds to a 'metal sheet' boundary, $\varepsilon_r = \infty$. The procedure described in Appendix A15, with which the data of Table 4.5 was obtained, is a modification of the classical Ewald Fourier sum. It takes the boundary as $\varepsilon_r = \infty$, and implies forming the sum in increasing values of a reciprocal space coordinate $d^* = 1/d$, where d is an interplanar spacing in real space (Appendix A20). Thus, the appropriate value for the parenthetical expression in Eq. (4.27) is correctly given as $\ln(2)$.

Table 4.5 *Madelung constants A and formal charge products for selected structure types*

	A	z_1z_2	Az_1z_2
CsCl	1.76267	1	1.76267
NaCl	1.74756	1	1.74756
α-ZnS	1.64132	4	6.56528
β-ZnS	1.63805	4	6.55220
CaF$_2$	2.51939	2	5.03878
TiO$_2^a$	2.38510	8	19.0808
β-SiO$_2$	2.20110	8	17.6088

[a] There is no unique r_e distance in the rutile structure type; A is calculated in terms of the shortest distance in the structure, 0.1945 nm.

4.4.2 Crystal energy calculation

The early calculation by Born and Landé in 1918 formulated the crystal energy by the equation [6]:

$$U(r) = -\frac{Az_1z_2e^2}{r}(1 - 1/n) \tag{4.31}$$

where n is an integer chosen usually as 9, the term $-1/n$ accounting for the rapid rise in U for values of r less than the equilibrium value r_e. Subsequently, the *repulsion term* $1/n$ was replaced by $B\exp(-r/\rho)$, where B is a constant for the structure type and ρ is a parameter for a particular compound and determined from compressibility data. The exponential term matches more closely the comparable term in quantum mechanics. Introducing current notation, the crystal energy is written as:

$$U(r) = -\frac{Az_1 z_2 e^2}{4\pi \varepsilon_0 r} + B\exp(-r/\rho) \tag{4.32}$$

For investigation, Eq. (4.32) may be written conveniently as:

$$U(r) = \frac{A'}{r} + B\exp(ar) \tag{4.33}$$

where $A' = -Az_1 z_2 e^2/(4\pi\varepsilon_0)$, $a = -1/\rho$ and B remains a constant for the structure type. The crystal energy is a minimum at $r = r_e$, *vide* Fig. 1.4; hence:

$$(\mathrm{d}U/\mathrm{d}r)_{(r=r_e)} = -A'/r^2 + aB\exp(ar) = 0 \tag{4.34}$$

so that

$$B = \frac{A'}{r_e^2 a \exp(ar_e)} \tag{4.35}$$

Substituting for B in Eq. (4.32), inserting the values of A' and a, and introducing L so that the result is in J mol^{-1}:

$$U(r_e) = -\frac{LAz_1 z_2 e^2}{4\pi\varepsilon_0 r_e}(1 - \rho/r_e) \tag{4.36}$$

The parameter ρ in the repulsion term is related, not surprisingly, to the *isothermal compressibility* of the crystal. The equation of state for an isotropic solid at 0 K (Appendix A16), is given by:

$$p = -(\partial U/\partial V)_T \tag{4.37}$$

from which it follows that

$$(\partial p/\partial V)_T = -(\partial^2 U/\partial V^2)_T$$

At any temperature T greater than 0 K, $p = -(\partial U/\partial V)_T + T\beta/\alpha$ (see Appendix A16). The isothermal compressibility κ is defined as

$$\kappa = -(1/V)(\partial V/\partial p)_T$$

and the expansivity β by:

$$\beta = (1/V)(\partial V/\partial T)_p$$

It follows that at 0 K:

$$1/(\kappa V) = d^2 U/dV^2 \tag{4.38}$$

and V is here the volume occupied in the crystal by a pair of oppositely charged ions. For an isometric (cubic) structure, this volume is proportional to r^3, or $V = cr^3$, where c is a constant. Hence:

$$dV/dr = 3cr^2 = 3V/r$$

and

$$d^2 V/dr^2 = 6cr = 6V/r^2$$

Now

$$dU/dV = (dU/dr)/(dV/dr)$$

and

$$\begin{aligned} d^2 U/dV^2 &= \frac{1}{(dV/dr)} \frac{d}{dr}[(dU/dr)/(dV/dr)] \\ &= \frac{(dV/dr)d^2 U/dr^2 - (dU/dr)(d^2 V/dr^2)}{(dV/dr)^3} \end{aligned} \tag{4.39}$$

From Eq. (4.34):

$$d^2 U/dr^2 = 2A'/r^3 + a^2 B \exp(ar)$$

and substituting appropriately in Eq. (4.39) gives

$$1/(\kappa V) = \frac{(3V/r)[2A'/r^3 + a^2 B \exp(ar)] - (6V/r^2)[-A'/r^2 + aB \exp(ar)]}{(3V/r)^3}$$

Eliminating B with Eq. (4.35) at the equilibrium distance:

$$9V/\kappa - 2A'/r_e = aA = aAr_e/r_e$$

Introducing the values for A' and a, and extracting ρ/r_e:

$$\rho/r_e = [Az_1 z_2 e^2/(4\pi\varepsilon_0 r_e)]/[9V/\kappa + 2Az_1 z_2 e^2/(4\pi\varepsilon_0 r_e)] \tag{4.40}$$

This result for ρ/r_e is used in Eq. (4.36) to express the molar crystal energy. Taking sodium chloride as an example, with $r_e = 0.5640$ nm, $\kappa = 4.1 \times 10^{-11}$ N^{-1} m^2 and $V = 2r_e^3$, ρ/r_e evaluates as 0.1125 and $U(r_e) = 764.1$ kJ mol^{-1}. Note that the units of V/κ are the same as those of $e^2/(\varepsilon_0 r_e)$. The result is approximately 20 kJ mol^{-1} (*ca.* 2%) more positive than the corresponding quantity from the thermochemical cycle. The main reason for the difference is that Eq. (4.40) is based on point-charge atoms, whereas they are regions of electron density, and a more refined model must take into account the mutual polarization of the electron density in the crystal.

4.4.3 Polarization in ionic compounds

Polarization was discussed in detail in Chapter 3, and a brief mention of ionic compounds was made in Section 3.7.1. Polarization may be regarded as a distortion of the electron density of a species by the electric field of its neighbours. This distortion creates a system of dipoles in each ion which, although they are symmetrical and therefore do not lead to a resultant dipole moment, the transients (Section 3.4.3), do give rise to a *dipole–dipole* attractive potential energy. This energy is proportional to $\alpha(M^+)\alpha(X^-)/r^6$, where α is the polarizability of a species, and r is the equilibrium M^+X^- interionic distance r_e. A *dipole–quadrupole* interaction, proportional to $1/r^8$, adds a small amount to the energy, approximately 1% of that derived from the dipole–dipole $1/r^6$ term. A quadrupole–quadrupole contribution is less still and quite negligible in relation to the precision of the other terms of the crystal energy equation.

The degree of polarization of an ionic species depends upon both the ease of distortion of its electron density, measured through its polarizability α, and the effectiveness (*polarizing power p*) of a neighbouring ion to induce polarization in the species and measured by its electric field: $p = ze^2/(4\pi\varepsilon_0 r)$. Table 4.6 list these two parameters for selected cations and anions, from which the dependence of total polarization of ionic size and charge can be judged.

Table 4.6 *Polarizability α and polarizing power p of common cations and anions*

	$10^{40}\alpha$/F m^2	$10^{18}p$/C^2 m^{-1}
Li$^+$	0.03	3.4
Na$^+$	0.30	2.3
Cs$^+$	3.4	1.3
Be^{2+}	0.01	13.2
Mg^{2+}	0.1	5.7
Ba^{2+}	2.8	3.4
F$^-$	1.0	1.7
Cl$^-$	3.4	1.2
I$^-$	7.3	1.1
O^{2-}	2.7	3.5
S^{2-}	6.1	2.5
Te^{2-}	10.0	2.2

4.4.4 Crystal energy with polarization

The polarization correction to the crystal energy produces a significant improvement over that given by Eq. (4.36). If terms are added to take account of dipole–dipole, dipole–quadrupole energies, the equation of state contribution and zero-point energy, the crystal energy can be written as

$$U(r) = -\frac{Az_1z_2e^2}{4\pi\varepsilon_0 r} + B\exp(-r/\rho) - C/r^6 - D/r^8 + \phi(T\beta/\kappa) + U_z \quad (4.41)$$

where C and D are lattice sums in the dipole–dipole and dipole–quadrupole interactions and ϕ is the equation of state function (Appendix A16) at a temperature $T > 0$. The zero-point energy of vibration given in J mol^{-1} by $9Lh\nu_{max}/4$, where ν_{max} is the frequency of vibration in the highest occupied energy state, which can be obtain from infrared *Reststrahlen* (residual rays) of the crystal. Following the analysis of Section 4.4.2 leads to the following equations for calculating the crystal energy:

$$\rho/r_e = \frac{[Az_1z_2e^2/(4\pi\varepsilon_0 r_e)] + 6C/r_e^6 + 8D/r_e^8 - 3VT\beta/\kappa}{[9V\Phi(T,p)/\kappa] + [2Az_1z_2e^2/(4\pi\varepsilon_0 r_e)] + 42C/r_e^6 + 72D/r_e^8}$$

$$\Phi(T,p) = 1 + [T(\partial\kappa/\partial T)_p/\kappa] + \beta(\partial\kappa/\partial p)_T + 2\beta/3$$

$$U(r_e) = -L\left\{\frac{Az_1z_2e^2(1-\rho/r_e)}{(4\pi\varepsilon_0 r_e)} + 6C(1-6\rho/r_e)/r_e^6 + 8D(1-8\rho/r_e)/r_e^8\right.$$

$$\left. + 3VT\beta(\rho/r_e)/\kappa\right\} + U_z \quad (4.42)$$

Table 4.7 *Crystal energies/kJ mol⁻¹ of the alkali-metal halides: $U(r_e)$, electro-static; U_c, thermodynamic; δ, percentage discrepancy*

	$-U(r_e)$	$-U_c$	$\lvert\delta\rvert/\%$
LiF	1031	1032	0.1
LiCl	846.0	850.2	0.5
LiBr	796.2	812.5	2
LiI	737.2	754.4	2
NaF	918.8	918.4	0.04
NaCl	781.0	781.5	0.06
NaBr	741.4	746.8	0.7
NaI	691.2	698.7	1.0
KF	813.0	813.0	0.0
KCl	712.5	712.1	0.06
KBr	684.9	682.8	0.3
KI	641.0	642.7	0.3
RbF	793.3	777.4	0.5
RbCl	692.9	684.5	1
RbBr	662.3	657.7	1
RbI	627.2	621.3	1
CsF	757.3	740.6	2
CsCl	669.4	668.6	0.1
CsBr	641.8	644.8	0.5
CsI	607.9	611.3	0.6

where U_z is the zero-point energy. These equations and the terms in C/r^6 and D/r^8 have been discussed fully in the literature [7–9]; neither they nor the zero-point energy are negligible where a precise value of crystal energy is sought. In the case of sodium chloride, for example, polarization contributes *ca.* −15 kJ mol⁻¹ to the crystal energy. A set of results for the lattice energies of the alkali-metal halides is given in Table 4.7. The reader may wish to write a program for $U(r_e)$, Eq. (4.42), and apply it to the data for sodium chloride, using data from the literature sources quoted.

4.5 Applications of crystal energies

The crystal energy is a measure of the stability of a compound with respect to its component ions in an ideal-gas phase. It can be used to investigate an interesting range of properties related to the energetics of formation of a crystal or its components, some of which will be considered here.

4.5.1 Deducing electron affinities and related quantities

When electrostatic calculations of crystal energies were carried out first, there were no experimental values for either dissociation energies or electron affinities. Among *MX*-type compounds, the other terms in Eq. (4.16), concerning metals and the ideal gaseous ions were known and were used to calculate the combined term $[\frac{1}{2}D_0(X_2) - E_X]$, which can be described also as $\Delta H_f^{\ominus}(X^-, g)$. The constancy found for this term for each given halogen was regarded as good evidence for the applicability of the electrostatic model.

As measurements of dissociation energies became practicable, values for the electron affinities of the halogens were determined from the average values of $[\frac{1}{2}D_0(X_2) - E_X]$. The values for the electron affinities given in Table 4.3 are experimentally determined. Single electron affinities for some species, including O⁻ and OH⁻, have also been measured, but data for species such as O^{2-}, S^{2-} and (CN)⁻, for example, are best obtained by comparing thermodynamic and electrostatic models for crystal energies of their compounds.

The electron affinity of oxygen for two electrons has been determined from detailed calculations on magnesium, calcium and barium oxides, all of which are NaCl-type structures. The result, $E(O^{2-}) = +748.9$ kJ mol⁻¹, has a precision of ±4% [10]. The uncertainty in the value of $E(O^{II})$ lies in the low accuracy of the compressibility data on the oxides, which has a direct bearing on the value of ρ/r_e.

With species such as $(NH_4)^+$, $(NO_3)^-$ and $(SO_4)^{2-}$ the parameters under the definitions of the enthalpies of sublimation, ionization, electron affinity and dissociation in the Born–Haber cycle Eq. (4.16) have no simple meaning. However, composite quantities such as $\Delta H_f(NO_3^-, g)$ and $\Delta H_f(NH_4^+, g)$, for example, take on meaning through appropriate thermodynamic cycles; $\Delta H_f(NO_3^-, g)$ may be defined by the theoretical process:

$$\tfrac{1}{2}N_2(g) + (3/2)O_2(g) +, e^- \xrightarrow{\Delta H_f[(NO_3)^-, g]} [(NO_3^-), g] \qquad (4.43)$$
(Elements in standard state)

and evaluated by a combination of the Born–Haber cycle and the electrostatic equation (4.35). Taking ρ/r_e as *ca.* 0.11 for 1:1 compounds (and *ca* 0.13 for more highly charged species) [11–15], $\Delta H_f(NO_3^-, g)$ evaluates as –82 kJ mol^{-1}.

4.5.2 Compound stability

Thermodynamic arguments can be used to discuss the probable existence or otherwise of chemical combinations and a few examples follow.

4.5.2.1 Energetics of inert gas compounds

Having discussed the energetics of the compound NaCl in detail, how is it that a compound of similar formula, NeCl, is not known? If it were to exist as an ionic compound, it would, as a 1:1 type, be expected to exhibit one of the three main types to be discussed in Section 4.6.2, probably the NaCl structure type. Then, from Eq. (4.16):

$$\Delta H_f = U_c(NeCl) + S(Ne) + I(Ne) + \tfrac{1}{2}D_0(Cl_2) + E(Cl) - nRT$$

Inserting the known data:

$$\Delta H_f/kJ\,mol^{-1} = U_c(NeCl) + 0 + 2080.7\,kJ\,mol^{-1} + 121.5\,kJ\,mol^{-1}$$
$$- 348.6\,kJ\,mol^{-1} - 5.0\,kJ\,mol^{-1}$$
$$= U_c(NeCl) + 1848.6\,kJ\,mol^{-1}$$

For NeCl to exhibit a thermodynamic stability, $\Delta H_f(NeCl)$ should be negative, so that $U_c(NeCl)$ must be less than –1848.6 kJ mol^{-1}. Using Eq. (4.36) for a sodium chloride structure type and taking ρ/r_e as 0.1, it follows that $(-218.52$ kJ mol$^{-1}/r_e$ nm$)^1$ must be less than –1848.6 kJ mol^{-1}, which implies an r_e of less than 0.1 nm. Since $r(Cl^-) \approx 0.17$ nm such a value for r_e must be significantly greater than this value. It follows that NeCl cannot form a stable, ionic compound; too much energy is needed to form Ne$^+$(g) than would be recoverable from the condensation of that ion-gas with Cl$^-$ ion-gas to form a solid. However, this result does not apply with all inert gases.

The ionization energy of xenon lies within 6 kJ mol^{-1} of that of molecular oxygen:

$$Xe(g) \rightarrow Xe^+(g) + e^- \quad I = 1170.4\,kJ\,mol^{-1}$$
$$O_2(g) \rightarrow O_2^+(g) + e^- \quad I = 1164.4\,kJ\,mol^{-1}$$

---

[1] -218.52KJ nm mol$^{-1} \equiv -LAe^2 \times 0.9/(4\pi\varepsilon_0 \times 10^{-9})$ *vide* Eq.(4.35)

The compound $O_2[PtF_6]$ was known from the gas phase reaction:

$$O_2(g) + PtF_6(g) \rightarrow O_2[PtF_6](s)$$

and was found from x-ray analysis to consist of O_2^+ and $[PtF_6]^-$ ions, with an r_e equal to 0.418 nm and a crystal energy of –523 kJ mol^{-1}. The experiment was repeated with xenon in place of oxygen: the gas was oxidized by PtF_6, forming the analogous, crystalline ionic compound $Xe[PtF_6]$ of $r_e(Xe - [PtF_6])$ distance 0.475 nm [16].

Example 4.2

If the compound $Xe[PtF_6]$ crystallizes with the sodium chloride structure type and $\rho/r_e \approx 0.1$ with the interionic spacing r_e of 0.475 nm, then the crystal energy is given by[1] -218.52 kJ nm mol^{-1}/0.475 nm $= -460$ kJ mol^{-1}.

The two crystal energies were calculated originally from the *Kapustinskii equation* which has a fixed value for the 'Madelung' constant independent of structure type. It is based on the Madelung constant for the sodium chloride structure type, modified by the number of ions in the formula. It assumes a fixed value of the repulsion parameter ρ for all structures and makes the unwarranted assumption that $r_e = r_+ + r_-$. It gives fairly satisfactory results with simple ionic structures by virtue of its construction. However, it is preferable always, where possible, to evaluate a true Madelung constant in terms of the parameters of the crystal structure of the species under consideration.

4.5.2.2   *Energetics of copper(I) fluoride and silver cyanide*

The existence of copper(I) fluoride has been challenged on the grounds that it is thermodynamically unstable with respect to its elements in the reference state; efforts to prepare it are reported to result in disproportionation to copper(II) fluoride and metallic copper:

$$2CuF(s) \rightarrow F_2(g) + 2Cu(s)$$

The criterion used in this example is not always all-important in preparative chemistry. For example, the enthalpy and free energy of formation of silver cyanide are positive ($\Delta G_f^\ominus = +164$ kJ mol^{-1}), but it can be prepared by mixing aqueous solutions silver nitrate and potassium cyanide. Once prepared and excluded from light, silver cyanide shows no tendency for spontaneous decomposition. It is thermodynamically unstable with respect to its elements in their reference state, but not with respect to the hydrated ions (Fig. 4.4). Hence, where doubt could arise, the term 'stability' is best used together with a statement of the reference level to which the term applies.

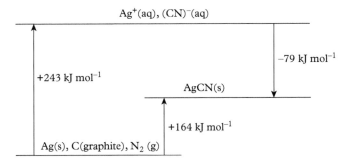

Fig. 4.4 Free-energy level diagram for silver cyanide. The solid is unstable with respect to its elements in the reference state but not with respect to the hydrated Ag^+ and $(CN)^-$ ions; hence, AgCN can be prepared by precipitation from a solution containing Ag^+(aq) and $(CN)^-$(aq) ions. [In terms of a Hess cycle, the reference state would specify $\frac{1}{2} N_2(g)$.]

4.5.2.3 *Unfamiliar chlorides of sodium*

Under certain conditions, some apparently inviolable rules of chemistry can break down, leading to the formation of unexpected compounds. Calculations have shown that under a pressure of *ca.* 250 GPa, Na_3Cl, Na_2Cl, Na_3Cl_2, $NaCl_3$ and $NaCl_7$ are theoretically stable compounds [14]. Over this pressure range NaCl itself is stable, so that other chlorides would be expected to disproportionate into NaCl and either sodium or chlorine in the absence of modifying factors. However, finely divided NaCl when heated in excess chlorine under 10–80 GPa produces the compound $NaCl_3$ which has been identified by x-ray analysis and Raman spectroscopy. In a similar experiment but in the presence of excess sodium, Na_3Cl was produced and remained stable down to 20 GPa. There would seem to be no reason why other unfamiliar alkali-metal halides are not waiting to be discovered.

4.5.3 Charges on polyatomic ions

By appropriate manipulations of Eqs. (4.16) and (4.36) it is possible to determine the distribution of charge on a polyatomic ion in certain compounds. This information is desirable in the calculation of the crystal energy of a compound, by either the thermodynamic or electrostatic model, because a polyatomic ion possesses an electrostatic self-energy of significant magnitude that must be taken into account in a calculation of crystal energy. In the following example, the determination of the distribution of charge on the cyanide ion by detailed calculations of the crystal energies of alkali-metal cyanides is considered [15].

At 298.15 K, potassium and sodium cyanides have the sodium chloride structure type (Fig 1.5). At 279 K, sodium cyanide and at 233 K potassium cyanide transform to the orthorhombic form (Fig. 1.6).

At a transition temperature, two polymorphs of a given compound have the same free energy, and at constant pressure the free-energy change at transition is given by:

$$\Delta G = \Delta U - T\Delta S + p\Delta V \qquad (4.44)$$

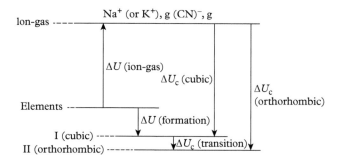

Fig. 4.5 Energy levels for two polymorphs of sodium and potassium cyanides. Except for ΔU(transition), the other terms labelled ΔU are actually U values because they refer to a process from reference levels of zero.

where ΔS is the change in entropy. At equilibrium $\Delta G = 0$, and since the volume change from one solid polymorph to the other is negligibly small:

$$\Delta U = T\Delta S \qquad (4.45)$$

The interrelationships of the energies of the two polymorphs of NaCN and KCN are shown in Fig. 4.5; the crystal energy change of each polymorph is determined by its entropy change according to Eq. (4.45),[2] and the ΔU_c values are 2.9 kJ mol^{-1} and 1.3 kJ mol^{-1} for NaCN and KCN, respectively.

The existence of the two cyanides in the sodium chloride structure type implies a disorder at the sites of the cyanide ions in the unit cell that simulates statistically a spherically symmetrical envelope for them. As such, their crystal energies can be calculated from Eq. (4.42) at their transitions temperatures, correcting the temperature dependent parameters r_e, β and κ to the values at the transition temperatures.

The single negative charge on the cyanide ion (CN)$^-$ is divided as C^{z_C} and N^{z_N}, where $z_C + z_N = -1$. The components of the electrostatic energy for a cyanide ACN are shown in Fig. 4.6: U_M is a crystal energy based on the ideal ion-gas A^+, C^{z_C}, N^{z_N}; U_C is the crystal energy based on the ion-gas A^+, (CN)$^-$, and the difference U_S is the electrostatic self-energy of the cyanide ion in the ideal gas phase, which refers to the process:

$$\text{C}^{z_C}(g) + \text{N}^{z_N}(g) \ \rightarrow \ (\text{CN})^-(g)$$

with the C–N distance taken as that in the crystal. The term U_S is defined by:

$$U_S = \sum_{\substack{i,j \\ (i \neq j)}} z_i z_j e^2 / (4\pi\varepsilon_0 r_{i,j}) \qquad (4.46)$$

where $r_{i,j}$ is the distance between the ith and jth species of numerical charges z_i and z_j, respectively. In the case of sodium and potassium cyanides, $r_{i,j}$ is r_{CN^-}. The term U_G refers to the process thermodynamic ion-gas $\rightarrow$ elements in the reference state but does not take part explicitly in these calculations.

[2] At the transition temperature: $\Delta S_t = \Delta H_t T_t$ for the Na and K cyanides.

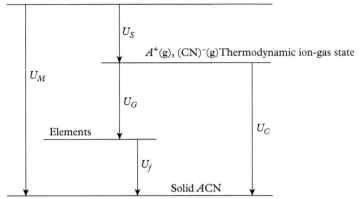

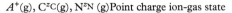

Fig. 4.6 Energy levels for sodium and potassium cyanides (not to scale), including the electrostatic self-energy of the $(CN)^-$ ion:

U_M: elemental ion-gas $\rightarrow$ crystal
U_S: C, N ion-gas $\quad\rightarrow (CN)^-$ ion-gas
U_G: A^+, $(CN)^-$ ion-gas $\rightarrow$ elements
U_f: elements $\quad\quad\quad\rightarrow$ crystal
U_C: A^+, $(CN)^-$ ion-gas $\rightarrow$ crystal

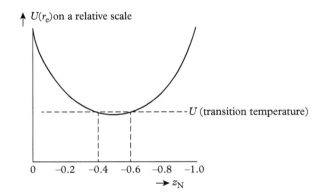

Fig. 4.7 Variation in $U(r_e)$ for potassium cyanide as a function of the charge z_N on the nitrogen atom; a similar curve can be drawn for sodium cyanide. The intersection U(transition temperature) is the crystal energy for the cubic form of potassium cyanide, leading to two choices for z_N; the value of -0.6 is preferred from a comparison of the values for the cohesive energies of KCN, NaCN and LiCN.

The energy U_M was calculated for the orthorhombic structures as a function of z_N from 0 to -1, using the value of ρ/r_e from the cubic structures, and Madelung constants calculated with the program MADC.

In Fig. 4.7, U(transition temperature) is the crystal energy for KCN in its cubic form; it relates to the energy of the orthorhombic form at two values of z_N, -0.4 or -0.6. On account of the symmetry of the structure, the energy is insensitive to spatial interchange of the C and N species in both KCN and NaCN. However, it is not so for the lithium cyanide structure which has a four-coordinate structure. The following results compare the crystal energies of these three compounds:

	LiCN	NaCN	KCN
$U_M(z_N = -0.4)$/kJ mol^{-1}	-682	-732	-669
$U_M(z_N = -0.6)$/kJ mol^{-1}	-791	-732	-669

Since electrostatic energy is proportional to $1/r$, the value of U_M for lithium cyanide cannot be less negative than that for sodium cyanide; hence, the value of z_N is chosen as -0.6. This work on the cyanides enabled the electron affinity of the cyanide ion to be determined as -379 kJ mol^{-1}. Confirmation of the charge distribution of z_N as -0.61 on the cyanide ion species was obtained by an *ab initio* calculation on the (CN)$^-$ ion. Similar calculations with ionic carbonates have shown $z_C \approx +0.82$ and $z_O \approx -0.94$ in the (CO$_3$)$^{2-}$ ion.

4.6 Crystal chemistry

Crystal chemistry is a subject of vast extent encompassing the structures of inorganic, organic, metallic and elemental species. Aspects of the structures of solids that can be termed covalent or that contain covalently bonded species (molecular solids) have been discussed in the Chapters 2 and 3.

Among inorganic materials, little was known about the structures of their crystalline states before the advent of x-ray diffraction techniques, although some interesting predictions had been made. Of particular note are Barlow's predictions in 1883 of the structures of the simple metals and inorganic structures, illustrated here:

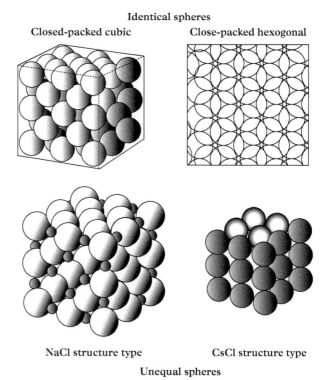

Identical spheres

Closed-packed cubic   Close-packed hexogonal

NaCl structure type   CsCl structure type

Unequal spheres

Crystal structures predicted by Barlow, 1883 [17]

He described the structures for metals in terms of cubic and hexagonal close-packed arrays of identical spheres, and sodium chloride and cesium chloride were pictured as cubic arrays of close-packed spherical atoms of unequal sizes. Although wholly speculative at that time, they were shown subsequently to be remarkably intuitive [17–19].

Crystal chemistry is concerned with the study of relationships between the physical and chemical properties of solids and their internal structure and bonding. It attempts to interpret properties in the light of known structures and, conversely, to associate certain structural characteristics with measured properties. A systematic approach to crystal chemistry began with Bragg's publication in 1920 of the interionic measurements of the alkali-metal halides, listed in Table 4.8 [20].

The table shows five sets of Δ-values for which the $\Delta(X_i^- - X_j^-)$ values are almost independent of the nature of the combined cation. Additionally, there are four sets of Δ-values for which the $\Delta(M_i^+ - M_j^+)$ values are almost independent of the nature of the combined anion. These results can be explained by a model in which the ions are regarded as spheres, each of a characteristic radius, with the sums of cationic and anionic radii equal to the corresponding interionic distances measured on the crystals. Taking sodium and potassium halides, for example,

$$r_e(KX) = r(K^+) + r(X^-)$$

and

$$r_e(NaX) = r(Na^+) + r(X^-)$$

Hence:

$$\Delta(KX - NaX) = r(K^+) - r(Na^+) \qquad (4.47)$$

Table 4.8 *Equilibrium interionic distances/nm in the alkali-metal halides*

	Li	Δ	Na	Δ	K	Δ	Rb	Δ	Cs	Mean Δ
F	0.201	(0.030)	0.231	(0.036)	0.267	(0.015)	0.282	(0.018)	0.300	
Δ	[0.056]		[0.050]		[0.047]		[0.046]		[0.056]	[**0.051**]
Cl	0.257	(0.024)	0.281	(0.033)	0.314	(0.014)	0.328	(0.028)	0.356	
Δ	[0.018]		[0.017]		[0.015]		[0.015]		[0.016]	[**0.016**]
Br	0.275	(0.023)	0.298	(0.031)	0.329	(0.014)	0.343	(0.028)	0.371	
Δ	[0.025]		[0.025]		[0.024]		[0.023]		[0.024]	[**0.024**]
I	0.300	(0.023)	0.323	(0.030)	0.353	(0.013)	0.366	(0.029)	0.395	
Mean Δ		(**0.025**)		(**0.033**)		(**0.014**)		(**0.026**)		

Table 4.9 *Radii/nm of ionic species, referred to coordination number 6*[a]

Ion	Ladd	Shannon & Prewitt	Pauling
Li^+	0.086	0.086	0.090
Na^+	0.112	0.102	0.116
K^+	0.144	0.138	0.152
Rb^+	0.158	0.152	0.166
Cs^+	0.184	0.167	0.181
NH_4^+	0.166	–	–
Ag^+	0.127	0.115	0.127
Tl^+	0.154	0.150	0.164
Be^{2+}	0.048	0.045	0.059
Mg^{2+}	0.087	0.089	0.086
Ca^{2+}	0.118	0.100	0.114
Sr^{2+}	0.132	0. 118	0.132
Ba^{2+}	0.149	0.135	0.149
H^-	0.139	–	–
F^-	0.119	0.133	0.119
Cl^-	0.170	0.182	0.167
Br^-	0.187	0.196	0.182
I^-	0.212	0.220	0.206
O^{2-}	0.125	0.140	0.126
S^{2-}	0.170	0.184	0.170
Se^{2-}	0.181	0.198	0.184
Te^{2-}	0.197	0.223	0.207

[a] The changes in ionic radius from coordination number 6 to coordination numbers 8 and 4 are approximately +3% and −5%, respectively.

which is independent of the nature of the halogen X, $(X = F, Cl, Br, I)$. A similar analysis shows an independence of the nature of M for any pair of halides such as bromides and chlorides: $\Delta(MBr - MCl) = r(Br^-) - r(Cl^-)$, independent of M $(M = Li, Na, K, Rb, Cs)$. The values of Δ involving cesium show less regularity because the cesium halides (Cl, Br, I) have the eight-coordinate cesium chloride structure, whereas all other halides listed in Table 4.8 have the six-coordinate sodium chloride structure. However, these three cesium halides give better agreement if adjusted for a change in coordination number from 8 to 6 (see footnote to Table 4.9).

4.6.1 Ionic radii

An interionic radius r_e is an experimentally measurable quantity. It is necessary to consider how it may be divided into its components, the *ionic radii*, a problem that has received much attention without the development of a totally unique set of ionic radii. The earliest method, by Landé [21], was based on structures in which close packing of ions was a reasonable assumption, namely, large anionic and small cationic species. Interionic distances for the sulphides and selenides of magnesium and manganese are shown here:

	r_e/nm		r_e/nm
MgO	0.210	MnO	0.222
MgS	0.260	MnS	0.261
MgSe	0.273	MnSe	0.273

The constancy of r_e in these sulphides and selenides, unlike the oxides, indicates close packing of the anions in these compounds. In the close-packed structure diagrams (Fig. 4.8), it is evident that $a\sqrt{2} = 2r_e\sqrt{2} = 4r_-$, whereupon $r_{Se^{2-}} = 0.193$ nm and $r_{S^{2-}} = 0.184$ nm. If the additivity of ionic radii is accepted, that is, $r_e = r_+ + r_-$, then other radii can be deduced.

Early tabulations of atomic radii were given by Wasastjerna [22] and Goldschmidt [23, 24], both of whom based their table on a value of 0.132 nm for the radius of the oxide anion. Pauling determined ionic radii on the basis of the relationship:

$$r_i = c_n/Z_{eff,i} \qquad (4.48)$$

where c_n is a constant for an isoelectronic series of ions and $Z_{eff,i}$ is the effective atomic number (Section 2.13.5) of the i ion [25]. It still requires the assumption that $r_e = r_+ + r_-$. From Eq. (4.48), for sodium fluoride:

$$r_{Na^+} = c/6.85$$
$$r_{F^-} = c/4.85$$
$$r_{Na^+}/r_{F^-} = 4.85/6.85 = 0.7080$$

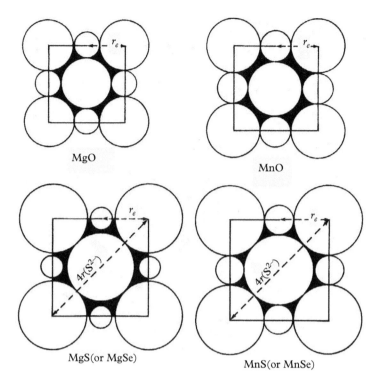

MgO

MnO

MgS(or MgSe)

MnS(or MnSe)

Fig. 4.8 Close-packed cubic arrays of ions, as seen in projection on to a face of a face-centred cubic unit cell. Close packing is present for the sulphide and selenide compounds, but not for the corresponding oxides because of the smaller size of $r_{O^{2-}}$ compared with $r_{Se^{2-}/S^{2-}}$.

From Table 4.7, $r_{Na^+} + r_{F^-} = 0.231$ nm; hence, $r_{Na^+} = 0.096$ nm and $r_{F^-} = 0.135$ nm, and a table of radii was drawn up on that basis. Ladd showed [26] that by applying Landé's method to lithium iodide, ionic radii were deduced that were equivalent to those produced from measurements on x-ray electron density maps of alkali-metal halides (Fig. 4.9) [27, 28]. Table 4.9 lists the radii deduced by Pauling and Ladd together with another set due to Shannon and Prewitt [29, 30]. It is evident that some uncertainty still attaches to the value of an ionic radius. However, the good correspondence between the Ladd and Pauling radii, obtained by quite different methods, would seem to indicate a preference for these data, the former being slightly more comprehensive.

4.6.2 Radius ratio and MX structure types

The *radius ratio* R_n, where n refers to the coordination number, may be defined as $r(\text{cation})/r(\text{anion})$ and may be used as a guide to predicting structure type. It is not successful in all cases, mainly because ionic radii are not defined precisely and their assumed additivity, that is, $(r_+ + r_-) = r_e$, may not be strictly true. The concept of radius ratio is examined first in structures of the general type *MX*.

Fig. 4.9 A portion of the x-ray electron density map for the sodium chloride structure, as seen in projection along a fourfold symmetry axis of the structure, the same viewpoint as in Fig. 4.8. The dashed lines are contours of zero electron density, and may be taken to represent the spatial limits of the ions [26–28].

Consider the cesium chloride structure type, represented by Fig. 1.11, in which the coordination pattern[3] is 8:8. In this mode, each ion is surrounded by eight ions of opposite sign forming the corners of a cube; a sharing of cube faces builds up a macro structure. Let the ions in this structure be of such sizes that the anions at the corners of the unit cell are simultaneously in contact with one another and with the central cation. Then, the unit cell side a is equal to $2r_-$, and the body diagonal $a\sqrt{3}$, or $2r_-\sqrt{3}$, is equal to $2r_- + 2r_+$. Hence, $R_8 = r_+/r_- = 0.732$, where R_8 is the radius ratio for 8:8 coordination.

As the cation is made smaller for a constant anionic radius, contact is lost between cations and anions. Work is done on the system in this separation leading to a less energetic system, but stability can be restored by a change in coordination to 6:6, as in the sodium chloride structure type. Now, each ion is surrounded by six ions of opposite sign at the apices of an octahedron, with a sharing of edges between these octahedra building up a complete structure. Referring to Fig. 4.8, it is clear that in the close-packed conformation (for sulphides and selenides) $2r_- + 2r_+ = 2r_-\sqrt{2}$, so that R_6 is 0.414. Continuing the argument to 4:4 coordination in zinc sulphide, R_4 evaluates as 0.225; each ion in this structure is surrounded by four of opposite sign at the apices of a tetrahedron, with linking by a sharing of apices.

The radius ratio results can be viewed in terms of the crystal energy equation (4.35). Let $U(r_e)$ be written as α/r_e, where $\alpha/\text{kJ mol}^{-1} \approx -218.5$ kJ nm mol$^{-1}/r_e$ (nm) (Section 4.5.2). In MX structures, $z_1 = z_2 = 1$. Assuming additivity of radii and keeping a fixed value of r_-, then for $n{:}n$ coordination:

$$U(r_e) = \alpha' (R_n + 1) \propto \alpha'/R_n \qquad (4.49)$$

where $\alpha' = \alpha/r_-$ and R_n is the radius ratio for $n{:}n$ coordination. The plot of $U(r_e)$ against R is shown in Fig. 4.10 for the three coordination limits discussed. Charges of unity have been used in the zinc sulphide structure in order to give a correct relative energy comparison.

At $R = 1$, the cesium chloride structure is the more stable of the three types illustrated. For $R \leq 0.732$, although the cation may be made smaller, $U(r_e)$ becomes constant because r_e remains constant, since the anions are in contact. The energy of the structure can be decreased further if a change is made to 6:6 coordination, which remains until $R = 0.414$; then, further stabilization requires a change to 4:4 coordination. Table 4.5 indicates the enforced closeness of the curves for the CsCl and NaCl structure types through the similarity of their Madelung constants; those for α-ZnS and β-ZnS (Fig. 4.11) are closer still.

The radius ratio limits for these three important MX structure types may be summarized as:

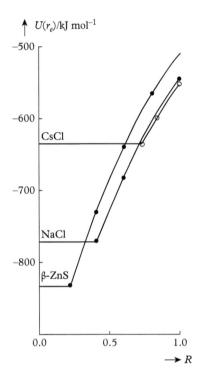

Fig. 4.10 Variation of the crystal energy of the point charge model with radius ratio R for a constant value of r_- for three common MX structure types; for correct comparison, singly-charged ions are used in the calculation of $U(r_e)$ for the β-ZnS (blende) structure type.

	R
CsCl	≥ 0.732
NaCl	0.414–0.732
α-ZnS	
β-ZnS	0.225–0.414

---

[3] $m{:}n$ coordination for a species $A_m B_n$ means that the species A is coordinated by m species B, and species B by n species A.

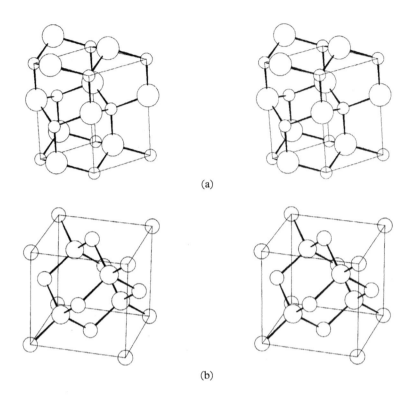

(a)

(b)

Fig. 4.11 Stereoview of the unit cell and environs of the structures of zinc sulphide; circles in decreasing order of size represent the S and Zn atoms. (a) Wurtzite α-ZnS. (b) Blende β-ZnS; this structure may be compared with those for α-Sn in Fig. 1.3 and diamond, Fig. 2.49.

Not all simple *MX* structures obey this pattern, as shown by Table 4.10. Eight values of *R*, those enclosed by the dashed line, although greater than 0.732 nevertheless exhibit the sodium chloride structure type under ambient conditions. There are three factors worthy of consideration.

The radius ratio is a geometrical concept based on the packing of spheres, an assumption that has limitations. The energy difference between the sodium chloride and cesium chloride structure types is *ca.* 4 kJ mol^{-1} at values of $R > 0.5$, *vide* Fig. 4.10. As the crystal energies are based on a point-charge model in this

Table 4.10 *Radius ratios for alkali-metal halides*

	Li	Na	K	Rb	Cs
F	0.72	0.94	0.83[†]	0.77[†]	0.67[†]
Cl	0.50	0.66	0.85	0.93	0.98[†]
Br	0.46	0.60	0.77	0.84	0.98
I	0.41	0.53	0.68	0.74	0.87

[†]r_-, r_+: permissible for *MX* structures with $m : m$ coordination.

analysis it is to be expected that polarization corrections are important, with extreme polarization bordering on covalency. Thus, a small covalent contribution is feasible: in the sodium chloride structure type, the p orbitals of adjacent ions are directed towards one another, thus facilitating orbital overlap. In the cesium chloride structure type, this condition does not obtain, because of the different coordination pattern.

A tendency towards a covalent contribution to crystal energy is more prominent in species with d electrons in the outermost shell. One measure of this increased interaction is a comparison of an interionic distance with the sum of the component radii based on the same coordination pattern, good examples being found among the silver halides:

Structure type		r_e/nm	$\sum_i r_i$/nm	Δ/nm	$10^{40}\alpha_{X^-}$/F m^2
AgF	NaCl	0.246	0.246	0.000	1.0
AgCl	NaCl	0.277	0.297	0.020	3.4
AgBr	NaCl	0.288	0.314	0.026	4.8
AgI	β-ZnS	0.281	0.322[a]	0.041	7.3

[a] A 5% reduction in r_i for the coordination change from 6:6 to 4:4 is included.

As the polarizability of the halide anion increases, so does the departure from additivity of the ionic radii. In the case of silver iodide, the more covalent, four-coordinated β-ZnS structure type is adopted. The thallium halides behave similarly, and these compounds will be encountered again in a discussion of the solubility of ionic solids.

Changes in temperature and pressure bring about polymorphic changes in the alkali-metal halides. Thus, rubidium chloride, for example, adopts the cesium chloride structure type at 83 K. This change can be appreciated from the temperature dependence of r_e. From linear thermal expansion, the decrease in r_e from 298 K to 83 K is *ca.* 0.008 nm, which would make $\underline{U}(r_e)$ from Eq. (4.42) 6–7 kJ mol^{-1} more negative, thus favouring the cesium chloride structure type. Polarization must be considered in a full explanation of the preference of the sodium chloride structure type in these compounds.

An increase in pressure tends to have an effect similar to that incurred by a decrease in temperature. For example, the potassium halides KF, KCl and KBr transform to the cesium chloride structure type at a pressure of 3–7 GPa. The increased pressure weakens the repulsion forces, thus lowering the energy and favouring the eight-coordinated structure type. Sodium chloride transforms similarly, but under a pressure of 30 GPa, the repulsion forces being stronger in this substance on account of the smaller size of the Na$^+$ ion.

Another structure type adopted by many *MX* compounds is that of nickel arsenide (Fig. 4.12). It is closely related to the sodium chloride structure type.

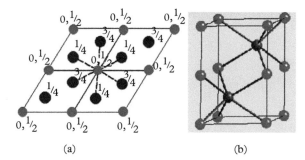

(a) (b)

Fig. 4.12 Structure of the unit cell and environs of nickel arsenide NiAs; circles in decreasing order of size represent the As and Ni atoms. (a) Four adjacent unit cells show the hexagonal packing of As around Ni; the fractions represent heights along the vertical direction in units of c, the unit cell vertical dimension. (b) The arsenic atom lies at the centre of a trigonal prismatic array of Ni atoms. In the hexagonal unit cell, $a = b$, $\angle \mathbf{a} - \mathbf{b} = 120°$, and c is the vertical length of the unit cell. [Reproduced by courtesy of Professor Chieh Chang, University of Waterloo.]

A close-packed hexagonal array of arsenic atoms contains nickel atoms in the octahedral holes of the hexagonal structure (a), whereas the arsenic atom lies at the centre of a trigonal prism formed by nickel atoms (b). The difference from the sodium chloride type is that only the nickel atom is in octahedral coordination. About fifty compounds are known to exhibit the nickel arsenide structure, including

- intermetallic compounds—MnS, AuS;

- borides and nitrides—PtB (anti-NiAs), NbB;

- sulphates, selenates and tellurates of transition-type metals—CoS, CrSe, NiTe.

4.6.3 Radius ratio and MX_2 structure types

Three important structure types of formula MX_2 are represented by fluorite CaF_2, rutile TiO_2 and β-cristobalite SiO_2; an oxide such as Na_2O, in which cations and anions have exchanged sites in a fluorite structure type, is termed an anti-fluorite. In the 8:4 coordination pattern of fluorite, a calcium cation is coordinated by eight fluoride ions at the corners of a cube, and a fluoride ion by four tetrahedrally disposed calcium cations (Fig. 4.13); this arrangement is similar to that of the cesium chloride type, but there are now two coordination polyhedra.

In rutile with 6:3 coordination, titanium is surrounded by a slightly distorted octahedron of oxygen ions: there are two slightly different Ti–O bond distances, 0.1952 nm and 0.1948 nm; the coordination of oxygen is that of an isosceles-triangular array of titanium atoms (Fig. 4.14).

In the 4:2 β-cristobalite structure, the basic structural unit is the $[SiO_4]$ tetrahedron, with the Si–O–Si bond linkage approximately linear (Fig. 4.15). The radius ratio limits for these three structure types are the same as those for the MX structure types, in the order 0.732, 0.414 and 0.225.

The radius ratio criteria are well obeyed among many structures in this group, and Table 4.11 illustrates some of them. A probable explanation for the structures following the radius ratio limits is that the differences in the Madelung energy are relatively much higher among these compounds, because of the more highly

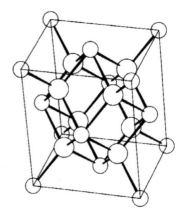
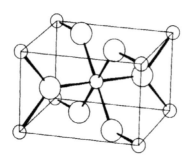

Fig. 4.13 Stereoview of the unit cell and environs of the structure of fluorite CaF_2; circles in decreasing order of size represent the F and Ca atoms; note the relationship to the cesium chloride structure type.

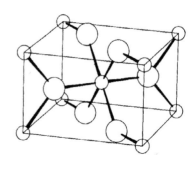

Fig. 4.14 Stereoview of the unit cell and environs of the structure of rutile TiO_2; circles in decreasing order of size represent the O and Ti atoms; note the relationship to the sodium chloride structure type. Two slightly different Ti–O distances and O–Ti–O angles are present in this structure.

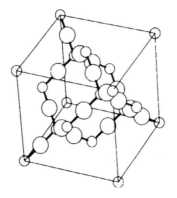
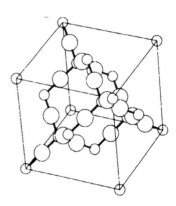

Fig. 4.15 Stereoview of the unit cell and environs of the structure of β-cristobalite (SiO_2); circles in decreasing order of size represent the O and Si atoms; note the relationship to the β-ZnS structure type. This structure consists of $[SiO_4]$ structural units with every oxygen atom shared with oxygen atoms of four other $[SiO_4]$ units.

charged species present, and polarization, a small percentage of the energy in ionic compounds, is insufficient to overcome the tendency for domination by the electrostatic component of the crystal energy.

Among the MX_2 structure types, an increase in polarization can lead to the formation of layer structures, among which cadmium chloride and cadmium iodide

Table 4.11 *Radius ratios for MX$_2$ structure types*

Fluorite		Rutile		β-Cristobalite	
CaF$_2$	0.99	TiO$_2$	0.54	β-SiO$_2$	0.21
BaF$_2$	1.25	MgF$_2$	0.73	BeF$_2$	0.23
SrF$_2$	1.11	CaCl$_2$	0.69		
SrCl$_2$	0.78	CaBr$_2$	0.63		
BaCl$_2$	0.88	ZnF$_2$	0.62		

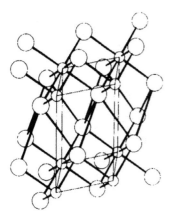 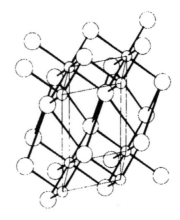

Fig. 4.16 Stereoview of the unit cell and environs of the structure of cadmium iodide CdI$_2$; circles in decreasing order of size represent the I and Cd ions. The structure comprises layers of cadmium cations sandwiched between two layers of iodide ions. A cadmium ion is coordinated by six iodine atoms forming a trigonal prismatic structure, whereas the iodide ions are coordinated by three cadmium ions; the nearest neighbours of any iodide ion all lie on one side of it.

are typical. Removal of one half of the nickel atoms symmetrically from nickel arsenide would lead to the cadmium iodide structure type (Fig. 4.16). Of the two planes containing the six neighbours of any cadmium ion, the three nearest neighbours all lie on one side of the plane. Cadmium iodide may be regarded also as a hexagonal close packed array of iodide ions with cadmium ions occupying one half of the octahedral holes in the structure (Fig. 4.17). Its layer-like composite structure consists of a layer of cadmium ions sandwiched between two close-packed layers of iodide ions. The bonding is ionic within the composite layers, but successive composites are linked by London forces, leading to the easy cleavage that is characteristic of a layer structure.

If polarization in the *MX*$_2$ structure types is decreased, dispersion energy is decreased and its effect on the predominantly ionic structure is greatly reduced. Conversely, if polarization is increased molecular *MX*$_2$ structure types arise, as with mercury(II) chloride, for example. The transformations among *MX*$_2$ structure types in terms of both radius ratio and polarization is indicated diagrammatically in Fig. 4.18: a high degree of polarization may be considered to merge into covalent character in the molecular compounds wherein attraction leads to significant atomic orbital overlap.

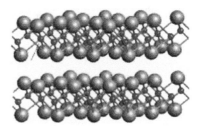

Fig. 4.17 Another view of the cadmium iodide structure; the close-packed ionic layers are linked by van der Waals forces, and define the cleavage planes of the structure.

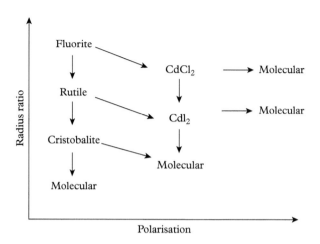

Fig. 4.18 Transformations among the MX_2 structure types in terms of both radius ratio and polarization. A decrease in radius ratio or an increase in polarization, or a combination of these two factors, leads ultimately to the molecular compounds associated with covalency.

4.6.4 Structures of silicates

In the early work on the x-ray analysis of crystal structures much attention was focussed on silicate minerals, and the detailed atomic arrangements in many types were determined [31]. In common with them all is the [SiO₄] tetrahedral structural building unit:

[SiO₄] structural unit

Silicate structures differ fundamentally in the manner in which the apical oxygen atoms in [SiO₄] units are shared with those of other tetrahedra, each mode of sharing being characterized by a *silicon: oxygen ratio*. In addition, silicon may be replaced by aluminium or other atoms of oxidation numbers less than IV, whereupon other cations such as Mg^{2+} and Fe^{3+} are incorporated so as ensure electrical balance; not all species exist in stoichiometric ratios. The more important sharing patterns are illustrated by Fig. 4.19; five classes of silicate structures can be recognized among this group of minerals.

- The [SiO₄] structural unit appears as a discrete entity in the olivine structures (Fig. 4.19a), among which magnesium silicate MgSiO₄, or forsterite, is the group example; the Si:O ratio is 1:4, and these compounds have been termed orthosilicates or *neosilicates* (Gk. *nêsos* = island). The magnesium sites, or some of them, can be replaced by other species, such as iron and

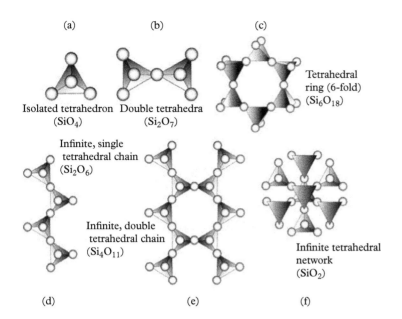

(a)   (b)   (c)

Isolated tetrahedron (SiO$_4$)   Double tetrahedra (Si$_2$O$_7$)   Tetrahedral ring (6-fold) (Si$_6$O$_{18}$)

Infinite, single tetrahedral chain (Si$_2$O$_6$)

Infinite, double tetrahedral chain (Si$_4$O$_{11}$)

Infinite tetrahedral network (SiO$_2$)

(d)   (e)   (f)

Fig. 4.19 The formation of silicate structures by sharing tetrahedral [SiO$_4$] structural units. As the extent of sharing of oxygen atoms increases, so the Si:O ratio increases, in stages, from 1:4 (olivine) to 1:2 (silica, cristobalite, tridymite). The formation of a sheet silicate structure (not shown) may be visualized by a side-to-side sharing of the double chains (e) to give Si:O = 2:5. [Reproduced by courtesy of Dr Phil Stoffer.]

manganese; olivine itself is (MgFe)SiO$_4$. An important olivine structure type is LiMPO$_4$, where M can be iron, cobalt or nickel; it has been developed as an effective storage device for sustainable energy. The LiFePO$_4$ compound has found applications in fuel cells for hybrid electric vehicles because of its good performance characteristics and low toxicity [32, 33]. Other typical olivines include phenacite Be$_2$SiO$_4$, zircon ZrSiO$_4$ and topaz (AlF)$_2$SiO$_4$.

The sharing of corners between two or more tetrahedra produces closed systems, such as double tetrahedra (Fig. 4.19b), or rings of three or six tetrahedra. Thortveitite Sc$_{1.5}$Y$_{0.5}$Si$_2$O$_7$, the primary source of scandium, and hemimorphite Zn$_4$(OH)$_2$Si$_2$O$_7$,H$_2$O, are double tetrahedra structures with the Si:O ratio of 2:7; these structures have been termed *sorosilicates* (Gk. *sôros* = heap).

Benitoite BaTiSi$_3$O$_9$ is a three-membered ring structure, and beryl Be$_3$Al$_2$Si$_6$O$_{18}$ is a ring structure comprising six linked [SiO$_4$] units (Fig. 4.19c); both three-membered and six-membered rings share two apices and have a Si:O ratio of 1:3; these structures are termed, not surprisingly, *cyclosilicates* (Gk. *kuklos* = circle).

Infinite chains or *inosilicates* (Gk. *inos* = fibre) are formed by a continuous linking of one oxygen atom of the [SiO$_4$] units: pyroxenes such as diopside [CaMg(SiO$_3$)]$_n$ form single chains (Fig. 4.19d) with Si:O = 1:3, and amphiboles like tremolite [Ca$_2$Mg$_5$(OH)$_2$Si$_8$O$_{22}$]$_n$, are obtained as infinite double chains (Fig. 4.19e), thus increasing the Si:O ratio to 4:11.

- Hexagonal sheets of linked tetrahedra are the *phyllosilicates* (Gk. *phyllon* = leaf) in which every tetrahedron shares three corners with other tetrahedra; the $(Si_2O_5)_n$ structure types have the Si:O ratio of 2:5. Species include the lamellar minerals such as talc $Mg_2(OH)_2Si_4O_{10}$, pyrophyllite $Al_2(OH)_2Si_4O_{10}$, which is a refractory material that has the unusual property of significantly zero contraction on firing as well as being machinable, and micas such as muscovite $KAl_2(OH)_2(Si_3Al)O_{10}$; these structures have hexagonal symmetry and show good cleavage parallel to the sheets. This silicate structure type is not shown in Fig. 4.19, but it may be visualized by an extended side-to-side sharing of the double chains in (e). Another example is shown by the white asbestos mineral chrysotile $Mg_3Si_2O_5(OH)_4$, now known to be a human carcinogen:

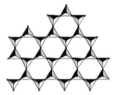

Chrysotile asbestos sheet

- In the phyllosilicates three basal oxygen atoms are shared with other tetrahedra, and the fourth oxygen atoms all point in the same direction, bonding octahedrally with the coordinating cations, forming plane, two-dimensional sheets.
- The final sharing principle with Si:O = 1:2 is obtained when all four corners of the $[SiO_4]$ tetrahedral units are shared with other similar units to build infinite three-dimensional networks, the *tectosilicates* (Gk. tekton = builder). These include the various forms of silica: quartz (Fig. 2.51), cristobalite (Fig. 4.15) and tridymite, together with other framework structures such as the rock-forming feldspars, like orthoclase $KAlSi_3O_8$ and anorthite $CaAl_2Si_2O_8$; they are based on an infinite number of tetrahedral units sharing all four apices (Fig. 4.19f).

The general formula for the more commonly occurring feldspars is $XAl_{(1-2)}Si_{(3-2)}O_8$, where X= Na, K or Ca. When X is singly charged, the formula contains Al and 3 Si, with aluminium taking the place of silicon in a random manner leaving the symmetry statistically unchanged. If X is doubly charged, then the formula will contain 2 Al and 2 Si. In every case the Si:O or (Si/Al):O ratio remains as 1:2.

Zeolite framework structures fall within this fifth class. They are microporous aluminosilicates that occur naturally in volcanic ash that has been deposited in alkaline waters. One of about fifty known examples is shown in Fig. 4.20. The channels in the structure are 0.3–0.6 nm in diameter, and can contain mobile

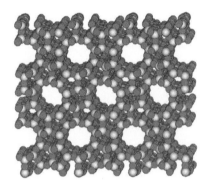

Fig. 4.20 Structure of a zeolite of composition $Na_nAl_nSi_{96-n}O_{192}.16H_2O$, where *n* can range from 0 to 27, subject to the structure remaining electrically neutral. [Public domain.]

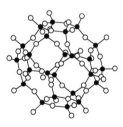

Fig. 4.21 Basic cage framework of the structure of ultramarine, an intensely blue coloured art pigment extracted from lapis lazuli rock. [Reproduced from Bragg WL *The Crystalline State*, Vol. I. G Bell and Sons Ltd, 1939.]

positive ions. Zeolites structures are able to act as catalysts in many organic processes, for example, in cracking and in hydrocarbon synthesis. Their use as ion exchange materials is well known: hydrated cations in the pores of the zeolite structure can exchange with other hydrated cations, and these materials are used in water softening by exchanging calcium and magnesium ions for sodium ions.

Other framework structures include the ultramarines (Fig. 4.21) and the naturally occurring semi-precious mineral lapis lazuli from which the ultramarine art pigments are made. Fig. 4.22a illustrates an example of natural lapis lazuli from the Sar-e-Sang mine in north-east Afghanistan, and a ring set with this material of gem quality is shown in Fig. 4.22b.

(a)

4.7 Structural and physical characteristics of ionic solids

Ionic bonds exist between an ion and an indefinite number of neighbours. Unlike the covalent bond, it has no particular directionality in space, and ionic structures tend to be governed to a first approximation by geometrical conditions subject to electrical neutrality of the structure a whole. Interactions based on polarization and covalency enhance the crystal energy, and may modify predictions based on radius ratios.

A refined electrostatic calculation, including polarization contributions, gives satisfactory crystal energy values where all necessary data are available. Such calculations on the more complex compounds are hampered by the paucity of data on compressibility, expansivity and polarizability, but a Born–Haber cycle procedure can sometimes be employed to overcome the lack of such parameters.

(b)

Fig. 4.22 Naturally occurring lapis lazuli. (a) A limestone base rock containing *inter alia* the intensely blue lazurite, $Na_{8-10}Al_6Si_6O_{24}S_{2-4}$; the blue colour of this substance is said to arise from the presence of the $(S_3)^-$ ion, unstable under normal conditions. (b) A gem quality ring made from a sample of semi-precious gemstone quality lapis lazuli from the Sar-e-Sang mine in north-east Afghanistan. (Pala International, CA 92028.]

Compounds in which a covalent contribution is evident, such as some of the silver halides, do not respond with high precision to the electrostatic model, even with the inclusion of London dispersion energy. However, these crystal energies can be determined from the Born–Haber cycle, with a precision governed by that of the parameters in the thermodynamic cycle, and generally give more exact values.

Molecules do not exist in ionic structures because electrons are localized in the atomic orbitals of the ions. In polyatomic ions such as $(NH_4)^+$ and $(SO_4)^{2-}$ covalent bonding is predominant within the ions themselves, with ionic bonding linking the cations and anions in the solid state. A typical structure involving a polyatomic species is shown in Fig. 1.12, which represents the isomorphic sodium nitrate and calcium carbonate (calcite) structures.

Ionic solids form hard crystal of low compressibility and expansivity, but of high melting point. They are electrical insulators in the solid but conduct in the melt, and in solution where a suitable solvent can be found. Magnesium oxide is as ionic a structure as is sodium chloride, but only sodium chloride dissolves in water. The reason lies in the relation between the crystal energy of the solid and that of the hydrated ions, as will be discussed shortly. Values of electrical conductivity of ionic solids distinguish them sharply from covalent and molecular solids.

Trends in physical properties can be related to crystal energy, principally because of its proportionality to $1/r$, as the following compilations show:

	BeO	MgO	CaO	SrO	BaO
r_e/nm	0.165	0.210	0.240	0.257	0.276
Hardness/Moh's scale	9.0	6.5	5.5	4.1	3.3
	NaF	NaCl	NaBr	NaI	
r_e/nm	0.231	0.281	0.298	0.323	
MP/K	1266	1074	1020	934	

In both the hardness and melting point series of data, the crystal energy decreases (becomes less negative) from left to right along the data.

4.8   Solubility of ionic compounds

The interesting property of solubility can be treated quantitatively by thermodynamic arguments, which show at the same time the relationship between solubility and crystal energy. Certain sources comment that 'ionic solids are soluble in water but covalent solids are not'; the second part of this statement is undoubtedly true, but the first statement needs qualification. Again, 'solubility decreases with an increase in the covalent character of the bonding'. But consider the following data:

$$AgF \quad \rightarrow \quad AgCl \quad \rightarrow \quad AgBr \quad \rightarrow \quad AgI$$
$$\longrightarrow \quad \text{Increasing polarization/covalent character}$$
$$\longrightarrow \quad \text{Decreasing solubility}$$

$$CaF_2 \quad \rightarrow \quad CaCl_2 \quad \rightarrow \quad CaBr_2 \quad \rightarrow \quad CaI_2$$
$$\longrightarrow \quad \text{Increasing polarization/covalent character}$$
$$\longrightarrow \quad \text{Increasing solubility}$$

Evidently, a more detailed study is required, but certain reference states need first to be defined.

4.8.1   Reference state for solubility

For convenience, this part of the discussion will be restricted to *MX* type compounds, although the ensuing results will be applicable *mutatis mutandis* in a general sense.

Consider the equilibrium:

$$MX(s) + aq \rightleftharpoons M^+(aq) + X^-(aq) \tag{4.50}$$

The equilibrium constant K is given by:

$$K = a(M^+aq)\, a(X^-, aq)/a(MX,\, s) \tag{4.51}$$

Since the activity a of a pure, crystalline solid in the reference state (Appendix A17) is defined as unity:

$$K = a(M^+, aq)a(X^-, aq) \tag{4.52}$$

or

$$K_c = c_\pm^2 f_\pm^2 \tag{4.53}$$

where $c_\pm$ is the mean concentration[4] of the solute in mol dm^{-3}, and $f_\pm$ is the mean activity coefficient for the electrolyte in its saturated solution; K_c is a constant at a given temperature, and 298.15 K will be assumed as a working temperature throughout this discussion.

The reference state for the solution is the infinitely dilute solution in which the ratio of the activity of the solute to its molar concentration is unity; the fact that the reference state is hypothetical does not invalidate the arguments which follow. The reference state may be regarded as a solution of mean concentration 1 mol dm^{-3} and unit mean activity, and in which the partial molar heat content of the solute is the same as that at infinite dilution (Appendix A17).

4.8.2   Solubility relationships

Generalizing Eq. (4.50):

$$\begin{array}{cc} A(s) & \rightleftharpoons B(\text{aq. ions}) \\ \text{solute} & \text{saturated solution} \end{array} \tag{4.54}$$

where the solute A is in equilibrium with the saturated solution B, and the equilibrium constant for this general case is given by:

$$K = a(B)/a(A) \tag{4.55}$$

The chemical potentials of the components in the system Eq. (4.54) are given by

$$\mu(A) = \mu^{\ominus}(A) + RT \ln a(A) \tag{4.56}$$

---

4 Strictly, *molality* m in mol kg^{-1}.

and

$$\mu(B) = \mu^{\ominus}(B) + RT \ln a(B) \qquad (4.57)$$

At equilibrium, the components A and B are at the same chemical potential. Hence, $\mu(A) = \mu(B)$, and the standard free energy change for Eq. (4.54) is:

$$\Delta G^{\ominus} = \mu^{\ominus}(B) - \mu^{\ominus}(A) = -RT \ln \frac{a(B)}{a(A)} = -RT \ln K' \qquad (4.58)$$

In an analogous manner for Eq. (4.53):

$$\Delta G_d^{\ominus} = -RT \ln K_c = -RT \ln c_{\pm}^2 f_{\pm}^2 \qquad (4.59)$$

This equation represents the free energy change for dissolution for the process *pure solid ⇌ ions* in their reference states (Appendix A17) and governs solubility.

Example 4.3

The solubility of silver iodide in water at 298.15 K is 1.02×10^{-8} mol dm^{-3}. At this concentration, $f_{\pm}$ is effectively unity (0.9999). Hence, from Eq. (4.59), $\Delta G_d^{\ominus} = -RT \ln c_{\pm}^2 = 91.2$ kJ mol^{-1}, an indication of the extreme insolubility in water of this compound.

Example 4.4

The solubility of lithium fluoride in water at 298.15 K is 0.09 mol dm^{-3}. Following the approach in Example 4.3 leads to $\Delta G_d^{\ominus} = 11.9$ kJ mol^{-1}. If the Debye limiting equation (Appendix A18) is introduced in order to estimate the mean activity coefficient, then $f_{\pm} = 0.703$, whereupon $\Delta G_d^{\ominus} = 13.7$ kJ mol^{-1}, a significant difference.

Then what further theory is needed?

The calculation in Example 4.4 requires data on the activity coefficient of the saturated solution. This was unnecessary in Example 4.3, where $f_{\pm}$ is effectively unity. Where solubility is less than *ca.* 0.01 mol dm^{-3} with a 1:1 electrolyte, $f_{\pm}$ can be calculated with good accuracy from the Debye limiting equation (Appendix A18), leading to a better value for the free energy of dissolution in the case of lithium fluoride. In concentrated solutions, however, this approach is very inaccurate. For example, sodium chloride is saturated at 6.146 mol dm^{-3}. At this concentration, the Debye limiting law gives $f_{\pm} = 0.0547$, whereas the experimentally measured value is 1.006. Unfortunately, there is a paucity of data on activity coefficients at saturation, and also it is desirable to analyse $\Delta G_{d}^{\ominus}$ in order to seek a clearer understanding of solubility. The Davies equation (A18.7) is an improvement on the Debye limiting law but cannot always be used to obtain reliable values for $f_{\pm}$ as it applies only to solutions of ionic strength no greater than *ca.* 0.05 mol dm^{-3}.

4.8.3 Solubility and energy

The important quantity in solubility $\Delta G_{d}\ominus$, the *free energy of dissolution*, is given by:

$$\Delta G_{d}\ominus = \Delta H_{d}^{\ominus} - T\Delta S_{d}^{\ominus} \qquad (4.60)$$

where $\Delta H_{d}^{\ominus}$ is the standard enthalpy change for the dissolution process referred to infinite dilution, and $\Delta S_{d}^{\ominus}$ is the corresponding change in entropy which may be expanded to:

$$\Delta S_{d}^{\ominus} = \sum_{i} \overline{S}_{i}^{\ominus} - S_{c}^{\ominus} \qquad (4.61)$$

where $\sum_{i} \overline{S}_{i}^{\ominus}$ is the sum of the standard *relative partial molar entropy* of the hydrated i ions (Appendix A17) and $S_{c}^{\ominus}$ is the standard entropy of the crystal under examination; these data are readily available in the literature [34, 35].

The thermodynamic relationships between solubility and energetics of the crystal and hydrated ion states are illustrated by the free-energy diagram of Fig. 4.23. The relevant equations are:

$$\Delta G_{d}^{\ominus} = \Delta G_{h}^{\ominus} - \Delta G_{c}^{\ominus}$$
$$\Delta G_{d}^{\ominus} = \Delta H_{d}^{\ominus} - T\Delta S_{d}^{\ominus} \qquad (4.62)$$
$$= \Delta H_{d}^{\ominus} - T\left(\sum_{i} \overline{S}_{i}^{\ominus} - S_{c}^{\ominus}\right)$$

and they show how the free energy of dissolution is related to the enthalpy and entropy parameters of the crystal and the hydrated ions, and how it may be determined in practice from thermodynamic data [36].

It is important that the enthalpies of dissolution be referred to infinite dilution because the process of dilution itself can involve a significant enthalpy change. For example, ΔH_{d} for cadmium sulphate in 200 mol water is -43.9 kJ mol^{-1}, but at infinite dilution it is -53.6 kJ mol^{-1}. The difference between two such data can, in some cases, be commensurate with $\Delta G_{d}^{\ominus}$ itself.

Solubility data for a number of 1:1, 2:1 and 3:1 compounds are listed in Table 4.12. The interaction between ions and water molecules on one hand and between ions in the crystal on the other both increase in magnitude as the ions become smaller and more highly charged, because coulombic energy, which is

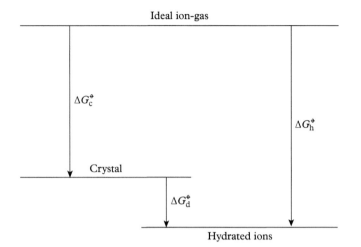

Fig. 4.23 Standard free energy levels for solubility at 298.15 K: ΔG_c, ideal ion-gas → crystal (free energy of the crystal); ΔG_h, ideal ion-gas → hydrated ions (free energy of hydration of ideal gaseous ions); ΔG_d, crystal → hydrated ions (free energy of dissolution of crystal).

Table 4.12 *Thermodynamic data[a] relating to solubility of selected halides in water at 298.15 K*

Halide	$\Delta H_d^{\ominus}$/kcal mol^{-1}	$\sum \overline{S}_i^{\ominus}$/cal mol^1 K^{-1}	$S_c^{\ominus}$/cal mol^1 K^{-1}	$T\Delta S_d^{\ominus}$/kcal mol^{-1}	$\Delta G_d^{\ominus}$/kcal mol^{-1}
LiF	1.1	1.1	7.0	1.8	2.9
LiCl	−8.9	16.6	13.2	1.0	−9.9
LiBr	−11.7	22.7	16.5	1.8	−13.5
LiI	−15.1	29.5	18.1	3.4	−18.5
NaF	0.1	12.1	14.0	−0.6	0.7
NaCl	0.9	27.6	17.3	3.1	−2.2
NaBr	−0.2	33.7	20.5	3.9	−4.1
NaI	−1.8	40.5	22.1	5.5	−7.3
KF	−4.2	22.2	15.9	1.9	−6.1
KC1	4.1	37.7	19.8	5.3	−1.2
KBr	4.8	43.8	23.1	6.2	−1.4
KI	4.9	50.6	24.9	7.7	−2.8
RbF	−6.3	27.4	17.4	3.0	−9.3
RbCl	4.0	42.9	22.6	6.1	−2.1
RbBr	5.2	49.0	25.9	6.9	−1.7
RbI	6.2	55.8	28.2	8.2	−2.0
CsF	−9.0	29.5	19.1	3.1	−12.1
CsCl	4.3	45.0	23.3	6.5	−2.2
CsBr	6.2	51.1	27.1	7.2	−1.0

Table 4.12 *continued*

Halide	$\Delta H_d^{\ominus}$ /kcal mol^{-1}	$\sum \overline{S}_i^{\ominus}$ /cal mol^1 K^{-1}	$S_c^{\ominus}$ /cal mol^1 K^{-1}	$T\Delta S_d^{\ominus}$ /kcal mol^{-1}	$\Delta G_d^{\ominus}$ /kcal mol^{-1}
CsI	7.9	57.9	30.0	8.3	−0.4
AgF	−4.9	15.4	20.0	−1.4	−3.5
AgCl	15.9	30.9	23.0	2.4	13.5
AgBr	20.1	37.0	25.6	3.4	16.7
AgI	26.7	43.8	27.3	4.9	21.8
TlCl	10.4	43.6	25.9	5.3	5.1
TlBr	13.7	49.7	28.6	6.3	7.4
TlI	17.7	56.5	29.4	8.1	9.6
MgF$_2$	−4.4	−32.8	13.7	−13.9	9.5
MgCl$_2$	−37.1	−1.8	21.4	−6.9	−30.2
MgBr$_2$	−44.5	10.4	29.4	−5.7	−38.8
MgI$_2$	−51.2	24.0	34.8	−3.2	−48.0
CaF$_2$	3.2	−17.8	16.5	−10.2	13.4
CaCl$_2$	−19.8	13.2	27.2	−4.2	−15.6
CaBr$_2$	−26.3	25.4	31.0	−1.7	−24.6
CaI$_2$	−28.7	39.0	34.7	1.3	−30.0
SrF$_2$	2.5	−14.0	21.4	−10.6	13.1
SrCl$_2$	−12.4	17.0	27.5	−3.1	−9.3
SrBr$_2$	−17.1	29.2	33.8	−1.4	−15.7
SrI$_2$	−21.6	42.8			
BaF$_2$	0.9	−1.6	23.1	−7.4	8.3
BaCl$_2$	−3.1	29.4	29.6	−0.1	−3.0
BaBr$_2$	−6.1	41.6	35.5	1.8	−7.9
BaI$_2$	−11.4	55.2	40.9	4.3	−15.7
AlF$_3$	−50.4	−81.8	15.9	−29.1	−21.3
AlCl$_3$	−79.3	−35.3	26.1	−18.3	−61.0
AlBr$_3$	−86.1	−17.0	43.0	−17.9	−68.2
AlI$_3$	−90.6	3.4	45.3	−12.5	−78.1

[a] Ladd MFC and Lee WH. *Trans. Faraday Soc.* 1958; 54: 34, listed in calorie units; they are converted to Joule units on multiplication by the Joule equivalent, 4.184 J cal^{-1}.

proportional to $1/r$, predominates in both situations. Both ΔG_d and ΔH_d depend upon the difference between two large quantities, one concerned with the solid and the other with the hydrated ions; Fig. 4.23 indicates this situation. Sparingly soluble halides have large positive values of $\Delta G_d^\ominus$ whereas negative values indicate ready solubility.

A decrease in ΔG_h tends to stabilize the hydrated ions state with respect to the ion-gas and so promotes solubility, whereas a decrease in either ΔG_c or ΔS_d tends to decrease solubility, the former by stabilizing the crystal with respect to the ion-gas, and the latter by making the hydrated ions state relatively less probable. The ΔS_d term becomes very significant with small or highly charged ions, such as Li^+, Mg^{2+} and F^-, and solubility cannot be discussed adequately on the basis of enthalpy changes alone.

The isothermal transfer of a gaseous ion into water involves two effects. There is a *structure-breaking* effect on the water itself owing to the interaction of the ions with water molecules, disrupting some hydrogen-bonded structure of the water, together with a *structure-making* effect as the ions become coordinated by water molecules and forming a hydration sphere around them. The first of these effects is important with large ions, as it tends to increase the value of the $\overline{S_i}$ parameters, as can be seen by comparing SrF_2 with BaF_2, for example. The second effect is prominent with small ions, since its action decreases $\overline{S_i}$, as exemplified by CaF_2 and MgF_2. Any given case involves the interplay of these factors, but whereas solubility may be often difficult to explain in molecular terms, except qualitatively, the thermodynamic analysis is quantitative. Where measured activity coefficients at saturation are known, results calculated by Eq. (4.59) agree well with the corresponding value in Table 4.12. For example, the sodium chloride concentration at saturation is 6.146 mol dm^{-3} and the measured activity coefficient at saturation is 1.006. Thus, ΔG_h evaluates from Eq. (4.59) as -9.03 kJ mol^{-1} (2.16 kcal mol^{-1}), showing excellent agreement with the value in Table 4.12. Such a result is to be expected since the solute and the saturated solution are at the one and the same chemical potential.

The trends in solubility with which this section opened can now be explained. The crystal energies of the calcium and silver halides, calculated by the thermodynamic model, are listed below in the usual units of kJ mol^{-1}; for the silver halides, the crystal energy calculations by thermodynamic and electrostatic models are compared, and their percentage variation listed:

	F	Cl	Br	I	
Ca	2609	2258	2176	2074	Eq. (4.16)
Ag	954	904	895	883	Eq. (4.16)
	941	833	816	778	Eq. (4.41)
	1.4	7.9	8.8	11.9	% discrepancy

The thermodynamic and electrostatic calculations of crystal energy for the calcium halides were found to agree well, and only the thermodynamic results for them are listed here. Along this series, ΔG_h decreases relatively more rapidly from fluoride to iodide than does ΔG_c,[5] with a progressive decrease in ΔG_d and consequent increase in solubility. With the silver halides, the reverse situation obtains: in the calculation of crystal energy, although polarization contributions have been included, the discrepancy increases from fluoride to iodide, the latter compound taking the four-coordinate β-ZnS structure type. It is evident that additional covalent contributions exist in silver chloride, bromide and iodide, becoming more significant in this order. It is this enhancement of their crystal energies that accounts for the observed solubility trends [36].

Compounds such as magnesium oxide and calcium oxide, for example, are highly ionic, as shown by crystal energy calculations [10], but nevertheless are insoluble in water. The crystal energies are numerically large, because of the doubly charged magnesium and oxide ions ($A_{NaCl} = 1.74756$; $A_{MgO} = 4 \times A_{NaCl}$), which quadruples the Madelung (electrostatic) component of the crystal energy for similar values of r_e as compared with the alkali-metal halides. The ionic bonding is too strong for disruption by the formation of hydrated ions;[6] if a solvent of sufficiently high permittivity were available, magnesium oxide would be soluble in it. The solvent N-methylacetamide $CH_3CON(H)CH_3$ has a relative permittivity of 179, and the solubility of silver chloride in this solvent has been reported as *ca.* 0.1 molar, whereas in water it is 1.3×10^{-5} molar.

4.9   Spectra of ionic compounds

All ionic solids absorb in the ultraviolet region of the spectrum; some compounds absorb also in the visible region. Absorption does not lead to opacity as it does in metals; often apparently opaque ionic crystals are simply highly coloured, and the transmission of light can be observed in thin section.

Ionic solids are coloured for two reasons: some contain ions which give rise to a characteristic colour through transitions involving d electrons, as in the $[Fe(H_2O)_6]^{2+}$ and $[Co(NH_3)_6]^{3+}$ complex ions. However, in the intensely coloured $[MnO_4]^-$ ion, for example, the colour arises through a charge transfer of electrons from the ligand ion into the d orbitals of the central atom. Other compounds may contain ions which, although colourless in solution, undergo polarization in the solid with the appearance of colour that arises from transitions among partially delocalized electrons, as exemplified by silver iodide and silver phosphate. If the absorption moves from the ultraviolet region just into the visible, the wavelengths absorbed would be in the blue region of the spectrum. Consequently, the crystalline solids that are coloured for this reason appear yellow to red, as in silver phosphate and silver chromate.

[5] Becomes more negative.
[6] $O^{2-} + H_2O \rightarrow 2(OH)^-$.

4.10 Heat capacity of ionic solids

At constant volume, the molar heat capacity $C_{V,m}$ is defined as:

$$C_{V,m} = (\partial U_m/\partial T)_V \qquad (4.63)$$

As the temperature of the solid is increased, its vibrational energy increases. If, as in cesium chloride, for example, there are no translational energy modes, and since rotation of a species about its own axis does not constitute a degree of freedom, all energy imparted to the solid in the form of heat enhances the thermal vibrations. As cesium chloride contains two ions, there are 6 vibrational degrees of freedom per CsCl species, each of which contributes kT (Appendix A5). Hence, the total vibrational energy at a temperature T is $6LkT$, or $6RT$, per mole. Thus, from Eq. (4.63) the molar heat capacity $C_{V,m}$ is $6R$, which is approximately 50 kJ mol^{-1}. For a monatomic species it is 25 kJ mol^{-1}, which is a statement of the Dulong and Petit law.

From statistical mechanics it can be shown that the principal molar heat capacities of a solid are related as:

$$C_{p,m} - C_{V,m} = \beta^2 TV_m/\kappa \qquad (4.64)$$

where the terms have the meanings as before. For sodium chloride at 298.15 K, $\beta = 1.1 \times 10^{-4}$ K^{-1}, $\kappa = 4.1 \times 10^{-11}$ N^{-1} m^2 and $V_m = 2.7 \times 10^{-5}$ m^3 mol^{-1}. Hence, $C_{p,m} - C_{V,m} = 2.37$ J K^{-1} mol^{-1}. The value for an ideal gas is R, or 8.3145 J K^{-1} mol^{-1}; the smaller value for solids arises because heat capacity is determined by vibrational energy, and this parameter does not vary appreciably between constant pressure and constant volume conditions.

According to the Planck quantum theory, a vibrating atom can acquire energy in quanta of $h\nu$, and the probability of a vibrating atom acquiring this amount of energy is proportional to $\exp(-h\nu/kT)$. At room temperature, almost all degrees of freedom are active, and $C_{V,m}$ tends to its limiting value (Fig. 4.24). As the temperature is decreased, vibrations of increasingly lower frequency cease to be excited. Thus, $C_{V,m}$ decreases with decreasing temperature because the average energy decreases. In the limit as $T \to 0$ K, $C_{V,m} \to 0$, as expressed in the third law of thermodynamics. From a curve such as that in Fig. 4.24, the entropy of the solid at a given temperature T can be determined, since a measure of entropy is $\int_0^T (C_{V,m}/T)\mathrm{d}T$, which is the area under a curve of $C_{V,m}$ against T between 0 and T.

The effect of a change in temperature on the vibrations can be gleaned from the following data: sodium chloride has a single absorption band at $\bar{\nu} = 164$ cm^{-1}, and the ratio of the probabilities that this oscillator will acquire the corresponding energy $hc\bar{\nu}$ at 500 K and 50 K is approximately 70; the energy absorbed is used in increasing the amplitudes of the vibrating atoms. The requirement for infrared activity is that a species contains an oscillating dipole, and the rate of change of the dipole determines the ability of the species to absorb in the infrared. In ionic solids each pair of vibrating ions of opposite sign is equivalent to an oscillating

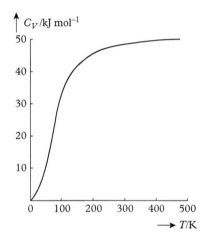

Fig. 4.24 Variation of molar heat capacity C_V with temperature T for crystalline sodium chloride; the portion from *ca.* 15 K to 0 K should be evaluated from the Debye formula. The limiting value is twice the Dulong and Petit value because two species are involved, Na$^+$ and Cl$^-$.

dipole, and its vibrations are excited by infrared radiation; the heavier the species, the smaller the wavenumber (energy) of the vibration.

The Debye temperature is an indicator of the degree of excitation of the vibrations in ionic solids. Thus, the Θ-value of 113 K for cesium iodide implies that its vibrations are excited under ambient conditions; the corresponding value for magnesium oxide is 774 K and its vibrations are only very slightly excited at room temperature.

4.11 Defects in ionic structures

Materials in general are not perfect: not every site required by the space group symmetry of the crystal is occupied; some structures have missing species or 'holes', other have inclusions that displace adjacent species from their equilibrium positions. Defects arise at point sites (zero-dimension defects), along lines (one-dimensional defects), involving planes of atoms (two-dimensional defects) or three-dimensional volume defects. The formation of a defect is the result of a competition between the internal energy U (cost) and entropy S (gain) of a system. Defects form initially for $|-T\Delta S| > |U|$ (Fig. 4.25); at equilibrium $\Delta A = \Delta U - T\Delta S = 0$, and subsequently the concentration of defects n_c increases if the internal energy is increased, for example, by applied heat.

4.11.1 Point defects

The simplest point defect is a vacancy at a site which, from the symmetry of the crystal, would be expected to be occupied. An ion or atom transferred from a site in the crystal structure to a position on its surface constitutes a *Schottky defect* (Fig. 4.26a). In a crystal at thermal equilibrium with its surroundings, a certain number of vacant sites always exist and contribute an increase in entropy compared to that of the ideal crystal with all sites correctly occupied. An expression for the number of sites unoccupied, or defect concentration, is obtained by a statistical procedure.

Consider a fixed volume of a crystal with n Schottky defects, or vacancies, distributed over N lattice sites ($n \ll N$). The first defect can be distributed in N ways, the second in $N-1$ ways, and defect n in $[N-(n-1)]$ ways. Thus, the total number of arrangements w of the n defects is:

$$w = N(N-1)(N-2)\ldots[N-(n-1)] = N!/(N-n)! \qquad (4.65)$$

Since the defects are indistinguishable from one another, there are n indistinguishable ways of obtaining the first defect, $(n-1)$ ways of obtaining the second, and in all $n!$. Thus, the total probability of the system W is now:

$$W = N!/[(N-n)!n! \qquad (4.66)$$

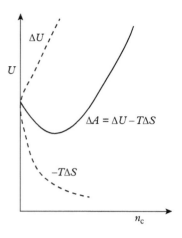

Fig. 4.25 Crystal defect concentration increases for $|T\Delta S| > |U|$; equilibrium exists when $\Delta A = 0$, that is, for $\Delta U = T\Delta S$. If the thermal energy is increased, by applied heat or by irradiation, U becomes the controlling factor for the defect concentration.

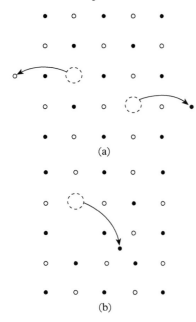

Fig. 4.26 Intrinsic point defects in ionic crystals; cations are filled circles, anions are open circles. (a) Pair of Schottky defects in potassium chloride; the ions are removed from the bulk crystal to its surface. (b) Frenkel defect in silver chloride; a cation has moved to an interstitial position within the bulk crystal.

If Δu represents the energy change for the formation of a single defect, then the change in work function (aka Helmholtz free energy) ΔA is

$$\Delta A = n\Delta u - T\Delta S \tag{4.67}$$

for the formation of the n defects. Using the Boltzmann equation $S = k\ln(W)$, Eq. (1.9), and since the ideal state of zero defects has a probability of unity, Eq. (4.67) becomes

$$\Delta A = n\Delta u - kT\ln(W) \tag{4.68}$$

Substituting for W, and using Stirling's approximation for factorials, $\ln X! = X \ln X - X$, ΔA becomes:

$$\begin{aligned} \Delta A &= n\Delta u - kT[N\ln(N) - N] - [(N-n)\ln(N-n) - (N-n)] - [n\ln(n) - n] \\ &= n\Delta u - kT[N\ln(N)] - [(N-n)\ln(N-n)] - [n\ln(n)] \end{aligned}$$

At equilibrium, the Helmholtz free energy change, $(\partial \Delta A/\partial n)_T = 0$, whereupon:

$$0 = \Delta u - kT\ln\frac{N-n}{n} \tag{4.69}$$

Since N is very much larger than n, the approximation $(N-n) = N$ is valid, and Eq. (4.69) rearranges to:

$$n = N\exp[-\Delta u/(kT)] \tag{4.70}$$

In ionic crystals, it is energetically favourable for Schottky defects to be formed in pairs, one of each sign in order to maintain the best electrical balance. Then W is squared, and Eq. (4.70) may be written as

$$n_{\pm}/N = \exp[-\Delta u/(2kT)] \tag{4.71}$$

and $n_{\pm}/N$ is the defect concentration.

The average nearest neighbour bond energy in a solid is of the order of 1 eV, and this amount of energy has to be expended to create a vacancy. The structure then readjusts itself around the vacancy and approximately two thirds of the energy expended is recovered in the process. The entropy change is positive and contributes to the driving force for the creation of defects, according to Eq. (4.67).

Example 4.5

In potassium chloride at 300 K, Δu for a pair of K^+ Cl^- defects is 1 eV, or 96.485 kJ mol^{-1}. Thus, the ratio of the number of defects to the number of sites, the defect concentration, is

$$n_{\pm}/N = \exp[-96485.5 \text{ J mol}^{-1}/(300 \text{ K} \times 8.3145 \text{ J K}^{-1} \text{ mol}^{-1})] = 1.59 \times 10^{-17}$$

If N is set equal to L, then $n_{\pm} = 9.56 \times 10^6$, or approximately 1 defect pair per 6×10^{16} sites. At 600 K this number is increased to 1 in 2.5×10^8.

Another zero-dimensional defect is the *Frenkel defect*. It can occur at either an anionic or a cationic site, anionic being the most usual. It implies the existence of a vacant cation site, the ion having been transferred to an interstitial position within the bulk crystal. An example of this defect (Fig. 4.26b) is found in silver chloride. It is necessary now to take into the probability equation the N' interstices, as well as the N sites and n defects. The probability can be evaluated by following the procedure for the Schottky defect concentration, and the result for the probability of Frenkel defects is

$$W = N!N'!/\{[(N-n)!n!(N'-1)!n!]\} \tag{4.72}$$

whereupon

$$n = (NN')^{\frac{1}{2}} \exp[-\Delta u/(2kT)] \tag{4.73}$$

for the creation of n Frenkel defects consisting of the *interstitial ion* and its *vacant site*, or hole. Frenkel defects do not have to be considered in pairs.

The reason that silver chloride exhibits Frenkel defects is probably related to its greater degree of covalent character in comparison with potassium or sodium chloride. The d electrons provide poorer screening than would be expected for their number, so that the tendency to covalency is enhanced *via* increased polarization, particularly as the cation is displaced preferentially. Moreover, the silver cation is under compressive stress in an interstitial position, a situation that would be expected to increase further the covalent character. Frenkel defects do not lead to a change in crystal density, whereas a high concentration of Schottky defects causes sufficient vacancies in the structure to be reflected in a decrease in bulk crystal density.

Example 4.6

In potassium chloride, 1 in 10^4 K^+ sites are replaced by Ca^{2+} ions. Assuming that all Cl^- sites are occupied, determine the stoichiometry of the compound. A first trial structure as $K_{0.99}Ca_{0.01}Cl$ proves unsatisfactory as the compound is no longer electrically neutral; the excess charge is +0.01 e-units. Therefore, it is necessary that one K^+ site becomes vacant for every Ca^{2+} added, leading to the formula $K_{0.98}Ca_{0.01}Cl$. This is satisfactory since the positive charge per cation site is $0.98(+1) + 0.01(+2) = +1$.

Example 4.7

The body-centred cubic unit cell of iron has an edge length of 286.7 pm. (a) Determine the density of iron. (b) Calculate the density if 0.15% of the iron atom sites become vacant.

(a)   The body-centred unit cell of iron contains 2 atoms, and the relative molecular mass of the atom is 55.845. Thus, the density D calculates as $2 \times 55.845 \times 1.6605 \times 10^{-27}$ kg$/(0.28665 \times 10^{-9}$ m$)^3 \approx 320 \times 0.027 \approx 8$; specifically, 7.874 kg m^{-3}.

(b)   Since $D \propto$ mass, the density of the defect iron structure is 7.874 – (0.0015 × 7.874 kg m^3) = 7.862 kg m^3.

4.11.2 Defects of higher dimensions

One type of one-dimensional defect involves a slip system consisting of a *slip plane* and a *slip direction*. It is observed in elemental metals, particularly those with close-packed planes. In the face-centred cubic metals, a slip plane can be any one of four (111) planes in a cube, and the slip direction any one of six [110] directions. The sequence of close-packed planes is *ABCABCA...* in the cubic close-packed structures and *ABABA...* in the hexagonal close-packed structures. A slip sequence may generate a combination, such as *ABCABCABABCA...* in a cubic structure.

The magnitude and direction of a defect is quantified by the *Burgers vector*. This vector is a measure of the lattice distortion caused by a defect, and is illustrated below by the edge dislocation diagram, which shows also the convention for measuring the Burgers vector.

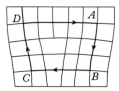

Distorted lattice

A circuit *ABCDA* is described around a *dislocation line* in a clockwise direction, with each step of the circuit connecting fully coordinated lattice sites. A trace is made around the dislocation plane from A to form a closed loop, and the total number of lattice vectors travelled along the sides of the loop recorded, 15 in this example. Then, the same path $A'B'C'D'A'$ is traced in an ideal lattice of the same type, moving the same number of lattice vectors along each direction as before. Since there is no dislocation in a perfect lattice, this circuit fails to close on itself, and the vector linking the end of the circuit of 15 translations to the starting point is the Burgers vector, $\mathbf{b} = VA$.

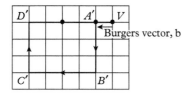

Ideal lattice, showing the Burger vector VA

For the (111) slip planes in the close-packed cubic structure type $|\mathbf{b}| = (a/2)\sqrt{U^2 + V^2 + W^2} = (a/2)\sqrt{2}$ for a [110] slip direction. The study of the extensive topic of crystal defects repays further attention, and the reader is directed to the literature [37, 38].

4.11.3   Colour centres

Many materials when exposed to the vapour of a component element or to high energy radiation exhibit colour. The colour is associated with defect inclusions in the structure, and the name colour centre or F centre (Ger. *Farbenzentrum =* colour centre) arises from the fact that early work on these materials was carried out in Germany.

4.11.3.1   F centres

Alkali-metal halides heated in alkali-metal vapour, or exposed to x-radiation, acquire a colour. The crystal contains an excess of cations and as a consequence some anion vacancies occur. A single electron from a chloride ion may remain delocalized at a vacant Schottky site, surrounded by six cations. Such an electron constitutes an *F centre*. The F-centre electron has a hydrogen-like character, and undergoes a $1s \rightarrow 2p$ transition. The electron energy can be calculated approximately by treating the system as an electron in a three-dimensional box. From Eq. (2.47) with $a = b = c$ for cubic potassium chloride ($a = 0.6295$ nm) and $n_1 = n_2 = n_3$, since $n = 1$ for the transition $1s \rightarrow 2p$, the energy is given by:

$$E = \frac{3h^2}{8m_e a^2}$$

which, on evaluation, gives $E = 4.561 \times 10^{-19}$ J, or 2.85 eV, which is in fair agreement with the experimental value of 2.29 eV. The corresponding wavelength is 435 nm, which is in the blue region of the spectrum, and the complementary colour is yellow, as observed.

Fig. 4.27 Specimens of 'smoky' quartz, SiO_2. (a) Naturally occurring mineral specimen; the hexagonal prismatic crystal habit is evident on the crystal. (b) Semi-precious, smoky quartz gem quality ring. [Reproduced by courtesy of Mineralminers.com.]

A naturally occurring mineral with a colour arising from F centres is 'smoky' quartz (Fig. 4.27a). The colour of this mineral arises from irradiation together with traces of elemental aluminium included in the crystal structure. Aluminium replaces silicon to form an $[AlO_4]$ structural group of nominal charge minus five. In order to compensate the imbalance of charge arising from this replacement, small singly-charged ions, such as H^+, Li^+ or Na^+, are built into the structure. Radiation releases an electron from the $[AlO_4]$ group which then reacts with H+ forming atomic hydrogen in the structure and creating a colour centre. High concentrations of hydrogen ion tend to interfere with this process, which takes place below *ca.* 40°C; thus, the smoky colour of the crystals appears long after crystal growth has taken place, probably several million years. The mineral can be formed into highly attractive semi-precious gemstones and jewellery (Fig. 4.27b).

4.11.3.2   H and V centres

As well as F centres, *H centres* and *V centres* are recognized. If an alkali halide crystal is heated in a halogen vapour, additional halogen ions are introduced into the structure; the accompanying cation vacancies trap holes like an anion vacancy traps an electron in an F centre. A single negative charge becomes spread over two atoms forming a $[Cl_2]^-$ species; it corresponds to a removal of a metal atom from the system. An H centre is, in a sense, the opposite of an F centre; hence, an F centre and an H centre can combine to annul the defect in the crystal structure.

Cl	Na	Cl	Na	Cl
Na	Cl	Na	e	Na
Cl	Na	Cl	Na	Cl

F centre

Cl	Na	Cl	Na	Cl
Na	(Cl-Cl)$^-$			Na
Cl	Na	Cl	Na	Cl

H centre

A series of colour centres that contain an excess of alkali-metal atoms have holes in place of electrons and are termed *V* centres. Crystals with these centres show several absorption bands which arise from the transitions of electrons into holes by a 'hopping' mechanism, somewhat different from the propagative motion of electrons in metals.

4.11.4   Electrical properties in defect ionic structures

Notwithstanding it has been indicated that ionic solids are electrical insulators, some ionic compounds exhibit electrical conduction that is dependent upon the nature and concentration of defects in the crystal. A defect concentration can be increased by heating the crystal to a high temperature and then quenching it rapidly. The defects at the high temperature become locked into the structure and, because ion mobility is related to defect concentration, enhanced electrical conduction results.

Point defects provide three mechanisms for electrical conduction by ion transport. In the *vacancy mechanism*, which can be pictured by referring back to Fig. 4.26a, ions move in jumps through the structure to the surface, where an electrical circuit can be set up to measure the conductance of the solid. A second process is the *hole mechanism*, or the movement of a hole, whereby an ion moves through the lattice by entering one vacant site after another. This, too, can be seen from Fig. 4.26a: if a hole exists in the structure, an ion can move into it, which is equivalent to the hole moving into the site previously occupied by the ion. Finally, there is an *interstitial mechanism* (Fig. 4.26b) in which the interstitial ion moves from one compression site to another carrying its charge with it.

The mobility of an ion is its drift speed under unit applied field strength; it is often quoted in the unit $cm^2\ s^{-1}\ V^{-1}$, which implies a rate of movement of a surface under an applied potential of 1 V. Studies by radioactive tracer techniques can be used to determine solid-state conductivities.

In a typical experiment with sodium chloride, a thin slice of $^{24}NaCl$ was placed between two plates of non-radioactive $^{23}NaCl$ and the composite maintained at a constant temperature. The defect concentrations throughout the composite were determined by cutting thin sections of the composite after given times and then counting the radioactivity in each section. The diffusion of the tracer is related to the concentration *c*, an elapsed time *t* and a distance *d* from the surface, by the equation:

$$\ln c = \alpha - d^2/(Dt) \tag{4.74}$$

where α is a constant and *D* is the *diffusion coefficient* for the substance under examination and varies with temperature according to the Boltzmann equation:

$$D = D_0 \exp[-E_d/RT)] \tag{4.75}$$

In this equation E_d is the energy of activation for diffusion, and may be determined by measuring D at several temperatures and then plotting it as a function of $1/T$. In sodium chloride, E_d is 173 kJ mol^{-1}, which includes the energy needed both to create the vacancy and to move an ion into the vacant site. For a single ion, and using data from Example 4.5:

$$E_d = E_\mu + 96 \text{ kJ mol}^{-1} \tag{4.76}$$

where E_μ is the energy needed to induce migration, 77 kJ mol^{-1} in this example.

Most simple ionic crystals have low defect concentrations at ambient temperature, but change significantly with a change of temperature, as the following results show:

T/K	$E_d = 60$ kJ mol^{-1}	$E_d = 300$ kJ mol^{-1}
300	5.7×10^{-6}	6.1×10^{-27}
1000	2.7×10^{-2}	1.4×10^{-8}

4.11.4.1 Fast-ion conductors

Certain mainly ionic crystals are endowed with unusually high electrical conductivity, in some cases almost the equivalent of 0.1 molar aqueous potassium chloride. Silver iodide, itself a conductor (*ca.* 130 S m^{-1}),[7] in the high temperature α-AgI phase reacts with alkali-metal iodides to form compounds $M\text{Ag}_4\text{I}_5$ ($M =$ K, NH$_4$, Rb) and also $\text{K}_{1/2}N_{1/2}\text{Ag}_4\text{I}_5$ ($N =$ Rb, Cs) which have high electrical conductivity and are termed *fast-ion conductors*.

These halides are isomorphous, with cubic unit-cell sides ranging from 1.113 to 1.125 nm. Madelung constants and crystal energies have been calculated for these halides [39], and $U(r_e)$ ranges from –487 kJ mol^{-1} to –492 kJ mol^{-1}. The compounds crystallize with the cubic space group $P4_132$, which is uncommon for ionic crystals, with four formula entities per unit cell [40–42].

In the crystal structures, the M ions take up a distorted octahedral coordination, with silver cations at the centre of iodide ion tetrahedra; these polyhedra share faces, thereby providing a mechanism for the diffusion of silver ions, and thus endowing the crystal with the property of electrical conductivity. In addition, the space group data indicates that not all silver atom sites demanded by space group rules are occupied. The Rb$^+$ and I$^-$ ions in RbAg$_4$I$_5$ are in fixed positions in the structure, but the Ag$^+$occupancy is random with site occupancies less than unity; this feature enhances electrical conduction as these ions are free to move through the structure, and do so on application of a potential gradient.

4.11.4.2 Doping

Defects can be created by introducing foreign ions into a structure. Sodium chloride can be *doped* by crystallization from the melt containing *ca.* 0.5% of calcium

---

[7] Written as mho cm^{-1} in earlier literature.

chloride. Because Ca^{2+} has a radius similar to that of Na^+, 0.118 and 0.112 nm, respectively, it can fit into the structure; electrical neutrality is maintained by ensuring one vacant Na^+ ion site for each Ca^{2+} ion replacing Na^+ in the crystal. The colours that arise from the defects produced by doping bring with them important changes in physical properties. In lithium niobate $LiNbO_3$ for example, the refractive index variations associated with the creation of F centres by doping are very large, and the material can be used for storing holographic images.

4.11.5 Image plates

Of particular importance in medical x-ray procedures and in recording data obtained from crystalline material by x-ray, neutron or electron diffraction is the *image plate*, a film-like radiation sensor. An image plate detector is shown diagrammatically in Fig. 4.28. It comprises several sections.

The imaging plate itself (a) consists of a barium halide phosphor doped with divalent europium BaFBr.Eu(II) which is held in a substrate of organic binder. On exposure to a high-energy radiation, a photon impinges on the image plate and *luminescent radiant* energy is stored in a Br^- 'hole', an F-centre defect, and oxidizes europium to the Eu(III) state (b) which remains metastable until further stimulated:

$$Eu(II) \xrightarrow{h\nu} Eu(III) + e^-$$

When next the image plate is stimulated by a helium-neon laser (c) of wavelength approximately 600 nm, a value longer than that of the luminescence wavelength from the first excitation, it reverses the excitation process and emits violet light of wavelength *ca.* 390 nm which is of intensity proportional to the absorbed energy:

$$Eu(III) + e^- \xrightarrow{laser} Eu(II) + h\nu \text{ (blue-violet)}$$

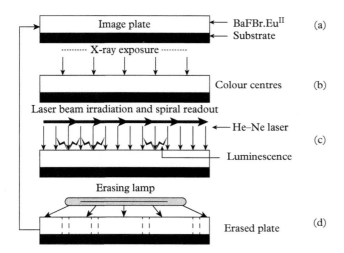

Fig. 4.28 Image plate exposure and cleaning cycle. (a) Resting state. (b) Exposure to radiation, leading to colour centres. (c) Scanning by laser light and measurement of emitted light by photomultiplier. (d) Cleaning plate for reuse. [*Structure Determination by X-ray Crystallography*, 5thed. 2014; reproduced by courtesy of Springer Science+Business Media, NY.]

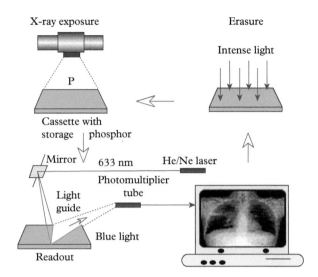

Fig. 4.29 Diagrammatic illustration of the use of the image plate in medical x-ray diagnosis, in this example a chest examination; the subject under examination is at *P*, between the x-ray source and the image plate. The plate is scanned, digitized and displayed on a monitor screen. Subsequently, it is exposed to intense yellow light ready for re-use. [Reproduced by courtesy of Dr K Maher.]

This light is then detected by a photomultiplier system, integrated and digitized. After the stored data has been processed, the plate is cleaned by exposure to bright yellow light and is then ready for reuse (d). The europium phosphor utilizes the process of *photostimulated luminescence* (PSL) which is different from both fluorescence and phosphorescence, but combines elements of both.

An important application of PSL is in the field of medical diagnosis. Here, the processed image may be printed either in colour or greyscale, or the digitized data fed into a computer system and viewed directly, thus providing a rapid diagnostic tool. A diagrammatic procedure is illustrated in Fig. 4.29; the subject under examination is at *P*, between the radiation source and the image plate.

In x-ray crystallography, which uses either x-rays or neutrons as the primary exciting radiation with a crystal, the digitized data are intensities of diffraction spectra from the crystal. They are processed to provide indexed crystallographic structure amplitudes, which are the first stage in determining the structure of the crystalline material on an atomic scale. The procedure is well documented in the literature [43, 44].

REFERENCES 4

[1] Born M. *Verhandl. Deut. Phys. Ges.* 1919; 21: 679.
[2] Fajans K. *ibid.* 21: 714.
[3] Haber F. *ibid.* 21: 750.
[4] Madelung E. *Phys. Z.* 1918; 19: 524.
[5] Evjen HM. *Phys. Rev.* 1932; 39: 675.
[6] Born M. and Landé A. *Verhandl. Deut. Phys. Ges.* 1918; 20: 210.

[7] Mayer JE. *J. Chem. Phys.* 1933; 1: 327.

[8] Ladd MFC. *J. Chem. Phys.* 1974; 60: 1954

[9] Shanker J *et al. J. Chem. Phys.* 1978; 69: 670.

[10] Ladd MFC and Lee WH. *Acta Crystallogr.* 1960; 13: 959.

[11] Ladd MFC and Lee WH. *Progress in Solid State Chem.* Vol 1 p 37. Elsevier B.V. 1964.

[12] *idem. ibid.* 1965; 2: 378.

[13] *idem. ibid.* 1967; 3: 265.

[14] Zhang W. *Science* 2013; 342: 1502

[15] Ladd MFC. *J. Chem. Soc. Dalton Trans.* 1977; 220.

[16] Bartlett N. *Proc. Chem. Soc.* (Lond.), 1962; 218.

[17] Barlow W. *Nature* 1883; 28: 186.

[18] *idem. Z. Krist. Min.* 1898; 29: 433.

[19] Bragg WH and Bragg WL. *X-rays and Crystal Structure.* G. Bell and Sons, 1915.

[20] Bragg WL *Phil. Mag.* 1920; 20: 169.

[21] Landé A. *Phys.* 1920; 1: 191.

[22] Wasastjerna JA. *Commun. Phys.-Math. Soc. Sci. Fenn.* 1923; 1: 1.

[23] Goldschmidt VM. *Skrifter Norske-Videnskaps-Akad. Oslo* 1926; 1: Mat. Natur.

[24] *idem. Naturwiss.* 1926; 21: 477.

[25] Pauling L. *Proc. Roy. Soc.* (*Lond.*) 1927; A114: 181.

[26] Ladd MFC. *Theor. Chim. Acta* 1968; 12: 333.

[27] Witte H and Wölfel E. *Z. Phys. Chem.* 1955; 3: 296.

[28] *idem. Rev. Modern Phys.* 1958; 30: 51.

[29] Shannon RD and Prewitt CT. *Acta Crystallogr.* 1969; B25: 925.

[30] *idem. ibid.* 1970; B26: 1046.

[31] Bragg WL. *The Crystalline State*, Vol. 1. Bell, 1933.

[32] Ellis B *et al. Faraday Disc.* 2009; 134: 119.

[33] Wang Y *et al. Energy Environ. Sci.* 2011; 4: 805.

[34] Rossini FD *et al. Selected Values of Chemical Thermodynamic Properties.* Circular No. 500 (and Supplements), National Bureau of Standards, 1952.

[35] Latimer WM *Oxidation Potentials*, 2nd ed. Prentice-Hall, 1952.

[36] Ladd MFC and Lee WH. *Trans. Faraday Soc.* 1958; 34: 54.

[37] Shockley W *et al. Imperfections in Nearly Perfect Crystals.* Wiley, 1952.

[38] Kelly AA and Knowles KM. *Crystallography and Crystal Defects.* Wiley, 2012.

[39] Ladd MFC and Lee WH. *Z. Krist.* 1969; 129: 57.

[40] Geller S. *Science* 1967; 157: 310.

[41] Bradley JN and Greebe PD. *Trans. Faraday Soc.* 1967; 63: 2516.

[42] Hull S *et al. J. Solid State Chem.* 2002; 165: 363.

[43] Ladd M and Palmer R. *Structure Determination by X-rays: Analysis by X-rays and Neutrons*, 5th ed. Springer, 2013.

[44] Clegg W *et al. Crystal Structure Analysis: Principles and Practice.* OUP/IUCr, 2001.

Problems 4

4.1. The convergence limit for the process:

$$Tl(g, \text{ ground state}) \rightarrow Tl^+(g) + e^-$$

is 49,250 cm^{-1}. Calculate the first ionization energy in (a) eV, and (b) kJ mol^{-1}.

4.2. The following values of vapour pressure p of molten lead were derived from measurements of the rate of effusion of the vapour, at different temperatures T, through a small hole into a vacuum:

T/K	895.4	922.1	964.5	1009.7	1045.5
$p/N\ m^{-2}$	0.0783	0.205	0.539	1.40	3.40

Determine the enthalpy of vaporization of lead over the given temperature range.

4.3. Refer to Problem 4.2. The molar heat capacity of lead may be represented by the equations:

298 K to 600 K:	$C_{p,m}/J\ K^{-1}\ mol^{-1} = 23.56 + 0.009757T$
600 K to 1200 K:	$C_{p,m}/J\ K^{-1}\ mol^{-1} = 32.43 - 0.003100T$

The enthalpy of fusion of lead at the melting point 600 K is 4.81 J mol^{-1}. If the value of ΔH_{vap} obtained already be taken to apply to the average temperature 970 K, obtain a value for $\Delta H^{\ominus}_{vap}$, the standard enthalpy of sublimation; it may be assumed that the vapour behaves ideally. It may help first to construct a thermodynamic cycle for the process, with an appropriate equation.

4.4. Given the following data, show how, on reaction of elemental magnesium with chlorine gas, formation of the compound $MgCl_2$ is preferred to that of MgCl. $I_1(Mg) = 6.09$ eV; $I_2(Mg) = 11.82$ eV; $S_M(Mg) = 149.0$ kJ mol^{-1}; $D_0(Cl_2) = 243.0$ kJ mol^{-1}; $E(Cl) = -348.6$ kJ mol^{-1}; $\Delta H_f(MgCl, s) = -221.8$ kJ mol^{-1}; $\Delta H_f(MgCl_2, s) = -641.8$ kJ mol^{-1}.

4.5. Use the following data to construct a thermodynamic cycle for the enthalpy of formation of solid ammonium chloride:

		$-\Delta H^{\ominus}$ kJ mol^{-1}
$\tfrac{1}{2}N_2(g) + \tfrac{3}{2}H_2(g)$	$\rightarrow NH_3(g)$	46.0
$NH_3(g) + aq$	$\rightarrow NH_4^+(aq) + (OH)^-(aq)$	34.7
$\tfrac{1}{2}H_2(g) + \tfrac{1}{2}Cl_2(g)$	$\rightarrow HCl(g)$	92.5
$HCl(g) + aq$	$\rightarrow H_3O^+(aq) + Cl^-(aq)$	74.9
$NH_4^+(aq) + (OH)^-(aq) + H_3O^+(aq) + Cl^-(aq)$	$\rightarrow NH_4^+(aq) + Cl^-(aq)$	52.3
$NH_4Cl(s) + aq$	$\rightarrow NH_4^+(aq) + Cl^-(aq)$	-15.1

4.6. (a) Use the spherical shell model (Section 4.4.1) to calculate an approximate value for the Madelung constant of sodium chloride; $a = 0.564$ nm. (b) Use Evjen's method on a cube of the sodium chloride structure of side $2a$. (c) How does the result in (b) relate to any of Pauling's empirical rules of crystal chemistry?

4.7. From the following data on magnesium oxide MgO, calculate the affinity of oxygen for two electrons:

Structure type	NaCl
Unit-cell side a	0.4212 nm
Madelung constant	1.74756 (remember the ionic charges)
ρ/r_e (for MgO)	0.143
I_1 (Mg)	737.7 kJ mol^{-1}
I_2 (Ca)	1451 kJ mol^{-1}
S_M (Ca)	148.1 kJ mol^{-1}
D_0 (O$_2$)	489.9 kJ mol^{-1}
$\Delta H_f^{\ominus}$ (CaO, s)	-601.8 kJ mol^{-1}

The terms ΔH_1 *and* ΔH_2 in Eq. (4.16) cancel each other in this example.

4.8. From the data below, together with data on the dissociation energies and electron affinities of the halogens, and the enthalpies of dissolution of the strontium halides, all given in the text, determine an average value for the standard enthalpy of hydration of the strontium ion Sr^{2+}(g).

	$-\Delta H^{\ominus}$ kJ mol^{-1}
I_1 (Sr)	-549.4
I_2 (Sr)	-1064.0
S_M (Sr)	-163.6
$\Delta H_f^{\ominus}$ (SrF$_2$, s)	1209.0
$\Delta H_f^{\ominus}$ (SrCl$_2$, s)	828.0
$\Delta H_f^{\ominus}$ (SrBr$_2$, s)	715.5
$\Delta H_f^{\ominus}$ (SrI$_2$, s)	569.4
$\Delta H_h^{\ominus}$ (F$^-$, s)	513.0
$\Delta H_h^{\ominus}$ (Cl$^-$, g)	371.1
$\Delta H_h^{\ominus}$ (Br$^-$, g)	340.6
$\Delta H_h^{\ominus}$ (I$^-$, g)	301.7

4.9. The sulphate ion may be considered as a regular tetrahedral arrangement of oxygen atoms equidistant around the sulphur atom. If the charge on oxygen is $-0.7e$ and the S–O bond length is 0.130 nm, calculate the electrostatic self-energy of the sulphate ion.

4.10. The radii $r(Cs^+)$ and $r(I^-)$ are 0.169 nm and 0.212 nm respectively based on 6:6 coordination, and the departure from additivity of ionic radii is +3% for 8:8 coordination. Assuming close contact between anions and cations, calculate the density of crystalline cesium iodide.

4.11. In rutile TiO_2, the coordination around titanium is sixfold, but not regular; the titanium atoms around each oxygen atom form an isosceles-triangular arrangement. Use the following data to calculate two unique Ti–O bond lengths and bond angles. Tetragonal crystal system; unit cell data $a = b = 0.4593$ nm, $c = 0.2959$ nm; two formula-entities per unit cell. Fractional coordinates of the atoms:

Ti at 0, 0, 0; ½, ½, ½

O at x, x, 0; $-x$, $-x$, 0; ½ + x, ½ - x, ½; ½ - x, ½ + x, ½, with $x = 0.3056$.

It may be helpful first to make a sketch of the rutile structure, from which two different Ti–O bond lengths and bond angles should become apparent.

4.12. (a) Show that the radius ratio R_4 lower limit for the wurtzite (ZnS) structure type is 0.225. (b) Using the data below, determine the shortest Zn–S distance; $a = 0.3823$, $c = 0.6261$. (c) What would be the ratio for an equilateral-triangularly, close-packed arrangement of ions of two species? Hint: The angle between $2\mathbf{r}_-$ and $\mathbf{r}_+ + \mathbf{r}_- = 30°$.

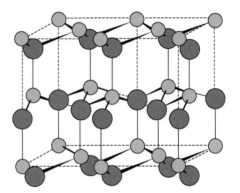

Two unit cells of the wurtzite α-ZnS crystal structure

Fractional coordinates of Zn and S:

Zn	0,	0,	0;	1/3.	1/3,	1/2
S	0,	0,	0.382;	1/3,	1/3,	1/2 + 0.375

4.13. In an experiment, precipitated silver iodide was dissolved in aqueous solution of potassium iodide and sodium iodide; the heats evolved were 9.45 kJ mol^{-1} and 7.46 kJ mol^{-1}, respectively. Next, finely divided silver was suspended in similar solutions of the alkali-metal iodides. On adding iodine, then silver dissolved rapidly to form silver iodide; the heats evolved were 72.0 kJ mol^{-1} (of AgI) in the potassium iodide solution and 70.3 kJ mol^{-1} in the sodium iodide solution. Set up equations to represent the chemical reactions taking place, and calculate an average value for the enthalpy of formation of silver iodide; all processes may be assumed to have taken place at 25°C.

4.14. (a) Determine the standard free energy of dissolution of magnesium fluoride, given that $\Delta H_d^{\ominus} (MgF_2) = -18.4$ kJ mol^{-1}, $S_c^{\ominus} = 57.3$ J K^{-1} mol^{-1} and $\sum S_i^{\ominus}$ for the hydrated ions is -137.2 J K^{-1} mol^{-1} with respect to the appropriate standard states. (b) The solubility of magnesium fluoride is 0.075 mol dm^{-3} at 25°C; calculate an approximate value for the mean ionic activity coefficient $f_{\pm}$ for this salt.

4.15. (a) Derive an expression for the number n of Frenkel defects in a crystal of silver chloride containing N ion sites and N' interstices. (b) If the energy needed to create a single defect (plus hole) in silver chloride is 1.5 eV, calculate the fraction of Frenkel defects as 500 K.

4.16. The probability of Schottky defects in potassium chloride is represented as $n/N = \exp[-\Delta H/(2kT)]$. What is the corresponding probability for calcium chloride?

4.17. Radioactive silver was allowed to diffuse through a silver/indium alloy at 100 K. The penetration depth x was determined after time intervals t of 6×10^4 s by measuring the radioactivity β_t. Show that the diffusion process follows the equation:

$$\beta_t = A \exp[-x^2/(Dt)]$$

Find also the constants A and D of the equation. The following data were recorded, where β_t are in dimensionless units:

x/mm	β_t	x/mm	β_t
0.000	600	0.329	88
0.084	540	0.376	50
0.132	450	0.425	25
0.183	360	0.470	12
0.230	250	0.520	5
0.279	160	0.568	2

4.18. In a further radioactive experiment, the values of the diffusion coefficient D were obtained as a function of temperature T. Given that D is related to T by the Boltzmann equation:

$$D = D_0 \exp[-U/(RT)]$$

determine the molar activation energy U for the diffusion process and its estimated standard deviation, using the following data:

T/K	878	1007	1176	1253	1322
$D/\text{m}^{-2}\,\text{s}$	1.6×10^{-18}	4.0×10^{-17}	1.1×10^{-15}	4.0×10^{-15}	1.0×10^{-14}

4.19. Calculate the standard molar entropy of nickel from the following data:

T/K	15.05	25.20	47.10	67.13	82.11	133.4	204.1	256.5	283.0	298.0
$C_{p,\text{m}}/\text{J K}^{-1}\,\text{mol}^{-1}$	0.1945	0.5994	3.532	7.639	10.10	17.88	22.72	24.81	26.09	26.24

Between 15.05 K and 0 K, the Debye approximation may be used: $C_{p,\text{m}} = aT^3$, where a is a constant.

4.20. Calculate the Bragg angle θ for the 300 reflection from cesium iodide, given the cell side $a = 0.4562$ nm and the wavelength λ of the X-radiation is 0.15418 nm. Would the intensity of this reflection be weak or strong?

4.21. Determine the equilibrium number n of Frenkel defects in 1 m^3 of copper at 1000 °C, given the following data: face-centred cubic unit cell with four atoms per unit cell of side a equal to 0.3615 nm, relative molar mass M_r is 63.546, energy for the creation of a single defect vacancy is 0.901 eV.

Metallic Compounds

Each metal has a certain power, which is different from metal to metal, of setting the electric fluid in motion.

Alessandro Volta

5.1 Introduction

The fourth section of those into which solids have been classified in this book encompasses metals, semiconductors, superconductors and alloys. Perhaps the feature that distinguishes metals most clearly from other classes of compounds is conduction, particularly that of electricity and heat. Conduction arises in a material because of a readiness in its structure to allow movement of its components under the application of appropriate stimuli, such as thermal or electrical gradients.

About three-quarters of the known elements are metals and their structure types are, in the main, few in number and geometrically simple. Little was known about these structures before the advent of x-ray diffraction, although some predictions had been made with intuitive correctness (Section 4.6).

It should be no surprise that metallic bonding involves an electrostatic attraction between positive and negative particles, and that the nature and distribution of these particles determine metallic bonding and its associated properties. The theory of metals, like those of other substances, evolved in stages, each one improving upon the applicability of its predecessor.

5.2 Drude free-electron theory

The earliest quantitative theory of metallic bonding, often called the *classical free-electron theory* was promulgated by Drude [1, 2] three years after the discovery of the electron. It draws upon certain ideas embodied in the kinetic theory of gases. Electrons in a metal were assumed to move freely among an array of effectively stationary metal atoms. The path of an electron was straight until colliding with

Bonding, Structure and Solid-State Chemistry. First Edition. Mark Ladd.
© Mark Ladd 2016. Published in 2016 by Oxford University Press.

a metal atom, whereupon a new path was followed with a speed governed by the temperature at the point of collision. The model takes on the following features:

- The inner electrons (core) remain bound to a lattice-like array of positive ions on fixed sites, but the valence electrons form an *electron gas or density distribution* (aka 'a sea of electrons') pervading the whole crystal;

- The electrons have an *average thermal energy* $(3/2)kT$ (Appendix A5) similar to that of an ideal gas;

- The electrons execute random motions through the metal, so that their *average velocity* $\bar{\mathbf{v}}$ *is zero* although $\sqrt{\overline{\mathbf{v}^2}}$ is greater than zero;

- Random motion of electrons results from *collisions with the positive ions*, but not with each other, and since the positive ions have a large mass compared with that of electrons, they are essentially static;

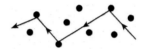

Random motion of free electrons

- The distance travelled by an electron between collisions is the *mean free path* λ and the time between successive collisions is the *relaxation time* (aka collision time) τ;

- The outer, free electrons are influenced by an applied field, be it electrical, thermal or optical, and the term *conduction electrons* is often used to describe these mobile electrons. As they are assumed to behave ideally, and their average kinetic energy is $(3/2)kT$; it follows that their average speed is $\sqrt{3kT/m_e}$, or approximately 1.2×10^5 m s^{-1} at 298 K ($\frac{1}{2}m_e v^2 = 3kT/2$; see also Fig. 5.2).

5.3 Electrical conductivity

In an applied field of strength $\mathbf{E}$, electrons acquire a *drift velocity* $\mathbf{v}_d$ that is superimposed on to that arising from their thermal motion. It is assumed that the probability that an electron collides with an ion during a time interval dt is proportional to dt/τ, τ being the relaxation time. From Newton's second law:

$$m_e \left(\frac{d\mathbf{v}_d}{dt} + \frac{\mathbf{v}_d}{\tau} \right) = -e\mathbf{E} \tag{5.1}$$

where e is the charge on an electron.

At a time $t = 0$ the field is set at zero, whereupon the drift velocity relaxes as $\mathbf{v}_d = \mathbf{v}_0 \exp(-t/\tau)$. In the steady state $\dot{\mathbf{v}}_d = 0$, so that from Eq. (5.1):

$$\mathbf{v}_d = (-e\tau/m_e)\mathbf{E} \tag{5.2}$$

where $(e\tau/m)$ is the drift velocity per unit field, or *drift mobility* μ. The velocity of an electron including both the thermal $\mathbf{v}'$ and the drift components is then

$$\mathbf{v} = \mathbf{v}' - e\tau\mathbf{E}/m_e \tag{5.3}$$

The *current density* $\mathbf{j}$ arising from the applied field is given by:

$$\mathbf{j} = \frac{1}{V}\sum_e -e\mathbf{v} \tag{5.4}$$

where the subscript 'e' implies the sum over all electrons in the system. For the thermal electrons (electrons flowing by heat activation), randomness implies $\sum_e \mathbf{v}_{Th} = 0$. Hence:

$$\mathbf{j} = \frac{N}{V}(-e)\left(\frac{-e\tau}{m_e}\right)\mathbf{E} = \sigma\mathbf{E} \tag{5.5}$$

and the *electrical conductivity* σ can be expressed variously as:

$$\sigma = n_c e^2 \tau/m_e = n_c e^2 \lambda/\left(m_e\sqrt{v^2}\right) = n_c e^2 \lambda/(3m_e kT)^{1/2} \tag{5.6}$$

where n_e is the electron concentration and $\sqrt{v^2}$ is the root mean square speed; n_c may be expressed also as N/V, where N is the number of electrons in a volume V of the species.

5.3.1   Drude–Lorentz model of metals

An enhancement of the Drude model by Lorentz [3] employed the Maxwell–Boltzmann distribution of velocities in place of Drude's assumption that every electron had the same speed. The argument is lengthy and the result obtained for electrical conductivity is:

$$\sigma = \frac{4n_c e^2 \lambda}{3(2\pi m_e kT)^{1/2}} \tag{5.7}$$

which differs from Drude's result only in a numerical factor of approximately 1.09 and need not be considered further in this discussion. The *Drude–Lorentz theory* held sway for about three decades.

5.3.2   Ohm's law

A current I flowing in a conductor is related to the potential drop V along it by Ohm's law:

$$V = IR \tag{5.8}$$

where R is the resistance of the conductor, and depends on its dimensions.

It is defined in an equivalent manner, but not involving the dimensions of the conductor, by Eq. (5.5), $\mathbf{j} = \sigma\mathbf{E}$. The current density vector $\mathbf{j}$ is parallel to the charge flow through the conductor, and measures the amount of charge per unit time passing through a unit cross section of the conductor. If a uniform direct current I flows along a wire of length l and cross-sectional area A, the magnitude j of the current density is I/A. The potential drop V is given in magnitude by El. Hence, from Eq. (5.8), $I/A = \sigma V/l$, whence:

$$R = l/(\sigma A) = \rho l/A \qquad (5.9)$$

where $\rho = 1/\sigma$ and is the *electrical resistivity* of the conductor. Both σ and ρ are parameters that are independent of the dimensions of the conductor.

5.3.3 Mean free path

The mean free path λ of an electron is its average distance of linear travel between collisions, and may be formulated as the produce of the average speed $<v>(=\sqrt{v^2})$ and the relaxation time:

$$\lambda = <v>\tau \qquad (5.10)$$

If the electrons in a metal are assumed to behave like an ideal gas, then the kinetic energy of an electron follows the equation:

$$\tfrac{1}{2}m_e v_d^2 = (3/2)kT \qquad (5.11)$$

from which the value for the average speed, reported above, is *ca.* 10^5 m s^{-1}; experimentally, this value is too small, and the energy term $\tfrac{1}{2}m_e v_d^2$ is better equated to three-fifths of a quantity E_F known as the Fermi energy, which will be discussed shortly.

Example 5.1

The Fermi energy for copper is 1.13×10^{-18} J; hence, $\sqrt{v_d^2} = 1.22 \times 10^6$ m s^{-1}, and the relaxation time can be determined if λ is known. From measurements of electrical resistivity of thin films of copper, λ has been determined as *ca.* 40 nm; hence, $\tau = 3.279 \times 10^{-14}$ s. In Eq. (5.6), $n_c = 8.5 \times 10^{28}$ m^{-3} (as will evolve shortly), whereupon $\sigma_{Cu} = 7.85 \times 10^7$ S m^{-1} (aka Ω^{-1} m^{-1}).

The experimental value for σ_{Cu} at 298.15 K is 6.00×10^7 S m^{-1}, so that the agreement with the early theory might be deemed good. It is a different matter, however, when examining certain other physical properties.

5.3.4 Hall effect

If an electric current I flows through a thin slice of conducting or semiconducting material in a magnetic field B, the field exerts a transverse force F_e on the

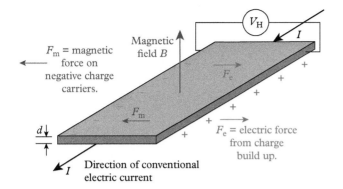

Fig. 5.1 Hall effect: a slice of conducting material in a magnetic field of strength B in the direction of the z-axis, with a current I flowing along the length of the slice (the x-axis). The Hall voltage V_H develops across the slice, perpendicular to both B and I, and is proportional to B. [Reproduced by courtesy of Professor R Nave, Georgia State University, Atlanta, GA 30302.]

moving charge carriers and causes them to move to one side of the material; the direction of movement follows Fleming's left-hand rule. A separation of charge is set up across the conductor so as to balance the magnetic effect, and this produces a voltage, the *Hall voltage* V_H, across opposite edges of the conducting slice (Fig. 5.1).

The time dependence of momentum of an electron is given by [4]:

$$\frac{d\mathbf{p}}{dt} = -e\left(\mathbf{E} + \frac{\mathbf{p}}{m_e} \times \mathbf{B}\right) - \frac{\mathbf{p}}{\tau} \tag{5.12}$$

where $\mathbf{p}$ is the momentum vector of the electron. In a steady state, the current is independent of time so that the components of momentum along x and y will be represented by

$$-eE_x - \omega p_y - p_x/\tau = 0$$

and $\tag{5.13}$

$$-eE_y - \omega p_x - p_y/\tau = 0$$

where $\omega = eB/(m_e)$. Utilizing Eqs. (5.5) and (5.6), and multiplying both equations by $-n_c e\tau/m_e$, it follows that:

$$\sigma E_x = \omega\tau j_y + j_x$$

and $\tag{5.14}$

$$\sigma E_y = \omega\tau j_x + j_y$$

The Hall field E_y has no transverse current since the component of the Hall field in that direction cancels the magnetic field, so that $j_y = 0$. Hence:

$$E_y = -(\omega\tau/\sigma)j_x = -(B/n_c e)j_x \tag{5.15}$$

The *Hall coefficient* R_H is defined by:

$$R_H = E_y/(j_x B) = -\frac{1}{n_c e} \tag{5.16}$$

Table 5.1 *Hall coefficients R_H for selected metals; electron concentrations n_c are calculated from the crystal structure density D*

	$10^{10}R_H/\text{m}^3\ \text{C}^{-1}$	$10^{-28}n_c/\text{m}^{-3}$	$10^{28}D^a/\text{m}^3$	$R_H n_c e$
Ag	−0.84	7.56	5.87	−1.02
Al	−0.30	2.12	6.02	−0.102
Na	−2.5	2.46	2.53	−0.99
Ni	−6.0	1.02	9.14	−0.98
Cd	+0.60	Negative	4.63	–
Fe	+0.25	Negative	8.49	–

[a] Density calculated from the crystal structure.

where n_c is the *electron concentration*. Since e is a constant, the Hall coefficient depends inversely on the density of charge carriers. From experimental measurements of the Hall coefficient, the quantity $R_H(n_c e)$ can be determined. Its ideal value should be the dimensionless quantity −1; the results for a selection of metals are listed in Table 5.1.

Agreements are good for some species and poor for others; more serious is the reversal of sign for some elements as this is not consistent with Eq. (5.16) and would imply positive charge carriers whereas in classical theory the charge carriers are electrons. Furthermore, values for the mean free path deduced from Hall data were found to be very large (70–200 nm) compared to the interplanar spacings in metal crystals (0.3–0.6 nm). Further aspects of the Hall effect will arise shortly, since it is evident that it cannot be explained fully by classical free electron theory.

5.4 Thermal properties of metals

The argument used by Drude for electrical conductivity was applied also to the *thermal conductivity* κ of metals, and the following equation developed [4]:

$$\kappa = (2/3)\tau v^2 c_V \tag{5.17}$$

where c_V is the constant volume heat capacity of the electron gas, and is equated to $(3/2)n_c k$ in the kinetic theory. Hence:

$$\kappa = \tau n_c k v_d^2 \tag{5.18}$$

and using Eq. (5.6) to eliminate v^2:

$$\kappa = 3\tau n_c k^2 T/m_e \tag{5.19}$$

Wiedemann and Franz [5] showed by experiment that the ratio κ/σ for good conductors was constant at a given temperature; from Eqs. (5.6) and (5.19) the *Wiedemann–Franz law* arises:

$$\kappa/\sigma T = 3 \left(\frac{k}{e}\right)^2 = 2.23 \times 10^{-8} \text{ W } \Omega \text{ K}^{-2} = \mathscr{L} \qquad (5.20)$$

where $\mathscr{L}$ is the *Lorenz*[1] *number* [4].[1] Since thermal conductivity at ambient temperatures is almost independent of temperature and electrical conductivity varies as $1/T$, the ratio $\kappa/(\sigma T)$ should be temperature-independent, as demonstrated by Lorenz. The following data summarize these findings for some example metals:

	Ag	Cu	Pb	Zn
$10^8 (\kappa/\sigma T)_{T=373\,K}/\text{W}\Omega\text{K}^{-2}$	2.37	2.33	2.56	2.33
$10^8 (\kappa/\sigma T)_{T=273\,K}/\text{W}\Omega\text{K}^{-2}$	2.31	2.23	2.47	2.31
$10^8 (\kappa/\sigma T)_{T=100\,K}/\text{W}\Omega\text{K}^{-2}$	1.95	1.90	2.00	1.81

Further work showed that the right-hand side of Eq. (5.17) was too large by a factor of two, but that other cancelling numerical errors led to the fortuitous, apparent agreement with $\mathscr{L}$ for the data listed above. A quantum mechanical calculation of $\mathscr{L}$ shows that the factor 3 in Eq. (5.20) should be replaced by $\pi^2/3$ giving $\mathscr{L}$ as $2.44 \times 10^{-8} \text{ W } \Omega \text{ K}^{-2}$. The agreement becomes less satisfactory at low temperatures, and the fall-off with decreased temperatures depends probably on the fact that thermal conduction involves both phonons[2] and electrons, whereas electrical conduction involves only electrons.

In Appendix A5 it was shown that the average kinetic energy of a system of n particles in equilibrium at a temperature T is given by

$$<E_K> = (3/2)nkT \qquad (5.21)$$

and if n is replaced by L the corresponding molar heat capacity at constant volume $C_{V,m}$ is given as:

$$C_{V,m} = (\partial E_K \partial T)_V = (3/2)Lk = (3/2)R \qquad (5.22)$$

In the Drude–Lorentz classical theory, the electron gas in a metal is assumed to follow the Maxwell–Boltzmann distribution of energies [6]. Hence, the fraction $\delta N/N$ of electrons in a system having energies lying between E and $E + \delta E$ is:

$$\delta N = N \frac{2\pi}{(\pi kT)^{3/2}} E^{1/2} \exp(-E/kT)\delta E \qquad (5.23)$$

[1] L. V. Lorenz, not H. A. Lorentz.
[2] Phonon: a quantum of acoustic or vibrational energy.

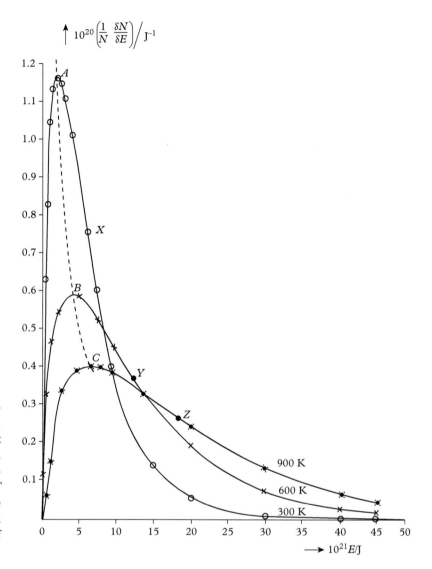

Fig. 5.2 Maxwell–Boltzmann classical distribution of kinetic energy. The maxima of the curves occur at $(1/2)kT$, and follow the path A, B, C with an increase in the temperature; the average energies $(3/2)kT$ are marked X, Y and Z. Since $E = \frac{1}{2}mv^2$, the distribution equation (5.23) can be re-cast in terms of speed v.

The variation of $\frac{1}{N}(\delta N/\delta E)$ with E is shown in Fig. 5.2. The maxima of the curves occur at $kT/2$, as can be verified by forming the differential dN/dE from Eq. (5.23) and setting it equal to zero; the positions on the curves corresponding to the average energies at each given temperature are marked A, B, C in order of increasing temperature.

The kinetic energy of a free electron of speed v is $\frac{1}{2}m_e v^2$, or in terms of momentum p:

$$E_K = p^2/2m_e \tag{5.24}$$

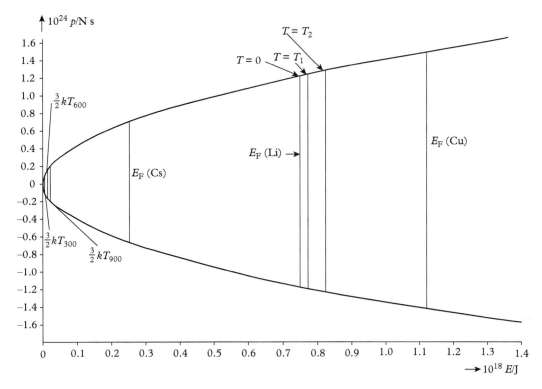

Fig. 5.3 Classical free-electron theory. The variation of momentum p with energy E is parabolic; other features on the curve are Fermi energies E_F (discussed later) and average classical energies $(3/2)kT$.

The variation of momentum with energy is shown in Fig. 5.3. As the energy is increased, the momentum moves along a parabola, its final position depending upon the balance between the applied thermal energy and the energy lost in collision with the positive ions. The parabolic form arises from the dependence of E on k^2 (*see* Section 5.5).

5.4.1   Classical solids

The geometrical basis for a crystal structure is its lattice, an arrangement of mathematical points corresponding to well-defined symmetry and orientation with respect to neighbouring points [7] and on or around them structural units are assembled in a regular manner. Many crystal structures, particularly elemental metals, consist of atoms located on these lattice sites, so that a crystal may be envisaged as a lattice-like array of atoms vibrating about their mean positions, the lattice sites. It is shown in Appendix A5 that the average energy of a one-dimensional oscillator is kT. In three dimensions, oscillations can be resolved

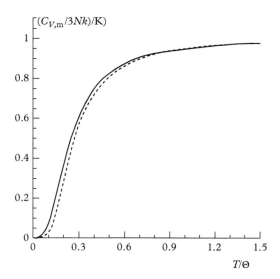

Fig. 5.4 Comparison of the molar heat capacity at constant volume $C_{V,\mathrm{m}}$ according to the Einstein theory (broken line) and the Debye theory (full line). Both equations are normalized to a maximum value of unity; the true maximum value for a monatomic solid is $3R$, or 24.94 J K^{-1} mol^{-1}. Plotting against T/Θ ensures that the curve is independent of the nature of the material, and the values on the ordinate must be halved for a monatomic species.

along three Cartesian axes and each component, or squared term, contributes the amount kT, so that the total average energy is $3nkT$, where n is the number of oscillators. If n is set equal to L, then the molar energy U_{m} and molar heat capacity at constant volume $C_{V,\mathrm{m}}$ are, respectively,

$$U_{\mathrm{m}} = 3RT \tag{5.25}$$

and

$$C_{V,\mathrm{m}} = (\mathrm{d}U_{\mathrm{m}}/\mathrm{d}T)_V = 3R \tag{5.26}$$

From this equation, the molar heat capacity of a monatomic solid is *ca.* 25 kJ K^{-1} mol^{-1}, a result enshrined in the Dulong and Petit law and obeyed well at high temperatures (Fig. 5.4). If a mole of electrons in a metal behaved as entirely free particles, their contribution to the heat capacity would, from Eq. (5.22), be $(3/2)R$ from the free electron 'gas' together with the vibrational contribution from Eq. (5.26), or $9R/2$ per mol in total. Experiment has shown that metals, like other solids, show little tendency to deviate from the Dulong and Petit limit under ambient conditions. Since there can be little doubt about the nature of atomic vibrations and their contribution to the heat capacity, it follows that electronic motion makes a negligibly small contribution to the heat capacity, a result clearly at variance with the classical theory, and often termed the *heat capacity paradox*.

5.4.2 Einstein and Debye solids

In the Einstein theory of heat capacity [8] the vibrations of N atoms in a lattice array were treated as $3N$ independent *harmonic oscillators* of frequency ν, but with the energy quantized in accordance with the Planck theory (Section 2.3). The distribution of energies is no longer continuous, and the average energy was given as:

$$<U> = 3N \frac{\sum\limits_{n=0}^{\infty} nh\nu \exp[-nh\nu/(kT)]}{\sum\limits_{n=0}^{\infty} \exp[-nh\nu/(kT)]} \tag{5.27}$$

Putting $x = -h\nu/kT$, Eq. (5.27) can be written as

$$<U> = 3Nh\nu \frac{d}{dx} \{\ln[1 + \exp(x) + \exp(2x) + \ldots]\} = \frac{3Nh\nu}{\exp[(-x) - 1]}$$

or

$$<U> = \frac{3Nh\nu}{\exp[(h\nu/kT) - 1]} \tag{5.28}$$

At high temperatures,

$$\exp[(h\nu/kT) - 1]^{-1} \text{ tends to } (kT/h\nu) \tag{5.29}$$

neglecting terms in $(h\nu/kT)^2$ and higher, so that

$$<U> = 3NkT \tag{5.30}$$

and

$$\left(\frac{d<U>}{dT}\right)_V = 3Nk \tag{5.31}$$

or $3R$, with N replaced with L, which is the classical value. At low temperatures, $\exp(h\nu/kT) \gg 1$,

$$C_V = \frac{d}{dT} \{3Nh\nu \exp[(h\nu/kT) - 1]^{-1}\}_V = \frac{3Nh\nu \exp[(h\nu/kT) - 1]^{-1}}{\exp[(h\nu/kT) - 1]^{-2}}$$

But since $\exp(h\nu/kT) \gg 1$, and with N replaced by L, the molar heat capacity at constant volume is then given as:

$$<U> = 3R \left(\frac{h\nu}{kT}\right)^2 \exp(-h\nu/kT) \tag{5.32}$$

The term $h\nu/kT$ is this equation is sometimes replaced by θ_E/T, where θ_E is known as the Einstein temperature, which expresses the frequency of atomic oscillations; thus, a high Einstein temperature represents an oscillator of high frequency.

From Eqs. (5.26) and (5.32) it is evident that C_V tends to 0 as T tends to 0 but it does so exponentially at low temperature, whereas experiment shows that C_V tends to zero as a T^3 variation at low temperatures. This problem was resolved satisfactorily by Debye.

He proposed that not all oscillators have the same frequency, and treated coupled vibrations of the atoms in a crystal in terms of $3N$ normal modes of vibration of the whole system [4], each with its own frequency. The vibrations of the atoms about their lattice sites (aka lattice vibrations) are therefore equivalent to $3N$ independent harmonic oscillators having these normal mode frequencies. At long wavelengths (low frequency vibrations), where the wavelength λ is much greater than the basic lattice translation $|\mathbf{a}|$, the crystal is treated as a homogeneous elastic medium and large volume regions in the crystal couple in their vibrational motion. At low temperatures, however, there are vibrations for which $h\nu \ll kT$ and for them the long wavelength vibrations are particularly significant. Thus, these vibrations make a classical contribution to the energy, thereby modifying the exponential dependence on T in the Einstein equation Eq. (5.32). Only for the low temperature region is the Debye modification significant. The Debye treatment is complex [9], and his equation for the average energy, setting $\frac{h\nu}{kT}$ to x for convenience, is:

$$<U> = 9NkT \left(\frac{T}{\Theta}\right)^3 \int_0^{\Theta/T} \frac{x^3}{[\exp(x) - 1]} dT \qquad (5.33)$$

The integral in Eq. (5.33) cannot be evaluated in that form, but high and low temperature limits can be assessed.

At high temperatures, $T \gg \Theta$ and x is then sufficiently small for the approximation $\exp(x) = 1 - x$ to be satisfactory, whence Eq. (5.33) becomes

$$<U> = 9NkT \left(\frac{T}{\Theta}\right)^3 \int_0^{\Theta/T} x^2 dx = 3NkT$$

and for $N = L$

$$C_{V,\mathrm{m}} = 3Lk = 3R \qquad (5.34)$$

which is the classical Dulong and Petit result.

At low temperatures, $T \ll \Theta$ and only long wavelength acoustic modes are thermally excited. Then, the Θ/T limit can be approximated to infinity and using the standard integral[3] $\int_0^\infty \frac{x^3}{\exp(x)-1} dx = \pi^4/15$, the average energy can be written as:

3 http://www.math10.com/en/university-math/definite-integrals/definite-integrals.html

$$<U> = 9NkT \left(\frac{T}{\Theta}\right)^3 \pi^4/15 = \frac{3\pi^4 NkT}{5} \left(\frac{T}{\Theta}\right)^3 \qquad (5.35)$$

By inserting the constants in the right-hand side of Eq. (3.35), $C_{V,m}$ becomes

$$C_{V,m} = \frac{12\pi^4 Nk}{5}\left(\frac{T}{\Theta}\right)^3 = 1943.8\left(\frac{T}{\Theta}\right)^3 \qquad (5.36)$$

The *Debye temperature* Θ is equal to $h\nu_{max}/k$, and the higher the value of Θ the lower the heat capacity at a given temperature. Some values for the Debye temperature are listed below:

	Be	C	Cu	Pb	Au	Na	NaCl
Θ/K	1440	2230	343.5	105	170	91	321

The wavelength λ_{min} corresponding to ν_{max} is given by $\lambda_{min} = h\nu_s/(k\Theta)$, where ν_s is the speed of sound in the material under consideration. For elemental copper $\nu_s = 4000$ m s^{-1} and $\lambda_{min} = 0.56$ nm, whence $\Theta_{Cu} = 342.8$ K; the currently accepted value is 343.5 K. At 14.68 K, $C_{V,m}$(Cu) from Eq. (5.36) is 0.1517 J K^{-1} mol^{-1}; by experiment, the value 0.1548 J K^{-1} mol^{-1} was obtained. The essential difference between the Einstein and Debye equations is brought out clearly by Fig. 5.4.

5.4.3 Summary of classical free-electron theory

So far, it has been shown that the Drude–Lorentz classical free-electron theory when applied to metals accounted satisfactorily for their electrical and thermal conductivities and, to some extent, also for the Wiedemann–Franz law. Heat capacity at ambient temperatures and above followed the Dulong and Petit law, but at low temperatures the enhancement embodied in the Debye theory was essential to a correct representation of the heat capacity. In summary, the following features of classical theory reveal a lack of general applicability:

- Electrical conductivity σ depends on the electron concentration n_c, therefore, a species such as Ti^{3+} should be a better conductor than a singly-charged ion like Na$^+$. Experiment shows:

	Ti^{3+}	Na$^+$
$10^{-7}\sigma$/S m^{-1}	0.23	2.1

- Electrical resistivity data, such as those listed in Section 2.25, do not follow from the theory. Furthermore, the electrical resistivity ρ of an alloy would be expected to lie between the values of those of its components. However, the copper–gold system exhibits electrical resistivities with maxima, lying well above the values for its components, as the plot below illustrates (see also Section 5.12.1):

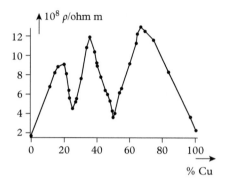

Electrical resistivity in the Cu–Au alloy system

It will transpire that the larger electrical resistivities occur in the disordered states of the system.

- The Hall coefficient is satisfactory for a number of metals, but some exhibit positive values for R_H, which is contrary to the view that the charge carriers are electrons. Electron concentrations calculated from the Hall coefficients are satisfactory for singly charged species but not for those of higher charge.

- Classical theory cannot explain why a majority of metals including silver are silvery in colour while gold is yellow.

- Lack of constancy of the Wiedemann–Franz law exists at low temperatures.

- For N species in a heat bath, classical theory requires each degree of freedom to contribute $(1/2)kT$, and the heat capacity would be expected to vary with valence:

Example		Energy		C_V
		Vibration contribution	Electron contribution	
Monovalent	Na^+	$6 \times (1/2)NkT$	$3 \times (1/2)NkT$	$(9/2)Nk$
Divalent	Ni^{2+}	$6 \times (1/2)NkT$	$6 \times (1/2)NkT$	$(3/2)Nk$
Trivalent	Al^{3+}	$6 \times (1/2)NkT$	$9 \times (1/2)NkT$	$(15/2)Nk$

Experimentally, the molar heat capacity found at ambient temperatures is *ca.* $3R$. Thus, the doubt about the negligible vibrational contribution of the electrons is substantiated.

In order to pursue the theory of metals further, and to explain the fundamental differences between conductors, semiconductors and insulators, and their temperature variations, recourse must be made to the application of wave mechanics and band theory to metallic bonding and properties of metals.

5.5 Wave-mechanical free-electron theory

It can be said that the fundamental difference between the classical and quantum electron theories of bonding lies in the *continuity* of the energy spectrum of classical particles as compared to the *discrete nature* of the energy spectrum of quantum particles: for metals, the condition $(E-E_F) \gg kT$, where E_F is the Fermi energy, marks the transition from discrete to continuous energy spectra.

In classical theory, free electrons interact with a lattice of positive ions only insofar as to use them to arrest or divert their motion. In the wave-mechanical treatment, electrons are still free with respect to any particular atom, but are bound collectively to the lattice of positive ions and interact with them over the volume of the whole crystal.

The application of wave mechanics to the Drude–Lorentz model was initiated by Sommerfeld [10]. In essence, classical mechanics was replaced by wave mechanics and the Boltzmann energy distribution gave way to the *Fermi–Dirac distribution* (Appendix A19), which is applicable to fermions. The electron-in-a-box treatment considered in Section 2.10 is utilized, but in its application to a metal crystal the potential acting on the electron must be periodic:

$$\text{Electron in a box } \psi(x, y, z) = 0 \begin{cases} x = 0 - a \\ y = 0 - a \\ z = 0 - a \end{cases} \tag{5.37}$$

$$\text{Electron in a metal } \psi(x, y, z) = 0 \begin{cases} \psi(x + A, y, z) \\ \psi(x, y + A, z) \\ \psi(x, y, z + A) \end{cases} \tag{5.38}$$

where A is the edge length of the cubical box in this argument. Following the discussion in Section 2.10, the solution of the wave equation for the energy evolves as:[4]

$$E_k = \hbar^2 k^2 / (2m_e) \tag{5.39}$$

where the square of a *wave vector* $\mathbf{k}$ has replaced the sum $(n_1^2 + n_2^2 + n_3^2)$ in Eq. (2.47). The $\mathbf{k}$ vector can be similarly allotted Cartesian components, such that

$$|\mathbf{k}|^2 = k^2 = k_x^2 + k_y^2 + k_z^2 \tag{5.40}$$

[4] In the context, distinguish between k (Boltzmann constant) and $k = |\mathbf{k}|$.

where

$$k_x = 2\pi n_x/A \tag{5.41}$$

and similarly for the y and z directions *mutatis mutandis*; E_k depends upon A through Eq. (5.39). The conditions expressed in Eq. (5.38) are satisfied by Eq. (5.41) through the following argument.

By analogy with the electron-in-a-box discussion, let the eigenfunctions for an electron at a vector distance $\mathbf{r}(x, y, z)$ be of the form:

$$\psi_{\mathbf{k}}(\mathbf{r}) = \exp(i\mathbf{k} \cdot \mathbf{r}) \tag{5.42}$$

where

$$\mathbf{k} \cdot \mathbf{r} = k_x x + k_y y + k_z z \tag{5.43}$$

Then, since n_x is integral,

$$\exp[ik_x(x + A)] = \exp(ik_x x)\exp(ik_x A) = \exp(ik_x x)\exp(i2\pi n_x) = \exp(ik_x x) \tag{5.44}$$

and similarly for the y and z directions, thus satisfying the conditions of Eq. (5.38).

The significance of Eq. (5.39) is that only discrete values for the energies are permitted. In a solid; these energy levels are closely spaced and approximate to a continuum or *band*. At 0 K electrons occupy energy levels up to a finite value, the *Fermi energy* E_F. If an average thermal energy is taken as $(3/5)E_F$ (as used in Section 5.3.3), then the equivalent temperature, the *Fermi temperature* T_F, or E_F/k, would be in the region of 50,000 K (see Problem 5.3). This result is not to be interpreted as the temperature of the electron gas. Its implication is that heat supplied to a metal crystal will be, normally, very much less than T_F and so will have negligible effect on the energy distribution of the electrons.

The wave vector $\mathbf{k}$ exists in *k-space*, which is similar to crystallographic reciprocal space (Appendix A20). In crystallography, a reciprocal distance $|\mathbf{d}^*|$ has the value $1/|\mathbf{d}|$, where d is the interplanar spacing in real space that corresponds to the distance $\mathbf{d}^*$ in reciprocal space. A vector $\mathbf{d}^*$ is labelled with the Miller indices of the family of planes giving rise to it; thus, the (hkl) family of planes determines the reciprocal point hkl distant $\mathbf{d}^*_{hkl}$ from the chosen origin of reciprocal space. In metals it is more convenient to define $\mathbf{k}$ by the equation (for the cubic box)

$$\mathbf{k} = \mathbf{i}\left(\frac{2\pi}{A}\right)(n_x + n_y + n_z) \tag{5.45}$$

through the combination of Eqs. (5.40) and (5.41). The electron of the wavefunction in Eq. (5.42) can be represented by a vector in $\mathbf{k}$-space. An example of a vector $\mathbf{k}$ is illustrated in Fig. 5.5: all points that are distant $|\mathbf{k}|$ from the origin O in $\mathbf{k}$-space have the same energy; the vector $\mathbf{k}$ here could be labelled $\bar{1}11$.

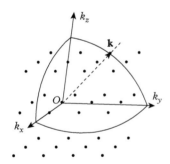

Fig. 5.5 System of N electrons represented as points on a triclinic reciprocal lattice in three-dimensional k-space; the components of $\mathbf{k}$ describe the periodicity of the wavefunction of an electron resolved along the three Cartesian axial directions of k_x, k_y and k_z; the point k may be labelled $\bar{1}11$.

Next, N electrons are introduced into the metal crystal, in accordance with the Pauli principle. For each energy state dictated by the *quantum numbers* n_x, n_y and n_z, one electron with a *spin* of either $+1/2$ or $-1/2$ can be accommodated. As **k** increases so does the energy and momentum represented by the **k**-space point. It may be noted that, although quantum numbers are positive integers, the quadratic form of E can be satisfied for n_x, n_y and n_z as positive or negative integers or zero, as is implicit in Fig. 5.5.

A system of N free electrons in the ground state may be represented by those points of a reciprocal lattice (Fig. 5.5) that lie within a sphere[5] of radius k_F. The energy at the surface of this sphere is the Fermi energy E_F, as given by Eq. (5.40); values of the Fermi energy for some metals are indicated on Fig. 5.3. Only those positions in **k**-space that correspond with acceptable wavefunctions of Eq. (5.42) are permitted, and **k**-space is filled when each of the possible **k** positions in Fig. 5.5 is surrounded by a cubic cell of side A, and thus of volume $(2\pi/A)^3$. This volume defines a primitive unit cell in **k** space and is identified with a single energy state; a single lattice point represents the volume of a primitive unit cell. Hence, in a sphere of radius k_F, the *total number* N of electron energy states is given by:

$$N = 2 \times \frac{4\pi k_F^3/3}{(2\pi/A)^3} \tag{5.46}$$

where the multiplier 2 arises because two energy states, corresponding to electron spins of $\pm 1/2$, are possible. The *electron* concentration[6] N/V (aka n_c), where $V = A^3$, is given as:

$$N/V = k_F^3/3\pi^2$$

or for metals crystallizing in the cubic system, as: $\qquad\qquad$ (5.47)

$$N/V = \eta/a^3$$

where η is the number of electrons in the unit cell of the metal of side a. Using the second expression from Eq. (5.47) for metallic copper, with a face-centred cubic (FCC) unit cell of side 0.3615 nm, $N/V = 4/(0.3615 \times 10^{-9} \text{ nm})^3$:

$$N/V(= n_c) = 8.467 \times 10^{28} \text{ m}^{-3}$$

Example 5.2

Using the above result, together with Eq. (5.39) and the first expression for N/V in Eq. (5.47), E_F may be written as:

$$E_F = \frac{\hbar^2}{2m_e}[3\pi^2(N/V)]^{2/3} \tag{5.48}$$

whence, $E_F = 1.127 \times 10^{-18}$ J, or 7.03 eV.

---

[5] The sphere (*Fermi sphere*) is the shape in **k**-space enclosing the wave vectors of the lowest energy electrons in the crystal.

[6] Also termed a *number density*, that is, a number of electrons per unit volume.

It may be noted that the Fermi energy is large in comparison to the energy of other interactions. For example, kT at 25 °C is only 0.026 eV; the implication is that thermal energy interacts with only a small fraction of the electrons, approximately (0.026/7.03) or 0.37% of the energy range. The majority of electrons are separated from the upper Fermi level by energy much greater than the thermal energy kT. Thus, electrons would not be expected to contribute significantly to the heat capacity at ambient temperatures. At very low temperatures the electron specific heat becomes significant, as will evolve in the next section; the molar heat capacity of electrons in copper under ambient conditions is only 0.7 mJ K^{-1} mol^{-1}.

5.5.1 Density of states

The *density of states function* $g(E)$ is defined as the rate of change of the number of energy states with change in energy. Thus, $g(E)dE$ represents the number of states in the energy range E to $E + dE$. Using Eq. (5.48) as an expression for E:

$$g(E) = dN/dE = 8\sqrt{2}V\pi\,(m_e/h^2)^{3/2}E^{1/2} \tag{5.49}$$

From the relationship between E and k, *vide* Eq. (5.39), the density of states is proportional to the number of points in a sphere in **k**-space lying between radii $|\mathbf{k}|$ and $|\mathbf{k}| + d|\mathbf{k}|$ (Fig. 5.6).

Strictly, m_e in Eq. (5.49), the rest mass of the electron, is not appropriate for all values of E. An electron in a periodic potential field of a crystal is accelerated relative to the crystal as though it had an effective mass m_e^*, given by

$$1/m_e^* = (1/\hbar^2)d^2E/dk^2 \tag{5.50}$$

The effective mass varies among the elements, but this correction term is not discussed herein.

A plot of the density of states function $g(E)$ is shown in Fig. 5.7. It is parabolic, like the curve in Fig. 5.2, but single-valued; at 0 K the energy levels are filled to a sharp cut-off at the Fermi energy, and higher energy states are unoccupied.

Consider raising the temperature of a metal from 0 K to a temperature T. According to classical theory, each particle would be increased in energy by the amount kT, which is 0.026 eV at 25°C. However, the energy distribution of conduction electrons extends over several electron volts, even at 0 K. An electron in this distribution will not, in general, be perturbed if it is less than *ca*. 0.1 eV below the Fermi level. Only those electrons within an energy range of approximately 0.1 eV of E_F will be excited thermally; otherwise they remain in lower energy levels with spins paired according to the Pauli principle. The electrons that are excited account for the rounding of the $g(E)$ curve (Fig. 5.7) at $T > 0$ K.

The wave mechanical theory of the thermal properties of conduction electrons, as compared to their classical behaviour, explains their small contribution to the heat capacity. Of a total of N electrons only a fraction, of the order of T/T_F, will be excited thermally at the temperature T, because only that fraction

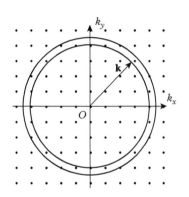

Fig. 5.6 The points in three-dimensional **k**-space lying between spherical annuli of radii $|\mathbf{k}|$ and $|\mathbf{k}| + d|\mathbf{k}|$, where $d|\mathbf{k}|$ is the width of the annulus, are proportional to the number of energy states lying between E and $E + dE$.

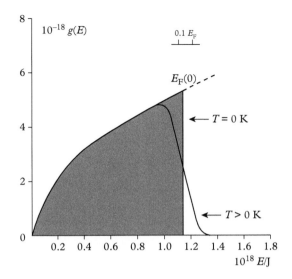

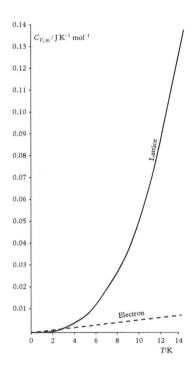

Fig. 5.7 Wave-mechanical free-electron theory density of states function $g(E)$ for metallic copper. The theoretical sharp cut-off at E_F at 0 K is modified as the temperature is increased.

of them lies within the amount kT of the top of the energy distribution (Fig. 5.7). The contribution of these electrons to the internal energy is $NkT(T/T_F)$, or in molar terms, RT^2/T_F, which is clearly proportional to T^2. Thus, $C_{V,m} = 2RT/T_F$. Bearing in mind the magnitude of T_F (Section 5.5), the electronic contribution to heat capacity becomes a significant contribution only at very low temperatures (Fig. 5.8).

The heat capacity of a metal at very low temperatures may be written as

$$C_{V,m} = \alpha T^3 + \gamma T \tag{5.51}$$

and experiment confirms that $C_{V,m}/T$ varies linearly with T^2. The term α is given in Eq. (5.36) and the Sommerfeld constant γ is a function of the electrons, given by:

$$\gamma = \pi^2 N k^2 / (2E_F) = \pi^2 N k / (2T_F) \tag{5.52}$$

where N is the number of conduction electrons in the specimen. The model makes a reasonable prediction of the electronic contribution to the molar heat capacity, as shown by the results in Table 5.2. It follows from Eq. (5.51) that a plot of $C_{V,m}/T$ against T would intercept the axis of the dependent variable axis at the value of γ.

5.5.2 Fermi–Dirac distribution

The Maxwell–Boltzmann (aka classical) distribution applies well to gaseous species under ordinary conditions. However, an electron has a much smaller mass than a molecule, and the concentration of valence electrons in a metal is

Fig. 5.8 Lattice and electronic contributions to the molar heat capacity of copper between 0 K and 14 K; only at $T < 5$ K does the electronic contribution become relatively significant.

approximately 10^4 times higher than molecules in a gas. The classical distribution is valid for a gas when the average spacing between molecules is large in comparison with the de Broglie wavelength $h/(m_e v)$.[7] In metals the average spacing between electrons is commensurate with the de Broglie wavelength, and the Maxwell-Boltzmann distribution does not furnish a satisfactory account of their energy distribution.

Fermi–Dirac quantum statistics treats all electrons as indistinguishable and requires that each energy state is either empty or fully occupied by a single electron specified by quantum numbers n_x, n_y and n_z, together with a spin quantum number m_s equal to $+1/2$ or $-1/2$. With a single electron per energy state, the lowest energy of a metal system may have many high quantum number states of individual electrons occupied. Under Maxwell–Boltzmann theory any number of electrons could have identical energies, the lowest state having zero energy.

Quantum statistics determines that the probability $f(E)$ for a single electron of energy E at 0 K is either zero or 1, and Fig. 5.9 indicates that the probability $f(E)$ for the energy E_i in the ith state will be unity for $E_i < E_F(0)$ or zero for $E_i > E_F(0)$.

A derivation of the Fermi–Dirac distribution is given in Appendix A19, where $f(E)$ is determined as:

$$f(E) = \frac{1}{\exp[E - E_F)/(kT)] + 1} \tag{5.53}$$

The behaviour of the exponent in Eq. (5.53) as T tends to zero depends on the sign $E - E_F$ as T tends to zero:

$$\frac{E - E_F(0)}{kT} = \begin{array}{l} -\infty \text{ if } E < E_F(0) \\[4pt] \infty \text{ if } E > E_F(0) \end{array} \quad \text{so that} \quad f((E) = \begin{array}{l} 1 \text{ if } E < E_F(0) \\[4pt] 0 \text{ if } E > E_F(0) \end{array}$$

Table 5.2 *Calculated and observed values of the Sommerfeld constant* γ

	$10^3 \gamma_{obs}/\mathrm{J\,mol^{-1}}$	$10^3 \gamma_{calc}/\mathrm{J\,mol^{-1}}$
Na	1.38	1.09
K	2.08	1.67
Cu	0.695	0.505
Ag	0.846	0.645

[7] Thermal velocity; $v \approx 10^5$ m s^{-1}.

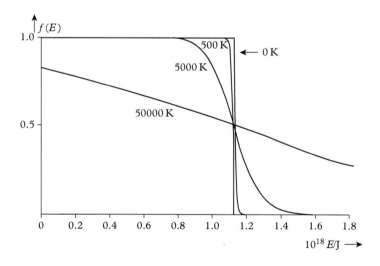

Fig. 5.9 Fermi–Dirac probability distribution function $f(E)$ for copper at different temperatures; the probability that a state of energy E will be occupied decreases as the temperature is increased. At 0 K the probability is either 1 or zero; E_F for copper is 1.13 eV. That the probability $f(E)$ is 1/2 for $E = E_F$ irrespective of temperature follows from Eq. (5.53).

At $T = 0$ K all states with energy E less than $E_F(0)$ are occupied and all states with E greater than $E_F(0)$ are empty; at this temperature, electrons occupy all the lowest energy states available. Thus, E_F is the sharp cut-off value to which reference has already been made and which is evident in Fig. 5.9. As the temperature is increased, $f(E)$ becomes less than unity for $E < E_F$, and greater than zero for $E > E_F$. The high energy 'Maxwellian tail' of the distribution for $T > 0$ corresponds to $(E - E_F) \gg kT$, or $(E - E_F)/(kT) \gg 1$. Then:

$$f(E) \approx \exp[E_F - E)/(kT)] \propto \exp[-E/(kT)] \tag{5.54}$$

which corresponds to the classical distribution. The curve for $T > 0$ K in Fig. 5.7 is just the density of states function at 0 K multiplied by the Fermi–Dirac distribution at a given temperature. At $E = E_F$, $f(E) = 1/2$ whatever the temperature, as can be seen in Fig. 5.9.

5.5.2.1 *Fermi energy and chemical potential*

The Fermi energy, that is, $f(E) = 1/2$, can be identified with the chemical potential of the electrons by the following argument. On expanding Eq. (A19.7) and with Eq. (A19.20):

$$S = k \sum_i [g_i \ln g_i - g_i \ln(g_i - N_i) + N_i \ln(g_i - N_i) - N_i \ln N_i]. \tag{5.55}$$

Recalling that $g_i/N_i = f(E)$ and setting $(E_i - E_F)/(kT)$ equal to x, rearrangement of Eq. (5.55) gives

$$S = k \left\{ \sum_i -g_i \ln[1 - f(E)] + N_i \ln\left[\frac{1}{f(E)}\right] - 1 \right\} \tag{5.56}$$

or

$$S = k \left\{ \sum_i g_i \ln\left[\frac{1 + \exp(x)}{\exp(x)}\right] + N_i \ln[\exp(x)] \right\}$$

which simplifies to

$$S = k \left\{ \sum_i g_i \ln[\exp(-x) + 1] + N_i x \right\}$$

Inserting the value of x and using Eq. (A19.6):

$$S = k \left\{ \sum_i g_i \ln\left[1 + \frac{\exp(E_F - E_i)}{kT}\right] + \frac{E}{T} - \frac{N_i E_F}{T} \right\}$$

A thermodynamic definition of chemical potential μ is

$$\mu = -T \left(\frac{\partial S}{\partial N} \right)_{E,V} \tag{5.57}$$

from which it follows that

$$\mu = E_F \tag{5.58}$$

namely, that the chemical potential of electrons is equal to their Fermi energy.

Example 5.3

The energy gap between the valence and conduction bands in diamond is 5.5 eV. What is the probability of electron promotion from the valence band to the conduction band at 20°C (k in eV = k/e = 8.6173 × 10^{-5} eV)? The Fermi level is that for $f(E) = 1/2$ (Fig. 5.9), and the probability of electron promotion is given by:

$$f(E) = \frac{1}{\exp[(\Delta E/2)/kT) + 1]}$$

$$= \frac{1}{\exp(2.75\,\mathrm{eV}/(8.6173 \times 10^{-5}\,\mathrm{eV\,K^{-1}}\,293\,\mathrm{K}) + 1]}$$

$$= 4.99 \times 10^{-48}$$

a result that confirms the insulating properties of diamond.

If germanium ($\Delta E = 0.670\,\mathrm{eV}$) is substituted for diamond under the same conditions then $f(E) = 1.73 \times 10^{-6}$ at 293 K, a result in accordance with the property of this species as a semiconductor.

5.6   Band theory of metals

The property of electrical conduction and its variation with temperature is very different in metals, semiconductors and insulators.

Pure or nearly-pure metals exhibit a linearity of resistivity with temperature, with electron–phonon interaction playing a key role. Thermal motion acts so as to scatter electrons and hinder their action as unidirectional charge carriers. At low temperatures, the activity of phonons (lattice vibrations) effectively ceases and the resistivity usually reaches a constant, low, residual value. In some materials the electrical resistivity at a sufficiently low temperature falls to zero, as discovered first for mercury (Fig. 5.10) by Kamerlingh-Onnes in 1911. Substances that exhibit this behaviour are termed superconductors (Section 5.8.4).

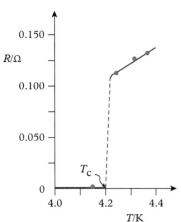

Fig. 5.10 Electrical resistance R of mercury in the 4 K temperature range; superconduction sets in with a discontinuity at 4.2 K at which temperature the measured resistance was less than $10^{-6}\,\Omega$. Within an increase of 0.1 K the resistance increases to 0.1 Ω.

Experimentally, electrical resistivity in intrinsic semiconductors is found to decrease exponentially, following the approximate relationship $\rho_T = \rho_0 \exp(-\alpha T)$, where ρ_0 is an initial value of resistivity and α is the temperature coefficient of resistivity. Extrinsic (doped) semiconductors show a more complicated variation with temperature. As the temperature is increased from 0 K, the resistivity decreases as the charge carriers leave the donors or acceptors; subsequently, the resistivity increases owing a reduced mobility of carriers, as is the case with pure metals. At higher temperatures, the doped semiconductors behave like the intrinsic semiconductors as the charge carriers now experience the effect of thermal motion.

Insulators are dielectric materials and non-conductors of electricity, that is, their electrical resistivities are very high, 10^{10}–10^{12} Ω m; conduction electrons are unavailable for a current carrying process. If a dielectric material is subjected to a sufficiently high voltaic potential, electrical breakdown occurs and current passes, often with a degradation of the material. Normally, the electrical resistivity of an insulator decreases approximately exponentially with an increase of temperature. These variations in the behaviour of materials will be explained by band theory shortly.

5.6.1 Wigner–Seitz cells and Brillouin zones

The fourteen conventional Bravais unit cells of the three-dimensional lattices are not all primitive, that is, they contain more than one lattice point per unit cell. For many applications including *inter alia* metallic bonding and spectroscopy it is desirable to work with representative primitive unit cells which contain only one lattice point per unit cell volume while retaining the full point group symmetry. It is achieved with the *Wigner–Seitz cells* (aka *Dirichlet domains* or *Voronoi cells*): they are true unit cells in that they can be stacked to fill space completely and exhibit the point-group of the lattice. In the majority of metal structures, atoms occupy the lattice points of the Bravais unit cells.

The fourteen Bravais lattices give rise to twenty-four Wigner–Seitz cells distributed unevenly among the crystal systems. In some systems they vary in shape according to the axial ratio of the system; thus, although there are just two tetragonal lattices with Bravais P and I unit cells, there are three tetragonal Wigner–Seitz cells, two being governed by the c/a ratio of the Bravais I cell.

The construction of a Wigner–Seitz cell may be visualized first in two dimensions. In Fig. 5.11 the real space square lattice is represented by a centred unit cell with translation vectors $\mathbf{a_r}$ and $\mathbf{b_r}$ and contains two lattice points per unit cell. The Wigner–Seitz cell is constructed by drawing lines from any lattice point P as an origin to its nearest neighbour lattice points; there are four such connections in this example. Lines are now drawn through the mid-points of these four lines and extended as necessary to form a closed polygon; it is clear that these cells fill space completely. The cell denoted here by the translation vectors $\mathbf{a_k}$ and $\mathbf{b_k}$ is the Wigner–Seitz cell of the original real space centred unit cell. It is a cell in

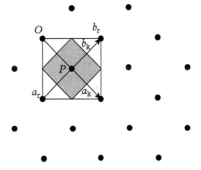

Fig. 5.11 First and second Brillouin zones constructed in a centred unit cell of a square two-dimensional lattice: $\mathbf{a_r}$ and $\mathbf{b_r}$ are translation vectors in real space; $\mathbf{a_k}$ and $\mathbf{b_k}$ are translation vectors in reciprocal space (k space). The two Brillouin zones have equal total area. (See also http://www.doitpoms.ac.uk/tlplib/brillouin_zones/zone_construction.php)

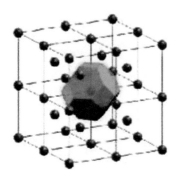

Fig. 5.12 Construction of the Wigner–Seitz cell from a Bravais body-centred unit cell in a cubic lattice. Eight unit cells are depicted, and the central lattice point of the array is the origin from which lines are drawn to the lattice points nearest to it. Bisecting these planes forms the surfaces of the Wigner–Seitz cell. In this example it is a truncated octahedron, a combination of the octahedron crystal form {111} and that of the cube {100}. [Reproduced by courtesy of Dr Nicolas Chamel, Université Libre de Bruxelles.]

reciprocal space, or **k**-space, has one lattice point per unit cell and shows the full symmetry of the array of points.

The cell that has been constructed in **k**-space defines the first Brillouin zone in two dimensions. A second such Brillouin zone is determined in a similar manner, by bisecting lines that are drawn to the next nearest points of the given origin point and forming another closed polygon. Each of the first, second, third and higher Brillouin zones have the same total area. Most interest in band theory centres on the first Brillouin zone, which is the geometrical equivalent in **k**-space of the Wigner–Seitz cell in real three-dimensional space.

In three dimensions, the construction is a little more complicated but similar principles apply. Consider a body-centred unit cell in a cubic lattice, which may be denoted *cI*; Fig. 5.12 shows a block of eight face-to-face body-centred unit cells of a cubic lattice. The Wigner–Seitz cell is obtained in the manner similar to that used for the two-dimensional example. In three dimensions, the mid-points of the lines emanating from a given central point are now bisected by *planes* that are extended to form a closed polyhedron. The square faces on this polyhedron arise from bisecting the vectors from the origin lattice point to the body-centring points of adjacent unit cells (fourfold symmetry directions), whereas the hexagonal faces are formed from the planes bisecting the vectors from the origin point to the corners of its unit cell (threefold symmetry directions).

The completed polyhedron is a *truncated octahedron*; it is of the same form as the first Brillouin zone of the *face-centred* unit cell of a cubic lattice. This relationship arises because a *cI* lattice in real space has a *cF* reciprocal lattice in **k**-space (reciprocal space). The relationships of these cubic cells are set out hereunder:

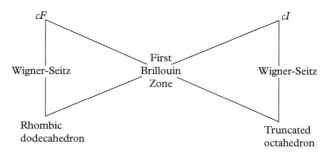

cF **and** *cI* **cells and their first Brillouin zones**

The complete filling of space by stacking the Wigner–Seitz truncated octahedral cells is illustrated in Fig. 5.13a, and Fig. 5.13b shows the Wigner–Seitz cells formed from the tetragonal *P* and *I* Bravais lattices.

The boundaries of Brillouin zones represent planes which can undergo Bragg reflection (diffraction) with constructive interference. The points of the first two-dimensional Brillouin zone, a unit cell in reciprocal space, can be reached from the origin point without crossing any boundary line (Bragg reflection plane in

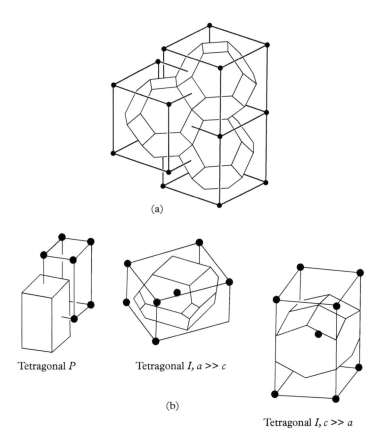

(a)

Tetragonal *P*   Tetragonal *I, a >> c*

(b)

Tetragonal *I, c >> a*

Fig. 5.13 (a) Stacking of Wigner–Seitz cells (Fig. 5.12) to fill space. (b) Single Wigner–Seitz cell from a tetragonal *P* Bravais unit cell, and two Wigner–Seitz cells from tetragonal *I* unit cells dependent on the *c/a* ratio. Each Wigner–Seitz cell contains only one lattice point while retaining the full tetragonal point group symmetry. [Burns G and Glazer AM. *Space Groups for Solid State Scientists*, 3rd ed. 2013; reproduced by courtesy of Elsevier.]

three dimensions), the second zone points are reached by crossing one boundary line and, in general, the *n*-zone points after crossing *n* – 1 boundaries.

Brillouin zones are important in chemistry for the progression from the theories based solely on bonds to those involving bands in the quantum mechanical theory of solids, and in the physics of lattice dynamics [11–14]. The working advantage of the Wigner–Seitz cell lies in its ability of presenting a space-filling unit cell containing one lattice point, and hence the smallest possible number of atoms of a structure while maintaining the full point-group symmetry of the structure under investigation.

5.6.2 Energy bands and the Fermi level

The differences between *conductors*, *insulators* and *semiconductors* are explained through band theory. Electron energy levels in solids are arranged in *bands*, which are closely spaced energy levels approximating to a continuum. Of particular importance are the conduction band and the valence band; electrical conduction requires the presence of electrons in the upper, conduction band. Between the bands are forbidden regions, or *energy gaps*, in which no energy states are allowed. Both the occupancy of the bands and the width of the gaps determine the electrical properties of solids. In insulators the electrons in the conduction band and valence band are separated by a large energy gap, whereas in metallic conductors the conduction and valence bands are very close or overlapping. In semiconductors the energy gap is such as to allow movement of electrons between the valence and conduction bands under the stimulus of heat or radiation. With small gaps the introduction of foreign atoms (doping) can have a dramatic effect on electrical conductivity. An important parameter in band theory is the *Fermi level*, which is the top of the available electron energy levels at low temperatures. The position of the Fermi level in relation to the conduction band is a crucial factor in determining electrical properties.

5.6.3 Schrödinger equation in a periodic potential

On the free-electron model, the permitted energy states are distributed quasi-continuously according to Eq. (5.39); the free-electron eigenfunctions are, in one dimension, of the form $\exp(ikx)$ and represent travelling waves of momentum $kh/2\pi$. Band theory starts from the Schrödinger equation, but incorporates the *periodic potential field* that arises from a lattice array of atoms.

A periodic potential may be written as:

$$V(\mathbf{r} + \mathbf{T}) = V(\mathbf{r}) \tag{5.59}$$

where the translation vector $\mathbf{T}$ is given as:

$$\mathbf{T} = n_x\mathbf{a} + n_y\mathbf{b} + n_z\mathbf{c} \tag{5.60}$$

where the n terms are integers and $\mathbf{a}$, $\mathbf{b}$ and $\mathbf{c}$ are the non-coplanar lattice translation vectors in real space.

The periodic nature of the potential means that it can be expressed as a Fourier series:

$$V(\mathbf{r}) = \sum_{\mathbf{k}} V_{\mathbf{k}} \exp(i\mathbf{k} \cdot \mathbf{r}) \tag{5.61}$$

where $V_{\mathbf{k}}$ are coefficients of the Fourier series. Since the n_i are integers, then from Eqs. (5.59) and (5.61):

$$\exp(i\mathbf{k} \cdot \mathbf{T}) = 1 \text{ so that } \mathbf{k} \cdot \mathbf{T} = 2n\pi \tag{5.62}$$

n being another integer. The vector $\mathbf{k}$ in reciprocal space is given by:

$$\mathbf{k} = m_x\mathbf{a}^* + m_y\mathbf{b}^* + m_z\mathbf{c}^* \tag{5.63}$$

where $\mathbf{a}^*, \mathbf{b}^*$ and $\mathbf{c}^*$ are translation vectors defining a unit cell in reciprocal space and the m terms are integers. It follows from Eqs. (5.60) and (5.62) that

$$\mathbf{a} \cdot \mathbf{a}^* = 1 \text{ and } \mathbf{a} \cdot \mathbf{b} = 0 \tag{5.64}$$

and similarly for other such scalar products. The nature of the reciprocal lattice shows that in $\mathbf{k}$-space there exists energy relationships $E(\mathbf{k} + \mathbf{T}) = E(\mathbf{k})$. However, such is the periodic nature of reciprocal space that an infinity of $\mathbf{k}$-space can be represented by a single primitive reciprocal unit cell, which is also the first Brillouin zone (see also Section 5.6.1).

The Brillouin zone always contains the same number of $\mathbf{k}$-states as the number of primitive unit cells in the crystal. The volume of a primitive unit cell of translation vectors $\mathbf{a}$, $\mathbf{b}$ and $\mathbf{c}$ is, from Appendix A20:

$$V = \mathbf{c} \cdot (\mathbf{a} \times \mathbf{b}) \tag{5.65}$$

and that of the reciprocal primitive unit cell:

$$V^* = \mathbf{c}^* \cdot (\mathbf{a}^* \times \mathbf{b}^*) \tag{5.66}$$

so that

$$VV^* = 1 \tag{5.67}$$

The Schrödinger equation for an electron of mass m_e in a periodic potential is written as:

$$-\left\{ \frac{\hbar^2 \nabla^2}{2m_e} + V(\mathbf{r}) \right\} \psi = E\psi \tag{5.68}$$

where $V(\mathbf{r})$, since it is periodic, may be represented by the Fourier series:

$$V(\mathbf{r}) = \sum_{\mathbf{k}} V_{\mathbf{k}} \exp(i\mathbf{k} \cdot \mathbf{r}) \tag{5.69}$$

The wavefunction ψ is the sum of plane waves that describe the motion of electrons *travelling* through the periodic potential and so contains the translation periodicity. It can be achieved by using the Born–von Karman boundary conditions [11, 12] to give the form:

$$\psi(\mathbf{r}) = \sum_{\mathbf{k}} C_{\mathbf{k}} \exp(i\mathbf{k} \cdot \mathbf{r}) \tag{5.70}$$

where $\mathbf{k}$ represents the allowed values of a reciprocal lattice vector, and $\mathbf{k}$ lies within the first Brillouin zone. The coefficients $C_{\mathbf{k}}$ in Eq. (5.70) are of this form and specify the form of the wavefunction. Hence, from Eq. (5.70):

$$\begin{aligned}
\psi_{\mathbf{k}}(\mathbf{r}) &= \sum_{\mathbf{k}} C_{\mathbf{k}} \exp[i(\mathbf{k} \cdot \mathbf{r})] \\
&= \exp[i(\mathbf{k} \cdot \mathbf{r})] \sum_{\mathbf{k}} C_{\mathbf{k}} \exp[-i(\mathbf{k} \cdot \mathbf{r})] \tag{5.71} \\
&= \exp[i(\mathbf{k} \cdot \mathbf{r})] u_{\mathbf{k}}(\mathbf{r})
\end{aligned}$$

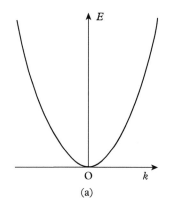

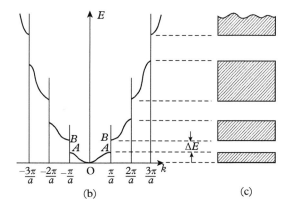

(a)

(b)   (c)

Fig. 5.14 Electron energies in solids as a function of $|\mathbf{k}|$. (a) Free-electron theory; energy parabolic and continuous. (b) Electrons in periodic potential field; gaps arise where the Bragg equation is satisfied at Brillouin zone boundaries. (c) Energy bands in a solid, showing energy gaps ΔE. [*Introduction to Physical Chemistry*, 3rd ed. 1998; reproduced by courtesy of Cambridge University Press.]

where $\mathbf{k}$ is an allowed wave vector for the electron under a constant potential and $u_q(\mathbf{r})$ is a function governed by the index $|\mathbf{k}|$ of the solution, always within the periodicity of the lattice, that is, $u_q(\mathbf{r} + \mathbf{T}) = u_q(\mathbf{r})$. The solution Eq. (5.70) was given by Bloch [13]. It may be stated as 'the eigenfunctions of the Hamiltonian in Eq. (5.68) take the form of a plane wave modulated by a potential function having the periodicity of the lattice, that is, $V(\mathbf{r} + \mathbf{T}) = V(\mathbf{r})$, and make no assumption about the value of that potential'.

The energy dependence is quadratic, as shown by Eqs. (5.39) and (5.40) and by Fig. 5.14a. For $k = \pm n\pi/a$, for the x direction, discontinuities (gaps) arise thus giving rise to the band structure shown by Figs. 5.14b and c. The Kronig–Penney model [11, 14] demonstrated that energy bands arise for an electron in a one-dimensional periodic potential field from which Figs. 5.14b and c derive.

Taking a one-dimensional case for simplicity, since the potential is periodic:

$$\psi_k(x + a) = u_k(x + a) \exp[\mathrm{i}k(x + a)$$
$$= u_k(x + a) \exp(\mathrm{i}kx) \exp(\mathrm{i}ka)$$

and because the potential is periodic:

$$\psi_k(x + a) = u_k(x) \exp(ikx) \exp(ika)$$

so that from Eq. (5.71) for the one dimension x:

$$\psi_k(x + a) = \psi_k(x) \exp(ika) \qquad (5.72)$$

From Eq. (5.45) it follows that the term $\exp(i|k|a) = \exp(i2\pi) = 1$. Thus, the probabilities, or electron density distributions are also periodic:

$$|\psi(x + a)|^2 = |\psi(x)|^2 \qquad (5.73)$$

Equation (5.72) suggests a band structure: each $u_{i,k}$ corresponds to a set of electron states E_k, which is equivalent to a band composed of many closely spaced levels. The number of wavefunctions in the band is given by $|k|$ the number of wave vectors in the first Brillouin zone, and the total states is twice this number because of the $\pm 1/2$ values for electron spins. The number of electrons in the energy bands governs the electrical behaviour of materials, as will evolve shortly.

For most values of k ($= |k|$) electrons behave very much like free particles. At $k = \pm n\pi/a$, however, the condition for Bragg reflection of electron waves in the x direction is realized. The relationship $k = \pm n\pi/a$ is equivalent to the Bragg equation in the form $2a \sin\theta = n\lambda$, where $k = 2\pi/\lambda$ and $\sin\theta = 1$. The first order reflection ($n = 1$) at $k = \pm\pi/a$ arises because waves reflected from adjacent atoms interfere constructively, their phase difference being 2π. The region in k-space between $+\pi/a$ and $-\pi/a$ is the *first Brillouin zone* (Fig. 5.15), and energy is quasi-continuous within a zone according to Eq. (5.39) but discontinuous at the zone boundaries.

As k increases towards $n\pi/a$, the eigenfunctions contain increasing amounts of the Bragg-reflected wave. At $k = \pi/a$, for example, the wave $\exp(i\pi x/a)$ reflects as $\exp(-i\pi x/a)$, and the resultant combinations are standing waves ψ_1 and ψ_2 of forms $\cos(\pi x/a)$ and $\sin(\pi x/a)$, respectively. The probability densities of the two waves are $|\psi_1|^2$ and $|\psi_2|^2$, whereas for the travelling wave the probability density is $\exp(i2\pi x/a)$, or $\exp(i2kx)$; Fig. 5.16 illustrates a one-dimensional potential field of period a and the wave probability functions.

A one-dimensional travelling wave distributes charge uniformly along the x-axis; $|\psi_1|^2$ and has peaks at $\pm ma$, where m is an integer, whereas $|\psi_2|^2$ has

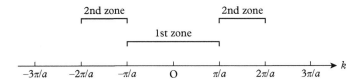

Fig. 5.15 The first two Brillouin zones of a one-dimensional lattice of periodicity a.

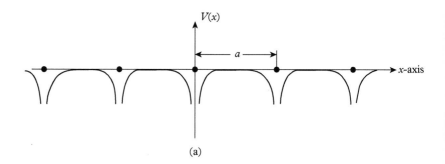

(a)

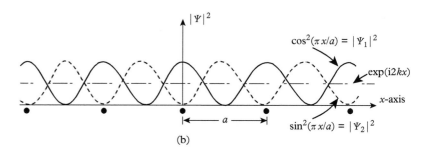

(b)

Fig. 5.16 One-dimensional lattice of periodicity *a*. (a) Periodic potential field $V(x)$. (b) Probability densities $\cos^2(\pi x/a)$, or $|\psi_1|^2$ and $\sin^2(\pi x/a)$, or $|\psi_2|^2$, and the plane wave $\exp(i2\pi kx)$. [*Introduction to Physical Chemistry*, 3rd ed. 1998; reproduced by courtesy of Cambridge University Press.]

peaks at $\pm(m+1/2)a$. It is clear, therefore, that the potential energies of the three distributions are in the order $|\psi_1|^2 < \exp(ikx) < |\psi_2|^2$. Hence, an energy gap $2u_k$ arises, where u_k is the potential energy function at the position corresponding to **k** in reciprocal space.

The ground state of N free electrons is determined by the occupation of all one-electron of energy levels **k** with energies $E_k(= \hbar^2 k^2/2m_e)$ less than the Fermi energy E_F. The Bloch theory labels the one-electron energy levels by the quantum numbers n and **k** (k_x, k_y, k_z), so that when the lowest levels are filled by N electrons two situations arise:

- A given number of bands are filled with the remainder empty, giving rise to a *band gap*. As will be discussed shortly, if the gap is much greater than kT at ambient temperatures, the solid is an insulator, whereas if the gap is commensurate with kT, the solid is an intrinsic semiconductor. Since the number of levels in a band is the same as the number of Wigner–Seitz primitive cells in the crystal, a band gap can arise only if N is an even number; and

- A number of bands may be partially filled, and the energy E_F of the highest occupied band lies within the range of one or more bands. For each partially filled band there exists a conceptual surface in **k**-space that separates occupied from unoccupied levels. The set of all such surfaces is termed the *Fermi surface*. Metallic properties require the existence of a Fermi surface in the solid.

Band theory modifies the density of states function. A simple example is shown in Fig. 5.17. The parabolic curve is still in evidence, but peaks arise from abrupt changes in slope where the Fermi surface approaches a Brillouin zone boundary, which are van Hove singularities (Section 6.2.1.1).

The argument can be extended to two and three dimensions. In three-dimensional space, the Brillouin zone boundaries are determined by reflections in reciprocal space where the Bragg equation is satisfied, that is, by the structure and symmetry of the crystal rather than by the nature of the crystal material.

The application of the Bragg condition can be explained further with the aid of Fig. 5.18a: the crystal under consideration is imagined at C, the centre of the Ewald sphere (Appendix A20), and the origin of reciprocal lattice is O; $\mathbf{k}$ is a vector in the direction of an incident wave and $\mathbf{k}'$ the corresponding vector in the direction of the reflected (diffracted) wave. Bragg reflection occurs when a reciprocal lattice point P intersects the Ewald sphere, and the condition for this intersection, and hence for Bragg reflection may be written as:

$$\mathbf{k}' - \mathbf{k} = \mathbf{G} \tag{5.74}$$

where $\mathbf{G}$ is the Bragg scattering vector of magnitude $2\pi/d_{201}$ in the example in the figure, which refers to the k_x, k_z reciprocal space level, a distance in reciprocal space as can be shown by the following analysis.

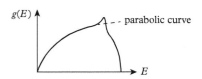

Fig. 5.17 Band theory density of states function $g(E)$ with a peak occurring where the Fermi surface approaches the Brillouin zone boundary.

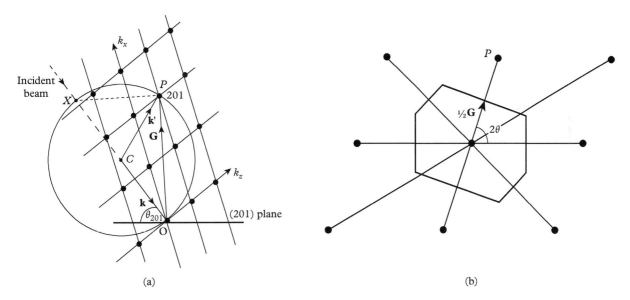

Fig. 5.18 (a) Ewald sphere (aka sphere of reflection) construction for Bragg reflection. The origin of the reciprocal lattice is at O and C is the crystal. The x-ray beam is incident on (201) planes at the glancing angle θ_{201}; P is the reciprocal lattice point 201 intersecting the Ewald sphere at which the Bragg equation is satisfied: $\mathbf{k} - \mathbf{k}' = \mathbf{G}$. (b) Construction of the first Brillouin zone in a two-dimensional reciprocal lattice.

That Eq. (5.74) has the form of the Bragg equation follows from the following argument. From Fig. 5.18a:

$$\frac{|\mathbf{G}|}{2|\mathbf{k}|} = \cos(\pi/2 - \theta) = \sin\theta \tag{5.75}$$

But $\dfrac{|\mathbf{G}|}{2|\mathbf{k}|}$ is also equal to $\dfrac{2\pi/d}{2 \times 2\pi/\lambda} = \dfrac{\lambda}{2d}$. Hence, the Bragg equation,

$$2d\sin\theta = \lambda \tag{5.76}$$

The Bragg equation may be written in the form $2(2\pi/\lambda)\cos\phi = d^*$, where $\phi = (90 - \theta)$ and d^* is the distance to a point in reciprocal space. Multiplying both sides by d^* gives $2[(2\pi/\lambda)d^*]\cos\phi = (d^*)^2$, which may be written as:

$$\mathbf{k} \cdot \mathbf{G} = (d^*)^2/2 (\equiv G^2/2) \tag{5.77}$$

which shows that whenever $\Delta\mathbf{k} = \mathbf{G}$, a Bragg reflection occurs at a scattering angle 2θ.

A Brillouin zone has been defined as the Wigner–Seitz cell in reciprocal space, and may be illustrated in terms of Fig. 5.18b in which the points represent a portion of a reciprocal lattice, P being the reciprocal lattice point of that notation in Fig. 5.18a. From a chosen origin the Brillouin zone is constructed along the lines discussed in Section 5.6.1; the outlined polygon is the first Brillouin zone, which contains all wave vectors $\mathbf{k}$ that can undergo Bragg reflection. Thus, the Bragg equation is satisfied when a $\mathbf{k}$ vector terminates on a plane normal to $\mathbf{G}$, or $\mathbf{d}^*$, at the mid-point $|\mathbf{G}|/2$; the terminations determine the boundaries of the Brillouin zones. Further reading on the topic of band theory may be found in the literature [4, 11–14].

5.6.4 Energy bands and electrical conduction

In order to explain more fully electrical conduction in different classes of solids, it is necessary to consider in which solid a certain fraction of electrons behaves as though free. The main types of solid electrical conductors are represented schematically in Fig. 5.19. In a *metallic conductor* the conduction band may be partially filled, as in (a): its electrons behave as though they were free electrons and can move into the conduction band under thermal or other stimulus provided that the energy gap is of the order of 1–2 eV. The solid will show metallic conduction also if a filled valence band and conduction band overlap, as shown in (b).

If a band is either nearly filled (c) or nearly empty (not shown) and the band gap ΔE is small then the solid will be a semiconductor. Since electrical conduction depends on the number of electrons in an almost empty band or on the number of holes in an almost full band, the conduction will be restricted, and the solid is

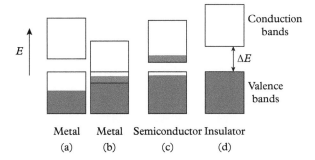

an *extrinsic* (*doped*) semiconductor. If all occupied bands are full but the energy difference between the uppermost filled band (valence band) and the band of next highest energy (conduction band) is small, then the solid will be an *intrinsic* semiconductor.

Finally, if all occupied bands are completely full there are no free electrons, the solid is then an *insulator*, as in (d); ΔE is large, in the region of 5–10 eV. Not surprisingly, gradations through these classes would be expected, depending on ΔE, and have been confirmed by experimental measurements of electrical resistivity:

	Cu/Ag/Au	Ge	Si	GaAs	Diamond	Quartz
ΔE/eV	0	0.66	1.2	1.4	5.5	9.0

5.6.4.1 Hall effect revisited

In discussing the Hall effect in Section 5.3.4, positive values of the Hall coefficient were noted for some species. The implication is conduction by movement of positive charge carriers (holes), which is an important concept in semiconductors and not explainable by free-electron theory where the charge carriers are electrons.

When electrons in a metal are subjected to both a magnetic field **B** and an electric field **E**, then Maxwell's equations show that the force on an electron of velocity **v** is given as $\mathbf{F} = -e[\mathbf{E} + (\mathbf{v} \times \mathbf{B})]$, rather than just $e\mathbf{E}$. The negative sign here is responsible for the negative sign of the Hall coefficient. However, it is shown [15] that when a band is very nearly filled, the time dependence of momentum, or force, is given as $+e[\mathbf{E} + (\mathbf{v} \times \mathbf{B})]$, and the positive sign now implies a positive value for the Hall coefficient; then, conduction is by movement of positive charge carriers or '*holes*'. In cadmium ($R_H = +6.0$), for example, there is a small amount of overlap in energy of the almost filled 5s band, derived from 5s energy states, and an almost empty 5p band. A competition for conduction exists between holes in the 5s band and electrons in the 5p band, and it turns out that hole conduction predominates. The following interesting analogy has been given by Blakemore [14]:

"Consider four glass tubes, each sealed at both ends. Let one be empty, another totally filled with water, the next one almost completely filled with water, and the fourth containing only a small amount of water. If one end of each tube be raised and lowered, there will obviously be no transport in the empty tube. Nor will there be transport in the full tube, for water molecules, like electrons, can then only interchange positions. However, *water* can be seen to run from one end to the other in the tube containing a little liquid, while *air bubbles* move in the opposite direction in the almost full tube. Air bubbles in a tube almost filled with water react in the same way as holes in an almost filled band. Holes rise to the highest attainable states of electronic energy, and move towards the negative pole in an electric field."

5.7 Energy bands and molecular orbital theory

The basis for determining electrical and other properties of solids rests on the distributions of electrons, for which there are two models. In the *nearly-free* electron approximation the assumption is made that the interaction between the conduction electrons and the ion cores can be modelled on a weak potential function. Hence, it appears to oppose coulomb attraction, which is strongest at small distances. However, the conduction electrons cannot approach the core electrons since the orbitals there are already fully occupied, in accordance with the Pauli principle. Thus, the conduction electrons are not close enough to the core to interact strongly with it. Additionally, the core electrons have a screening effect on the other electrons so that they experience only a small effective nuclear charge (Section 2.13.5). The result is that the conduction electrons have a freedom of movement which is reflected in certain properties, such as the electrical conductivity of monovalent compounds.

The second model, the *tight-binding* approximation, is more in accord with discussions in Chapter 2; in fact, the Bloch theory is essentially a molecular orbital model of solids. The important feature of molecular orbital theory applied to a chemical species is that each electron moves in the potential field created by all other atoms in the species, including the core field. It uses one-electron Hamiltonian operators and the solutions of the corresponding wavefunctions are obtained by an LCAO approximation. In benzene (Section 2.18.3) strong overlap of π orbitals in the molecule produces a delocalization of these electrons over the whole molecule. In the case of a metal, the number of atoms is infinite, or at least very large, and the electrons, initially associated with the atoms, are now delocalized over the whole metal crystal. The LCAO molecular orbitals are now the conduction orbitals, and metallic properties depend upon their degree of overlap.

The model will be applied to a one-dimensional solid, an infinite row of atoms. Although somewhat hypothetical, it will lead to the concepts for an understanding of the conductivity of metals and semiconductors, and also aid the description of long structures such as the carbon nanotubes to be discussed in the next chapter.

It is assumed that the atom has one s orbital available for constructing LCAO molecular orbitals by the addition of further atoms in succession, according to the Aufbau principle (Section 2.14), up to some value N. On adding a second atom it overlaps with the first atom to create two atomic orbitals, bonding and antibonding. Addition of a third atom leads to strong overlap with the second atom and a weaker overlap with the first, and bonding, antibonding and non-bonding molecular orbitals are formed. A fourth atom produces a fourth orbital which overlaps the third strongly and earlier orbitals less strongly, and so on. This process is continued to N atoms, where N is a very large number, albeit not quite infinity. The ultimate effect of adding more and more atoms is a fanning out of the range of energies spanned by the molecular orbitals while filling in the range of energies with more and more molecular orbitals. After the addition of N atoms there are then N molecular orbitals within a band of finite width. A Hückel $N \times N$ determinantal equation for this configuration is:

$$
\begin{vmatrix}
\alpha - E & \beta & 0 & 0 & 0 & \cdots & 0 \\
\beta & \alpha - E & \beta & 0 & 0 & \cdots & 0 \\
0 & \beta & \alpha - E & \beta & 0 & \cdots & 0 \\
0 & 0 & \beta & \alpha - E & \beta & \cdots & 0 \\
\cdot & \cdot & \cdot & \cdot & \cdot & & \\
\cdot & \cdot & \cdot & \cdot & \cdot & & \\
0 & 0 & 0 & 0 & 0 & \cdots & \alpha - E
\end{vmatrix} = 0 \qquad (5.78)
$$

where β is the (s,s) resonance integral. The solution of the determinantal equation is:

$$
E_n = \alpha + 2\beta \cos \frac{n\pi}{N+1} \quad (n = 1, 2, 3, \ldots, N) \qquad (5.79)
$$

where n is an integer ranging from 1 to N. The reader is invited to compute E_n for $N = 4$ and to compare the result with that for 1,3-butadiene (see Problem 2.20; remember that β is a negative parameter). Since $-1 \leq \cos\left(\dfrac{n\pi}{N+1}\right) \leq 1, (\alpha + 2\beta) \leq E_n \leq (\alpha - 2\beta)$; thus, molecular orbitals lie within the amount 4β of the original atomic orbital. When N is infinitely large the difference in energy of any n and $(n + 1)$ levels is infinitely small but still of finite width, as the following analysis shows.

Consider the limit:

$$
\lim_{N \to \infty} E_1 = \alpha + 2\beta \cos \frac{\pi}{N+1} \to \alpha + 2\beta
$$

whereas the limit:

$$
\lim_{N \to \infty} \alpha + 2\beta \cos \frac{N\pi}{N+1} \to \alpha - 2\beta
$$

Thus, the difference is 4β, as determined above.

As N increases, the HOMO–LUMO separation decreases so that the photon energy needed to excite an electron from the HOMO level to the LUMO level decreases also. As N tends to infinity, the separation of any two energy levels, while remaining discrete, becomes infinitely small, and all such levels effectively merge to give a continuous band of energy levels of width 4β. In general, the band width depends upon the degree of overlap of the individual atomic orbitals, and governs the strength of interaction and the magnitude of the resonance integral β. As the HOMO–LUMO separation tends to zero electrical conduction can occur readily.

The band consisting of N molecular orbitals formed from s atomic orbitals is referred to as an s band; if formed from p atomic orbitals then it is termed a p band, and so on. Some aspects of this discussion are illustrated by Fig. 5.20. In the s band the lower level band orbitals are fully bonding, whereas the higher level orbitals are antibonding; a similar structure arises for the p band. The figure shows the band gap, which is a range of energies with no corresponding orbitals. In another situation the bands may be adjacent or even overlap, as with the $3s$ and $3p$ orbitals of magnesium or the $5d$ and $6s$ orbitals of gold (see Fig. 5.31).

In metallic lithium $(1s)^2(2s)^1$, a cube of crystal of side 1 mm contains approximately 4.5×10^{19} atoms. Thus, there are 9×10^{19} $2s$ energy states that form a quasi-continuous series of energy states; a continuum is the limiting case of infinite degeneracy.

At $T = 0$ K only the lowest $N/2$ molecular orbitals are occupied and the HOMO is the Fermi level. Above this level there are empty orbitals and if the

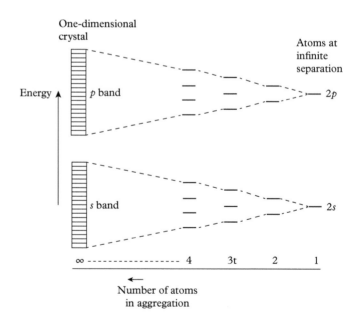

Fig. 5.20 Schematic representation of the formation of s and p energy bands, consisting of very closely spaced energy levels. The overlap of s atomic orbitals forms the s band, and the p atomic orbitals give the p band. The distance between the top of the s band and the bottom of the p band is the energy gap ΔE for these bands.

HOMO–LUMO energy gap is not too large (Fig. 5.20a and b) electrons can be excited into a mobile state in an upper conduction band. At temperatures above 0 K it is possible for electron excitation to occur through the thermal motion of other electrons in the metal and so cause the metal to be electrically conducting.

The electrical conductivity σ of a metal decreases with an increase in temperature. Notwithstanding an increase in temperature raises the degree of thermal excitation, which would be expected to increase conductivity, it also increases the number and frequency of collisions. Of the two competing processes, that of collision has the greater effect so that the electrons become less effective as charge carriers and the conductivity decreases.

5.8 Semiconductors

Semiconductors at ambient temperatures have electrical resistivities ρ of 10^{-4} to $10^7\,\Omega$ m, whereas metallic conductors are *ca.* $10^{-8}\,\Omega$ m and insulators 10^{12} to $10^{18}\,\Omega$ m in resistivity. The reciprocal of electrical resistivity is electrical conductivity, ranges of which are shown below for the materials under discussion:

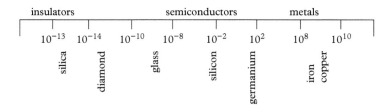

Typical electrical conductivities, σ/S m^{-1}

At 0 K pure crystals of semiconducting material behave as insulators, and semiconduction arises either through thermal agitation of the material or by the presence of defects in it, such as impurities or lattice defects. Among the many semiconducting materials, silicon, germanium and gallium arsenide are well-known examples.

Thermal energy excites electrons across a band gap in a semiconductor, so that an increase in temperature increases the number of electrons with kinetic energy sufficient to enter the conduction band. The electrical resistivity ($\rho = 1/\sigma$) of a semiconductor therefore decreases rapidly with an increase in temperature. Experimental data on the electrical resistivity of a doped germanium semiconductor are listed in Table 5.3 and illustrated below. The plot is very closely linear; Pearson's r is 0.9996 for the equation $\rho_T = \rho_0[1 + \alpha(T - T_0)]$, where $\rho_0 = 0.250\,\Omega$ m, $\alpha = -0.693\,\Omega$ m and $T_0 = 290$ K. The resistivity is very dependent upon the nature and extent of the doping.

Table 5.3 *Electrical resistivity of a doped germanium semiconductor*

T/K	ρ/Ω m
290.0	0.250
300.0	−1.401
310.0	−3.412
320.0	−5.240
330.0	−6.603
340.0	−8.746
350.0	−10.001
360.0	−12.154
370.0	−13.602
380.0	−15.403
390.0	−17.114

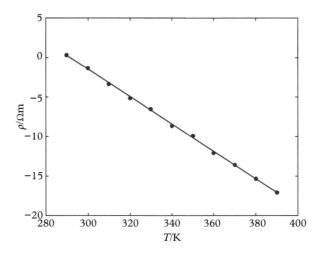

Electrical resistivity ρ $v.$ temperature T for doped germanium

An electron travels through a metal in response to an applied potential difference, but it travels only a short distance before it collides with a metal ion. In response to an increase in temperature the frequency of these collisions increases, the amount of travel and overall movement of electrons through the metal is reduced and the resistivity increased in consequence. If the temperature of the solid is increased further, the metal ion vibrations about their mean positions increase in amplitude. Thus, the electrons encounter metal ions more frequently which also brings about an increase in resistivity. The changes of electrical resistivity with temperature in metals are significantly less than the corresponding changes in semiconductors, as the following typical data indicate:

	$\rho/\,\Omega\,\text{m}$	
	500 K	1000 K
W	10.34×10^{-8}	24.43×10^{-8}
Si	39.91	4.75×10^{-2}

5.8.1   Intrinsic semiconductors

In an intrinsic semiconductor, conduction is a property of the band structure, whether it be in a pure element like silicon, or in a compound such as gallium arsenide. At 0 K the conduction band is empty and separated from a filled valence band by a small energy gap ΔE of 1–2 eV. As the temperature is increased, thermally excited electrons are promoted across the Fermi level and enter the conduction band. The electrical resistivity varies with temperature, and follows the equation:

$$\rho = A\exp[\Delta E/(2kT)] \qquad (5.80)$$

Hence, the energy of activation ΔE for the conduction process may be found from a plot of $\ln\rho$ against $1/T$; some values for the energy gap in well-known intrinsic semiconducting materials are listed below.

	Si	Ge	GaAs	InAs	CdS
ΔE/eV	1.1	0.70	1.4	0.33	2.4

5.8.2 Extrinsic semiconductors

Conduction in an extrinsic semiconductor takes place on account of a replacement of some atoms in the structure by atoms of another species, to an extent of approximately 1 in 10^9. The addition of small amounts of impurity substances (doping) has a significant effect on the electrical properties of semiconducting materials. Both silicon and germanium have the diamond structure (Fig. 2.51). On the one hand, if silicon $(Ne)(3s)^2(3p)^2$ be doped with a donor species such as arsenic $(Ar)(4s)^2(4p)^3$ or phosphorus $(Ne)(3s)^2(3p)^3$, then the donor atom replaces a silicon atom in its tetrahedral structural unit, not in an interstitial position. The structure then has an *excess electron* from each donor atom introduced and is positive at the donor site; it becomes a negative carrier and is termed an *n*-type semiconductor (Fig. 5.21a). On the other hand, if boron $(He)(2s)^2(2p)^1$ or aluminium $(Ne)(3s)^2(3p)^1$ (acceptor species) be used as dopant in place of a donor species, each boron or aluminium site introduced becomes negative and a *positive hole* is available for conduction; this semiconductor is *p*-type (Fig. 5.21b). At 25 °C, the electrical resistivities of doped silicon and germanium are approximately 2×10^3 Ω m and 0.4 Ω m, respectively.

5.8.3 *p–n* Junction semiconductors

If an *n*-type and a *p*-type semiconductor are placed in contact and a potential difference set up in the direction shown in Fig. 5.22a, then electrons in the *n*-type portion of the composite are attracted towards the positive electrode, and the holes move to the negative electrode; in this condition, there is *no flow* of electric current. If the direction of the applied potential difference is reversed, the positive and negative electrodes are *p*-type and *n*-type semiconductors, respectively. Now, *charge flows* across the *p–n* junction, as indicated by Fig. 5.22b. A *p–n* junction device provides control over the magnitude and direction of current flow, and consequently is of considerable importance in transistor- and diode-operated electric circuits. While the potential difference is directed as in Fig. 5.22b, positive and negative charge carriers continue to flow.

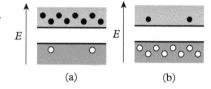

(a)   (b)

Fig. 5.21 Extrinsic semiconductors; the dark circles in the conduction band are electrons and open circles in the valence band are holes. (a) *n*-Type: electrons are the majority charge carriers; electron concentration > hole concentration. (b) *p*-Type: holes are the majority charge carriers; hole concentration > electron concentration. Thus, *n*-doped silicon has one 'extra' electron per doped atom whereas *p*-doped silicon is 'missing' one electron per doped atom, leading to a positive 'hole'. [Reproduced under Creative Commons Attribution-ShareAlike 3.0 Unported License from https://en.wikipedia.org/wiki/Extrinsic_semiconductor.]

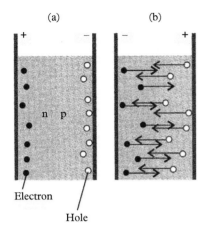

Electron

Hole

Fig. 5.22 Juxtaposition of *n*-type and *p*-type extrinsic semiconductors: (a) With a positive potential applied to the *n* portion no current flows; (b) With a positive potential at the *p* portion, current flows as shown. Thus, a *p–n* junction can form a current controlling device. [Reproduced under Creative Commons Attribution-ShareAlike 3.0 Unported License from https://en.wikipedia.org/wiki/Extrinsic_semiconductor.]

A continuation of this current flow generates heat in many electronic devices, such as computers, circuit-boards and electric-car batteries. Cooling can be achieved by various means, such forced air or Peltier cooling plates. High-power transistors and laser installations can be cooled by diamond sinks, using industrial diamonds. Most materials of high thermal conductivity also have high electrical conductivity: diamond is notable for its high thermal conductivity (*ca.* 3 kW m^{-1} K^{-1}) coupled with low electrical conductivity (100 GΩ m to 1 EΩ m), and is incorporated into some transistor devices where overheating is a serious problem. But diamonds are not forever: under high temperature over long periods of time they transform slowly to the thermodynamically stable graphite allotrope.

5.8.4 Superconductors

A *superconducting* species has zero resistance below a transition temperature, but behaves as a normal conductor above that temperature, as shown by Fig. 5.10 for mercury; transformation takes place with no enthalpy change. The discontinuity in heat capacity is a second order transition, involving no latent heat, from the normal electronically disordered state to one of high order and consequent lower entropy. An example is shown by the heat capacity curve for gallium in Fig. 5.23, where it is compared with that of a normal conductor over the same temperature range.

Superconductivity occurs in approximately thirty elements at ambient pressure, and in about another fifteen under high pressure. The transition temperatures for selected elements and superconducting binary alloys are listed in Table 5.4.

The interior of a superconductor is not penetrable by a weak magnetic field; this flux exclusion is termed the *Meissner effect* [16]. If the applied magnetic field becomes very large, superconduction breaks down; superconductors are classified as Type I or Type II according to the mode of breakdown.

In Type I, superconduction is destroyed when the applied field strength exceeds a critical upper limit, and is generally exhibited by pure metals. A Type II

Fig. 5.23 Molar heat capacity curves $C_{p,m}$ for a normal conductor and a gallium superconductor; the superconducting transition temperature for gallium is 1.13 K.

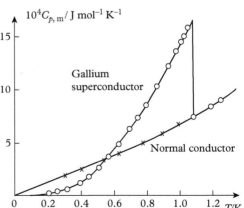

superconductor is characterized by the formation of magnetic vortices above a critical value of field strength. As the magnetic field strength is increased the vortex density increases, until the field strength is sufficiently high to destroy the superconductivity.

In the Meissner effect, the magnetic field penetrates to a depth λ given by [11]:

$$\lambda = \sqrt{\frac{m_e c^2 \varepsilon_0}{e^2 n_s}} \qquad (5.81)$$

where n_s is the density of electrons in the superfluid condition of the superconducting material. Superconducting electrons do not move singly, but in spin-paired couples, known as Cooper pairs from the *BCS theory*[8] [17, 18].

The coherence length, or spatial range ξ, of the wavefunction for a pair is given as:

$$\xi = \frac{2\hbar v_F}{\pi \Delta} \qquad (5.82)$$

where v_F is the Fermi velocity;[9] Δ is the superconductor energy band gap and is given approximately by $(7/2)kT_t$, where T_t is the superconductor critical transition temperature. These parameters λ and ξ are used to differentiate between Type I and Type II superconductors: for Type I, $0 < \lambda/\xi < 1/\sqrt{2}$, and for Type II, $\lambda/\xi > 1/\sqrt{2}$.

Type I superconductors are usually pure metals, such as aluminium, lead and mercury. Type II superconductors are generally metal alloys, such as niobium–tin and niobium–titanium, elemental niobium and vanadium or else complex oxide ceramics, of which $La_{1.85}Ba_{0.15}CuO_4$ and $YBa_2Cu_3O_7$ (YBCO) are well known; superconductors for high-temperature operation are of Type II.

Strong superconducting electromagnets have important applications, in non-invasive medical magnetic resonance imaging (MRI) scanning equipment, in scientific nuclear magnetic resonance (NMR) imaging machines, particle accelerators and 'petaflop' computers[10] in transport, such as trains operating on frictionless rails of powerful superconducting magnets, as in the Japanese MagLev system; a speed of *ca.* 300 mph has been reported for these trains.

Table 5.4 *Transition temperatures T_t for superconducting elements and alloys*

	T_t /K		T_t /K
W	0.01	$ZrAl_2$	0.30
Ti	0.40	TiCo	0.71
Cd	0.56	AuBe	2.64
Al	1.19	NiBi	4.25
Hg	4.15	NbN	16.0
Pb	7.18	Nb_3Ga	20.3
Nb	9.46	Nb_3Ge	23.2

[8] So-called from the authors' surnames, Bardeen, Cooper and Schrieffer.
[9] The Fermi velocity is $\sqrt{2E_F/m_e}$ and is in the range *ca.* $(0.8–2.3) \times 10^6$ m s^{-1}.
[10] 1 petaflop = 10^{15} floating point operations per second.

Japanese 'MagLev' levitation train [courtesy evworld.com]

Superconduction is an active field of current research. Among the many topics abounding is the use of computer models to design commercial materials. Density functional theory[11] (DFT) predicted correctly that at high pressure and low temperature silicon would act as a superconductor [19]. Another computer prediction is the formation of iron boride, FeB_4, which was found to be diamagnetic below 3 K, an indication of superconductivity; this compound has the strength and hardness of diamond [20].

5.9   Fuel cells, batteries and solar cells

Fuel cells and batteries are devices that convert chemical energy into electrical energy. Fuel cells operate from a continuous supply of chemical reactants in order to maintain a constant electric current. The essential difference between a battery and a fuel cell is that the battery is a *complete electrical storage unit* incorporating its own reactants, whereas a fuel cell requires the *continual replenishing* of its reacting materials.

5.9.1   Fuel cells

The concept of a fuel cell was implicit through the work on electrolysis by Humphry Davy in 1802: water was decomposed by an electric current into hydrogen and oxygen; therefore, the combination of these two gases should generate electricity. Schönbein was credited with the development of the first fuel cell 1838, but he is better known for discovering the explosive properties of nitrocellulose. It was Grove who produced the first 'gas voltaic' fuel cell in 1839 by the platinum catalysed combination of hydrogen and oxygen [21, 22]. Bacon in 1932 modified the system used by Langer and Mond (1889) and produced a 5 kW fuel cell system. A few years later a 15 kW Bacon-type cell was used to power an Allis-Chalmers agricultural tractor, and a number of industrial machines powered by fuel cells was developed subsequently.

The many types of fuel cells have in common an *anode*, a *cathode* and an *electrolyte*. An individual fuel cell produces a small voltaic potential, and cells are stacked in series to increase the available electromotive force. The energy efficiency of a fuel cell is *ca.* 60%, or effectively more if the 'waste' heat is harvested for another purpose.

A fundamentally important fuel cell is the *hydrogen cell*, illustrated in Fig. 5.24. The anode and cathode plates are often compressed carbon powder containing an oxide catalyst or finely divided platinum or silver. Many substances have been uses as electrolytes: liquids include sodium hydroxide or potassium hydroxide, phosphoric acid or molten carbonates; a solid electrolyte is usually a *proton exchange membrane* (PEM), and these fuel cells have the distinct advantage of clean working.

In each type of hydrogen cell the reaction at the anode is an oxidation of hydrogen to protons and electrons. The electrons travel via the electrode to the external

[11] DFT is a computational quantum mechanical modelling method used in physics, chemistry and materials science to investigate the electronic structure of atoms, molecules and condensed phases.

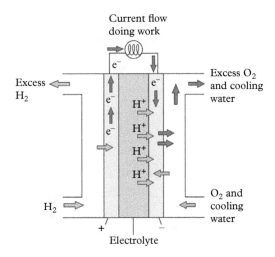

Fig. 5.24 Hydrogen gas fuel cell: the electrodes are compressed finely divided nickel, or compressed carbon with a catalyst material, and the electrolyte in the early cells was sodium hydroxide; the potential difference developed is *ca.* 1.2 V.

circuit to do mechanical or other work, and the protons travel through the electrolyte to the cathode where they reduce molecular oxygen to form water. The fundamental half-cell reactions are:

$$H_2(g) \rightarrow 2H^+(aq) + 2e^- \qquad E° = 0.00 \, V$$

and

$$\tfrac{1}{2}O_2(g) + 2H^+(aq) + 2e^- \rightarrow H_2O(l) \qquad E° = 1.23 \, V$$

The overall reaction is exothermic so that the process is less productive at elevated temperatures, and water cooling of the cell is necessary in some cases. A descriptive diagram of a PEM fuel cell is shown in Fig. 5.25.

In addition to these hydrogen-type cells, there are hydrocarbon fuel cells that employ diesel, methanol and hydrides; the waste products with these fuel cells are carbon dioxide and water.

5.9.2 Batteries

Many successful batteries have been based on lead-acid, nickel-metal oxides or lithium-ion types. The lead-acid battery was invented by Planté in 1859 and was the first rechargeable-type battery. A single cell provides a potential of about 2 V, and its operation is now well known. It is able to supply a large surge current, which is necessary in such applications as mechanical starter motors. As it presents problems in toxicity and corrosion, other sources of electrical energy have been sought.

The nickel-metal hydride battery, known as NiMH, has a nickel oxyhydroxide anode, as in the now obsolete NiCd battery, and the cathode is a

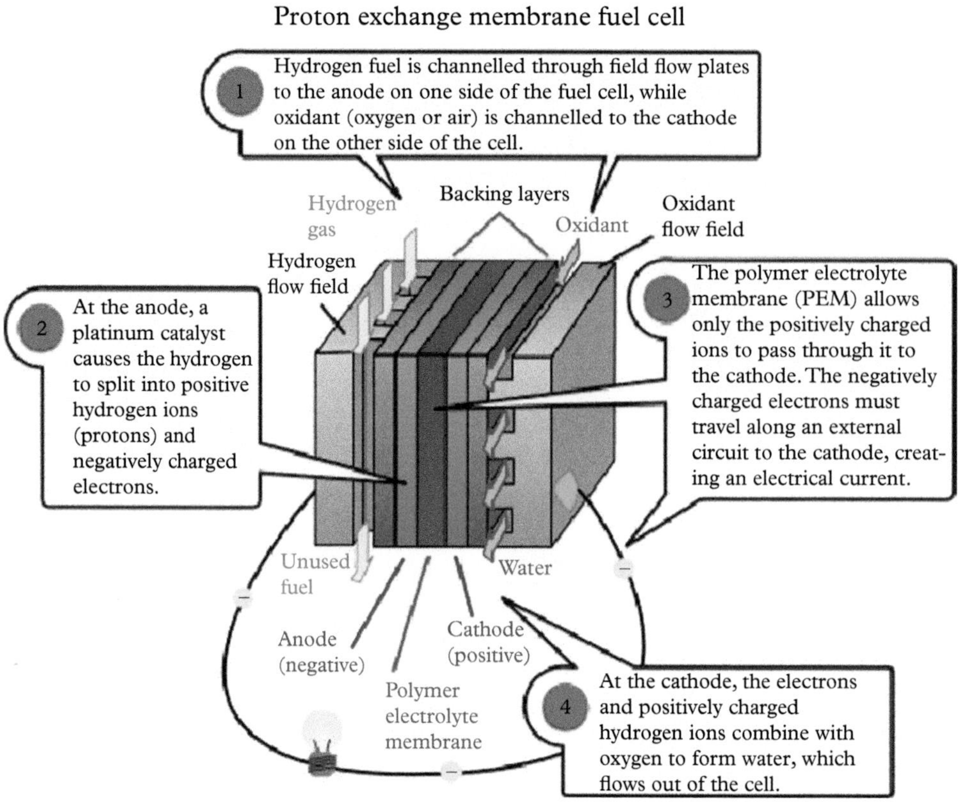

Proton exchange membrane fuel cell

Fig. 5.25 Explanatory diagram of a proton exchange membrane fuel cell. [Public domain.]

hydrogen-absorbing alloy capable of forming metal hydrides on charging. The chemical process can be represented thus:

$$H_2O(l) + M(s) + e^- \underset{\text{charge}}{\overset{\text{discharge}}{\rightleftarrows}} OH^-(aq) + MH(s)$$

The NiMH battery develops a potential of 1.2 V per cell, so that stacking is necessary in order to obtain a useful working voltage. Although the NiMH has had wide use and has powered some automobiles, lithium-ion type batteries have proved more successful and less toxic.

The lithium-ion battery uses a lithium cobalt oxide $LiCoO_2$ as the anode and a specially crystallized form of graphite compressed to form the negative electrode. The electrolyte is a lithium salt, such as $LiCl\,O_4$, $LiB\,F_4$ or $LiP\,F_6$ in an organic solvent such as ethylene carbonate or diethyl carbonate. The essential chemical reaction is:

$$LiCoO_2 + C \underset{\text{charge}}{\overset{\text{discharge}}{\rightleftarrows}} Li_xCoO_2 + Li_{1-x}C$$

Table 5.5 *Characteristics of nickel and lithium battery types*

	NiCd	NiMH	Li-ion	Li–S
Rating / W-h kg^{-1}	50	70	120	500
Fast charge / h	1–2	2–3	3–4	–
Cycle life	1500	500	500–1000	1500

On discharge, electrons are set free at the anode to travel through the external work circuit and lithium cations travel through the electrolyte to form a carbon-lithium compound. The process is reversed when the cell is recharged. Lithium-ion batteries have proved to be useful in a range of industrial applications, and research in this field is on-going. Improvements have been made by incorporating a graphene (Section 6.3.4) anode into the lithium-ion cell. A comparison of some properties of battery types is given in Table 5.5.

A lithium–sulphur combination has led to rechargeable, high energy-density batteries. On account of the low atomic masses of lithium and sulphur, these batteries are light in weight compared to most other types; in addition, their high energy density and cheapness recommends them over the lithium-ion batteries, as shown in Table 5.5. A recent development in lithium–sulphur batteries reports a much improved rating and cycle life, and has suggested that battery-powered travel of up to 300 miles should be attainable.

The basic chemistry is an overall two-electron reaction:

$$2Li^+(aq) + S(s) + 2e^- \underset{\text{charge}}{\overset{\text{discharge}}{\rightleftarrows}} Li_2 S(s)$$

and a lithium–sulphur battery has a rating of 167 mA-h g^{-1}. This value has been achieved by employing a modified sulphur–graphene oxide cathode and a lithium metal anode protected by lithium nitrate in the organic electrolyte. The loss of sulphur as polysulphides remains a challenge for maintaining a long life, currently standing at 1500 cycles [23].

Recent reports [24] have focused on the improvement of the anode in those lithium cells that use graphite for this function. A new material TiNb$_2$O$_7$ has been developed as a nanoporous assembly[12] with a packing density twice that of graphite and so can store more energy for the same size cell. The interconnected channels of the nanoporous anode facilitate electrolyte diffusion and provide a continuous conduction path. The cell has a rating of 281 mA-h g^{-1}, and can be recharged to 70% of full capacity in three minutes. A downside is the high cost of separating niobium from tantalum, with which it occurs in the most common ore columbite (Fe,Mn)(Nb,Ta)$_2$O$_6$.

The problem with accumulation of polysulphides in batteries has been ameliorated to some extent by the introduction of DNA into the cell. The four

[12] 'Nanoporous' implies a porosity less than 100 nm.

nucleotide base pairs and phosphate groups attract sulphur by an adsorption process, and a 1% addition of DNA derived from salmon sperm produces a significant improvement in performance [25].

A new rival to the lithium-ion battery is the aluminium-ion battery. Aluminium is stored at the cathode as $[AlCl_4]^-$. On discharge, ions migrate to the graphite anode forming $[Al_2Cl_7]^-$. The battery develops 2 V on discharge and is rechargeable in less than one minute. The overall processes may be represented as a reversible intercalation reaction:

$$Al + 7[AlCl_4]^- \rightleftharpoons 4[Al_2Cl_7]^- + 3e^-$$

Three electrons are liberated in the reaction in contrast to only one for the lithium-ion battery [26].

5.9.3 Solar cells

Solar cells convert incident light radiation into electricity by means of the photoelectric effect. Silicon plates in the solar cells absorb photons, dislodging electrons and harvesting them into an electric current by means of p–n type semiconductors (Section 5.8.3). The excess electrons in the n-type silicon fill the holes in the p-type silicon, thus creating an electric field across the solar cell that can be harnessed to do work.

Currently, about twenty-six different types of solar cell are recognized. The most conventional solar cells use silicon as the photovoltaic material, but although it is cheap to produce it is not an efficient absorber because it is an indirect band gap semiconductor (Section 5.9.3.1) and a thick layer is required to be prepared for the cell to be useful. A silicon solar cell develops a potential of 0.75 V, so that three such cells are needed to decompose water into hydrogen and oxygen.

The electrolysis of water, known as *water splitting*, is an important method of achieving clean, renewable energy systems. The water splitting procedure can be enhanced by means of a photovoltaic catalyst, such as finely-divided titanium dioxide and platinum, making the production of electricity a one-step process instead of the combination of photovoltaic and electrolytic systems.

An interesting use of perovskites arises in the context of photovoltaic cells. Perovskite is the general name of a material of the type ABO_3, where A is an alkaline earth or rare earth element and B is a transition metal; they are usually cubic structures, perovskite itself being $CaTiO_3$, identified first in 1829.

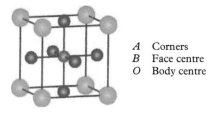

A Corners
B Face centre
O Body centre

Perovskite ABO_3 structure type

Perovskite is a direct band gap material: its crystals harvest light efficiently, requiring much less material for the process than do other materials, and transport electric charge in the form of ions. A perovskite solar cell that uses lead methylammonium triiodide $CH_3NH_3PbI_3$ as an active material achieves a potential of *ca.* 1 V, so that only two cells are required for decomposing water, and their operating efficiency is *ca.* 18%. Recent advances with perovskite solar cells should make it a dynamic area of research offering new and promising opportunities for renewable energy [27]. However, queries have been raised about the level of efficiency on account of hysteresis in the current–voltage measurements [28, 29].

"The sun will be the fuel of the future." Anon. *ca.* 1876

The water splitting process has an improved efficiency *via* a photocatalysis technique that uses a N, N'-dicyanocarbodiimide(C_3N_4)–carbon nanodot composite as catalyst. The water is split into hydrogen and hydrogen peroxide with a conversion efficiency of 2%. A more efficient catalyst is cobalt(II) oxide, but it loses its 5% efficiency within 1 hour [30].

5.9.3.1 Direct and indirect band gap semiconductors

If the energy E of a photon incident on to a semiconductor is greater than the band gap energy ΔE, the incident radiation is absorbed strongly. If, on the one hand, the minimum energy of the conduction band lies at or close to the same value **p** of momentum as the maximum energy of the valence band (Fig. 5.26a), an electron in the conduction band shares the same value of momentum **p** as the hole and releases excess energy as a photon. This process occurs with a *direct band gap* semiconductor, such as perovskite or gallium arsenide. On the other hand, where the above minimum and maximum are separated by a significant momentum Δ**p** in an *indirect band gap* semiconductor (Fig. 5.26b) the sharing cannot occur since photons do not carry crystal momentum, and the conservation of crystal momentum cannot be violated. For sharing to occur in an indirect band gap material, such as silicon, a phonon of momentum Δ**p** must be absorbed or emitted, depending upon the relative positions of the two extrema.

Recently, a new orthorhombic allotrope of silicon has been prepared. First the compound Na_4Si_{24} was synthesized under high pressure [31]. Subsequently,

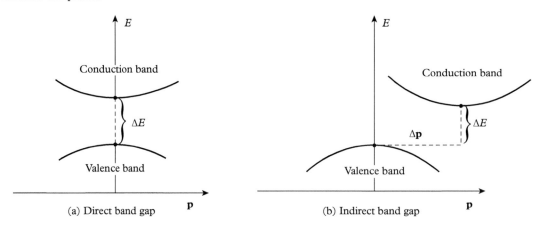

Fig. 5.26 (a) Direct band gap semiconductor (GaAs, InP, InAs): the minimum in the conduction band and the maximum in the valence band coincide in momentum. (b) Indirect band gap semiconductors (Si, Ge, AlAs): the minimum in the conduction band and the maximum in the valence band differ in momentum by the significant quantity $\Delta \mathbf{p}$.

Fig. 5.27 A channel parallel to the a-axis formed by 6- and 8-membered rings of silicon atoms in the Si_{24} allotrope. [Reproduced by courtesy of Dr Timothy Strobel.]

removal of the sodium by thermal degassing formed Si_{24}, and x-ray diffraction showed that the new allotrope crystallized with space group *Cmcm* and twenty-four silicon atoms in the unit cell. The structure exhibits channels parallel to the a-axis, formed by six- and eight-membered rings of tetrahedral silicon atoms (Fig. 5.27). This new allotrope acts as a direct band gap species with $\Delta E \approx 1.3$ eV, less than that of gallium arsenide (1.43 eV). The combined experimental and theoretical study has expanded the known allotropy for element 14 and demonstrated the value of high-pressure precursor synthesis as a potential for the development of new materials [32].

5.10   Structures of metals

The metallic bond is without directional character, so that metals take up structures that are determined to a large extent by space-filling criteria. A majority of metals adopt one or more of three relatively simple structures: close-packed cubic (Fig. 5.28a), close-packed hexagonal (CPH) (Fig. 5.28b), or body-centred cubic (BCC) (Fig. 5.29).

The close-packed structures represent equally efficient ways of filling space with identical, spherical atoms. A first layer of the close-packed structures is obtained by placing spheres in contact such that the centres of any three in close contact form the apices of an equilateral triangle (Fig. 5.30a, layer A). A second similar layer is added so that the spheres of the second layer rest in the depressions of the first (Fig. 5.30b, layer B). A third layer can be added in two ways: either the spheres in the third layer lie over the spaces in both the first and second layers

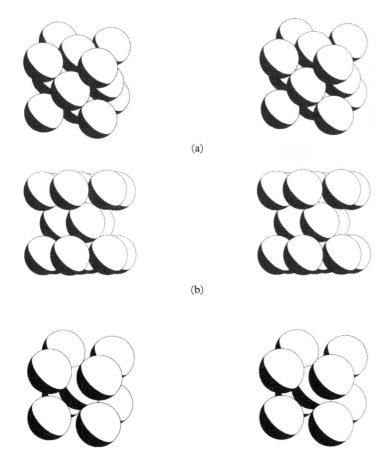

(a)

(b)

Fig. 5.28 Stereoviews of the unit cell and environs of the structures of close-packed metals of coordination number 12: (a) Close-packed cubic; examples include calcium, copper, silver, cerium. (b) Close-packed hexagonal; examples include beryllium, cobalt, zinc, thallium.

Fig. 5.29 Stereoview of the unit cell and environs of the structure of less closely-packed, body-centred cubic metals of coordination number 8; examples include sodium, chromium, iron, tungsten.

(Fig. 5.30c, layer C), in which case the close-packed cubic structure is obtained with the sequence $ABCABCA\ldots$, or the spheres in the third layer lie sphere for sphere exactly above those in the first layer (layer A) (Fig. 5.30d), in which case the close-packed hexagonal array obtains with the sequence $ABABA\ldots$.

The close-packed cubic structure (aka FCC) is referred conventionally to a face-centred cubic unit cell of side a equal to $2r\sqrt{2}$, where r is the radius of atom. The volume of that portion of the cell containing one atom is $(16r^3\sqrt{2})/4$, or $4r^3\sqrt{2}$. The volume of an atom is $4\pi r^3/3$, so that the efficiency of packing, or packing fraction, is $[(4/3)\pi r^3]/(4r^3\sqrt{2})$, or 0.740. The same value holds for the equally close-packed hexagonal structure. In the case of the less closely-packed body-centred cubic structure (aka BCC), (Fig. 5.29), the cube side a is $4r/\sqrt{3}$, and it is easily shown that the packing fraction is now 0.68.

Close packing leads to high density. The close-packed planes are $\{111\}$ in the close-packed cubic structure, and (0001) in the close-packed hexagonal structure. From the relationships between unit cell side and atomic packing, it is possible

Fig. 5.30 Close packing of identical spheres. (a) First layer: full lines, denoted *A*. (b) Second layer: long dashes, denoted *B*. (c) Third layer: short dashes, denoted *C* with the spheres lying over the octahedral holes created by the packing of the first two layers; the close-packed cubic sequence is *ABCABCA....*. (d) Alternative third layer: very short dashes, denoted by *A* with the spheres lying exactly above the spheres in the first layer; close-packed hexagonal sequence is *ABABA....*.

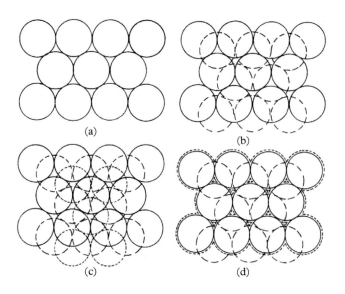

(a)

(b)

(c)

(d)

Table 5.6 *Selected metallic radii/nm*

Li	0.157	Be	0.112	Cu	0.128
Na	0.191	Mg	0.160	Ag	0.144
K	0.235	Ca	0.197	Au	0.144
Rb	0.250	Sr	0.215	Fe	0.126
Cs	0.272	Ba	0.224	Co	0.125
Fr	0.293	Ra	0.229	Ni	0.125
Zn	0.137	Cd	0.152	Hg	0.155

to develop a set of metallic (atomic) radii, by the procedure with which the ionic radii were deduced in Section 4.6.1; they are listed in Table 5.6.

Some metals occur in polymorphic (allotropic) modifications. From an experimental study of these metals, the following relationship between radius and coordination number has been established:

Coordination number	12	8	6	4
Relative atomic radius	1	0.97	0.96	0.88

5.11 Structural and physical characteristics of metallic solids

The metallic bond is spatially undirected, giving rise principally to the structure types discussed in Section 5.10; metal structures generally have high coordination numbers, frequently 12 or 8. Metals have variable strength. Deformation by gliding is common in metals, and it takes place preferentially in directions parallel to close-packed planes; four (111) planes in close-packed cubic and one, (0001), in close-packed hexagonal structures. Consequently, metals such as copper and silver are more malleable and ductile than are the body-centred cubic structures of iron or tungsten. Metals have sharp melting points but they vary widely, Hg 234 K, W 3584 K, and the liquid interval in the melt is often very long, Ga

2370°, Hf 3400°. Perhaps the most distinctive properties of metals are their high electrical conductivities (low resistivities) and thermal conductivities, to which detailed reference has been made in earlier sections.

Electrons near the Fermi level of a metal can absorb energy quanta and are thus raised to higher states. If there is but little interaction between these electrons and the lattice ions, the energy gained is radiated away without change of phase and the crystal is transparent. Metals interact in this way with radiation of wavelengths less than that of the ultraviolet region of the electromagnetic spectrum.

Colour in metals depends upon the absorption of photons by the conduction electrons, principally d electrons. In silver, for example, the $4d \rightarrow 5s$ transition involves a large energy gap that corresponds to photons in the ultraviolet spectral region; frequencies in the visible region have insufficient energy for absorption, and all visible frequencies are reflected, that is, the material shows no colour. This situation applies to most metals, which look white or silvery in white light. In contrast, the $5d \rightarrow 6s$ transition for gold is much smaller in energy and the electron transition occurs with frequencies just into the visible region from the ultra-violet region; hence, the metal exhibits the complementary colour, yellow.

Shapes and energies of bands may be determined in electron spectroscopy by exposing a specimen to x-radiation of known energy E_ν, sufficient to induce electron emission from the metal. The kinetic energy E_K of an emitted electron is of the order of 1 keV, and can be counted accurately by a β-ray spectrometer. The energy E needed to remove an electron from an atom is given as:

$$E = E_\nu - E_K \qquad (5.83)$$

The overlap of the $5d$ and $6s$ bands shown in Fig. 5.31 was determined by electron spectroscopy on a single crystal of gold, using Al $K\alpha$ radiation. The result applies

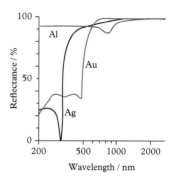

Reflectance *v.* wavelength for Al, Ag and Au

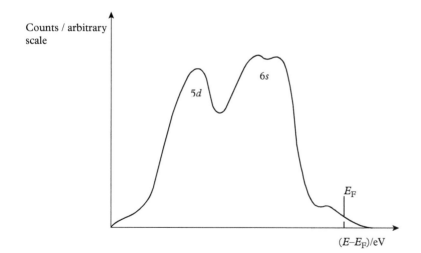

Fig. 5.31 Electron spectrum of elemental gold, showing an overlap of the $5d$ and $6s$ energy bands.

strictly to the outer layers of the metal, as electron emission can take place only through approximately 2 nm distance; the corresponding energies deeper within the metal have differing values [33].

5.12   Alloy systems

Metals form alloys readily, and their study has played an important part in the development of the theory of the metallic state. Alloys are numerous in type and exhibit variable, non-stoichiometric compositions. They are formed generally by melting the constituents together, a feature that is consistent with the picture of the metallic bond. In this section the structures and properties of two binary alloys will be studied.

5.12.1   Copper–gold alloys

Copper and gold are chemically similar and have similar atomic radii (Table 5.5); these elements provide a good example of an order–disorder system. If gold is added to copper and the molten alloy quenched rapidly, a face-centred cubic structure is obtained which shows random replacement of copper by gold (Fig. 5.32).

A decrease in the unit-cell dimension is found, in accordance with Vegard's law, which may be written concisely for cubic crystals as:

$$a \propto c \tag{5.84}$$

where a is the cubic unit cell dimension and c is the concentration of the added alloying metal. The following data were given by Vegard [34, 35]:

Copper–0.23 atomic % Au: $a = 0.3616$ nm

Copper–2.8 atomic % Au: $a = 0.3632$ nm

Copper–10.0 atomic % Au: $a = 0.3669$ nm

Fig. 5.32 Stereoview of the unit cell and environs of the face-centred cubic structure of a random solid solution of gold in copper; each sphere represents, statistically, a certain fraction c of copper and $(1 - c)$ of gold.

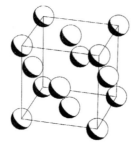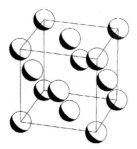

to which can be added the zero value:

Copper–0.00 atomic % Au: $a = 0.3615$ nm.

A linear least-squares fit to these data has a root mean square error of 1.2×10^{-4} and a Pearson r of 0.998, thus confirming the validity of the law for this system. The space group of the solid solution phases are $Fm\overline{3}m$.

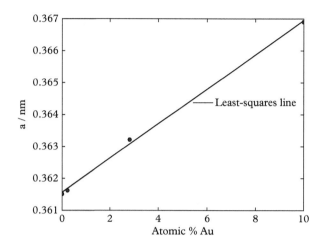

Confirmation of Vegard's law for the α phase of the Cu–Au alloy system

5.12.1.1 *Order-disorder structures*

A different situation is obtained if the alloy specimen is annealed. Random re-placement continues until the composition reaches the formula Cu_3Au, which has a pseudo face-centred structure (Fig. 5.33). The space group is now $Pm\overline{3}m$ with $a = 0.3749$ nm, and the unique atomic positions are:

Au at 0, 0, 0;

Cu at 0, 1/2, 1/2; 1/2, 0, 1/2; 1/2, 1/2, 0

Continued replacement of copper by gold followed by quenching ultimately reveals a tetragonal structure with the composition CuAu (Fig. 5.34); the space group is $P\frac{4}{m}mm$ with $a = 0.3963$ nm and $c = 0.3710$ nm; the atomic coordinates are: Au 0, 0, 0 and 1/2, 1/2, 0; Cu 0, 1/2, 1/2 and 1/2, 0, 1/2.

The Cu_3Au and CuAu ordered structures are termed *superstructures* (aka, in-correctly, *superlattices*). The electrical resistivity of each quenched copper–gold alloy shows a smooth variation from that of pure copper to that of pure gold. The annealed specimens, however, show pronounced minima in resistivity for the two superstructures (Section 5.4.3).

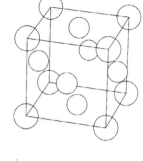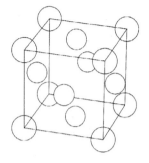

Fig. 5.33 Stereoview of the unit cell and environs of the structure of the ordered pseudo-face-centred cubic structure, Cu_3Au; circles in order of decreasing size represent Au and Cu.

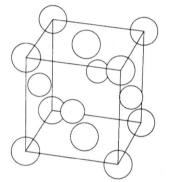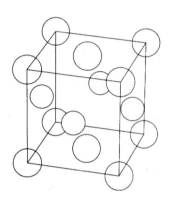

Fig. 5.34 Stereoview of the unit cell and environs of the structure of the ordered tetragonal structure, CuAu; circles in order of decreasing size represent Au and Cu.

If the metallic radii of two components of an alloy are very similar, the strain introduced by replacement will be correspondingly small. Possible superstructures under annealing conditions will then be no more stable than those of the random solid solutions. Thus, silver and gold, of equal atomic radii, owing to the lanthanide contraction,[13] form a continuous series of solid solutions but no superstructures. If there is little or no strain for a given composition such as Ag Au, ($r_{Au} = 0.1442$ nm, $r_{Ag} = 0.1445$) then the free energy change for the reaction:

$$\text{AgAu (superstructure)} \rightarrow \text{AgAu (solid solution)}$$

will be governed by the change in entropy, so that a random solid solution would be favoured.

5.12.2 Silver–cadmium alloys

X-ray diffraction has proved a valuable tool in the study of alloys and their phase diagrams. Powder photographs of the silver–cadmium system are shown in Fig. 5.35.

From 100% silver, cadmium replaces silver in a face-centred cubic solid solution in the α phase of silver to just over 40% Cd, after which cadmium replaces silver at random sites in the crystal (Fig. 5.36).

[13] Poor shielding by the *4f* electrons causes a drop of 0.03 nm in radius with increase in atomic number from Ce to Lu.

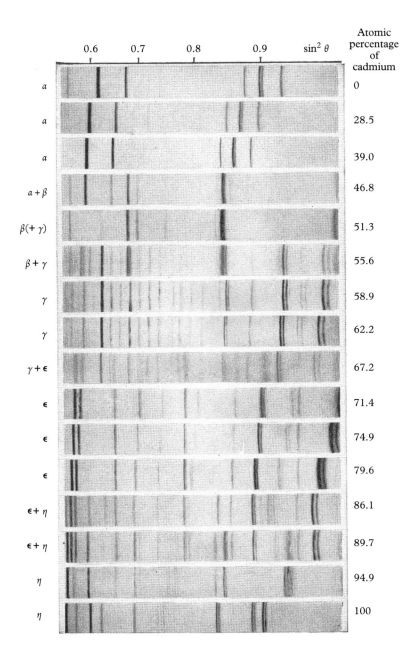

Fig. 5.35 X-ray powder photographs of the silver-cadmium system. The variation in composition of the α–β solid solution finally giving way to the complex γ structure can be traced on the x-ray photographs; similarly for the ε and η phases. [Westgren A and Phragmén G. *Metallwirtschaft.* 1928; 7: 700.] Note how the simplicity of the elemental structures is revealed by the small number of diffraction lines on the photographs of the pure α and η phases.

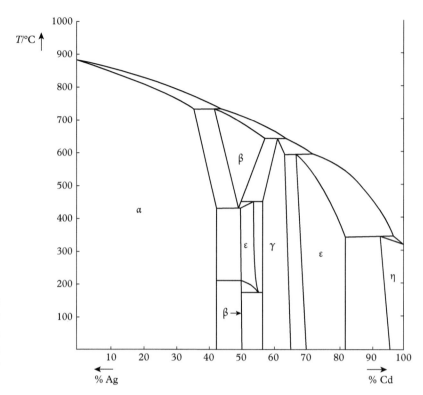

Fig. 5.36 The silver–cadmium phase system, as deduced from the x-ray diffraction study (Fig. 5.33) [Westgren A. *Angew. Chem.* 1932; 45; 33; reproduced by courtesy of John Wiley and Sons.]

The shift of equivalent x-ray spectral lines in the diagram is a manifestation of Vegard's law in this system; the movement of a given reflection to lower values of $\sin\theta$, equivalent to higher d values, with increase in the percentage of cadmium arises because of the larger radius of this species compared with that of silver (Table 5.5). By 46.8% Cd, lines from the β-phase appear, those from α disappearing completely by 51.3%; other phase changes can be traced in a similar manner.

The β-phase it is not a true body-centred cubic unit cell: it may be called pseudo body-centred, but strictly it is a primitive cubic unit cell containing two atoms, with silver atoms at the corners of the unit cell and cadmium at the centre, or *vice versa*. The γ-phase that appears next is a complicated cubic structure containing fifty-two atoms in the unit cell. This structure occurs in many alloys, and has a characteristic hardness and brittleness.

Continued replacement of silver by cadmium leads to the close-packed hexagonal ε-phase, showing with moderate intensity together with the γ-phase at 67.2% Cd and developing fully with further increase in the cadmium concentration. At higher cadmium concentrations it enters into solid solution with the η-phase that represents pure cadmium which also a close-packed hexagonal

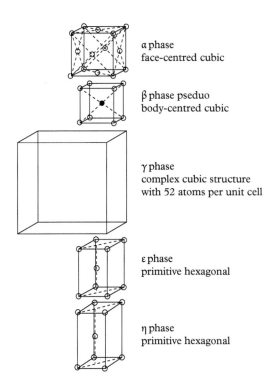

α phase
face-centred cubic

β phase pseduo
body-centred cubic

γ phase
complex cubic structure
with 52 atoms per unit cell

ε phase
primitive hexagonal

η phase
primitive hexagonal

Fig. 5.37 The crystal structures of the phases $\alpha, \beta, \gamma, \varepsilon$ and η in the silver–cadmium alloy system. [Reproduced from Bragg WL. *The Crystalline State*, Vol. I. G Bell and Sons Ltd, 1939.]

structure; the unit cells for the five phases in this system are shown in Fig. 5.37. The sum total of x-ray information permitted the complete phase diagram to be plotted, which demonstrates the power of the x-ray diffraction in the study of alloy systems.

5.13   Hume-Rothery rules for alloys

From an extensive study of alloys, Hume-Rothery drew up a set of rules governing binary alloy systems [36, 37]. Subsequently, the *Hume-Rothery rules* were explained in terms of the theory of metallic bonding; they are now accepted features of binary alloys and display the following features:

- *The diameters of the solvent and solute atoms must differ by no more than ca. 15% for complete solid solution to occur.*

The free energy of formation ΔG_{AB} of an alloy of metals A and B in the proportions c and $(1-c)$, respectively, may be written as

$$\Delta G_{AB} = G_{AB} - [cG_A + (1-c)G_B] \tag{5.85}$$

and is approximately one tenth of G_{AB}. If the enthalpic component of ΔG_{AB} tends to zero, the entropy of mixing ensures a random solid solution, as with the silver–gold system. A difference in atomic size opposes the entropic effect, and a difference in diameter of more than *ca.* 15% restricts the range of solid solution formation.

- *The solvent and solute should have similar electronegativities.*

A significant difference in electronegativity leads to a negative value for ΔG_{AB}, and the components A and B tend to form an intermetallic compound rather than a solid solution. These rules imply also that solid solution is favoured by species similar in both valency and crystal structure.

- *The electron–atom ratio correlates well with a range of binary alloys of given compositions.*

This rule can be examined in the light of the copper–zinc (brass) alloy system, the phase diagram for which is shown by Fig. 5.38

The phases of most interest in the Cu–Zn system are α (pure copper, Cu), β (Cu$_3$Zn$_2$), γ (CuZn), ε (CuZn$_3$) and η (pure zinc, Zn). Hume-Rothery pointed

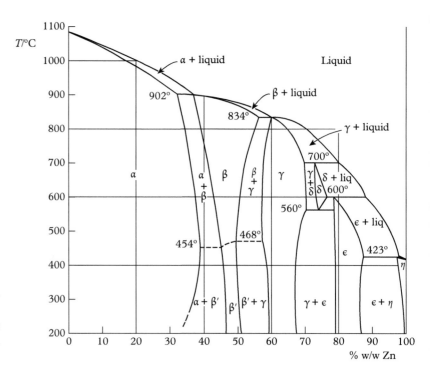

Fig. 5.38 The $\alpha, \beta, \gamma, \varepsilon$ and η phases in the copper–zinc (brass) system. [Reproduced from Bragg WL. *The Crystalline State*, Vol. I. G Bell and Sons Ltd, 1939.]

Table 5.7 *Electron:atom (e/a) boundary ratios for binary alloy phases; the bold values in parentheses are the Hume-Rothery values*

	FCC	BCC (minimum) (3/2 = 1.5)	γ-Phase (21/13 = 1.62)	CPH (7/4 = 1.75)
Cu–Zn	1.38	1.48	1.58–1.66	1.78–1.87
Cu–Sn	1.27	1.49	1.60–1.63	1.73–1.75
Ag–Zn	1.38	1.49	1.58–1.63	1.67–1.90
Ag–Cd	1.42	1.50	1.59–1.63	1.65–1.82

out that the alloy phases have boundaries that are determined by an electron/atom ratio, e/a. Thus, pure copper (face-centred cubic) exists until $e/a = 3/2$, which corresponds to the composition of the β-phase ('body-centred' cubic). This phase gives way to the complicated γ-phase, with 52 atoms per unit cell and $e/a = 21/13$, which is followed by the ε-phase (close-packed hexagonal), $e/a = 7/4$, finally arriving at pure zinc (close-packed hexagonal). Table 5.7 lists ideal and experimental e/a ratios for selected binary alloys.

The ratios were found to persist over a wide range of alloys, and were given an explanation in terms of band theory [12]. The observed limit of the face-centred cubic α-phase is very close to the electron concentration of 1.36 for which the inscribed Fermi sphere just touches the surface of the Brillouin zone of the face-centred cubic lattice (Fig. 5.39), the Wigner–Seitz cell of a *body-centred* unit

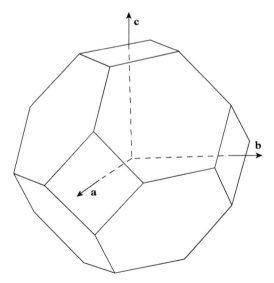

Fig. 5.39 Truncated octahedron: the Wigner–Seitz primitive cell of a body-centred unit cell of a cubic Bravais lattice referred to the conventional crystallographic axes (see also Fig. 5.12).

cell of a cubic lattice. The points of contact are the centres of the hexagonal faces of the first Brillouin zone (Fig. 5.40) [36]. For the onset of the β-phase CuZn, the observed e/a ratio of 1.48, at which value the Fermi sphere just touches the Brillouin zone for the body-centred cubic lattice (the Wigner–Seitz cell of a *face-centred* unit cell of a cubic lattice), and is very close to the Hume-Rothery empirical value of 3/2. The contact ratios for the γ-phase and the ε-phase are very close to the Hume-Rothery values of 21/13 and 7/4, respectively [37]).

The situation is somewhat akin to that discussed in Section 4.6.2. In this case, however, once filled states make contact with a Brillouin zone boundary, it can become energetically unfavourable to add electrons above the energy gap as the number $N(E)$ of energy states would fall off rapidly as a function of energy [4]. However, a structural change to one with a larger Fermi surface enables further states to be populated, until that structure comes to a new limit, and so on (Fig. 5.41). Thus, the empirically deduced Hume-Rothery rules are given an expression in terms of band theory.

5.13.1 Intermetallic compounds

Intermetallic compounds are formed from two or more types of metal atoms, and exist as homogeneous substances that differ in a structurally discontinuous manner from their constituent metals. They are termed, perhaps more correctly, intermetallic phases since their physical and structural properties differ from those of their components. Thus, they form distinct crystalline compounds of stoichiometric compositions that are separated by phase boundaries from their components or the solid solutions thereof. One component is frequently more strongly metallic than the other, such as in the pairs Cu–Zn, Ag–Cd and Mg–Sn. The melting points of intermetallics are normally higher than those of their

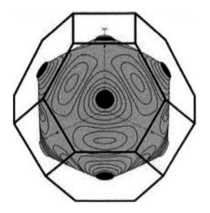

Fig. 5.40 Fermi surface of copper according to Pippard [36]; the contours mark constant distances from the origin.

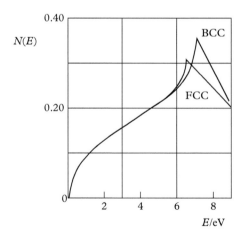

Fig. 5.41 Number $N(E)$ of states of energy states E for face-centred cubic (FCC) and body-centred cubic (BCC) lattice structures. The transition FCC (α) to BCC (β) allows an increase in the number of energy states for the β-phase until a further transition occurs.

components; thus the compound Mg_2Sn melts at 1053 K, whereas magnesium and tin melt at 923 K and 503 K, respectively.

Experimental results from x-ray diffraction studies in the copper–zinc system, for example (Fig. 5.38), have shown the existence of intermetallic compounds of definite structure and composition; they differ from the alloys across phase boundaries in terms of order–disorder relationships, magnetic properties, bond type and deformation behaviour. The phases of composition $CuZn$, $CuZn_3$ and Cu_5Zn_{13} are good examples in this alloy system.

The subject of metals and intermetallics is well documented in the literature, to which the reader may wish to turn for more detailed studies [38–41].

....

REFERENCES 5

[1] Drude P. *Ann. Phys.* 1900; 306: 566.

[2] *idem. ibid.* 308: 369.

[3] Lorentz HA. *Proc. Acad. Amsterdam* 1905; 7: 684.

[4] Kittel C. *Introduction to Solid State Physics*, 8th ed. John Wiley and Sons, 2004.

[5] Wiedemann G and Franz R. *Ann. Phys.* 1853; 165: 497.

[6] Gupta MC. *Statistical Thermodynamics*, 2nd ed. New Age International, 2003.

[7] Ladd M and Palmer R. *Structure Determination by X-ray Crystallography*, 5th ed. Springer Science+Business Media, 2013.

[8] Einstein A. *Ann. Phys.* 1907; 22: 180.

[9] Debye P. *Ann. Phys.* 1912; 39: 789.

[10] Sommerfeld A. *Z. Phys.* 1928; 47: 1.

[11] Ashcroft NW and Mermin ND. *Solid State Physics*. Philadelphia, PA: Saunders College, 1976.

[12] Singleton J. *Band Theory and Electronic Properties of Solids*. Oxford University Press, 2001.

[13] Bloch F. *Z. Phys.* 1928; 52: 555.

[14] Blakemore JS. *Solid State Physics*, 2nd ed. Cambridge University Press, 1985.

[15] Shockley W. *Electrons and Holes in Semiconductors*. van Nostrand, 1950.

[16] Meissner W and Ochsenfeld R. *Naturwiss.* 1933; 21: 787.

[17] Bardeen J, Cooper LN and Schrieffer JR. *Phys. Rev.* 1957; 106: 162.

[18] *idem. ibid.* 108: 1175.

[19] Chang KJ and Cohen ML. *Phys. Rev.* 1984; B30: 5376.

[20] Gou H *et al. Phys. Rev. Lett.* 2013; 111: 157002.

[21] Grove WR. *Phil. Mag. J. Sci.* 1839; 14: 127.

[22] *idem. ibid.* 1842; 21: 417

[23] Song M-K, Zhang Y and Cairns EJ. *Nano Lett.* 2013; 13: 5891.

[24] Guo B *et al. Energy Environ. Sci.* 2014; DOI: 10.1039/C4EE00508B.

[25] Li Q. *J. Mater. Chem. A* 2015; 3: 7241.

[26] Lin M-C *et al. Nature* 2015; 520: 325.

[27] Grätsel M. *Nat. Mater.* 2014; 13: 837.

[28] Unger EL *et al. Energy Environ. Sci.* 2014; 7: 3690.

[29] Tress W. *ibid.* 2015; 8: 995.

[30] Liu J *et al. Science* 2015; 347: 6225.

[31] Kurakevych OO *et al. Cryst. Growth Des.* 2012; 13: 303.

[32] Kim DY *et al. Nat. Mater.* 2015; 14: 169.

[33] Tarantino SC *et al. Phys. Chem. Min.* 2010; 37: 769.

[34] Vegard L. *Z. Phys.* 1921; 5: 17.

[35] *idem. Z. Krist.* 1928; 67: 239.

[36] Pippard AB. *Phil. Trans. Roy. Soc.* 1957; 250: 325.

[37] Hume-Rothery W and Powell HM. *Z. Krist.* 1935; 91: 23.

[38] Hume-Rothery W. *Atomic Theory for Students of Metallurgy.* Institute of Metals, 1969.

[39] Cotterell A. *Introduction to the Modern Theory of Metals.* Institute of Metals, 1991.

[40] Blakemore JS. *Semiconductor Statistics.* Dover, 2002.

[41] Yu PY and Cardona M. *Fundamentals of Semiconductors.* Springer, 1999.

Problems 5

5.1. The electrical resistivity ρ of elemental lithium at 273 K is 8.55×10^{-8} Ω m, its body-centred unit cell dimension is 0.3510 nm and the rms electron speed is 1.30×10^6 m s^{-1}. Calculate (a) the mobility and (b) the mean free path of the valence electrons in lithium under the given conditions.

5.2. The anharmonic vibrations of atoms about their mean positions in a solid may be represented approximately by the potential energy function $V(x) = ax^2 - bx^3$, where a and b are constants. By determining an expression for the mean displacement $<x>$ of atoms, show that $<x>$ is directly proportional to the absolute temperature, that is, it is consistent with the solid expanding on heating. *Note.* Since x is small, $\exp(bx^3) \approx (1 + bx^3)$.

5.3. Calculate from the wave mechanical free-electron theory (a) the Fermi energy in eV, and (b) the Fermi temperature in °C for lithium; the unit cell dimension of the body-centred cubic unit cell of this metal is 0.3510 nm.

5.4. Determine $g(E_F)$ for lithium.

5.5. Use the density of states function to calculate the average thermal energy of electrons between 0 and E_F.

5.6. (a) Draw a block of several primitive unit cells of a two-dimensional, square lattice of translation vector **a** and outline the boundaries of the first three Brillouin zones. (b) What is the first Brillouin zone for a two-dimensional lattice referred to a primitive unit cell of side $|\mathbf{a}|$?

5.7. Calculate the packing fraction for a close-packed hexagonal structure of identical spheres of radius r.

5.8. The close-packed cubic structure of spherical atoms of equal radius contains tetrahedral holes and octahedral holes. (a) How many holes of each type are unique to one face-centred unit cell and what are the fractional coordinates of their centres? (b) What is the radius of the sphere that just fits a tetrahedral hole?

5.9. (a) What is the volume V_I of the largest sphere that can be fitted into the Wigner–Seitz cell of a cubic lattice referred to a body-centred unit cell of side a (BCC)? (b) What is the ratio $V_I : V_F$, where V_F is the volume of the largest sphere that can be fitted into the Wigner–Seitz cell of a cubic lattice referred to a face-centred unit cell (FCC) of the same size?

5.10. A crystal has the band structure shown at (A): dark regions represent occupied bands, so that the lower band is completely filled by electrons. (a) What are the expected electrical properties of this crystalline solid? (b) What electrical property would result if all electrons from the upper band were removed? What chemical agent would effect such a change? (c) What would be the electrical property if electrons were added so as to fill completely the upper band? What chemical agent would be used to bring about this change?

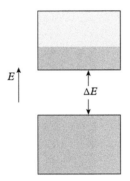

Semiconductor A

5.11. A solid has the band structure shown at (B): the lower band is half-filled with electrons and the upper band is empty. (a) What are the expected electrical properties of this crystalline solid? (b) What electrical property would result if all electrons from the lower band were removed? (c) What would be the electrical property if electrons were added so as to fill completely the lower band? What chemical agent would be used to bring about this change?

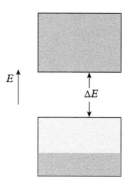

E

ΔE

Semiconductor B

5.12. Surface effects are important with very small crystals. (a) Construct arrays of body-centred cubic unit cells of elemental sodium in order to find that array for which there are just more interior atoms than surface atoms. (b) Given that the cubic unit cell dimension of sodium is 0.4291 nm, what is the size of the final crystal? (One isolated body-centred unit cell provides a total of nine atoms, as shown by Fig. 5.27.)

5.13. What is the maximum kinetic energy and average speed of emitted electrons when photons of wavelength 250 nm are incident upon the surface of a metal having a work function of 4.25 eV?

5.14. An intrinsic semiconductor has a band gap ΔE of 1.55 eV. Determine the wavelength of the photon that could promote an electron from the top of the valence band into the conduction band.

5.15. A pure crystalline material appears red in transmitted light. (a) What is the spectral nature of this material? (b) What is an approximate value in eV for the finite band gap ΔE for this material?

5.16. A given crystalline solid transmits light of frequency of *ca.* 1.25×10^{14} Hz. Sketch a likely band structure and indicate a value for the band gap ΔE.

5.17. An excited electron in a sodium atom emits radiation of wavelength 589.5 nm on returning to its ground state. The average time for the transition is 20 ns. Calculate (a) the uncertainty in the energy, and (b) the spread width of the emission line, and (c) the photon length.

5.18. (a) The intensity of sunlight falling on the earth's surface on a clear day is 1360 W m^{-2}. What is the energy per unit volume in the sunlight? (b) An efficient 60 watt tungsten lamp converts 15% of its power into light of average wavelength 600 nm (the remainder is lost as heat). If 3/4 of the light falls on to a table surface, how many photons reach that surface in 1 second?

5.19. The band gaps ΔE for cadmium sulphide CdS and zinc sulphide α-ZnS have been given as 2.39 eV and 3.61 eV, respectively. What are the colours of these compounds?

5.20. (a) Exposure of rutile TiO_2 to ultraviolet radiation of 340 nm promotes an electron from the valence band to the conduction band. What is the energy gap between these bands? (b) Are the following extrinsic semiconductors n-type or p-type: germanium doped with (i) arsenic and (ii) gallium? (c) A p-type and an n-type extrinsic semiconductor are placed in contact with a source of emf completing the circuit. At which semiconductor must the positive terminal of the emf source be connected in order for a current to flow in the circuit?

5.21. (a) One mole of a lead–zinc alloy was found to contain 20.76 g Pb. What is the weight percent of zinc in the alloy? (b) Analysis of an annealed alloy of silver and cadmium reported 62.51% by weight of cadmium. What is the probable formula for the alloy?

5.22. In phosphorus-doped silicon at 20 °C (ΔE = 1.11 eV), the Fermi level is shifted upward by 0.1 eV through the application of heat to the n-type semiconductor. What is the probability of electron promotion from the valence band to the conduction band?

5.23. Elemental copper has a face-centred cubic crystal structure, with a = 0.3615 nm. If the electrical resistivity is 1.68×10^{-8} Ω m, calculate the mean time between electron collisions.

5.24. A precise unit cell dimension for elemental sodium is 0.42906 nm. Calculate E_F, v_F and k_F.

5.25. The following data were recorded with a pure germanium thermistor:

$T/°C$	0	25	50	75	100
$R/k\Omega$	27	11	5.2	1.7	1.0

Determine the band gap for germanium.

5.26. Formulate Eq. (5.23) into a distribution in terms of molecular speeds v.

6

Nanoscience and Nanotechnology

The principles of physics, as far as I can see, do not speak against the possibility of manoeuvring things atom by atom.

Richard Feynman, 1959.

6.1   Introduction

The idea of a nanotechnology (Gk. *nános* = dwarf) emanated from Feynman during a discussion in 1959 on the possibility of carrying out reactions with individual atoms. Drexler (1986) conceived the building devices of varying complexity atom by atom, a nanoscale level. Thus, nanotechnology has now emerged as a viable field of activity, as illustrated by the scanning tunnelling microscope, discussed in Section 2.10.1.1, which enabled the manipulation of individual atoms, and graphene tubes for application in nanoscale electronics, to name but two.

The terms nanoscience and nanotechnology are closely connected, but a useful distinction defines *nanoscience* as the manipulation of materials of atomic sizes, their properties being very different from those of the same material in bulk, and *nanotechnology* as the design and precise production of devices from nanoscale materials.

One nanometre is 10^{-9} m: the molecule of C_{60} has a diameter of *ca.* 0.7 nm (Fig. 6.1), and a human hair has a width of approximately 80,000 nm. Atoms themselves have sizes *ca.* 0.1 to 0.2 nm, and the range of greatest interest lies between particle sizes 0.2 nm and 100 nm. At these values, materials have properties very different from those in the bulk material, because quantum and surface effects become paramount.

Nanoscience involves the understanding of the influence of size and surface effects on the preparation, properties and reactions of these materials, whereas nanotechnology is concerned with the application of these effects in novel-sized electrical and mechanical devices. From this large and rapidly expanding field of scientific endeavour, it is possible here to treat only a small section of topics on science and technology of these materials.

Bonding, Structure and Solid-State Chemistry. First Edition. Mark Ladd.
© Mark Ladd 2016. Published in 2016 by Oxford University Press.

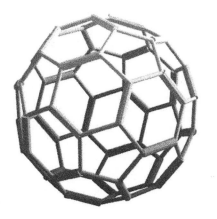

Fig. 6.1 Stereoscopic illustration of the C_{60} molecule. [Joris Mooij (2003) *The Vibration Spectrum of Buckminsterfullerene*, Master's Thesis, Radboud University of Nijmegen; reproduced by courtesy of Dr Gert Heckman.]

6.2 Physics of small systems

Quantum mechanics is applicable to systems generally, but where large numbers of species or large objects are involved, then classical mechanics provides frequently a satisfactory and simpler model. Nanoscience involves both classical and quantum mechanics in its study and application.

As the name suggests, nanoscience indicates the importance of a size factor, the size of the individual crystals, if they can truly be so called. Fluctuations in energy are of particular significance since the total energy of a system depends on the size of the system. For a given size of fluctuation, the smaller the system the more important does a fluctuation become. In a system containing an Avogadro number of particles, small fluctuations are negligible, but when the energy of a fluctuation arising, say, from thermal agitation, becomes commensurate with a bond energy, a change or breakdown will be initiated. Such processes exist in a vast number of physical and biochemical systems.

In the discussion of dipolar forces in Chapter 3 it was noted that an increased thermal agitation in a polar fluid led to dipolar alignment, with consequent effects on the magnetic properties of the material. Thermal fluctuations may be considered as the driving force from one state of a system to another, and the balance of such forces leads to dynamic equilibrium conditions.

The ATP-dependent motor protein myosin II that is involved in muscle action operates in conjunction with another protein, actin. Power is released by ATP hydrolysis; at pH 6 the overall reaction may be written as:

$$[ATP]^{3-} + H_2O \overset{p\text{H}=6}{\rightleftharpoons} [ADP]^{2-} + [H_2PO_4]^-$$

The power stroke occurs at the release of phosphate from the myosin molecule after hydrolysis of ATP, while myosin is tightly bound to the globular protein actin that participates in muscle contraction. The effect of the power release is a conformational change in the myosin molecule that pulls against the actin.

The subsequent release of the ADP molecule and binding of a new ATP molecule then releases myosin from actin. The ATP hydrolysis within the myosin species will cause it to bind to actin again to repeat the cycle. The combined effect of numerous such power strokes causes muscle contraction [1]. ATP is nature's smallest rotary motor.

Statistical mechanics has shown that while the total energy of a system increases with a rise in the number N of particles that it contains, fluctuations (deviations) from the average energy increases as $\sqrt{N}$, and the corresponding relative expression is $\sqrt{N}/N$. For a system of 100 particles $\sqrt{N}/N$ is 0.10 and for 10 particles it is much more significant at 0.32; but for 10^{23} particles $\sqrt{N}/N$ is *ca.* 10^{-12} which is insignificant in terms of thermal fluctuations. Most fundamental nanoscale processes involve 100 or less particles and, since an average atomic size is 1 nm, particle sizes of 100 nm or less are implied in this context.

6.2.1   Quantum dots

A metal or a semiconductor can carry a current because delocalized electrons are present in the material. For this reason, useful results are obtainable from the electron-in-a-box model (see also Sections 2.10 and 5.5). A metal wire corresponds to a box of large dimension, and Eq. (2.47) shows that for this situation the separation of energy levels is so small that they form a continuum within an energy band. In a crystal, bonding electrons are in a lower *valence band*, which could be an s band or a p band. Two electrons with opposed spins occupy each energy level, and Eq. (2.47) shows that there are several ways of forming a given energy by the sum $(n_1^2 + n_2^2 + n_3^2)$; the system is highly degenerate (Fig. 6.2a). If energy is supplied to the material, an electron can move from the valence band to the conduction band leaving a positive 'hole' in the valence band. The material then becomes electrically conducting if a potential difference is applied to it. An expelled electron and its positive electron hole together constitute an *exciton*.

As the three-dimensional box of semiconductor material is made smaller in size the electron levels decease in number and, the energy gap between the valence

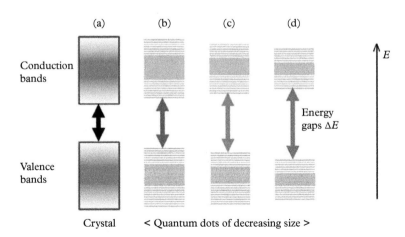

Fig. 6.2 Energy levels of the valence and conduction bands: (a) for a crystal and (b–d) for quantum dots of progressively decreasing size. As the quantum dots become smaller in size, the energy states decrease in number and the band gap ΔE widens. Typical sizes for (b)–(d) could be 30, 15 and 5 nm, corresponding to crystals of, say, 50, 25 and 10 atoms.

band and the conduction band increases (Fig. 6.2b–d). The crystals have now entered the nanoscale size range; in molecular orbital terms, the energy gap is the difference between the HOMO and LUMO levels, and such nanometre-sized crystals of semiconductors are known as *quantum dots.*

Quantum dots were reported first in 1981 from experiments on small crystals of copper(I) chloride. Discussion centred on particles in the 20–100 nm size range, with *effective masses* $m_e^* = 0.44m_e$ for electrons and $m_h^* = 3.6m_e$ for holes [2]. A more recent illustration (Fig. 6.3) shows cadmium selenide dots of approximately 6 nm size recorded by transmission electron micrography. On account of the size of quantum dots, electrons within them are contained in a small space.

As the difference in energy, contingent upon crystal size, between the valence band and conduction band increases, it follows that more energy is needed progressively for excitation. In turn more energy is released when the electron returns to the valence band, and the transition is accompanied by a spectral emission of colour dependent upon energy change involved.

Quantum dots can arise by virtue of the photoelectric effect discussed in Section 2.5.1; the energy of an electron at an nth level is given through Eq. (2.37). By experiment, it has been shown that crystals of semiconducting cadmium selenide ranging from 0.5 nm to 4.0 nm in size exhibit photoluminescent emission from the blue region of the visible spectrum to the red. Over this range the photon energy decreases from 3.0 eV to 1.9 eV, approaching the band energy gap of 1.73 eV. Initially, a quantum dot electron is tightly bound in the valence band. Once the electron has received the gap energy E_g, which raises it to a free state, the kinetic energy given by Eq. (2.37) is added to give the total, or exciton, energy E_{tot}:

$$E_{tot} = E_g + \frac{n^2 h^2}{8 m_e a^2} \tag{6.1}$$

It follows from Eq. (6.1) that there is a minimum energy needed to raise an electron from the valence band to the conduction band. This energy itself is dependent upon the dot size, which is represented by the term a in Eq. (6.1). The spectral range of cadmium selenide/zinc sulphide quantum dots of composition $CdSe_x ZnS_{1-x}$ is illustrated by Fig. 6.4; the colour is 'tuneable' by varying the composition parameter x.

Fig. 6.3 Transmission electron micrograph of cadmium selenide quantum dots of *ca.* 7.5 nm diameter. [Public domain.]

Example 6.1

What colours are associated with the first two photoluminescent emissions for a CdSe dot of size 3 nm? Since E_g has been given as 1.73 eV and the emission wavelength λ by hc/E_{tot}, it follows from Eq. (6.1) that for the lowest energy wavelength ($n = 1$) $\lambda = 699.8$ nm, which corresponds to the red region of the visible spectrum. For $n = 2$, $\lambda = 653.5$ nm which is in the orange region.

Fig. 6.4 Photoluminescence of CdS_x Se_{1-x}/ZnS quantum dots of 6 nm diameter in size. The material emits light of differing colour by tuning the composition variable x. [Reproduced by courtesy of the Sigma-Aldrich Corporation.]

What colour is associated with the third level spectral emission from this system and what is its wavelength?   [Yellow; 588.7 nm.]

Example 6.2

Assume that a given substance behaves like a one-dimensional box with respect to energy. At what size of box would it exhibit the energy of bulk material at 25°C? For the material to exhibit bulk behaviour, the energy must be less than kT. From the particle-in-a-box discussion it follows then that $\dfrac{h^2}{8m_e a^2} < kT$. At the limit $E = kT$, and $a = \sqrt{\dfrac{h^2}{8 \times m_e \times k \times T}} = \dfrac{h}{\sqrt{8 \times m_e \times k \times T}}$, which evaluates to 3.826 nm. Thus, the required 'box' size must be greater than *ca.* 3.83 nm.

6.2.1.1 *Dimensionality and density of states function*

A crystal (bulk material) has no restriction of size in any particular direction. A sheet, which corresponds to a well, is small in one dimension and a wire in two, whereas a dot is small in all three mutually perpendicular dimensions.

In three dimensions the density of states function Eq. (5.49), together with Eqs. (5.39) and (6.1), and introducing m^*, is given, per unit volume, as

$$\frac{1}{V}\frac{dN_{3D}}{dE} = g_3(E) = \frac{8\pi\sqrt{2}}{h^3} m^{*3/2}\sqrt{E - E_{min}} \quad (E \geq E_{min}) \tag{6.2}$$

Thus, the density of states for a crystal is proportional to $\sqrt{E}$, and has the form shown by the rising curve of Fig. 5.7. In two dimensions the number of **k**-states per unit area A is:

$$\frac{1}{A}\frac{dN_{2D}}{dk} = \frac{k\pi}{\pi^2} = \frac{k}{\pi} \tag{6.3}$$

and from Eq. (5.39):

$$dE/dk = \hbar^2 k/m^* \tag{6.4}$$

so that

$$\frac{1}{A}\frac{dN_{2D}}{dk} = \frac{1}{A}\frac{dN_{2D}/dk}{dE/dk} = g_2(E) = \frac{k/\pi}{\hbar^2 k/m^*} = \frac{4\pi m^*}{h^2} \tag{6.5}$$

The density of states is now independent of E, and increases in steps of constant $g_2(E)$ each over a range δE corresponding to a particular atomic orbital.

In the case of one dimension per unit length L, a similar analysis shows that now

$$\frac{1}{L}\frac{dN_{1D}}{dE} = \frac{1}{L}\frac{dN_{1D}/dk}{dE/dk} = g_1(E) = \frac{1/\pi}{\hbar^2 k/m^*} = \frac{m^*}{\pi \hbar^2 k} = \frac{\sqrt{2\pi m^*}}{h\sqrt{E-E_{min}}} \quad (E > E_{min}) \tag{6.6}$$

where L is the dimension of the wire and $g_1(E)$ has in the form shown in Fig. 6.5, proportional to $E^{-\frac{1}{2}}$ and with associated with van Hove singularities.

A van Hove singularity is a non-smooth region in the density of states function. The wave vectors corresponding to these singularities are known as critical points of the Brillouin zone. They present as irregularities in the density of states function, and in three-dimensional crystals they take the form of kinks in the $g_3(E)$ curve (Fig. 5.41). These singularities arose in van Hove's 1953 analysis of optical absorption spectra.

The $g_0(E)$ function in the zero-dimensional case returns the result for atoms; it has the form of a delta function[1] at the eigenvalues for the system, and shows a proportionality to $\sqrt{E}$. There is an increasing concentration and sharpness of the energy as the dimensionality decreases, as illustrated below:

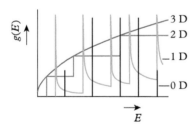

Density of states function in 0, 1, 2 and 3 dimensions
For $E = 0$ or $E < E_{min}, E(k) = 0$

6.2.1.2 Applications of quantum dots

The small size of quantum dots allows them to travel faster than larger particles, thus enabling electronic processes to operate quickly in devices such as optical

[1] $\int_{-\infty}^{\infty} \delta(x)\,dx = 1,\; x = 0.$
$\delta(x) = 0,\; x \neq 0.$

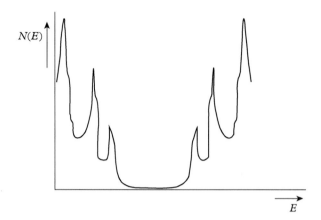

Fig. 6.5 Density of states $N(E)$ as a function of energy E for a quantum wire. The peaks represent van Hove singularities (Section 5.6.3), and correspond to specific exciton energy levels. The region of E for $N(E) = 0$ represents the energy gap between the two bands.

switches, logic gates and quantum computers. The tuneable electric properties of quantum dots afford them wide application in manufacturing processes, as well as an intrinsic scientific and biomedical interest. Their range of bright, pure colours together with high efficiencies and long lifetimes makes them eminently suitable for light-emitting diodes, solid-state lighting displays and photovoltaic devices.

A light-emitting diode (LED) consists of a two-lead semiconductor light source, usually indicated in electronic circuitry by the symbol

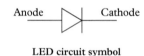

LED circuit symbol

In essence it is a *p–n* junction that emits light when activated by an applied voltage. A potential difference applied such that conventional current flows from anode to cathode causes charge carriers, the electrons, to move in the opposite direction towards the *p*-type slice of the semiconductor composite. There, they combine with the holes (Fig. 5.22b) and light photons are released. It is in effect the reverse of the photoelectric effect described in Section 2.5.1 and may be described as *electroluminescence*. The colour of the emitted light is determined by the band gap of the semiconductor, as indicated in Example 6.1. It follows that in order to excite luminescence in an LED, energy greater than that of the band gap must be supplied in terms of an applied voltage.

The diode element is frequently a doped gallium aluminium arsenide *p–n* junction, with the *n*-plate of the junction connected to a source of negative potential. The generation of light takes place without the heating effect common to incandescent lamps, and the LED is very much less costly to operate than are conventional forms of lighting. A typical LED unit is illustrated below:

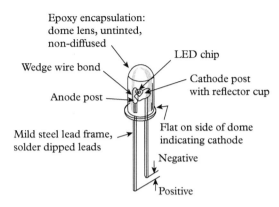

Typical LED element [Courtesy of FiberOptic.com]

Quantum dots have found important application in medical diagnosis. Because of their small size they can act as biological markers in the human body, and with their brightness and spectral range of colours they are an important alternative to fluorescence based biosensors that depend upon organic dyes of broad spectral width with limited effectiveness. Nanocrystals for this work can be used in either aqueous or organic media according to the application.

Imaging by x-rays, computed tomography, ultrasound, radionuclide and magnetic resonance are used widely in screening for carcinoma. These techniques are limited to some extent by insufficient sensitivity for the detection of small numbers of malignant cells in primary or metastatic sites, and by an inability to detect specific carcinoma cell-surface markers which could assist in the diagnosis and staging of carcinoma. These limitations demand new imaging probes that are highly sensitive and biospecific. Quantum dot probes provide improvement in imaging techniques for *in vivo* carcinoma screening, particularly in visceral organs.

The toxicity associated with cadmium compounds has led to the development of other quantum dot diagnostic materials. Some success has attended the use of silicon quantum dots, and they are currently undergoing detailed preparatory studies. This field of quantum dots and their applications is one of intensive present activity.

6.2.2 Mechanical properties of nanocrystals

Conventional polycrystalline materials have particle sizes in the range 25 nm to 250 μm, whereas nanocrystalline substances range from 2 nm to 250 nm in size. Consequently, nanocrystals have a large volume fraction of particle boundaries, which can alter significantly their physical and chemical properties in comparison with the larger particulate materials. Size plays a vital role in the mechanical behaviour of all materials, particularly so in the nanosize range. In macrocrystals, deformation is determined by processes such as migration of defects and dislocations or by gliding, particularly across close-packed planes. Not surprisingly, the migration of dislocations becomes negligible in nanocrystals, and deformation occurs mainly along particle boundaries. Ductility in crystals increases with a decrease in particle size, but in nanocrystals ductility is limited by the porosity of a specimen, and by instability under tensile stress or shear action. Simulation studies have shown that for small nanocrystals a shear mechanism is the most probable mode of deformation (Fig. 6.6).

On account of the large surface to volume ratio in nanocrystals, the surface atoms will be less strongly bound than those in its bulk. The ratio of the surface cohesive energy E_s to that in the body of the nanocrystal E_b is given by:

$$E_s/E_b = \frac{S-s}{S} \tag{6.7}$$

where s is the atomic size and S the size of the nanocrystal. Equation (6.7) shows that for $S \gg s$, the cohesive energy of a nanoparticle approaches that of its bulk material. Figure 6.7 illustrates the ratio T_s/T_b for the melting point for gold as a function of particle size; only at values of S less than about 40 nm does T_s, begins to depart significantly from T_b.

A discussion of the Lindemann criterion of melting [3] gives the result for a metal as:

$$\sqrt{<u^2>} = \delta_L s \tag{6.8}$$

where $\sqrt{<u^2>}$ is the rms amplitude of thermal vibrations of the atoms, s is now the nearest neighbour distance and δ_L is a parameter, often denoted the Lindemann constant, for the substance under investigation. For the alkali halides, since there are two atomic vibrators, Eq. (6.8) becomes:

$$\sqrt{<u^2>} = 2\delta_L s \tag{6.9}$$

Fig. 6.6 Shear deformation lines in a crystal of iron of 250 nm in size.

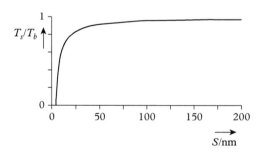

Fig. 6.7 The ratio T_s/T_b of melting-point temperatures of a particle of size S to that of bulk size b as a function of crystal size S for gold; the ratio begins to fall significantly from unity for $S < 40$ nm.

The parameter δ_L is approximately 0.1 and some actual results are listed below:

	$\sqrt{<u^2>}$/nm	$\frac{\sqrt{<u_+^2>}+\sqrt{<u_-^2>}}{2}$/nm	s/nm	$\delta_{L,expt}$	$\delta_{L,calc}$
Al	0.0218	–	0.2910	0.075	0.072/0.0738
Cu	0.0219	–	0.2608	0.084	0.069/0.0745
LiF	–	0.0235	0.2061	0.114	
NaCl	–	0.0328	0.2928	0.112	

6.3   Carbon and its nanomaterials

Carbon is a well-known material, particularly in the forms of graphite and diamond, but also as carbon black, and more recently as fullerenes, graphene, nanowires and nanotubes; carbon intercalation compounds involve graphite in which metal atoms are contained between the layers of the graphite structure.

6.3.1   Carbon black

Carbon black is obtained from the incomplete combustion of petroleum products and other residual oils. It is *ca.* 98% pure elemental carbon, different from soot, and presents as grape-like masses to examination under the electron microscope.

Carbon black powder   Electron micrograph cluster of carbon black

It is a versatile pigment with many applications, the most important of which is the reinforcement of rubber products, such as car tyres. It is used also as a black pigment in inks, paints and photographic toners. The performance of carbon black depends upon fundamental properties such as particle size and surface activity. It has also various applications in electronics: as a filler in plastics it provides electrical conduction through the material, and on account of this property it is a useful antistatic additive.

6.3.2   Graphite

The structure of graphite has been discussed in Section 2.25 and illustrated in Fig. 2.52. The uses of finely divided graphite in 'lead' pencils and as a lubricant are well known. However, its lubricant properties are limited by its corrosive nature in contact with construction materials such as steel and aluminium. In a finely divided form it has catalytic and gas absorbent activity.

Natural graphite [Reproduced by courtesy of Tiny Pencil.com]

The properties of graphite are highly anisotropic in acoustic and thermal conduction, since phonons can travel rapidly along the hexagonal sheets of the structure but less readily from layer to layer. Electrical conduction in the plane of the hexagonal sheet is enabled by the delocalized state of electrons within the layers.

Tin-coated nanosize graphite composites employed as anodes in lithium-ion batteries have been found to increase the battery performance. Although tin reacts with lithium forming a lithium-tin alloy, this reaction is offset by forming the composite in a graphite matrix which is able to compensate for the reaction products.

The electrical conductivity of graphite can be best understood in reference to diamond and benzene. The very low electrical conductivity of diamond follows from its band structure (Fig. 5.19d), where the energy gap between the valence and conduction bands is *ca.* 5.5 eV. In benzene, each trigonal carbon atom forms σ and σ^* molecular orbitals: the unused p_z orbitals form six π molecular orbitals with four discrete energy levels, their separation being much smaller than that of the σ and σ^* separation. In graphite each of the levels is broadened into bands, and the π orbital separation is actually –0.04 eV, that is, they overlap (Fig. 5.19b). Hence, this continuous, half-filled band enables graphite to exhibit electrical conduction within the layers of the structure (Fig. 6.8). The following data on electron mobility compare graphite and diamond:

	Graphite		Diamond
	In layer	Layer-to layer	
μ / m^2 V^{-1} s^{-1}	2.0	0.010	0.18

From the earlier discussion, it follows that the band overlap will increase as the size of the graphite particles becomes smaller.

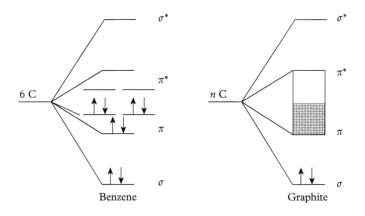

Fig. 6.8 Molecular orbital energy-level diagrams for benzene and graphite; in each substance the σ level is a $(1s)^2$ core. Graphite exhibits a much higher degree of delocalization of the π electrons than does benzene.

6.3.3   Fullerenes and buckyballs

When the products of a high-energy laser interaction with graphite in a helium atmosphere were subjected to mass spectrometric examination, fragments were discovered that had a carbon atom content varying in number according to the pressure of helium [4]. A very stable carbon atom cluster was found to correspond to the formula C_{60}, a new allotrope of carbon. The stability arises because a sheet of carbon atoms large enough to form a ball-like structure can satisfy fully the valence requirements for a resonance form of carbon without imposing mechanical strain on the structure as a whole.

The geometrical structure of C_{60} is similar to that of a geodesic dome (Fig. 6.9). The dome is a spherical or near spherical lattice type surface formed

Fig. 6.9 *Spaceship Earth* at Disneyland, Florida; an example of a geodesic dome. [Reproduced by courtesy of Benjamin D Esham.]

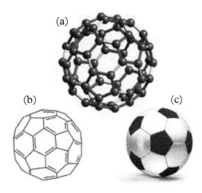

Fig. 6.10 Molecular structure of the C_{60} buckyball in three representations. (a) Ball and spoke model. (b) Resonance structure. (c) Football-style. [Harrison P and McCaw C *Educ. Chem.* 2011; 48: 113; reproduced by permission of The Royal Society of Chemistry.]

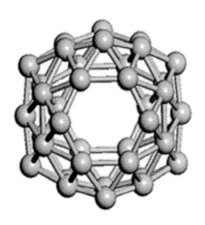

Fig. 6.11 A boron buckyball B_{38} of point group symmetry D_{2h}.

by a network of great circles, or geodesics, on a sphere. The geodesics intersect to form a rigid, stress-free triangular edifice.

The C_{60} molecule is similar in structure, but with interlocking pentagons and hexagons that form a complete sphere (Fig. 6.10). This structure was named buckminsterfullerene [5] after R. Buckminster Fuller who devised the mathematics for the dome and described it in detail. There are no double bonds in the five-membered rings so that delocalization is not well developed; the compound behaves something like an electron-deficient alkene, and reacts readily with electron-rich reagents such as the halogens, forming compounds such as $C_{60}Cl_6$ and $C_{60}Br_{24}$. Fullerenes are soluble in non-polar solvents, such as benzene and toluene. The solution in toluene is red in colour and on evaporation deposits pure crystalline carbon.

A molecule that consists entirely of carbon atoms in the form of hollow spheres, ellipsoids or tubes is known by the general name of *fullerene*. Spherical or near-spherical fullerenes have been given the name *buckyballs*, the simplest, stable structure being that of buckminsterfullerene. This buckyball structure has 32 faces, comprising carbon atoms arranged in 20 hexagons and 12 pentagons, all related through icosahedral symmetry. In theory there is no limit to the number of possible fullerenes. The C–C bond lengths in C_{60} are 0.1458 nm for the bonds fusing five-membered rings with six-membered rings and 0.1401 nm for the bonds fusing the six-membered rings one to the other. Buckyballs and buckytubes are topics of current research both in pure chemistry, in which fullerenes are manipulated to form various organic compounds, and in technological applications, such as carbon nanotubes.

Recent theoretical calculations have predicted a stable boron buckyball, a boron analogue of fullerene, with the formula B_{38} [6]. This boron species, if it can be made, would have the structure of a Mackay icosahedron [7] comprising 56 triangles and 4 hexagons (Fig. 6.11); heptagonal holes can also be detected in the structure. A hexagonal arrangement of boron atoms would lead to a large band gap, *ca.* 2.25 eV, and a high degree of aromatic character. No doubt more will be heard of this and similar compounds in the future.

A new structure corresponding to B_{40} has been produced which is slightly different from that predicted. The 40-atom species has been termed borophene, and its molecular structure comprises two hexagons, seven heptagons and forty-eight triangles arranged in two shapes, one being almost planar and the other a hollow, pseudo-spherical molecule (Fig. 6.12). It is similar to the C_{60} buckyball but is not as smooth since boron is only trivalent; loss of planarity induces strain on the B–B bonding. The reactivity of B_{40} makes it difficult to isolate but it can bond with hydrogen, which opens a way for possible mechanism for hydrogen storage [8].

6.3.4 Graphene

Graphene is pure, single-crystal carbon in the form of a sheet of one atom thickness, first prepared in 2004; it is a new allotrope of carbon, similar in form to a single layer of graphite, from which substance it was first prepared [9, 10]. Its thickness of 0.335 nm is approximately 10^5 times thinner than a human hair. Its original preparation involved peeling away layers from a graphite crystal by means of Scotch tape (aka Sellotape) until a single layer, or 'two-dimensional carbon molecule', remained (Fig. 6.13). Graphene is a feature of much current research on account of its great mechanical strength, as also are carbon nanotubes which might be regarded as convoluted forms of graphene.

Graphene is a better conductor of electricity than copper, as electrons are able to flow through graphene more easily; its electrical resistivity is 35% less than that of copper. The electrons travel through the graphene sheet as if they were massless, and as fast as one hundredth the speed of light. The charge carriers make hop-like movements from a negative electron-like state to a positive hole-like state, a process that has been likened to Klein tunnelling.

Klein tunnelling is a process by which a relativistic electron[2] begins to penetrate a potential barrier if its height V_0 is greater than $m_e c^2$, the rest energy of the electron. In these circumstances the transmission probability P_T depends only weakly on the height of the barrier. This feature is in strong contrast to the tunnelling of non-relativistic electrons, where P_T decays exponentially as the value of V_0 increases (see Section 2.10.1). The relativistic effect occurs because a strong potential that is repulsive for electrons creates positive areas within the barrier that can attract and combine with electrons, a process that is similar to the movement of electrons into positive holes in a semiconductor device. Klein tunnelling, which has remained a theoretical topic for many years, has been now tested and validated for the first time with single and bi-layer graphene sheet [11]. The mechanical strength of graphene, as determined by atomic force microscopy, is about three hundred times greater than that of steel; it has also a hardness greater than that of diamond. It is ductile to an extent of 20%, and can be used to form strong composite materials. Graphene springs have been made by wrapping fibres of wet graphene around a glass tube followed by annealing at 500 °C. The springs can be elongated by *ca.* 500% and are very durable. They find applications in electronic devices, such as magnetostrictive switches and actuators [12].

Graphene is visible to the naked eye and appears blue or purple; it transmits approximately 97% of visible light, falling to 89% through five layers of the material. The absorption of light has been increased by *ca.* 50% by spraying the graphene surface with nanosize crystals of lead sulphide. These crystals are quantum dots of 3 nm to 4 nm in size and enable the material to be used in night-vision applications. The interaction of quantum dots with graphene is another manifestation of the mobility of its delocalized electrons.

The self-assembly of nanosize cubes of magnetite to form larger structures occurs by exposing the material to a magnetic field. In a magnetic field of slowly increasing strength the cubes form into strips, whereas under a constant field helical structures obtain; the value of the structures as magnetic materials is yet to be evaluated [13].

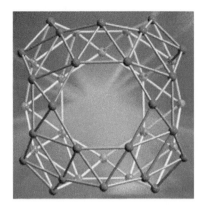

Fig. 6.12 A pseudo-spherical molecule of composition B_{40}, borophene. [Reproduced by courtesy of Professor L-C Wang.]

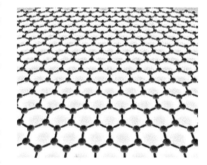

Fig. 6.13 Graphene: a single-crystal sheet of fused hexagonal carbon rings, similar in structure to graphite, but of one atom in thickness (0.335 nm); the C–C bond length is 0.142 nm.

---

[2] An electron moving with a speed $v > 0.1c$.

6.3.4.1 *Chemical sensing*

Graphene can be used as a sensor of gases, such as NH_3, CO and NO_2. The vibrations of a suspended sheet of graphene carrying an electric current are modified in the presence of gases; even a single molecule can be detected. The gas undergoes adsorption/desorption processes at pores and interstitial sites on the graphene surface, interacting through its π-bonding system. The amplitude of a vibration of the graphene sheet, rather than its frequency, is sensitive to the mass of a species. The detection of NO_2 and NO can be achieved at sub-ppm concentrations, which is of importance in the nitric acid industry on account of the toxicity of these photochemical oxidants.

6.3.4.2 *Graphene oxide*

Graphene oxide is formed as a yellow material by treating graphite with a strong oxidizing mixture, such as potassium chlorate and *ca.* 100% nitric acid. It has a monomolecular structure similar to that of graphite. Membranes of graphene oxide, when immersed in water, act as molecular sieves and can block the passage of molecular species of size greater than about 0.9 nm. The effect is attributed to nanocapillaries that open up in the hydrated state and accept only species that fit [14].

The interaction of graphene oxide with soil is a feature arising from the discharge of nanomaterials into the environment. In a recent study, soil samples were exposed for *ca.* 90 days to the action of pure graphene oxide (PGO). It was found that the soil-modified graphene oxide (SMGO) became bacterially richer and more diverse than that the control soil sample PGO. In particular, nitrogen-fixing genera were selectively enriched. Nitrogen-containing groups, such N–O, and elements including Mg, Al, Si, K and Fe were detected in SMGO. A higher chemical activity than in PGO was noted, including structures with unpaired electrons and others exhibiting disorder. This work highlights the need for investigation of the probable long-term effects and risks arising from the exposure of soil to waste nanomaterials [15, 16].

6.3.5 Carbon nanotubes and nanowires

Nanosize carbon sheet can be processed into tubes by wrapping or rolling the sheet, and differing properties of the tubes arise according to the chosen mode of wrapping. Single-wall nanotubes (SWNTs) have diameters between 0.5 nm and 5 nm. The C–C bond length for trigonal carbon is 0.142 nm, and the nearest non-bonded C $\cdots$ C distances are 0.246 and 0.284 nm. Multi-wall nanotubes (MWNTs) range from 1.5 nm to 15 nm for the inner wall, and from 2.5 nm to 50 nm for the outer wall, and with a distance of *ca.* 0.35 nm between walls.

The nature, or *chirality*, of a nanotube is determined by the manner of wrapping. The allowed wrapping directions are based on the two vectors a_1 and a_2 as shown in the diagram, which represents a portion of a graphene sheet:

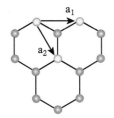

Portion of graphene sheet showing the basic wrapping vectors a_1 and a_2

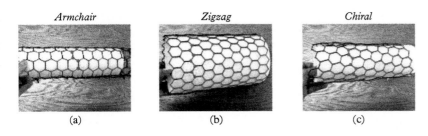

Fig. 6.14 Single-wall carbon nanotubes (SWNTs) of the 'armchair', 'zigzag' and 'chiral' wrapping modes. [Reproduced by courtesy of Professor H Dai, Stanford University.]

Three directions of wrapping of the sheet to form SWNTs are important (Fig. 6.14). They are designated *armchair*, *zigzag* and *chiral*, and are based on wrapping vectors **v** given by the equation:

$$\mathbf{v} = n\mathbf{a}_1 + m\mathbf{a}_2. \tag{6.10}$$

The nanotubes are then designated concisely as (n, m) where n and m are integers, including zero. For $n = m$, that is, a vector $(n\mathbf{a}_1 + n\mathbf{a}_2)$, or $n(\mathbf{a}_1 + \mathbf{a}_2)$, the armchair SWNT is formed; the configuration $m = 0$, that is, the wrapping vector is $n\mathbf{a}_1$, denotes the zigzag structure; and the vector $(n\mathbf{a}_1 + m\mathbf{a}_2)$ produces the chiral nanotube.

A nanotube is formed physically by wrapping the graphene sheet around the given vector until the points $n\mathbf{a}_1$ and $m\mathbf{a}_2$ meet. A tube can be capped by a fullerene molecule if desired, and a singly-capped armchair SWNT is shown here:

Singly-capped armchair SWNT

The electrical conduction of a carbon SWNT is governed by the chiral angle, that is, the angle between a hexagon and the axis of the tube. It can be related to the values of n and m by the following generally applicable conditions:

- for $n = m$ the nanotube will be metallic in character, so that the SWNT of armchair mode function as metals for electrical conduction;

- for $(n-m) = 3r$ ($r = 0, 1, 2$, etc.) the SWNT will be a good semi-conductor with a small band gap, like silicon;

- otherwise, the nanotube will function only moderately as a semi-conductor.

The density of states functions for carbon nanotubes exhibit van Hove singularities, the number thereof dependent upon the nanotube conformation.

Carbon nanotubes have very great mechanical strength: Young's modulus, which is a measure of the ratio of the stress along an axis to the strain in the same direction, has been measured up to 10^3 GPa; steel at its best is *ca.* 60 GPa.

Carbon nanotubes and nanowires are finding numerous applications in science and technology. Their mechanical strength makes them suitable for mechanical construction materials, body armour and prosthetics, and they alloy with aluminium to produce a material of the strength of steel. Applications arise also in the field of microelectronics, solar cells, and lithium batteries. Desalination of brine can be achieved by passing it through electrically controlled nanotubes of nanoporosity size.

In medicine, carbon nanotubes are used in biomedical research; tumours can be targeted and drugs delivered *in vivo*. Nanotubes have the advantage in this field over the heavier metal quantum dots by their much less hazardous nature, but long-term toxicity is yet to be evaluated. Single wall nanotubes can be tracked *in vivo* by Raman scattering, luminescence and other optical methods. Recently a computer has been made from many thousands of carbon nanotubes, which can store and execute simple programs and output results.

Example 6.3

It is instructive to try to form the nanotube structures. If nanotube material itself is not available, hexagon outlines are easily drawn and reproduced on paper; for something more durable, chicken wire is a good substitute. To construct models of the *armchair, zigzag* and *chiral* nanotubes, the reader's attention is directed to the following sites:

- https://web.stanford.edu/group/cpima/education/nanotube_lesson.pdf

- http://education.mrsec.wisc.edu/IPSE/educators/activities/supplements/carbon-NanotubeGuide.pdf

---

6.3.6 After graphene

The authors of the graphite exfoliation process that first produced graphene [9, 10] have shown that interesting two-dimensional crystals can be produced from transition-metal dichalcogenides, such as molybdenum or niobium, bonded

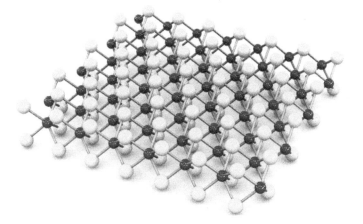

Fig. 6.15 The nanosheet structure of molybdenum disulphide MoS_2. The molybdenum transition-metal atoms form a layer between two layers of sulphur atoms. It is slightly less strong than graphene, but has important electrical and catalytic properties; it has been used also as an engine lubricant.

with sulphur or selenium. These materials differ from graphene in forming layer structures, with the transition-metal layer sandwiched between two layers of a chalcogen (Fig. 6.15) [17]. In certain electrical properties these layers structures behave similarly to graphene.

The molybdenum disulphide MoS_2 structure behaves in like manner but its faster vibration makes it a more sensitive detector [18]. Sheets of molybdenum disulphide or tungsten disulphide have been shown to catalyse hydrogen evolution reactions, particularly if nanoparticles of gold are deposited on the surfaces of the sheets. The catalytic activity is believed to be due to gold nanoparticles enhancing charge transport between the sheets [19]. Other work is in progress on defect structures and magnetic properties of these new materials [20].

Computer simulation studies have predicted the existence of a pentagonal graphene termed 'penta-graphene'. It is predicted to be stronger than graphene and to withstand temperatures up to 1000 K. It is semiconducting with a direct band gap length of *ca.* 2.35 eV, forming carbon nanotubes with properties that are independent of chirality. The material would be auxetic: if graphene is stretched it expands along the direction of stretching but contracts in a perpendicular direction, whereas penta-graphene when stretched it should expand in both directions in its monomolecular plane. Work is in progress to synthesize this proposed carbon allotrope; currently, the only known carbon nanomaterial composed solely of pentagons is C_{20} [21].

6.3.7   Graphite intercalation compounds

Graphite intercalation compounds are structurally ordered compounds involving metal atoms sandwiched between the layers of the graphite structure (Fig. 2.52).

They have the form MC_n, where M can be potassium, rubidium or cesium, in which case n is often 8 or 16, or M is lithium, calcium or strontium and n is 6 or higher.

Intercalation compounds with graphite were reported first in 1926, having been prepared by heating rubidium or cesium under finely-divided graphite [22]. Compounds such as MC_8 and MC_{16} were isolated, and these and similar intercalation compounds have been characterized by x-ray diffraction. Improved methods of preparation involve the vapour-phase reaction of potassium at 250 °C with graphite at 250–600 °C. X-ray powder analysis has identified several phases; KC_{24}, for example, was found to have a cubic unit cell of side 0.876 nm, with an interlayer separation of 0.541 nm. Not every layer of the graphite structure need be populated in order to form an intercalation compound.

Potassium graphite KC_8 is well known as a chemical reducing agent, particularly in organometallic chemistry, and can be prepared by reacting molten potassium with graphite in either an inert atmosphere or a sealed tube. In its crystal structure, graphite acts as an acceptor of electrons; the potassium atoms reduce graphite by donating valence electrons to the orbitals of its empty π^* bands, and the K^+ ions enter the interlayer spaces of the graphite structure (Fig. 6.16). Since the π^* electrons are mobile, the intercalation compound has a high but strongly anisotropic electrical conductivity: $\sigma_a = 2.5 \times 10^5$ S m^{-1} and $\sigma_c = 5.9 \times 10^2$ S m^{-1}, where the subscripts refer to directions parallel to the a and c hexagonal axes. At very low temperature, KC_8 becomes superconducting, with a transition temperature of *ca.* 0.15 K.

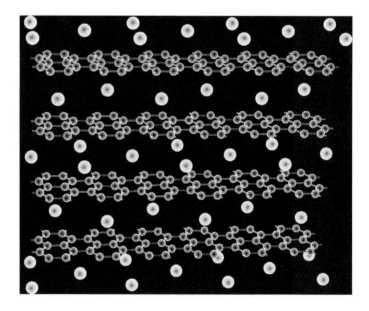

Fig. 6.16 Structure of the intercalation compound KC_8. The population of the interlayer spaces by potassium ions increases the layer spacing from 3.35 Å in graphite to 5.41 Å in this structure. The difference of 2.06 Å is less than twice the radius of the potassium ion (1.44 Å), indicating that the K^+–benzene interaction that is more than mechanical.

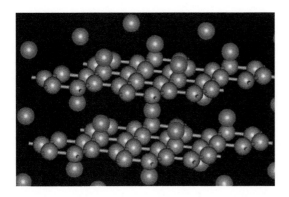

Fig. 6.17 Crystal structure of calcium graphite CaC_6; space group $R\bar{3}m$; interlayer distance 4.52 Å ($r_{Ca^{2+}} = 1.18$ Å); the superconducting transition temperature of 11.5 K at ambient pressure increases to 15.1 K at 8 GPa. [Emery N *et al.* Synthesis and superconducting properties of CaC_6. *Sci. Tech. Adv. Mater.* 2008; 9: 044102. DOI:10.1088/1468-6996/9/4/044102.]

In another example of the reducing properties of KC_8, liquid silicon tetrachloride reacts with it to produce nanosize silicon and graphite. The compacted solid material finds application as the cathode in lithium-ion batteries. The charge retention is better than that with a cathode prepared by compacting normally comminuted graphite and silicon powder, which suggests an increased bonding interaction with the nanosize materials.

Calcium graphite CaC_6 has been prepared by reacting liquid calcium under pyrolysed hydrocarbons. The interlayer distance in this compound is *ca.* 0.452 nm, with an increase in the C–C bond length to 0.144 nm (Fig. 6.17). The calcium graphite compound becomes superconducting at 11.5 K under ambient pressure [23]. It may be noted that whereas intercalation with potassium increases the interlayer spacing of graphite by 0.206 nm, a change of $1.4\, r_{K^+}$, the corresponding change with CaC_6 is 0.117 nm, equivalent to $r_{Ca^{2+}}$ (0.118 nm), indicating a stronger interaction with the divalent cation.

6.4 Magnetism in nanosize materials

Nanosize crystals are commensurate with the sizes of domains of magnetic materials, Fe 15 nm and Co 70 nm, for example. Not surprisingly, nanomaterials have interesting magnetic properties; first, however, a brief resume of the magnetic properties of materials.

6.4.1 Diamagnetic compounds

All substances exhibit *diamagnetism*. Where other magnetic effects are also present, the diamagnetic property is severely overshadowed. Bergman observed in 1778 that elemental **bismuth** and **antimony** were repelled by magnetic fields, the term *diamagnetism* was given to these materials by Faraday. Diamagnetic materials include **water**, organic compounds, such as petroleum and certain plastics, and elemental metals, such as copper, gold and mercury. Diamagnetic susceptibilities χ_v are usually 10^{-6} to 10^{-5}. The magnetic moment vectors, symbolized by $\rightarrow$, in a diamagnetic substance are arranged at random, but in an applied magnetic field they align in opposition to the direction of the field.

Diamagnetic: random magnetic moment vectors

6.4.2 Paramagnetic compounds

The existence of unpaired electrons in a substance, with concomitant electron spin, results in *paramagnetism*. In the absence of a field the individual magnetic moment vectors remain randomly orientated, but application of a magnetic field tends to align the vectors with the field. The susceptibility χ_v ranges from 10^{-3} to 10^{-2}. Not all individual moment vectors are necessarily aligned, but at a decreased temperature, when thermal motion is reduced, more vectors will turn into the field if the material is sufficiently fluid.

Paramagnetic: partially oriented magnetic moment vectors

6.4.3 Ferromagnetic, antiferromagnetic and ferrimagnetic compounds

The three classes under this heading differ in the value of their magnetic moments. In a *ferromagnetic*, such as iron or cobalt, the magnetic moment vector μ has a magnitude much greater than unity. Ferromagnetic materials are characterized by long-range order of magnetic moment vectors, which implies a lining up of unpaired electrons in nanosize regions known as *domains*.

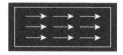

Ferromagnetic: strongly oriented magnetic moment vectors

A bulk ferromagnetic material is normally unmagnetized because the domains are set at random orientations within the material. An applied magnetic field causes the domains to line up thus greatly enhancing the magnetic permeability of the material. Ferromagnets exhibit *hysteresis*, that is, a significant magnetism remains after the applied field is removed.

If the temperature of a ferromagnet is increased, magnetism falls to zero at a temperature known as the Curie temperature, owing to the increase in thermal agitation. The Curie law expresses the fact that the susceptibility χ_v of a compound varies inversely as its temperature. For iron the Curie temperature is 1043 K; its thermal energy at this temperature is 0.14 eV compared with 0.03 eV at room temperature.

In an *antiferromagnet* the magnetic moment vectors, which are related to the unpaired electrons in the substance, align in a regular manner, but with neighbouring **spins** aligned in opposing directions.

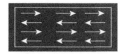

Antiferromagnetic: oriented but opposed magnetic moment vectors

The order in an antiferromagnetic or a ferrimagnetic compound is destroyed at an increased temperature known as the Néel temperature, and above this temperature the material becomes paramagnetic. Examples of antiferromagnetic compounds, with their Néel temperatures, are: CoO, 291 K; haematite Fe_2O_3, 680 K; chromium, 308 K.

Elemental chromium is interesting as a simple antiferromagnetic compound: it has the electronic configuration $(Ar)(4s)^1(3d)^5$, and the crystal structure is shown in Fig. 6.18. Illustration (a) indicates that under x-ray diffraction the structure is body-centred cubic. Illustration (b) shows its structure by neutron diffraction: although the individual atoms are similarly placed in the unit cell, magnetic interactions with the neutron beam reverse the direction of the magnetic moment vector of atom at the centre of the unit cell. Then, it is no longer body-centred but primitive cubic, with two atoms of chromium, with differing magnetic orientation, in the unit cell.

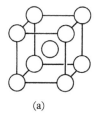

(a)

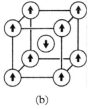

(b)

Fig. 6.18 Unit cell and environs of the crystal structure of elemental chromium as seen by (a) x-rays and (b) neutrons; the arrows in (b) represent the direction of the magnetic moment vectors of the chromium atoms that interact with the neutron radiation. [*Structure Determination by X-ray Crystallography*, 5th ed. 2013; reproduced by courtesy of Springer Science+Business Media, NY.]

Finally there is the *ferrimagnetic* class, materials that are populated by more than one species, with opposing spins that represent differing values of magnetic moment.

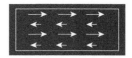

Ferrimagnetic: oriented, partially opposed magnetic moment vectors

Well known examples are the cubic ferrites, such as magnetite Fe_3O_4, garnets, one formulation being $Ca_3Al_2[SiO_4]_3$, and spinels (Lat. *spina* = thorn ≡ acicular crystals); hexagonal ferrites of the type of $Ca_2(AlFe)O_5$ and $BaFe_{12}O_{19}$ are ferrimagnetic ceramic materials.

Spinel structures have the general formula AB_2O_4, where A and B are metals in oxidation states II (Mg, Fe, Ni, ...) and III (Al, Fe, Cr, ...) respectively; some of these structures were determined first by Bragg in 1915. Spinel itself is $MgAl_2O_4$: it is cubic, space group $Fd\bar{3}m$, with a unit cell dimension of 0.8083 nm and eight formula entities in the unit cell. The structure is illustrated by Fig. 6.19: spinel is an almost perfect close-packed cubic structure (see also Fig. 5.28a).

The magnesium species in this spinel lie in the tetrahedral holes, coordinated by four oxygen atoms, with aluminium occupying the octahedral holes. Given the space group, the single fractional coordinate x for oxygen (0.365) defines

Fig. 6.19 Unit cell and environs of the crystal structure of spinel, general formula $A^{II}B^{III}O_4$; the atom labelling shows the four-coordinated A species (Mg^{2+}) occupying tetrahedral holes in the almost close-packed array of oxygen atoms, and the six-coordinated B species (Al^{3+}) occupying the octahedral holes. The unit cell contains eight formula-entities, and the atoms O, Mg and Al are represented in decreasing order of size.

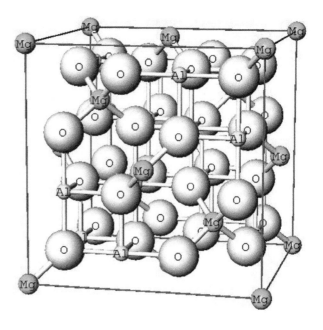

completely the crystal structure; had x been 0.375 the structure would have been a perfect close-packed cubic array of oxygen atoms. Another view of the spinel structure is shown in Fig. 6.20.

As well as spinel structures containing varying types of A and B species, there are also the inverse spinels, in which the A and B species interchange sites. Perhaps the best known of them is magnetite Fe_3O_4, better written as $Fe^{III}(Fe^{II}Fe^{III})O_4$. Its space group is $Fd\bar{3}m$, with $x = 0.379$, now very nearly close-packed, in a unit cell of side 0.8394 nm. It is a lower entropy form than the normal Fe_3O_4 spinel, and is well known for its magnetic properties as the mineral lodestone. The following scheme is typical of normal and inverse spinels:

Normal spinel	Inverse spinel
Example: $MgAl_2O_4$	Example: $Co^{II}(Co^{II}Fe^{III})O_4$
Mg^{II} in *tetrahedral holes*	Co^{III} in *tetrahedral holes*
Al^{III} in *octahedral holes*	Co^{II}, Fe^{III} in *octahedral holes*
O^{2-} *close-packed cubic*	O^{2-} *close-packed cubic*

Nanosize spinels have been used to improve the storage properties of lithium-ion electrodes in batteries. On account of the size of nanoparticles, surface effects change significantly the storage properties of batteries. Calculations indicate that storage depends on surface orientation: the lowest energy surface, (110) for a spinel structure type, is energetically favourable for addition of Li^+ ions into the vacant Wyckoff (c) space-group sites. The results obtained indicate the considerable value of nanosize materials in the construction of battery electrodes.

Copper chromite or nickel chromite at nanosize provide for a clean and uncomplicated catalytic procedure in organic degradation processes. The nickel chromite nanosize crystals are eco-friendly and replace other more hazardous reagents in such procedures.

Nanosize crystalline spinels such as $CoFe_2O_4$ or $LiNi_xMn_{2-x}O_4$ ($x \leq 0.5$) can be prepared by carbohydrate-combustion: one such process reacts cobalt(II) nitrate and iron(III) nitrate with glycine, and combustion takes place at *ca.* 200 °C. The compounds have been characterized by x-ray diffraction and consist of particles 20–25 nm size. The sample of particular composition $LiNi_{0.5}Mn_{1.5}O_4$ forms another useful cathode material in lithium-ion batteries, which exhibit the high rating of *ca.* 133 mA-h g^{-1} and recycles to 99.7%.

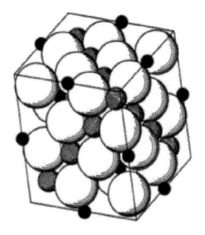

Fig. 6.20 Another view of the unit cell and environs of $MgAl_2O_4$; again, spheres in decreasing order of size represent the O, Mg and Al atoms; the unit cell origin is shifted by $a/2$ compared with the standard orientation for space group $Fd\bar{3}m$ in order to show more clearly the tetrahedral and octahedral holes occupied by Mg and Al respectively. [Verwey EJW and Heilmann EL. *J. Chem. Phys.* 1947; 15: 174. Reproduced by permission of the American Institute of Physics.]

6.5 Nanopolymers

Nanopolymers comprise a polymer into which nanosize particles of various shapes are dispersed, forming *polymer nanocomposites* (PNCs). The physical properties of the polymer are changed significantly because of an increase in the total surface area, which is dependent on the nanoparticle size. The smaller the particles the greater the surface area, and the more the particles then become the controlling factor on the properties of the composite. At a 10% loading of nanoparticles, the following data have been reported:

Particle size	Surface area	Average interparticle distance
1 μm	0.6 m^2	8 nm
1 nm	60 m^2	800 nm

An important application of nanopolymers involves the regeneration of human tissue that has been marred by accident or other misfortune. The PNC provides a means for adding and thus enabling growth of cells that have been damaged. Nanopolymers in the form of wires form a particularly useful scaffold for cells of fibril tissue structures which attach readily to nanowires. Not only carbon but many other materials, such as silicon and metals, have been fabricated as nanosize particles with interesting and far-reaching applications of this nature.

6.6   Fabrication of nanosize materials

The production of nanosize materials has been mentioned briefly in some of the foregoing discussions. There are two basic general methods for preparing nano-materials, usually referred to as the 'bottom-up' and 'top-down' procedures, that is, either the building up or breaking down of atomic bonds; both processes can lead to materials of nanosize particles:

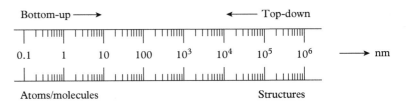

Bottom-up/top-down size scale

In the *bottom-up* method, the nanomaterial is assembled effectively atom by atom through chemical processes or physical means. Chemical properties of single species or small structural entities are caused (a) to self-organize or self-assemble into some useful conformation, or (b) to rely on positional assembly. These procedures utilize molecular self-assembly and molecular recognition, a field known as supramolecular chemistry. Examples of this procedure are employed by biological systems, as in the hydrogen-bonding of DNA to form the double helix.

Chemical methods under various conditions of temperature and pressure may form other bottom-up approaches. Precision, nanosize graphene for electronics can be prepared by the bottom-up method, rather than by a top-down method such as graphite exfoliation. Linear chains of carbon atoms can be converted under carefully controlled conditions into extended hexagonal sheets. In a particular preparation, oligo-en-diyne (Fig. 6.21a), composed of three benzene rings linked by carbon atoms, placed on a silver surface was induced to react by raising the temperature above 90 °C.

The progress of the reaction was monitored by non-contact atomic force microscopy. The sharp-tip probe detects not only individual atoms, but also the forces of covalent interatomic bonds. The resulting images are readily interpreted as chemical moieties, and a distinction between single and multiple bonds is discernible (Fig. 6.21a–c).

Another technique for bottom-up production of nanomaterials is based on the use of ultrasound. The sol–gel method is a wet chemical process that can be applied, for example, to metal oxides and involves the sequence

$$\text{sol} \rightarrow \text{hydrolysis} \rightarrow \text{condensation} \rightarrow \text{gelation}$$

under ultrasonic control. The particle size of the product is governed by chemical conditions, such as pH, temperature and pressure, as well as by the intensity of the ultrasound.

Chemical reactions in bottom-up processes can be induced by electrodeposition. It is relatively cheap and can be carried out at relatively low temperatures. In the process a conductive substrate is coated with a layer of metal of nanosize thickness controllable by the amount of charge passed. The technique is used in producing microelectronics under control of insulating masks to direct the circuitry. This nanoscale technique has application also in biological systems, for

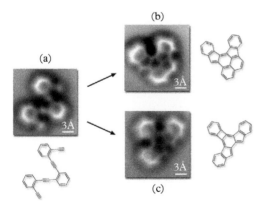

(a)

(b)

(c)

Fig. 6.21 Bottom-up syntheses of a nanomaterial. (a) Starting material, oligo-en-diyne. (b)–(c) Initial products from the reaction on a silver surface. A distinction between bond orders is perceptible, and the benzene rings show evidence of electron delocalization.

example, in the use of proteins to control processes where nanoscale materials are laid down by electrodeposition, an area of union between biology and nanoscale chemistry.

Among *top-down* methods, as well as the original exfoliation technique, small size particles can be produced by grinding, dispersive, irradiative and chemical methods. In the grinding procedure, material is passed through stainless steel or tungsten carbide coated rollers in a sealed system, from which nanosize, powdered material is obtained. It is necessary to be aware that any mechanical comminution process may introduce impurity from the material of the rollers and from strain inherent in the grinding process itself.

Grinding can be combined with chemical reaction to obtain nanosize material such that the desired material is obtained in a matrix that can be eluted subsequently. Examples are:

$$CoCl_2 + 2K \xrightarrow{\text{Milling}} Co + 2K^+Cl^-$$

$$CdCl_2 + K_2S \xrightarrow{\text{Milling}} CdS + 2K^+Cl^-$$

Subsequent treatment with water removes the potassium chloride in each case. Alternatively, nanosize particles can be obtained by vaporization and condensation on to a clean, non-reacting surface, such as gold foil. In another technique, bulk material can be sized by an electron beam or laser-produced x-rays under vacuum, and the particle size is governed here by the wavelength of the radiation.

An important top-down procedure uses a form of photolithographic process, particularly for manufacturing miniature electronic devices such as computer chips. In one technique, a flake of oxidized silicon is coated with an organic photoresistant material. It is then exposed to ultraviolet radiation in a designated pattern. The pattern is developed and the procedure repeated so as to build up a sufficient pattern. Alternatively, the pattern may be etched by an electron beam. There results a pattern of better resolution on account of less diffraction interference than with ultraviolet radiation, because of the shorter electron wavelengths used.

More recent and less costly are mechanical methods of lithography. A silicon flake is coated with a soft polymer through which the desired pattern is imprinted mechanically. Even higher quality resolution, down to *ca.* 20 nm, can be achieved by a gallium ion beam *in vacuo* focussed by electromagnets, which effectively cuts the desired pattern.

Although many products of nanoscience and nanotechnology are now in general use, much remains in the theoretical and research stages. Areas of development would be expected to include optics, engineering, analytical chemistry, transport, biochemistry and medicine, a large growth area in solar power and energy storage. The increase in the world market, measured in billions of US dollars, over the period 2002–2014 is illustrated by Fig. 6.22, and a brief account of the current and anticipated industrial preparation of graphene has been given recently [24].

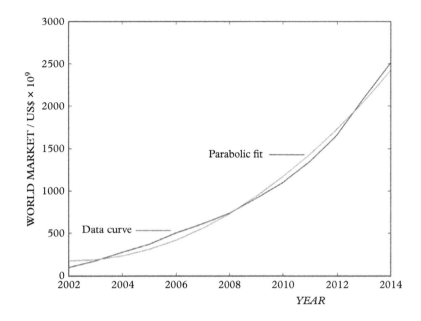

Fig. 6.22 Growth of the world market in nanotechnology from 2002 to 2014, measured in billions of US$; the growth is very closely parabolic.

The market growth shows a good parabolic fit over the past twelve years rather than the exponential increase that has been suggested. In the literature field it is notable that there are now about twenty journals dedicated to reporting research and applications of nanoscience and nanotechnology, together with a rapidly growing textbook literature of general and specialized natures, through which the subjects of nanoscience and nanotechnology can be pursued in depth [24–31].

....

REFERENCES 6

[1] Holmes KC. *Novartis Foundation Symp.* 1998; 213: 76.
[2] Ekimov AI and Onushchenko AA. *JETP Lett.* 1981; 34: 345.
[3] Martin CJ and O'Connor DA. *J. Phys. C: Solid State Phys.* 1977; 10: 3521.
[4] Kroto H *et al. Nature* 1985; 318: 162.
[5] Hedberg K *et al. Science* 1991; 254: 410.
[6] Lv J, Wang Y and Ma Y. *Nanoscale* 2014; June.
[7] Mackay AL. *Acta Crystallogr.* 1962; 15: 916.
[8] Wang L-S *et al. Nat. Chem.* 2014; 6: 727.
[9] Geim AK *et al. Science* 2004; 308: 666.
[10] Geim AK and Novoselov KS. *Nat. Mater.* 2007; 6: 187.
[11] Robinson TR. *Am. J. Phys.* 2012; 80: 141.
[12] Lee J-H *et al. Science* 2014; 344: 286.
[13] Cheng H *et al. Nanoscale* 2014; DOI 10.1039/c4nr03409k.

[14] Klajn R *et al. J. Am. Chem. Soc.* 2014; 136: 11276.

[15] Yang K *et al. Small* 2013; 9: 492.

[16] Du J *et al. RSC Adv.* 2015; 5: 27009.

[17] Joshi RK *et al. Science* 2014; 343: 752.

[18] Novoselov KS *et al. Proc. Natl. Acad. Sci. USA* 2005; 102: 10451.

[19] Lee J *et al. ACS Nano* 2013; 7: 6086.

[20] Kim J *et al. J. Phys. Chem. Lett.* 2013; 4: 1227.

[21] Zhang S *et al. Proc. Natl. Acad. Sci. USA* 2015; 112: 2372.

[22] Evans J. *Chem. World* 2014; 11: 44

[23] Fredenhagen K and Cadenbach G. *Z. Anorg. Allgem. Chem.* 1926; 158: 249.

[24] Chalmers M. *Phys. World* 2015; 28: 15.

[25] Murty BS *et al. Textbook of Nanoscience and Nanotechnology.* Springer, 2014.

[26] Lindsay SM. *Introduction to Nanoscience.* Oxford University Press, 2010.

[27] Binns C. *Introduction to Nanoscience and Nanotechnology.* Wiley Online Library, 2010

[28] Varadan VK *et al. Nanoscience and Nanotechnology in Engineering.* World Scientific, 2010.

[29] Wolf EL. *Nanoscience and Nanotechnology, 2nd* ed. Wiley-VCH, 2006.

[30] Rietman EA. *Molecular Engineering of Nanosystems.* Springer-Verlag, 2001.

[31] Bridge Technology, 2013. http://four-point-probes.com/sheet-resistance-and-the-calculation-of-resistivity-or-thickness-relative-to-semiconductor-applications/

Problems 6

6.1. The density of silicon is 2329.0 kg m^{-3}. If a computer chip is 250 mm square, how many transistors of a size containing 100 atoms can be contained by on to the chip? The silicon atoms can be assumed to be spherical.

6.2. In a photovoltaic cell based on a silicon semiconductor, the energy required to cross the gap between the valence and conduction bands is 1.31 eV. What irradiating wavelength would be needed to activate this transition, and in which part of the electromagnetic spectrum would it be found?

6.3. If a quantum dot is doubled in size, what would be the effect on the frequency of its emitted fluorescence?

6.4. The charge carriers in semiconductors are electrons and holes. What are their relative mobilities under the influence of an applied electric field? If the electric field is reduced to zero, how will the mobilities of electrons and holes be affected?

6.5. The density of silicon is 2329 kg m^{-3}. (a) What is the mass of an atomic force microscope silicon cantilever of length 0.125 mm, width 0.03 mm and thickness 4.0 μm? (b) If the force constant f of the cantilever spring is 41.0 N m^{-1}, what is the resonant frequency of the cantilever?

6.6. For each of the following types of SWNT indicate their conformation and whether they are metallic or semiconducting:

(a) $(5a_1 + 4a_2)$, (b) $4(a_1 + a_2)$, (c) $(3a_1 + 3a_2)$, (d) $(6a_1 + 0a_2)$, (e) $(11a_1 + 2a_2)$, (f) $(-2a_1 + 0a_2)$

6.7. For this problem and Problem 6.8, the web site http://nanotube.msu.edu/nt05/abstracts/NT05tutor-Dresselhaus.pdf (page 14) is desirable. [It is worthwhile to read the whole article.]

(a) On an enlarged copy of the illustration of a portion of a graphene sheet hereunder, draw the wrapping vector $v = (4a_1 + 3a_2)$, or concisely (4, 3).

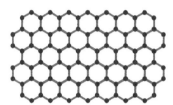

Portion of a graphene sheet

(b) If the diameter of a carbon SWNT is given by 7.83×10^{-2} nm $\times \sqrt{n^2 + m^2 + nm}$, determine the diameter of the (4, 3) nanotube.

6.8. (a) Calculate the chiral angle θ for each of the vectors in Problem 6.6; θ is the angle between a vector (n, m) and a_1.

6.9. Euler's polyhedron formula is V(vertices) $+ F$(faces) $= E$(edges) $+ 2$. (a) Determine the number of pentagons in a fullerene. (b) What is the smallest fullerene molecule? (c) Represent the molecule from (b) by a plane drawing (a fullerene map) that shows the number of vertices, faces and edges. (d) How many edges are there in the fullerene from (b)?

6.10. Show by means of a fullerene map, that a fullerene with a single hexagonal face is not possible. Give a reason for this situation. (*Hint:* Begin by drawing a hexagon.)

6.11. An intrinsic semiconductor of small energy gap has electrons in the conduction band. The conductance G measured at five temperatures gave the following results:

T/K	273	298	323	348	373
G/S	0.1198	0.04450	0.1411	0.3489	0.7942

On account of the exponential factor in the Fermi–Dirac distribution, the density of charge carriers and, thus, the conductance of the semiconductor, follows the Arrhenius-type formula $G = G_0 \exp(-\Delta E/2kT)$. Determine the energy gap in eV for the semiconductor.

6.12. The equilibrium constant K_c for the forward direction of the ATP hydrolysis reaction given in Section 6.2 at 37°C and $\underline{p}H \approx 6$ is 5.8×10^5. If an aqueous solution of ATP of concentration 1 mmol dm^{-3} is allowed to attain equilibrium under the given conditions, what are the equilibrium concentration of ATP and the free energy change for the forward reaction?

6.13. An exciton has an average separation between the electron and its hole; this distance is termed the exciton Bohr radius r_{ex} and is given by:

$$r_{ex} = \frac{\varepsilon a_0}{m_r^*/m_e}$$

where a_0 is the Bohr radius and m_r^* is the reduced mass of an exciton. The relative permittivity of silicon is 11.9 and the reduced masses of the electron and hole are $0.26m_e$ and $0.36m_e$, respectively. Calculate the exciton radius for silicon.

6.14. A monomolecular graphene sheet has a resistance per square R_s of 5.9 $k\Omega\,sq^{-1}$ measured by a four-point probe technique [27]. What is its electrical conductivity? A datum from the text will be required. (*Resistance per square* is strictly dimensionless, but the inclusion of 'square' avoids confusion between R and R_s.)

6.15. (a) Carbon, silicon and germanium are in group 14 (IVA) of the periodic table, and their electron configurations are, respectively, $(He)(2s)^2(2p)^2$, $(Ne)(3s)^2(3p)^2$ and $(Ar)(3d)^{10}(4s)^2(4p)^2$. Which, if any, of the following statements about band gaps E_g is correct?

(i) $(E_g)_{Si} > (E_g)_{Ge} > (E_g)_C$

(ii) $(E_g)_C < (E_g)_{Ge} = (E_g)_{Si}$

(iii) $(E_g)_C > (E_g)_{Si} > (E_g)_{Ge}$

(iv) $(E_g)_C = (E_g)_{Si} > (E_g)_{Ge}$

(v) $(E_g)_C < (E_g)_{Si} < (E_g)_{Ge}$

(b) A p–n photodiode has a band gap of 1.73 eV. Can it be activated by a signal of wavelength 2250 nm radiation?

6.16. (a) Calculate the density of states for a silicon chip of volume 1.0×10^5 nm^3 and energy 0.1 eV above the edge of the conduction band. Assume $m^* = 1.05m_e$. (b) Check the units of the expression that you use.

Computer-Aided Chemistry

7

The real danger is not that computers will begin to think like men, but that men will begin to think like computers.

Sydney J. Harris

7.1 Introduction

From time to time reference has been made to the Web Program Suite, a set of computer programs that has been designed around the text. Computing is an essential feature of any modern scientific endeavour, but computers do not teach; rather, they assist in both the study of scientific textual material and the solving of problems. The collection of programs accompanying this book can be accessed via the publisher's website http://www.oup.co.uk/companion/ladd/.

The plotting programs use the Python interpreter [1–4], whereas the other programs in the Suite are written in FORTRAN [4–6] and supplied as IBM-compatible .exe files. It is suggested that the complete Suite be first loaded from the website into a personal program folder, and a back-up copy be made, although the Suite can always be downloaded again. The programs are executed in a command window and their operation is mostly self-explanatory; however, the following sections may help with their best use.

Note. There are two complete sets of programs, one for 32-bit operating systems and the other for 64-bit systems. In use, the four-letter program name listed here and throughout the text is applicable, but with the addition of either 32 or 64 as appropriate. Thus, LSLI is called into execution as either LSLI32 or LSLI64, and so on.

7.2 Plotting with Python

Those readers who are familiar with the Python computing language should have no problems in working with the programs provided; only straightforward applications of Python are used herein.

Python is not a compiler but a line-by-line interpreter. Its coding format is *case-sensitive* and it is necessary to work Python commands always in *lower case*, except

Bonding, Structure and Solid-State Chemistry. First Edition. Mark Ladd.
© Mark Ladd 2016. Published in 2016 by Oxford University Press.

for certain situations such as those involving labelling or other strings where upper case may be desirable. Data files also are best set in lower case although some implementations of Python will accept upper case file names.

The *Enthought* distribution may be used for obtaining a copy of the Python interpreter:

1. Go to https://www.enthought.com/products/canopy/ As a student or academic, sign up for an Academic licence: click on the Create Account link at the top of the Canopy page and follow the instructions; your e-mail address must be used when registering for the licence. Enter a username and a password for registering the software later.

2. Download Canopy by clicking on Get Canopy and choose the Express version, which is sufficient for present purposes; be patient here as the download is lengthy; the version 1.5.2 was released in January 2015.

3. Install Python GUI (graphical user interface) by following the on-screen instructions; it is simplest to use the default location. In Windows, the Windows Installer is needed to open to downloaded *.msi* files.

4. Open the Canopy window and sign in with your username and password. Once you have logged in, you will be remembered on subsequent occasions when opening Canopy.

5. At the bottom right of the Welcome screen the Canopy version is listed; install any update that is available.

There is now a sufficient Python environment on your computer. It is recommended that you set up a dedicated work folder named PYTHON (or a preferred name) for work with the interpreter, and copy the file *grfn32.py* (or *grfn64.py*) and the data file parb.txt to it.

A correct installation should result in the Canopy icon appearing in the taskbar. If not, generate one and install it there for subsequent easy access. Click on the Python logo in the taskbar: a "Hi 'your name' welcome to Canopy" will appear. Click on Editor and then minimize the Welcome window.

If the Editor should show only a single window, click on View and tick Python in the drop-down menu; the latter procedure should be utilized whenever the Python shell (lower window) is not in evidence for any reason. The Editor (upper pane) is used for organizing programs; the process of typing correct Python instructions is assisted by the appearance of a red line below any erroneous entry. Note that this discussion is not concerned with writing programs, nor teaching Python, but rather with using correctly those programs supplied in the Suite. After use, the Navigation (left-hand) pane will carry a name 'e.g. John' and a list of recent files used. If the Navigation pane is not present Click View and tick File Browser.

For the present work Click on *Select files from your computer*; Click on *Create a new file* only if writing a new program. Then DClick (double Click) on the program name *grfn32.py*. (Alternatively, DClick on the program name in the dedicated Python work folder.) It will appear in the Editor (upper) pane. As it is a working program, proceed to the lower pane (Python shell) and press Enter. The cursor should appear in the shell. If only the upper pane shows, Click on View and tick Python (as above). Now press Enter in the shell and look for the flashing cursor line there. It should appear as

'In [2] | '

or other number depending on how many times Enter is pressed.

Notes:

(a) Pressing Enter in the Editor pane will simply add an unwanted blank line to the program.

(b) Any keystroke will go first into the Editor pane, and an error could be introduced into the program.

If the cursor is not in the shell then:

1. Click the ∇ on the lower right-hand side of the shell and ensure that 'Keep Directory Synced to Editor' is ticked; or

2. Click Run $\rightarrow$ Restart Kernel; or

3. Click File $\rightarrow$ Close and recall the program; or

4. Click File $\rightarrow$ Exit and start again.

Assuming now that the program has been called and the cursor is flashing in the shell, carry out the following commands (lower case), noting the format carefully. The ENTER key is required after each command. Bold font is used here only to highlight the commands, and **grfn32** can be replaced by **grfn64**, and similarly for all other programs in the Suite;

import grfn32 (Enter)
grfn32.graf(['parb']) (Enter)

A successful plot will be indicated by /\/\/ in the taskbar, and will be that of the parabola $y = x^2 + x + 1$; a file *grfn32.pyc* will have been generated in the Python work folder by the procedure and is of no concern in this discussion; it can be deleted if desired. More details and examples are contained in this chapter and in other sections of the text where plotting examples have been presented.

If it is desired to plot the points on the graph the following procedure can be followed. Suppose three *x, y* points are given, as below. Then the following commands will plot them as filled circles to the graph just plotted; if other shapes are desired consult

help(plot):
x = [11, 47, 91]
y = [111, 2163, 8191]
plot(x, y, 'o')

The three points appear on the plot of the parabola.

7.2.1 Plotting program (GRFN)

The program *grfn32* plots a graph from an x, y data set contained in a .txt file; the data file can be given any name and is not case sensitive. It must carry the suffix .txt, but the suffix is *not* entered with the data file name; the program system assumes (and requires) that data are in .txt files.

If an error is made in typing a command, it will be indicated. Use the *up-arrow* key and then the *back-arrow* to return to the fault and then correct it. The axes of a plot are labelled X and Y by default, but may be changed by appropriate label commands. To change a label, the following commands are a sample of some of the instructions that can be used.

xlabel(r'T') gives T as an abscissa label.
xlabel(r'$(1/T)/ \rm K^{-1}$') gives $(1/T)/K^{-1}$ as an abscissa label.
ylabel(r'$C_p/ \rm J\,mol^{-1}\,K^{-1}$') gives $C_p/J\,mol^{-1}K^{-1}$ as an ordinate label.

Notes on the above examples:

1. The default font for a label is italic.
2. \rm followed by a *spacebar space* converts to Roman font; \it followed by a *spacebar space* reverts to italic font.
3. \, is a *space within text* as in J\,K{−1} above in order to separate J and K^{-1}. Spacebar spaces can be added for clarity of view but are not interpreted as spaces on output; thus, {−1}\,**mol** and {−1}\, **mol** produce the same result.
4. '$.....$' writes everything written between the $ signs.
5. ^ for superscript; _ for subscript; additionally {..} must be used if the subscript/superscript contains two or more characters.
6. If two or more sets of data are to be plotted on one graph then the appropriate command is **grfn32.graf(['set1', 'set2'])** plots the two x, y data sets on one and the same graph, annotated P1 and P2.

7. **grfn32.graf(['set1', 'set2'], True]** plots the two data sets on one graph and their sum on another, *provided* that the data sets cover the same ranges of x and y; **True** is the Python indicator to sum the sets and plot the result; the default value is **False** and indicates no sum.

8. Following this input command with a label command as above adjusts the labels X and Y as needed.

9. **plt.plot([p,q], [r,s])** plots a straight line from p, r to q, s; thus **plt.plot([0,p], [0,0])**, where p marks the end of the plotted x-axis, draws the line $y = 0$, whereas **plt.plot([1,2}, [3,4])** draws a line from the point 1,3 to 2,4.

10. Ensure that 'Keep Directory Synced to Editor' (see above) is always ticked.

11. If a problem arises, it may be necessary to reset the shell, Click View → Restart Kernel.

12. Further notes on Python plotting are given at those points in the text or the problems where its use is applicable, and also in the references given in Section 7.1, or by means of the Python help facility, for example, the command **help(plot)**.

7.2.2 Contouring program (ANFN)

Contour line plots of sections through the angular functions of atomic orbitals, normal to either the X or the Z axes may be done with the program ANFN. The program requires *inter alia* an input of the value n for an $n \times n$ square matrix of numerical values of the orbital function which may be generated by the program PLOT (Section 7.2.3). Typical input files an2s.txt and an3s.txt, for hydrogenic $\psi(2s)$ and $\psi(3s)$ atomic orbitals, are included in the Program Suite. For an alternative three-dimensional presentation of the angular wavefunctions see Section 7.8.

The Python commands in the shell for this program are, with the data set an2s:

```
import anfn32
anfn32.anfn('an2s')
```

7.2.3 Angular functions field figures (PLOT)

The program PLOT produces field figures of the angular functions for s,p and d orbitals in a form suitable for input to ANFN. The program is opened by clicking on PLOT, and the directions on the monitor screen should be followed. The matrix for *field figures* is square and of maximum dimension 201; the value 101 is recommended and it must be an *odd* integer. The *transformation* symbol XYZ should be selected for orbitals $1s$, $2s$, $2p$, $3s$, $3p$, $3d_{x^2-y^2}$ and $3d_{xy}$. For $3d_{z^2}$

the transformation YZX is required since the XYZ transformation is a view along axis of Z; this plot should present as follows (labels adjusted):

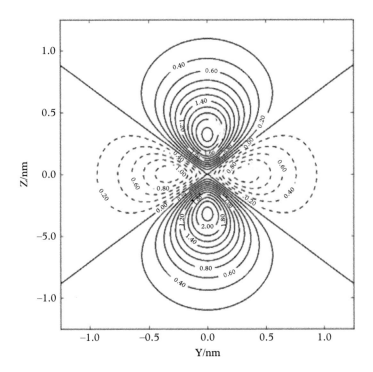

Hydrogen-like $3d_{z^2}$ atomic orbital, as seen along the X-axis

Only one example of the $2p$ orbitals is given as they are identical except for their spatial orientation with respect to Cartesian reference axes; the $3p$ orbitals are treated similarly. Of the $3d_{xy}$, $3d_{xz}$ and $3d_{yz}$, only one example is given as these three differ only in their relation to the reference axes.

There are several variables at the choice of the user. In particular, the *input plot-size parameter* is best related to the quantum number n, as indicated here:

n	Plot-size parameter
1	0.5–1.0
2	*ca.* 2.0
3	2.0–2.5

The *offset parameter* can be set at zero; if it is increased, it may be necessary to increase the plot-size parameter. Two example data sets, *an2s.txt* and *an3s.txt*, are provided with the Suite.

The user can experiment with different values of the parameters. It is not essential to use the *effective* atomic number (see Table 2.7); the atomic number leads to the same plot with only a small difference in the numerical values of the wavefunction. The usefulness of these plots is that they reveal the gradation in ψ or $|\psi^2|$ functions through the orbitals and also how they change with distance from the origin.

Note. The heading-lines reference information on the output from PLOT must not be removed as ANFN is programmed to pass over it.

7.3 Hückel molecular orbitals (HUCK)

Hückel molecular orbital theory (Section 2.18ff) uses the linear combination of atomic orbitals to determine the π molecular orbitals of conjugated hydrocarbons. Aromatic hydrocarbons and extension to heterocyclic systems are treated by the program; the associated h and k parameters for heterocyclic systems are listed in Section 7.3.1.

Characteristics of the Hückel procedure are:

- Only π molecular orbitals are included, as they determine the general properties of conjugated molecules; σ-electrons are ignored, which is justified by the orthogonality of π and σ orbitals in the planar molecules to which the theory applies.

- The method determines the energy levels that exist for the molecule under examination, indicates degeneracy and lists the energies of the molecular orbitals in terms of the parameters α and β discussed in the text. In addition it calculates coefficients of the molecular orbital wavefunctions, bond orders, atomic charges, free valence indexes and delocalization energies.

The program is entered with the program name HUCK, and data preparation is straightforward. Input is at the keyboard, and output to a user-named file. Two example data schemes are provided in a HUCK–README document, namely, 1,3-butadiene,

1,3-Butadiene

and the heterocyclic molecule pyrrole,

Pyrrole

The numbering refers to carbon and heterocyclic atoms; hydrogen atoms are not involved in the procedure.

Note: HUCK does not subtract the charge distribution from unity, unlike Eq. (2.153).

7.3.1 Hückel extension: *h* and *k* parameters

Atom	Bond type	Example	π-Electrons/atom	h	k
C	–C=C–	Benzene	1	(0.0)	(1.0)
N	–C=N–	Pyridine	1	0.5	1.0
N	=C–N<	Pyrrole	2	1.5	0.8
N	–N=N–	Azo	1	1.0	1.0
O	–C=O	Carbonyl	1	1.0	1.0
O	=C-O–	Furan	2	2.0	0.8
S	=C–S–	Thiophene	2	1.5	0.8

7.4 Linear least squares (LSLI)

The program LSLI determines the best-fit straight line $y = Ax + B$ to a series of x, y data points that must number at least 3 and, in this program, cannot exceed 100. Data are entered from a user-named file, such as LSSQ.TXT, and a sample data file has been included with the Program Suite. Execution follows input of the program name LSLI; unit weights are assumed for each observation. It is implicit that errors in x_i are significantly smaller than those in y_i. Output is to both the monitor screen and a file named LSLN.TXT. The goodness-of-fit is reflected by the values of the estimated standard deviations in the slope $\sigma(A)$ and the intercept $\sigma(B)$, and in Pearson's r coefficient for which a value in excess of 0.95 is regarded as satisfactory.

If the errors in the parameters A and B are to be propagated to another quantity, then the procedure in Section 7.4.1 should be followed.

7.4.1 Transmission of errors

Let q be a function of several variables p_i $(i = 1, 2, 3, \ldots, n)$ each with its own estimated standard deviation (esd), $\sigma(p_i)$. If the variables are independent, the esd σ of q is the positive square root of the variance σ^2:

$$\sigma^2(q) = \sum_i \left(\frac{\partial q}{\partial p_i} \sigma(p_i) \right)^2 \tag{7.1}$$

Example 7.1

This example of the application of Eq. (7.1) uses bond length data for a crystal. Consider a bond between two atoms lying along the vertical edge c of a tetragonal unit cell, with $c = 1.006(1)$ nm. Two atoms A and B have z fractional coordinates 0.3712(3) and 0.5418(2), respectively. A fractional coordinate z is Z/c, where both quantities are expressed in one and the same length unit, and the figures in parentheses are σ-values that apply to the final digits of the attached parameter. The bond length r_{AB} is given by:

$$d_{AB}/\text{nm} = \sqrt{(z_1 - z_2)^2 c^2} = 0.17162$$

Applying Eq. (7.1):

$$\begin{aligned} \sigma^2(d_{AB}) &= (0.5418 - 0.3712)^2 (0.001\,\text{nm})^2 + (1.006\,\text{nm})^2 (0.0002)^2 \\ &\quad + (1.006\,\text{nm})^2 (0.0003)^2 \\ &= (1.60667 \times 10^{-7}\,\text{nm})^2 \end{aligned}$$

Hence, $\sigma(r_{AB}) = 0.0004008$ nm, and d_{AB} is 0.1716(4) nm.

7.5 Madelung constants (MADC)

This program calculates the Madelung constant of a structure type from a set of crystal data; all crystal systems other than triclinic are accommodated. Program execution begins by entering the name MADC; crystal structure data are required for the program. A set of specimen input data, applicable to the CsCl structure type, is provided with the Suite. The β-angle should be entered as 90.0 except for the monoclinic and hexagonal systems when the true value should be used together with System Number 2. In addition, for hexagonal crystals, and trigonal crystals referred to hexagonal axes, a transformation must be applied to the data such that the angle γ is transformed to β, that is, $\mathbf{a}' = \mathbf{b}$, $\mathbf{b}' = -\mathbf{c}$ and $\mathbf{c}' = \mathbf{a}$, where the primes refer to the transformed unit cell; atomic coordinates must be transformed accordingly [6].

The result for the CsCl structure type is 1.7622 (2π), 1.7626 (3π) and 1.76266 (5π), which is the currently accepted value. Further information on the method is given in the references to Appendix A15.

7.6   Matrix operations (MATS)

The program MATS performs operations on one 3×3 matrix or on two, according to choice. The output is to the file MATOUT.TXT. If operations are required on only one matrix $\mathbf{A}$, neglect all reference to $\mathbf{B}$ in the output; the $\mathbf{B}$ matrix will have been set to unity by the program. Thus, meaningful results will then be the determinant, transpose, cofactor and inverse of $\mathbf{A}$. There will be no inverse of a matrix with a determinant value less than or equal to zero. For operations on two matrices, as well as the results above, there will be the sum, difference and product of $\mathbf{A}$ and $\mathbf{B}$.

7.7   Radial wavefunctions (RADL)

The program RADL calculates values of the radial wavefunction $R(r)$ for the $1s$, $2s$, $2p$, $3s$, $3p$, $3d$, $4s$, $4p$, $4d$ and $4f$ hydrogen-like radial functions. Input is at the keyboard, and output is to a user-named file and to three files named R1.TXT, R2.TXT and R3.TXT for $R(r)$, $|R(r)|^2$ and $4\pi r^2 |R(r)|^2$. These three files may be used to plot the radial functions with the program GRFN (*see* Section 7.2ff) either separately or all on a single graph. Example data sets *rd2s.txt* and *rd3s.txt* are provided.

7.8   Angular wavefunctions ('Orbitron')

For viewing atomic orbitals in three dimensions, the reader's attention is directed to the excellent display produced by Professor Mark Winter of Sheffield University, which can be found on the website http://winter.group.shef.ac.uk/orbitron, referred to herein as 'Orbitron'. Atomic orbitals from $1s$ to $7g$ are represented, together with s, p and d, s, p combinations, molecular orbitals for hydrogen and nitrogen molecules and other information of structural and chemical interest.

7.9   Gaussian quadrature (QUAD)

This numerical integration program calculates the area under a curve by Gaussian quadrature; the data are pairs of x, y coordinates of points on the curve.

There must be at least 4 data pairs and up to 100 pairs can be accommodated by the program QUAD. The input is from the keyboard or, preferably, from a prepared user-named file. The data consists of a title, the number n of data pairs and the n pairs of x, y values. The output is to the monitor screen.

7.10   Roots of polynomials (ROOT)

The program ROOT calculates the real and complex roots of a polynomial equation by Bairstow's method of quadratic factorization. The input consists of a title, the degree n of the polynomial and three starting parameters: the quadratic factors v_0 and w_0, both often set at unity in initially, and a precision level parameter *phi*, often 10^{-4} initially, all of which can be adjusted interactively according to requirements. The nth degree polynomial coefficients are then input *starting from the $(n-1)$th coefficient*; the program requires the nth coefficient to be unity. A specimen data set referring to the equation

$$x^5 + 17x^4 - 21x^3 + 3x^2 - 14x + 23 = 0$$

(input coefficients 17 −21 3 −14 23) is contained in the program; the coefficient of x^5 is unity.

The following results are obtained with a 10^{-6} degree of precision:

REAL ROOTS	IMAGINARY ROOTS
−0.482952E+00	0.863692
−0.482952E+00	−0.863692
0.106672E+01	0.393675
0.106672E+01	−0.393675
−0.181675E+02	0.000000

and they are output to both the monitor screen and a file named ROUT.

7.11   Inverse r^6 and r^{12} curves (LJON)

The program LJON produces the r, $V(r)$ data for curves of the form $V(r) \propto 1/r^6$ and $V(r) \propto 1/r^{12}$ named LJR6 and LJR12, respectively. The results are used in the Plotting exercise 3.1 (Section 3.8).

7.12   Electron-in-a-box (BOXS)

This program generates the electron-in-a-box wavefunction numerical values and energies for principal quantum numbers n of 1 to 15. A facility exists also for determining several features of conjugated systems of carbon atoms. The program is executed by clicking on the program name BOXS and then following the instructions on the monitor screen.

In the 'box' mode, as well as the output of the function values to the user-named file, output files are generated as BOX1–BOX15 ($n = 1 – 15$), any or all of which may be used in conjunction with the plotting program GRFN (Section 7.2.1) both singly and in sum.

In the 'conjugated' mode, the program calculates the energies E_n of the electron levels n, together with the energy and wavenumber $\bar{\nu}$ of the HOMO $\rightarrow$ LUMO transition.

7.13 Waveforms (WAVE)

The program WAVE enables studying, with varying values of n, the effect of summing sine functions, cosine functions and (sine + cosine) functions, the latter being equivalent to a harmonic function of the form $\exp(ikx)$. The superposition is made through the program GRFN (Section 7.2.1), and the value of n is from 1 to N, where the value of N is limited only by time. Some examples of plots of these functions can be seen in Appendix A21; Fig. A21.2d was obtained by summing 10^6 harmonic waves, and represents a particle of effectively exactly known position but of indeterminate momentum.

....

REFERENCES 7

[1] Lutz M. *Learning Python*, 5th ed. O'Reilly Media, 2013.
[2] Dawson M. *Python Programing*, 3rd ed. Course Technology PTR, 2010.
[3] Swaroop CH. *A Byte of Python*. Free E-book, 2012. http://www.swaroopch.com/notes/python/
[4] Dowling B. *Interfacing Python with Fortran*. http://www.ucs.cam.ac.uk/docs/course-notes/unix-courses/pythonfortran/files/f2py.pdf
[5] Chapman SJ. *Fortran 90/95 for Scientists and Engineers*. WCB/McGraw-Hill, 1998.
[6] Ladd M and Palmer R. *Structure Determination by X-ray Crystallography*, 5th ed. Springer Science+Business Media, NY, 2013.

Problems 7

Although the programs are used extensively throughout the text, the following additional problems are provided for practice and as a check on the correct installation and working of the programs.

7.1. Write a program to calculate the function $y = x^3 - 3x^2$ for $x = -2$ to $+4$ in steps of 0.01 in x. Plot the function with program GRFN and with the mouse pointer over the graph determine the x, y coordinates of (a) the maximum, (b) the minimum. Check the results with those determined from the derivatives of the function.

7.2. How many nodes would be expected in the radial distribution functions for the $5s$, $6p$ and $7d$ atomic orbitals?

7.3. Determine the β-energies, bond orders and free valence indexes for cyclobutadiene, using the program HUCK.

7.4. The diffusion coefficient D of carbon in α-iron as a function of temperature follows the equation $D = D_0 \exp(-U/RT)$. Determine the activation energy U for the diffusion from the following data:

T/K	300	500	700	900	1100
D/m^2 s^{-1}	4.73×10^{-21}	3.35×10^{-15}	1.08×10^{-12}	2.66×10^{-11}	2.05×10^{-10}

Note: If a least-squares fit is done, it is possible to plot both the least-squares line and the real data points.

7.5. Calculate the Madelung constant for sodium chloride with program MADC, given the following data: Face-centred cubic unit cell, with $a = b = c = 0.56402$ nm; take α as 5.

7.6. Determine manually (a) the determinant value $|\mathbf{A}|$, (b) the cofactor matrix $\mathrm{Cof}(\mathbf{A})$ of $\mathbf{A}$ and (c) the inverse matrix $\mathbf{A}^{-1}$, given $\mathbf{A} = \begin{pmatrix} 1 & 0 & 1 \\ 2 & 1 & \bar{1} \\ 0 & 1 & 2 \end{pmatrix}$.

Check the results with the program MATS.

7.7. Use RADL to obtain the $2p$ radial distribution function. Transfer the output file (R3.TXT) to the PYTHON work folder and plot it with GRFN. With the mouse pointer record the r/a_0 value of the maximum. Check the result with the data of Table 2.2.

7.8. The following data approximate to the equation $f(x) = x^2 + x + 1$.

x	y
0.00	1.00
0.50	0.80
1.00	1.00
1.50	1.80

x	y
2.00	3.00
2.50	4.50
3.00	7.00
3.50	9.70
4.00	13.00
4.50	16.70
5.00	21.05
5.50	25.80
6.00	31.00
6.50	36.70
7.00	43.00
7.50	49.80
8.00	57.05
8.50	64.70
9.00	73.05
9.50	81.80
10.00	91.00

Use them with the program QUAD to find the area under the curve $f(x)$ from $x = 0$ to $x = 10$. Compare the result with an analytical value for the same area.

7.9. Determine the roots of the equation $x^3 + 3x^2 - x - 4 = 0$ using program ROOT.

7.10. Use programs WAVE and GRFN to obtain and plot the sum of the first 500 harmonic functions.

7.11. Use program BOXS to calculate ψ_n, $n = 1$–5 Transfer BOX1–BOX5 to the Python folder and plot both the five functions and their sum with GRFN. (Note: BOX6–BOX15 will be output as blank files in this run.)

7.12. Use program BOXS to calculate the wave number for the ground state transition in hexatriene; assume a mean bond length of 0.14 nm.

Hexatriene

A1
Stereoviewing and Stereoviewers

Stereoscopic views, or *stereoviews*, have been used to illustrate the three-dimensional character of objects for more than a century and the technique is now commonplace. Computer programs are available for preparing the views needed for producing the desired three-dimensional images of crystal and molecular structures [1–3].

Two diagrams of an object are needed and they may be prepared by computer, photography or drawing; an angular separation of 6° between the two positions of the object is required. The diagrams are mounted for viewing at approximately 63 mm apart between centres. In order to form a three-dimensional image, each eye should see only the appropriate half of the complete illustration.

The simplest procedure for viewing is with a *stereoviewer*, whereupon the three-dimensional image appears centrally between the stereoscopic pair of diagrams. A supplier of a relatively inexpensive stereoviewer (*ca.* $4) is 3Dstereo.com Inc., 1930 Village Center Circle, #3-333, Las Vegas, NV 89134, USA.

In another technique the unaided eyes can be trained to defocus, with each eye seeing only the appropriate stereoview. The eyes must be relaxed and look straight ahead: it may help to close the eyes for a moment, then open them wide and allow them to relax without consciously focusing on the diagrams. The viewing process may be aided by holding a white card edgeways between the two diagrams.

Finally, a simple stereoviewer can be constructed. A pair of biconvex or plano-convex lenses, each of focal length approximately 100 mm and diameter approximately 30 mm, is mounted between two opaque cards, with the centres of the lenses approximately 63 mm apart, as shown in Fig. A1.1. The card frame is shaped such that the lenses may be brought close to the eyes. It may be helpful to obscure a segment S of each lens, closest to the nose region N, of width approximately 25% of the lens diameter.

....

REFERENCES A1

[1] Oak Ridge National Laboratory, *ORTEP* III, 2000. http://www.chem.gla.ac.uk/~louis/software/ortep3/

[2] National University of Ireland, *OSCAIL*, 2015. http://www.nuigalway.ie/cryst/dload.html

[3] Collaborative Computational Project, CCP14. http://www.ccp14.ac.uk/solution/structuredrawing/

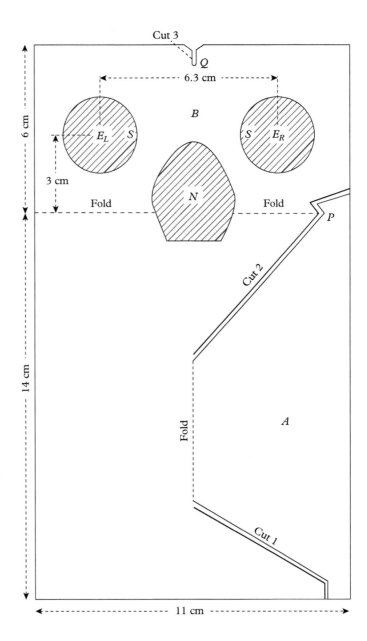

Fig. A1.1 Two pieces of thin card are prepared as shown in the diagram; the shaded portions are then cut out and discarded. Cuts are made along the double lines, marked 'Cut 1', 'Cut 2' and 'Cut 3'. The two cards are glued together with the lenses E_L and E_R set in position. If plano-convex lenses are employed their convex faces should set to lie forwards, in the direction away from the eyes. Fold the portions A and B backwards, along the lines marked 'Fold'. Set the projection P into the 'Cut 3' at Q. Strengthen the folds with Sellotape, if necessary. The stereo diagrams are viewed with the side marked B facing the diagram. [*Crystal Structures: Lattices and Solids in Stereoview*, 1999, Ellis Horwood Limited, UK; reproduced by courtesy of Woodhead Publishing, UK.]

A2
Miller Indices

In studying the *morphology*, or external form, of a crystal, it is convenient to refer it to three non-coplanar reference axes *x*, *y* and *z* that need not be orthogonal but, by convention, form a *right-handed* set (Fig. A2.1). A crystal face that intercepts all three axes is chosen as the *parametral plane* and is allocated integral *Miller indices* (*hkl*), normally (111). Any other face can be described then in terms of its Miller indices, *h*, *k* and *l*, which are generally *small* numbers on most crystals. The Miller indices of a plane specify its orientation uniquely with respect to the reference axes and the parametral plane.

The plane *ABC* in Fig. A2.1 has intercepts on the *x*, *y* and *z* axes at unknown values *a*, *b* and *c*, respectively. This plane is chosen as the parametral plane and is labelled (111). Another plane *LMN* makes intercepts *a/h*, *b/k* and *c/l* along the *x*,

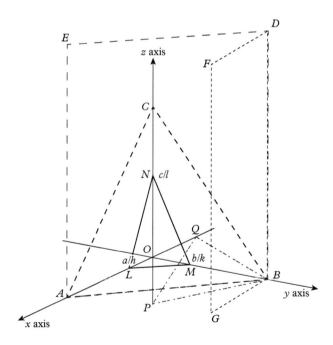

Fig. A2.1 Illustration of Miller indices (*hkl*); *ABC* is the parametral plane (111). The plane *LMN* has been drawn such that $OL = OA/4$, $OM = OB/3$ and $ON = OC/2$; thus, the Miller indices of this plane are (432). The plane in the lattice parallel to *BDFG* but with intercepts 0, *b/2*, 0 is labelled (020); such an identifier would not arise in morphology as the plane would be designated (010). Morphology does not reveal values for *a*, *b* and *c*, only their ratios. [*Structure Determination by X-ray Crystallography*, 5th ed. Springer Science+Business Media, 2013; reproduced by courtesy of Springer Science+Business Media, NY.]

Fig. A2.2 Rational character of Miller indices. (a) Rectangular unit cell of a crystal as seen along the *z*-axis. (b) Conventional crystallographic, right-handed reference axes. (c) Possible crystal shape obtained by the regular stacking of unit cells; the bounding planes are, in anti-clockwise sequence from the bottom line, (100), (110), (010), ($\bar{1}$10), ($\bar{1}$00), ($\bar{1}\bar{1}$0), (0$\bar{1}$0) and (1$\bar{1}$0). The shaded steps are of the order of 10^{-6}–10^{-5} mm and would not be observed on a crystal of macroscopic size. [*Structure Determination by X-ray Crystallography*, 5th ed. 2013; reproduced by courtesy of Springer Science+Business Media, NY.]

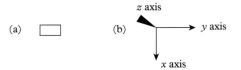

(a)   (b)

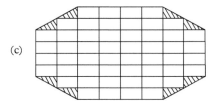

(c)

y and *z* axes; its Miller indices are expressed as the ratios of the intercepts of the parametral plane to those of the plane *LMN*. The diagram has been drawn such that $OL = a/4$, $OM = b/3$ and $ON = c/2$; hence, $h = a/(a/4)$, $k = b/(b/3)$ and $l = c/(c/2)$ whereupon the plane *LMN* is designated (432). If fractions remain in the indices after their evaluation, they are cleared by multiplying throughout by the lowest common denominator. If *LMN* had been chosen as (111), then *ABC* would have had the Miller indices (346). Experience has shown that a correct choice of parametral plane leads to indices of crystal faces that are rarely numerically greater than five.

The plane *ABDE* is parallel to the *z* axis, and its intercept on that axis may be said to be at infinity; hence, the Miller indices are (110). Similarly, *BDFG* is the plane (010). A plane that makes a negative intercept with an axis has a corresponding negative Miller index. Thus, the plane *QPB*, which makes intercepts $-a/2$, *b* and $-c/3$, is designated ($\bar{2}$1$\bar{3}$), which is read as 'bar-2 1 bar-3' ('2-bar 1 3-bar' in the USA).

The rational nature of the Miller indices of crystal planes is illustrated by Fig. A2.2. Crystal faces are the terminations of unit cells, which are the building blocks of crystals. In x-ray crystallography, the symbols *a*, *b* and *c* represent the lengths of the crystal unit cell. Since these lengths are of the order of 0.5–5 nm, the areas of the crystal shown by shading on the diagram will be of a size undetectable to the eye (or the microscope), so that the faces appear plane. Further discussion on Miller indices may be found in the literature [1, 2].

Exercise A1

Given that face *r* is (111), allocate possible Miller indices to the visible faces of the cubic crystal below. (Answer at end of Appendices)

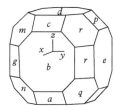

Cubic crystal

....

REFERENCES A2

[1] Ladd M. *Symmetry of Crystals and Molecules*. Oxford University Press, 2014.
[2] Ladd M and Palmer R. *Structure Determination by X-ray Crystallography*, 5th ed. Springer Science+Business Media, 2013.

A3
Bragg Reflection of X-rays from Crystals

A crystal of volume 1 mm^3 will contain approximately 10^{18} unit cells of average dimension 1 nm, stacked in a manner similar to that shown in Fig. A2.2(c). Even if the stacking is not perfect, over the large number of unit cells in the crystal specimen the symmetry of the x-ray diffraction spectra from the crystal will conform statistically to that of its ideal unit cell of that symmetry.

A plane of Miller indices (hkl) is one of a *family* of parallel, equidistant planes of spacing d_{hkl}. Bragg's treatment of x-ray diffraction as though it were reflection from the crystal planes was based on an experimental observation: when a crystal in a diffracting position was rotated through an angle ϕ to the next diffracting position, the direction of the diffracted beam was observed to have been rotated through 2ϕ. The comparable situation with the reflection of visible light from a plane mirror is well known in optics, and is illustrated by Fig. A3.1.

Consider a parallel beam of x-rays incident at a glancing angle, θ_{hkl} (the Bragg angle),[1] on an (hkl) family of planes of interplanar spacing d_{hkl} and reflected from them at an equal angle (Fig. A3.2). The incident x-rays are in phase along the incoming wavefront BQ and again at the outgoing wavefront BS. The path difference δ for reflection from the first two planes is $PD + DR$; hence,

$$\delta = 2d_{hkl} \sin \theta_{hkl} \tag{A3.1}$$

since $\angle PBD = \angle RBD = \theta_{hkl}$. For constructive interference between the reflected rays, the path difference must equal the wavelength λ of the radiation; thus:

$$2d_{hkl} \sin \theta_{hkl} = \lambda \tag{A3.2}$$

[1] The complement of the angle of incidence in geometrical optics.

Fig. A3.1 Reflection of light from a plane mirror in an initial position M_1 and after rotation of the mirror through an angle ϕ to M_2. The reflected beam has turned through the angle 2ϕ; the distinction from Bragg reflection is that the *glancing* angle, shown as θ_{hkl} in Fig. A3.2, must satisfy Eq. (3.2) for x-ray reflection from a crystal.

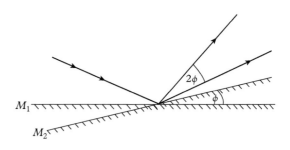

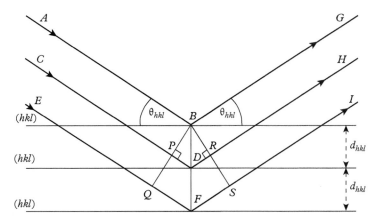

Fig. A3.2 Bragg reflection from crystal planes; three planes of an (*hkl*) family of interplanar spacing d_{hkl} are shown; 'rays' *A*, *C* and *E* represent the incident sinusoidal rays and *G*, *H* and *I* the corresponding reflected rays. [*Structure Determination by X-ray Crystallography*, 5th ed. 2013; reproduced by courtesy of Springer Science+Business Media, NY.]

An obvious extension of the argument shows that if Eq. (A3.2) is satisfied for the first and second planes, it will be satisfied also for the second and third, and so on for the whole (*hkl*) family.

The *Bragg equation* (A3.2) was written originally as

$$2d_{hkl} \sin \theta_{hkl} = n\lambda, \tag{A3.3}$$

where *n* is an integer. However, in x-ray crystallography all families of planes are defined uniquely by their Miller indices, including multiples such as (*nh, nk, nl*) which would be listed morphologically as (*hkl*); d_{020}, for example, is equal to $d_{010}/2$, and in applying Eq. (A3.3) *n* would be equal to 2; conventional practice employs the formulation of Eq. (A3.2).

The values of d_{hkl} can be related to *h*, *k* and *l* through Fig. A3.3 as follows; for simplicity, the reference axes will be assumed to be orthogonal. Let *ABC* be the first plane in the (*hkl*) family from the origin *O*. In the right-angled triangle *ONA*, $ON = d_{hkl} = (a/h) \cos\alpha$, where $\cos\alpha$ is the direction cosine of d_{hkl} with respect to the *x*-axis. Similar equations apply to the intercepts on the *y* and *z* axes *mutatis mutandis*.[2] Hence:

$$(h^2/a^2)d_{hkl}^2 + (k^2/b^2)d_{hkl}^2 + (l^2/c^2)d_{hkl}^2 = \cos^2\alpha + \cos^2\beta + \cos^2\gamma \tag{A3.4}$$

Since the sum of the squares of the direction cosines is unity [1]:

$$1/d_{hkl}^2 = (h^2/a^2) + (k^2/b^2) + (l^2/c^2) \tag{A3.5}$$

In a cubic unit cell $a = b = c$, whereupon Eq. (A3.5) may be formulated as:

$$1/d_{hkl}^2 = (h^2 + k^2 + l^2)/a^2 = N/a^2 \tag{A3.6}$$

[2] 'the necessary changes having been made.'

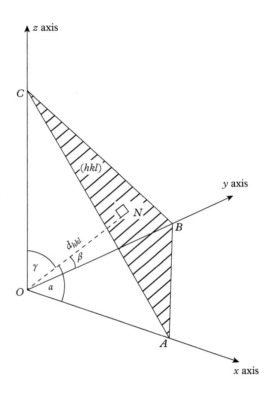

Fig. A3.3 Relationship between the Miller indices (*hkl*) of a crystal plane *ABC* and the interplanar spacing d_{hkl}. Its intercepts at *A*, *B* and *C* are a/h, b/k and c/l, respectively; $\cos\alpha$, $\cos\beta$ and $\cos\gamma$ are the direction cosines of the normal d_{hkl}. For orthogonal axes $1/d_{hkl}^2 = (h^2/a^2) + (k^2/b^2) + (l^2/c^2)$. [*Structure Determination by X-ray Crystallography*, 5th ed. 2013; reproduced by courtesy of Springer Science+Business Media, NY.]

where *N* is the sum $(h^2 + k^2 + l^2)$. From Eqs. (A3.2) and (A3.6) it follows that in the cubic crystal system:

$$\sin\theta_{hkl} = (\lambda/2a)(h^2 + k^2 + l^2)^{1/2} = (\lambda/2a)\sqrt{N} \tag{A3.7}$$

Exercise A2

The unit-cell dimension *a* of a cubic crystal is 0.7500 nm. What is the value of $\sin\theta_{040}$ if the wavelength of x-rays used in a diffraction experiment with this crystal is 0.1542 nm? (Answer at end of Appendices.)

....

REFERENCE A3

[1] Ladd M and Palmer R. *Structure Determination by X-ray Crystallography*, 5th ed. Springer Science+Business Media, 2013.

A4
Boltzmann Distribution from the Hypsometric Formula

Consider a rectangular column of ideal gas of mass m, cross sectional area A and height z from ground level (Fig. A4.1). For an external pressure p and temperature T, the gas law applies:

$$pV = RT \qquad (A4.1)$$

where V is the molar volume. The gas constant R is equal to Lk, where L is the Avogadro constant and k the Boltzmann constant. Since $L/V = N$, the number of molecules in unit volume of gas

$$p = NkT \qquad (A4.2)$$

At a height $z + dz$ the pressure is $p + dp$. The gravitational force on the segment of width dz is $ADg\,dz$, where D is the density of the gas and g is the gravitational acceleration. The pressure difference across the segment is $-\frac{dp}{dz}dz$, the negative sign indicating that the gas pressure decreases with increasing height z.

The hypsometric force on the segment is $-A\frac{dp}{dz}dz$, and at equilibrium the two forces are balanced. Thus:

$$-A\frac{dp}{dz}dz = ADg\,dz \qquad (A4.3)$$

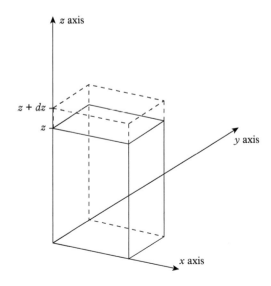

Fig. A4.1 Construction for the hypsometric formula; the x, y plane represents ground level.

or

$$dp = -Dg\,dz \tag{A4.4}$$

From Eq. (A4.2):

$$dp = kT\,dN \tag{A4.5}$$

Since $D = mN$, and combining Eqs. (A4.4) and (A4.5):

$$\frac{dN}{N} = -\frac{mg}{kT}dz \tag{A4.6}$$

Integrating Eq. (A4.6):

$$\ln N = -\frac{mgz}{kT} + \text{constant} \tag{A4.7}$$

At $z = 0$ let $N = N_0$, whereupon the constant becomes N_0, the value of N at $z = 0$. Hence:

$$N = N_0 \exp(-mgz/kT) \tag{A4.8}$$

Now, mgz is the gravitational potential energy per molecule of gas; thus, $Lmgz$ is the potential energy per mole U, which leads to the *Boltzmann distribution equation*:

$$N = N_0 \exp(-U/RT) \tag{A4.9}$$

A more general derivation of this equation may be obtained through statistical mechanics [1].

Example A4.1

The mean molar mass of air is 0.02897 kg mol^{-1}. At ground level ($z = 0$), $N/N_0 = 1$. An average value of the gravitational acceleration may be taken as 9.812 m s^{-2} (London). At 11 km (6.8 mile) above ground level, where T is $-56.5°C$, the value of (U/RT) is $\approx \frac{3 \times 10^3}{1.8 \times 10^3}$, or ≈ 1.7[a], and $\exp(-1.7) = 0.18$. Evaluating the expression precisely, $N/N_0 = 0.169$, that is, the relative number of molecules per unit volume of air at the 11 km level is 0.176 of the value at ground level; in other words, the density of air at 11 km is approximately one sixth of its value at ground level.

[a] The result is dimensionless: $J \equiv kg\ m^2\ s^{-2}$.

REFERENCE A4

[1] McQuarrie DA. *Statistical Mechanics*. University Science Books, 2000.

A5

Mean Classical Thermal Energy and the Equipartition Theorem

A5.1   Mean kinetic energy

Consider a system of particles, each of mass m but with differing values of speed v; the *kinetic energy* of each particle is $mv^2/2$, which may be resolved along mutually perpendicular x, y and z axes. The mean value $<X>$ of any distribution of the form $\phi(X)$ is given by:

$$<X> = \frac{\int X\phi(X)\mathrm{d}X}{\int \phi(X)\mathrm{d}X}, \tag{A5.1}$$

where the integration is taken over the range of the variable. Assuming that the energies of the particles follow a Boltzmann distribution, then from Appendix A4, the mean value for the kinetic energy u_K for a single particle is:

$$<u_K> = <mv^2/2> = \frac{\int_{-\infty}^{\infty}\int_{-\infty}^{\infty}\int_{-\infty}^{\infty} (mv^2/2)\exp(-mv^2/2kT)\,\mathrm{d}v_x\mathrm{d}v_y\mathrm{d}v_z}{\int_{-\infty}^{\infty}\int_{-\infty}^{\infty}\int_{-\infty}^{\infty}\exp(-mv^2/2kT)\,\mathrm{d}v_x\mathrm{d}v_y\mathrm{d}v_z} \tag{A5.2}$$

Since positive and negative directions of v are equally probable, introducing spherical coordinates from Appendix A6 and replacing r of that discussion by v, Eq. (A5.2) may be now written as:

$$<u_K> = \frac{2\int_0^{\infty}(mv^2/2)\exp(-mv^2/2kT)v^2\mathrm{d}v\int_0^{\pi}\sin\theta\,\mathrm{d}\theta\int_0^{2\pi}\mathrm{d}\phi}{2\int_0^{\infty}\exp(-mv^2/2kT)v^2\mathrm{d}v\int_0^{\pi}\sin\theta\,\mathrm{d}\theta\int_0^{2\pi}\mathrm{d}\phi} \tag{A5.3}$$

which simplifies to:

$$<u_K> = \frac{(m/2)\int_0^{\infty}v^4\exp(-mv^2/2kT)\,\mathrm{d}v}{\int_0^{\infty}v^2\exp(-mv^2/2kT)\,\mathrm{d}v} \tag{A5.4}$$

Making the substitution $t = pv^2$, where $p = m/2kT$, and following the argument of Example A7.1 in

Appendix A7, Eq. (A5.4) becomes:

$$< u_K > = \frac{(m/4)p^{-5/2}\,\Gamma(5/2)}{(1/2)p^{-3/2}\Gamma(3/2)} = \frac{(m/2)p^{-1}(3/2)\,\Gamma(3/2)}{(3/2)\,\Gamma(1/2)}$$

$$= \frac{(m/2)p^{-1}(3/2)^2\Gamma(1/2)}{(3/2)\,\Gamma(1/2)} = 3m/4p \tag{A5.5}$$

$$= (3/2)kT$$

and for the system of n particles, the average kinetic energy $< U_K >$ is given by:

$$< U_K > = \frac{3}{2}nkT \tag{A5.6}$$

A5.2   Equipartition theorem

A molecule possesses energy in terms of translation, rotation and vibration. The *equipartition theorem*, which derives from the Boltzmann distribution (Appendix A4), states that for a system of particles at thermal equilibrium under a temperature T, the mean value of each quadratic contribution to the energy of the system is $kT/2$. A quadratic contribution is one that can be expressed as the *square* of a property such as position or velocity.

Translational energy is kinetic and involves three quadratic terms:

$$U_K = (1/2)\,mv_x^2 + (1/2)\,mv_y^2 + (1/2)\,mv_z^2 \tag{A5.7}$$

with each quadratic term in this expression equal to $kT/2$. Hence, the total classical kinetic energy is $(3/2)kT$; it is a mean value, since all directions of translation are equally probable.

A rigid rotor has two perpendicular axes about which it can rotate, each leading to a quadratic kinetic energy contribution; rotations about the axis of the rotor have no effect on the energy. The key term in a rotor is the moment of inertia; it has the form mr^2 in the simplest case and constitutes a quadratic term. In accordance with the equipartition theorem, each such term contributes $kT/2$ to the energy; thus, the total mean rotational energy is kT for two quadratic terms. The vibrational energy of an atomic oscillator involves quadratic terms in both kinetic and potential energy; hence, there are again two quadratic terms and a total mean vibrational energy of kT.

A monatomic species such as argon possesses only energy of translation of total mean value $(3/2)RT$ per mole. A diatomic molecule like oxygen may rotate about two directions perpendicular to each other and to the molecular axis, with each rotation contributing to the energy of the molecule. Rotation about the molecular axis does not contribute to the energy of the molecule which has the total mean value of $(5/2)RT$ per mole, that is, $(3/2)RT$ for translation plus RT for rotation.

A6
Spherical Polar Coordinates

A6.1   Coordinates

In Fig. A6.1, the *spherical polar coordinates* (aka polar coordinates) r, θ and ϕ are given by:

$$\left.\begin{array}{l} x = r\sin\theta\cos\phi \\ y = r\sin\theta\sin\phi \\ z = r\cos\theta \end{array}\right\} \qquad (A6.1)$$

where

$$r^2 = x^2 + y^2 + z^2 \qquad (A6.2)$$

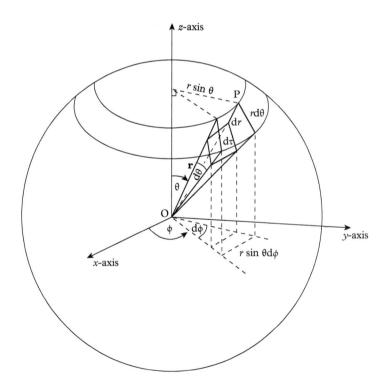

Fig. A6.1 Conversion of Cartesian coordinates to spherical polar coordinates. The distance OP is $|\mathbf{r}|$, where $\mathbf{r}$ is the vector from the origin O to the point $P(x, y, z)$.

A6.2   Volume element

The *volume element* $d\tau$ in Fig. A6.1 corresponds to the infinitesimal quantity $dx\,dy\,dz$ in Cartesian coordinates; it is straightforward to show the relationship:

$$d\tau = r^2 \sin\theta\, dr\, d\theta\, d\phi \qquad (A6.3)$$

The limits of the polar variables that correspond to x, y and z each between $-\infty$ and $+\infty$ are:

$$\left. \begin{aligned} 0 &\le r \le +\infty \\ 0 &\le \theta \le \pi \\ 0 &\le \phi \le 2\pi \end{aligned} \right\} \qquad (A6.4)$$

Exercise A3

Convert the following Cartesian coordinates to spherical polar coordinates. (Answer at end of Appendices.)

$$x = -1 \text{ nm}, \; y = 1 \text{ nm}, \; z = -\left(\sqrt{2}/2\right) \text{ nm}.$$

A7
Gamma Function $\Gamma(n)$

The *gamma function* [1, 2] is useful in handling integrals of the type:

$$\int_0^\infty x^n \exp\left(-ax^2\right) dx \qquad (A7.1)$$

where a is a constant and n is a positive number. The integrals occur in studying *inter alia* quantum chemistry and atomic scattering factors. The gamma function may be defined by the equation:

$$\Gamma(n) = \int_0^\infty x^{n-1} \exp\left(-ax^2\right) dx \qquad (A7.2)$$

The following results are useful.

1. For $n > 0$ and integral:

$$\Gamma(n) = (n-1)! \qquad (A7.3)$$

2. For $n > 0$ and not necessarily integral:

$$\Gamma(n+1) = n\Gamma(n) \qquad (A7.4)$$

   but if n is also integral, then:

$$\Gamma(n+1) = n! \text{ (which follows also from 1)} \qquad (A7.5)$$

3. $$\Gamma(1/2) = \sqrt{\pi} \qquad (A7.6)$$

Example A7.1

Consider the solution of the integral:

$$\mathcal{G} = \int_0^\infty x^4 \exp\left(-x^2/2\right) dx$$

Let $x^2/2 = t$, so that $x = (2/t)^{1/2}$ and $dx = (2t)^{-1/2}dt$. Then:

$$\mathcal{G} = 2\sqrt{2} \int_0^\infty t^{3/2} \exp\left(-t\right) dt = 2\sqrt{2}\,\Gamma\,(5/2) = 3\sqrt{\pi/2}$$

The reduction formula can also be useful:

$$\int x^n \exp(ax)\, dx = \frac{x^n \exp(ax)}{a} - \frac{n}{a} \int x^{n-1} \exp(ax) dx \qquad (A7.7)$$

....

REFERENCES A7

[1] Margenau H and Murphy GM. *The Mathematics of Physics and Chemistry.* van Nostrand, 1943.
[2] eFunda Inc. Γ–function finder, 2015. http://www.efunda.com/math/gamma/findgamma.cfm

A8
Reduced Mass

The *reduced mass* is an effective inertial mass of a system, considered here as a two-particle problem in classical mechanics. Let the particles of masses m_1 and m_2 be at position vectors $\mathbf{r}_1$ and $\mathbf{r}_2$ with respect to an origin at the centre of mass C; r is the linear distance between m_1 and m_2 (Fig. A8.1).

Then:

$$r = r_1 + r_2 \tag{A8.1}$$

and

$$m_1 r_1 = m_2 r_2 \tag{A8.2}$$

Now:

$$r_2 = r - r_1 \tag{A8.3}$$

Hence:

$$m_1 r_1 = m_2 (r - r_1) \tag{A8.4}$$

so that

$$r_1 = m_2 r / (m_1 + m_2) \tag{A8.5}$$

and

$$r_2 = m_1 r / (m_1 + m_2) \tag{A8.6}$$

The moment of inertia I of the system is given by

$$I = m_1 r_1^2 + m_2 r_2^2 = m_1 \left(\frac{m_2 r}{m_1 + m_2} \right)^2 + m_2 \left(\frac{m_1 r}{m_1 + m_2} \right)^2$$

$$= \frac{m_1 m_2^2 r^2}{(m_1 + m_2)^2} + \frac{m_1^2 m_2 r^2}{(m_1 + m_2)^2} = \frac{m_1 m_2 (m_1 + m_2) r^2}{(m_1 + m_2)^2} \tag{A8.7}$$

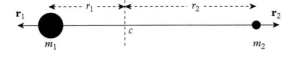

Fig. A8.1 Two-particle system of masses m_1 and m_2 vibrating about the centre of mass C.

$$I = \frac{m_1 m_2 r^2}{(m_1 + m_2)} = \mu r^2 \qquad \text{(A8.8)}$$

where μ is the *reduced mass* of the system:

$$\mu = \frac{m_1 m_2}{(m_1 + m_2)} \qquad \text{(A8.9)}$$

A similar expression arises for the moment of inertia of a diatomic molecule, in which case m_1 and m_2 are the masses of the two atoms and r is the length of the bond between them.

A9
Solution of a Differential Equation

Let the differential equation be:

$$\frac{d^2y}{dx^2} + \kappa^2 y = 0 \qquad\qquad\text{(A9.1)}$$

where κ is a constant. It may be recast as:

$$D^2 y + \kappa^2 y = 0 \qquad\qquad\text{(A9.2)}$$

where $D = \frac{dy}{dx}$. Consider next the equation:

$$[(D - p_1)(D - p_2)]y = 0 \qquad\qquad\text{(A9.3)}$$

Expanding Eq. (A9.3):

$$D^2 y - (p_1 + p_2)Dy + p_1 p_2 y = 0 \qquad\qquad\text{(A9.4)}$$

Comparing Eqs. (A9.2) and (A9.4), it is evident that they become comparable if $p_2 = -p_1$. Hence:

$$p_1^2 = -\kappa^2 \qquad\qquad\text{(A9.5)}$$

so that

$$p_1 = \pm i\kappa \qquad\qquad\text{(A9.6)}$$

Considering the terms in Eq. (A9.3):

$$(D - p_1)y = 0 \qquad\qquad\text{(A9.7)}$$

or

$$dy/y = p_1\,dx \qquad\qquad\text{(A9.8)}$$

Integrating:

$$\ln y = p_1 x + K \qquad\qquad\text{(A9.9)}$$

Setting the constant K equal to $\ln A$ and using Eq. (A9.6):

$$y = A\exp(i\kappa x) \tag{A9.10}$$

Continuing in a similar manner from Eq. (A9.3), another solution is

$$y = B\exp(-i\kappa x) \tag{A9.11}$$

where B is another constant; then, the complete solution is

$$y = A\exp(i\kappa x) + B\exp(-i\kappa x) \tag{A9.12}$$

Double differentiation to give $\frac{d^2y}{dx^2}$ confirms the correctness of the solution *vide* Eq. (A9.1).

A10
Determinant Expansion, Matrices and Matrix Properties and Hückel Molecular-Orbital Coefficients

A10.1 Expansion of a determinant

Consider the determinant:

$$D = \begin{vmatrix} y & 1 & 0 & 0 \\ 1 & y & 1 & 0 \\ 0 & 1 & y & 1 \\ 0 & 0 & 1 & y \end{vmatrix} \tag{A10.1}$$

The *value D* of the determinant is obtained by forming the sum of the product of each element in a given column (or row) with its *cofactor* (see Section A10.2.9). The cofactor F_{ij} of the ij element in D is the determinant of one order lower, obtained by striking out the ith row and jth column of D and multiplying by $(-1)^{(i+j)}$. Thus from Eq. (A10.1), using the first row $i = 1$ (there is a manual practical advantage in choosing a row or column with some zero elements):

$$D = a_{11}F_{11} - a_{12}F_{12} + a_{13}F_{13} - a_{14}F_{14}$$

The terms $F_{1,j}$ above are *minors* (third order determinants in this example) of D. Expanding:

$$D = y \begin{vmatrix} y & 1 & 0 \\ 1 & y & 1 \\ 0 & 1 & y \end{vmatrix} + (-1) \begin{vmatrix} 1 & 1 & 0 \\ 0 & y & 1 \\ 0 & 1 & y \end{vmatrix} + 0 + 0$$

$$= \left((y^2) \begin{vmatrix} y & 1 \\ 1 & y \end{vmatrix} + (-y) \begin{vmatrix} 1 & 1 \\ 0 & y \end{vmatrix} \right) - (1) \left(\begin{vmatrix} y & 1 \\ 1 & y \end{vmatrix} - (-1) \begin{vmatrix} 0 & 1 \\ 0 & y \end{vmatrix} \right)$$

$$= y^2 (y^2 - 1) - y^2 - y^2 + 1 = y^4 - 3y^2 + 1 \tag{A10.2}$$

Some useful properties of determinants are:

- A matrix and its transpose (Section A10.2.6) have the same determinant value.
- Interchanging two columns of a determinant multiplies its value by −1.
- If any row or column is zero, the determinant value is zero.

A10.2   Matrices

A matrix is an ordered rectangular array of numbers or variables. The matrix may be symbolized by $\mathbf{A}$ and its size as $m \times n$. Its elements a_{ij} run from $i = 1$ to m down the columns and from $j = 1$ to n along the rows.

A10.2.1   General matrix

Thus, in a 4×3 matrix $\mathbf{A}$:

$$\mathbf{A} = \begin{pmatrix} a_{11} & a_{12} & a_{13} \\ a_{21} & a_{22} & a_{23} \\ a_{31} & a_{32} & a_{33} \\ a_{41} & a_{42} & a_{43} \end{pmatrix} \qquad (A10.3)$$

where i runs from 1 to 4 and j runs from 1 to 3. In most cases, the matrices of interest herein will be 3×3 size matrices; a matrix in which $m = n$ is a *square* matrix.

A10.2.2   Row matrix

A row matrix consists of a single line array, dimensions $1 \times n$. The array

$$\mathbf{R} = (a_{11}\ a_{12}\ a_{13}) \qquad (A10.4)$$

is a row matrix (aka row vector) $\mathbf{R}$ of size 1×3.

A10.2.3   Column matrix

The matrix

$$\mathbf{L} = \begin{pmatrix} a_{11} \\ a_{21} \\ a_{31} \\ a_{41} \end{pmatrix} \qquad (A10.5)$$

is a column matrix (aka column vector) $\mathbf{L}$ of dimension 4×1.

A10.2.4   Symmetric, skew-symmetric, equal and identity matrices

A *symmetric* matrix $\mathbf{S}$ has elements $s_{ij} = s_{ji}$ for all i and j; otherwise it is *skew-symmetric*; the matrices $\mathbf{A}$ and $\mathbf{B}$ below are both skew-symmetric. Note that a symmetric matrix is equal to its transpose, that is $\mathbf{S} = \mathbf{S}^T$ (see Section A10.2.6). *Equal* matrices have both the same dimensions and equal corresponding elements; they need not be square. The *identity* matrix $\mathbf{E}$ has diagonal elements of unity and zero elements otherwise. Examples of skew-symmetric matrices and the identity matrix are:

$$\mathbf{A} = \begin{pmatrix} 1 & 0 & 1 \\ 0 & 1 & 2 \\ \overline{2} & 1 & 1 \end{pmatrix} \quad \mathbf{B} = \begin{pmatrix} 1 & 1 & 2 \\ 0 & \overline{1} & 0 \\ 1 & 0 & 0 \end{pmatrix} \quad \mathbf{E} = \begin{pmatrix} 1 & 0 & 0 \\ 0 & 1 & 0 \\ 0 & 0 & 1 \end{pmatrix} \quad \text{(A10.6)}$$

For neatness, the negative sign of an element is placed above the digit to which it refers, as with Miller indices.

A10.2.5 Addition and subtraction of matrices

Two matrices $\mathbf{A}$ and $\mathbf{B}$ may be added or subtracted if $\mathbf{A}$ has both the same number of columns and same number of rows as $\mathbf{B}$. Thus:

$$\underbrace{\begin{pmatrix} 1 & 0 & 1 \\ 0 & 1 & 2 \\ \overline{2} & 1 & 1 \end{pmatrix}}_{\mathbf{A}} + \underbrace{\begin{pmatrix} 1 & 1 & 2 \\ 0 & \overline{1} & 0 \\ 1 & 0 & 0 \end{pmatrix}}_{\mathbf{B}} = \underbrace{\begin{pmatrix} 2 & 1 & 3 \\ 0 & 0 & 2 \\ \overline{1} & 1 & 1 \end{pmatrix}}_{\mathbf{A}+\mathbf{B}} \quad \text{and} \quad \underbrace{\begin{pmatrix} 1 & 0 & 1 \\ 0 & 1 & 2 \\ \overline{2} & 1 & 1 \end{pmatrix}}_{\mathbf{A}} - \underbrace{\begin{pmatrix} 1 & 1 & 2 \\ 0 & \overline{1} & 0 \\ 1 & 0 & 0 \end{pmatrix}}_{\mathbf{B}} = \underbrace{\begin{pmatrix} 0 & \overline{1} & \overline{1} \\ 0 & 2 & 2 \\ \overline{3} & 1 & 1 \end{pmatrix}}_{\mathbf{A}-\mathbf{B}}$$

A10.2.6 Transposition

A transposed matrix has a_{ij} replaced by a_{ji} for all i and j. Thus, if a matrix $\mathbf{A}$ is

$$\mathbf{A} = \begin{pmatrix} 1 & 0 & 1 \\ 0 & 1 & 2 \\ \overline{2} & 1 & 1 \end{pmatrix} \quad \text{(A10.7)}$$

its transpose $\mathbf{A}^T$ is given by

$$\mathbf{A}^T = \begin{pmatrix} 1 & 0 & \overline{2} \\ 0 & 1 & 1 \\ 1 & 2 & 1 \end{pmatrix} \quad \text{(A10.8)}$$

A10.2.7 Multiplication with matrices

If a matrix is multiplied by a scalar quantity s, then all elements a_{ij} of the matrix are transformed to sa_{ij}. Two matrices $\mathbf{A}$ and $\mathbf{B}$ can be multiplied if and only if $\mathbf{A}$ has the dimensions $m \times p$ and $\mathbf{B}$ has the dimensions $p \times n$, so that their product $\mathbf{C}$ has the dimensions $m \times n$. The element $c_{i,j}$ of $\mathbf{C}$ is given by multiplying the ith row of $\mathbf{A}$ and the jth column of $\mathbf{B}$ element by element. In general:

$$c_{ij} = \sum_{k=1}^{p} A_{ik} B_{kj} \quad \text{(A10.9)}$$

For a 3×3 matrix, $c_{12} = a_{11}b_{12} + a_{12}b_{21} + a_{13}b_{23}$. Thus, if $c_{i,j}$ (also written as c_{12}) is formed from matrices $\mathbf{A}$ and $\mathbf{B}$ in Eq. (A10.6), $c_{12} = (1 \times 1) + (0 \times \overline{1}) + (1 \times 0) = 1$. Note that the position of c_{ij} in $\mathbf{C}$ is the junction of the ith row and jth column.

The multiplication of $\mathbf{A}$ and $\mathbf{B}$ from Eq. (A10.6) to give $\mathbf{C}$ is written concisely as:

$$\mathbf{C} = \mathbf{A}\,\mathbf{B} \tag{A10.10}$$

It is important to note that, in general, $\mathbf{BA} \neq \mathbf{AB}$. Thus, using the matrices from Eq. (A10.6):

$$\mathbf{A}\,\mathbf{B} = \begin{pmatrix} 2 & 1 & 2 \\ 2 & \overline{1} & 0 \\ \overline{1} & 3 & \overline{4} \end{pmatrix} \qquad \mathbf{B}\,\mathbf{A} = \begin{pmatrix} \overline{3} & 3 & 5 \\ 0 & \overline{1} & 2 \\ 1 & 0 & 1 \end{pmatrix} \tag{A10.11}$$

A10.2.8 Multiplicative properties of matrices

The following properties of matrices are useful from time to time; the list is not exhaustive:

$$\left.\begin{array}{l} \mathbf{A}(\mathbf{BC}) = (\mathbf{AB})\mathbf{C} \\[4pt] (\mathbf{A} + \mathbf{B})\mathbf{C} = \mathbf{AC} + \mathbf{BC} \\[4pt] s(\mathbf{A} + \mathbf{B}) = s\mathbf{A} + s\mathbf{B} \\[4pt] \text{where } s \text{ is a scalar constant} \\[4pt] (\mathbf{A} + \mathbf{B})^{\mathrm{T}} = \mathbf{A}^{\mathrm{T}} + \mathbf{B}^{\mathrm{T}} \\[4pt] (\mathbf{AB})^{\mathrm{T}} = \mathbf{B}^{\mathrm{T}} + \mathbf{A}^{\mathrm{T}} \end{array}\right\} \tag{A10.12}$$

A10.2.9 Inverse of a matrix

The inverse of a *square* matrix may be obtained through its *cofactor* matrix. Using matrix $\mathbf{A}$ from Eq. (A10.6) as an example, the *cofactor* matrix $\mathbf{F}$, is obtained with elements f_{ij} given by:

$$f_{ij} = (-1)^{i+j} M_{ij} \tag{A10.13}$$

where M_{ij} is the *minor* determinant of $\mathbf{A}$, obtained by striking out the row and column containing the ij element; thus:

$$f_{12} = - \begin{vmatrix} 0 & 2 \\ \overline{2} & 1 \end{vmatrix} = 4$$

Proceeding in this manner:

$$\mathbf{F} = \begin{pmatrix} \overline{1} & \overline{4} & 2 \\ 1 & 3 & \overline{1} \\ \overline{1} & \overline{2} & 1 \end{pmatrix} \tag{A10.14}$$

The transpose of the cofactor matrix is the *adjoint* matrix $\mathbf{A}^\dagger$, where $a_{ij}^\dagger = f_{ji}$ but there is no need to use it explicitly here. Finally, the inverse of $\mathbf{A}$, written as $\mathbf{A}^{-1}$, is given by

$$\mathbf{A}^{-1} = \frac{1}{\det(\mathbf{A})}\mathbf{D}^{\mathrm{T}} = \frac{1}{1}\begin{pmatrix} \bar{1} & 1 & \bar{1} \\ 4 & 3 & \bar{2} \\ 2 & \bar{1} & 1 \end{pmatrix} = \begin{pmatrix} \bar{1} & 1 & \bar{1} \\ 4 & 3 & \bar{2} \\ 2 & \bar{1} & 1 \end{pmatrix} \tag{A10.15}$$

A simpler, equivalent procedure is to form the elements of $\mathbf{A}^{-1}$ directly as

$$a_{ij}^{-1} = (-1)^{i+j} M_{ji} \tag{A10.16}$$

where M_{ji} is the minor determinant formed by deleting now the row and column containing the ji element of the matrix $\mathbf{A}$. Thus, the a_{23}^{-1} element of the inverse matrix is given by Eq. (A10.13), but with M_{ji} obtained by striking out the row and column of the a_{32} element $[\det(\mathbf{A}) = 1]$:

$$a_{23}^{-1} = \frac{1}{\det(\mathbf{A})}(-1)^5 \begin{vmatrix} 1 & 1 \\ 0 & 2 \end{vmatrix} = -2$$

A10.2.10   Orthogonal and unitary matrices

A matrix that fulfils the condition

$$\mathbf{A}^{-1} = \mathbf{A}^{\mathrm{T}} \tag{A10.17}$$

is an *orthogonal* matrix. All orthogonal matrices are square but not necessarily symmetrical, as in the following example:

$$\mathbf{A} = \begin{pmatrix} \cos\theta & \overline{\sin\theta} & 0 \\ \sin\theta & \cos\theta & 0 \\ 0 & 0 & 1 \end{pmatrix} \quad \mathbf{A}^{-1} = \begin{pmatrix} \cos\theta & \sin\theta & 0 \\ \overline{\sin\theta} & \cos\theta & 0 \\ 0 & 0 & 1 \end{pmatrix} \quad \mathbf{A}^{\mathrm{T}} = \begin{pmatrix} \cos\theta & \sin\theta & 0 \\ \overline{\sin\theta} & \cos\theta & 0 \\ 0 & 0 & 1 \end{pmatrix}$$

A matrix $\mathbf{A}$ is *unitary* if its adjoint is equal to its inverse:

$$\mathbf{A}^\dagger = \mathbf{A}^{-1}$$

that is, $a_{ij}^\dagger = a_{ij}^{-1}$ for all i and j: matrix $\mathbf{A}$ below is unitary:

$$\mathbf{A} = \begin{pmatrix} 1 & 0 & 0 \\ 0 & 0 & \mathrm{e}^{\mathrm{i}} \\ 0 & \mathrm{e}^{-2\mathrm{i}} & 0 \end{pmatrix} \quad \mathbf{A}^\dagger = \begin{pmatrix} 1 & 0 & 0 \\ 0 & 0 & \mathrm{e}^{2\mathrm{i}} \\ 0 & \mathrm{e}^{-\mathrm{i}} & 0 \end{pmatrix} \quad \mathbf{A}^{-1} = \begin{pmatrix} 1 & 0 & 0 \\ 0 & 0 & \mathrm{e}^{2\mathrm{i}} \\ 0 & \mathrm{e}^{-\mathrm{i}} & 0 \end{pmatrix}$$

If $\mathbf{A}$ is a complex matrix, the adjoint matrix $\mathbf{A}^\dagger$ is complex conjugate of the transpose, $(\mathbf{A}^{\mathrm{T}})^*$, as can be seen in the matrices above.

A10.2.11 Matrices, rows and columns

In certain transformations, row and column matrices are involved. Thus, the multiplication of a row $\mathbf{x}$ by a matrix $\mathbf{A}$ with the result $\mathbf{x}'$ would be written as:

$$\mathbf{xA} = \mathbf{x}'$$

For example:

$$\underset{\mathbf{x}}{(1\ 2\ 3)} \underset{\mathbf{A}}{\begin{pmatrix} 1 & 0 & 1 \\ 0 & 2 & 1 \\ 1 & 3 & \bar{1} \end{pmatrix}} = \underset{\mathbf{x}'}{(4\ 13\ 0)}$$

whereas if $\mathbf{x}$ is a column:

$$\underset{\mathbf{A}}{\begin{pmatrix} 1 & 0 & 1 \\ 0 & 2 & 1 \\ 1 & 3 & \bar{1} \end{pmatrix}} \underset{\mathbf{x}}{\begin{pmatrix} 1 \\ 2 \\ 3 \end{pmatrix}} = \underset{\mathbf{x}''}{\begin{pmatrix} 4 \\ 7 \\ 4 \end{pmatrix}}$$

Note also that:

(i) $\mathbf{A}^{\mathrm{T}}\mathbf{x} = \mathbf{x}'$

(ii) $(\mathbf{A}^{\mathrm{T}})^{-1} = (\mathbf{A}^{-1})^{\mathrm{T}}$

A10.3 Hückel molecular orbital wavefunction coefficients

The *secular equations* from which Eq. (A10.1) was derived (see Problem 2.20) were:

$$\begin{aligned} c_1(\alpha - E) + c_2\beta &= & 0 \\ c_1\beta + c_2(\alpha - E) + c_3\beta &= & 0 \\ c_2\beta + c_3(\alpha - E) + c_4\beta &= 0 \\ c_3\beta + c_4(\alpha - E) &= 0 \end{aligned} \tag{A10.18}$$

One of the determined values of E is chosen, say, $\alpha + 1.618\beta$, and substituted into the secular equations, the left-hand side of which forms the *secular determinant D*:

$$\begin{aligned} -1.618c_{1,1} + c_{1,2} &= 0 \\ c_{1,1} - 1.618c_{1,2} + c_{1,3} &= 0 \\ c_{1,2} - 1.618c_{1,3} + c_{1,4} &= 0 \\ c_{1,3} - 1.618c_{1,4} &= 0 \end{aligned} \tag{A10.19}$$

These equations may be solved by back substitution which can be tedious, and a more general procedure will be now described.

Choose a row, say, row 1 and calculate the *apparent* cofactors A_{ij}:

$$A_{ij} = (-1)^{i+j} M_{ij} \tag{A10.20}$$

where M_{ij} is the *minor determinant*, obtained by striking out the ith row and jth column of D. Thus, referring to Eq. (A10.1):

$$A_{11} = (-1)^{1+1} \begin{vmatrix} y & 1 & 0 \\ 1 & y & 1 \\ 0 & 1 & y \end{vmatrix} = y(y^2 - 1)$$

$$\tag{A10.21}$$

$$A_{12} = (-1)^{1+2} \begin{vmatrix} 1 & 1 & 0 \\ 0 & y & 1 \\ 0 & 1 & y \end{vmatrix} = 1 - y^2$$

Similarly, $A_{13} = y$ and $A_{14} = 1$. The *true* cofactors are now obtained by multiplying by the roots of D. Taking $y_1 = -1.618$, then: $C_{11} = -1.000$, $C_{12} = -1.618$, $C_{13} = -1.618$, $C_{14} = -1$. Proceeding in this manner and tabulating the results, first normalizing by division with $\sqrt{\sum_j (C_{ij})^2}$, which is 2.690 in this example:

ij	C_{ij}	c_{ij}
1 1	1.000	0.3717
1 2	1.618	0.6015
1 3	1.618	0.6015
1 4	1.000	0.3717

For convenience, the values of C_{ij} have been made positive, a trivial alteration influencing only the orientation of the π system in space.

A11

Electrostatics and Volume of Molecules in a Gas

A11.1 Law of force

The force F between two charges q_1 and q_2 at a distance d apart *in vacuo* is given by:

$$F = \frac{q_1 q_2}{4\pi \varepsilon_0 r^2} \tag{A11.1}$$

where ε_0 is the *electric constant* (aka the *permittivity of a vacuum*). If q_1 and q_2 include magnitude and sign, then a negative value for F indicates a force of attraction. Experimentally, it has been shown that the deviation of the exponent of r from 2 is less than one part in 10^{15}, but the law cannot be proved *a priori*. If q is in the units C and r is in m, then the units of F are J m^{-1}, or N. In order to show the vector nature of force, Eq. (A11.1) may be written as:

$$\mathbf{F} = \frac{q_1 q_2 \mathbf{r}}{4\pi \varepsilon_0 r^3} \tag{A11.2}$$

A11.2 Coulomb's law

If a charge q_1 is brought from infinity to a distance r from another charge q_2, the potential energy V *in vacuo* is the work done in that process, and may be thought of as $V_r - V_\infty = V_r$. Since energy is force $\times$ distance, it follows from Eq. (A11.1) that:

$$V = \frac{q_1 q_2}{4\pi \varepsilon_0 r} \tag{A11.3}$$

which is the *Coulomb potential energy* of the system of the two charges separated by the distance r. A force F is the negative gradient of the potential, that is, $F = -dV/dr$; thus, from Eq. (A11.3):

$$F = -\frac{dV}{dr} = -\frac{d}{dr}\left(\frac{q_1 q_2}{4\pi \varepsilon_0 r}\right) = \frac{q_1 q_2}{4\pi \varepsilon_0 r^2} \tag{A11.4}$$

which conforms to Eq. (A11.1); for an attractive force, one of q_1 and q_2 must be of negative sign.

Exercise A4

Two point charges of values 2×10^{10} and -3×10^{10} are separated by a distance of 1×10^{-2}m. Calculate (a) the force that the charges exert on each other, and (b) their Coulomb potential energy. (Answers at end of Appendices.)

A11.3 Permittivity

If two charges q_1 and q_2 are surrounded by any medium, their potential energy is reduced from that *in vacuo* to:

$$V = \frac{q_1 q_2}{4\pi \varepsilon r} \qquad\qquad (A11.5)$$

where ε is the *permittivity* of the medium. It is convenient to work with a dimensionless, relative permittivity ε_r (aka *dielectric constant*):

$$\varepsilon_r = \varepsilon/\varepsilon_0 \qquad\qquad (A11.6)$$

The permittivity has an important effect on the behaviour of electrolytes. At 25°C, the relative permittivity of water is 78 and that of methanol 33; the molar conductivity at infinite dilution Λ_0 for manganese chloride is 2.05×10^4 S m^{-1} mol^{-1} in water,[3] and 0.95×10^4 S m$^-$ mol^{-1} in methanol. In non-conducting materials, the permittivity has the effect of decreasing the electric field acting on a medium; relative permittivity is large for polar or highly polarizable species.

A11.3.1 Measurement of relative permittivity

Relative permittivity may be determined by comparing the capacitance of a capacitor filled with a dielectric with that of the same capacitor in the absence of the dielectric and *in vacuo*; the measurement is carried out by a heterodyne-beat technique. Two similar oscillators of frequency approximately 10^6 Hz are coupled to an amplifier and an output measuring device. One oscillator has a fixed frequency, while the other is coupled to a standard variable capacitor connected in parallel with the experimental cell. By adjusting the variable capacitor, the frequency of the variable oscillator can be brought into coincidence with that of the fixed oscillator. The beat frequency, equal to the difference in frequency of the two oscillators, is detectable to better than 1 Hz, resulting in a precision of at least 1 in 10^6.

Data from the standard capacitor are collected with the experimental cell first *in vacuo* and then containing the dielectric under examination; the ratio of the two capacitances is the relative permittivity of the substance. In practice, corrections must be applied for the capacity of the connections, and for end-effects of the capacitor where the applied field is non-uniform. A calibration technique may be employed with carbon dioxide or benzene as a standard substance.

[3] 205 in the units ohm^{-1} cm^{-1} mol^{-1}.

A11.4   Potential and field of a charge

A force **F** acting on a test charge q, without indicating the source of the charge responsible for the force, implies the existence of a field **E** at q, where

$$\mathbf{E} = \mathbf{F}/q \tag{A11.7}$$

In order that q should not itself perturb the field, Eq. (A11.7) may be recast as:

$$\mathbf{E} = \lim_{q \to 0} \mathbf{F}/q \tag{A11.8}$$

but in practice the formulation of Eq. (A11.7) will generally suffice; it follows that:

$$\mathbf{E} = \frac{q\mathbf{r}}{4\pi\varepsilon_0 r^3} \tag{A11.9}$$

Let a unit test charge be moved from A to B under the influence of a charge q at O (Fig. A11.1). In moving the infinitesimal distance ds from P to Q, the field exerts a force **E** and does work **E**ds:

$$\mathbf{E}\mathrm{d}s = \frac{q\cos\theta\,\mathrm{d}s}{4\pi\varepsilon_0 r^2} = \frac{q\,\mathrm{d}r}{4\pi\varepsilon_0 r^2} \tag{A11.10}$$

The total work done V by the field in moving the unit test charge from A to B is:

$$V = \int_A^B \frac{q\,\mathrm{d}r}{4\pi\varepsilon_0 r^2} = \frac{q}{4\pi\varepsilon_0 r_B} - \frac{q}{4\pi\varepsilon_0 r_A} \tag{A11.11}$$

The terms on the extreme right-hand side of Eq. (A11.11) define the *difference in electric potential* between A and B due to the charge q. The work done V' in moving the test charge is thus:

$$V' = -\left\{ \frac{q}{4\pi\varepsilon_0 r_B} - \frac{q}{4\pi\varepsilon_0 r_A} \right\} \tag{A11.12}$$

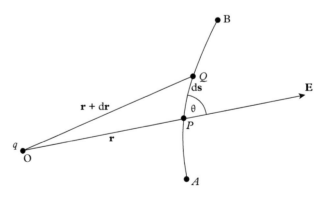

Fig. A11.1 Electrostatic field **E**, with a point P at a vector distance **r** from a charge q at the origin O.

If A is at infinity, the potential V at a point r distant from a charge q is the work done in bringing a unit charge from infinite distance to that point:

$$V = \frac{q}{4\pi\varepsilon_0 r} \tag{A11.13}$$

The difference in potential between the infinitesimal distance from P to Q is:

$$dV = V_Q - V_P = -\text{work done by field } (-\mathbf{E}\,d\mathbf{s}) \tag{A11.14}$$

With respect to the distance AB, the potential difference is:

$$V_B - V_A = -\int_A^B \mathbf{E}\,d\mathbf{s} \tag{A11.15}$$

The vector $\mathbf{E}$ is resolved into Cartesian components along the x, y and z axes, whence

$$dV = -(E_x dx + E_y dy + E_z dz) \tag{A11.16}$$

and

$$E_x = -\left(\frac{\partial V}{\partial x}\right) \quad E_y = -\left(\frac{\partial V}{\partial y}\right) \quad E_z = -\left(\frac{\partial V}{\partial z}\right) \tag{A11.17}$$

or concisely

$$\mathbf{E} = \mathbf{grad}(V) \tag{A11.18}$$

where the operator $\mathbf{grad}$ is given by:

$$\mathbf{grad} = \mathbf{i}\frac{\partial}{\partial x} + \mathbf{j}\frac{\partial}{\partial y} + \mathbf{k}\frac{\partial}{\partial z} \tag{A11.19}$$

and $\mathbf{i}$, $\mathbf{j}$ and $\mathbf{k}$ are unit vectors along the x, y and z Cartesian reference axes. From Eq. (A11.13),

$$\mathbf{grad}(V) = \frac{q}{4\pi\varepsilon_0}\mathbf{grad}\left(\frac{1}{r}\right) = \frac{q}{4\pi\varepsilon_0}\left(-\frac{1}{r^2}\right)\mathbf{grad}(r). \tag{A11.20}$$

Since $r^2 = x^2 + y^2 + z^2$, and using Eq. (A11.19):

$$\mathbf{grad}(r) = \left(\frac{1}{r}\right)(\mathbf{i}x + \mathbf{j}y + \mathbf{k}z) = \frac{\mathbf{r}}{r} \tag{A11.21}$$

whence

$$\mathbf{grad}(V) = -\frac{q\mathbf{r}}{4\pi\varepsilon_0 r^3} = -\mathbf{E} \tag{A11.22}$$

Exercise A5

A fly accumulates 3.5×10^{-10}C of positive charge while flying through the air. Calculate the magnitude and direction of the electric field at a location 2.5 cm away from the fly?

A11.5   Potential of an ideal dipole

The dipole represented in Fig. A11.2 consists of charges $\pm q$ separated by a distance a, and the positive direction of the dipole vector μ is indicated here by the arrowhead (in chemistry, the opposite direction is often used).

The potential V at the point P is given by:

$$V = \frac{1}{4\pi\varepsilon_0} \left(\frac{q}{R} - \frac{q}{r} \right) \tag{A11.23}$$

which is $(q/4\pi\varepsilon_0)$ multiplied by the change in $(1/r)$ from Q to S; in other words:

$$V = \left(\frac{qa}{4\pi\varepsilon_0} \right) \frac{\partial}{\partial s} \left(\frac{1}{r} \right) \tag{A11.24}$$

where $\partial/\partial s$ denotes differentiation at Q in the direction QS in three-dimensional space. From the foregoing, and since qa is the dipole moment magnitude μ, it follows that

$$V = \left(\frac{1}{4\pi\varepsilon_0} \right) \mu \cdot \mathbf{grad}_Q \left(\frac{1}{r} \right) = -\left(\frac{\mu \cdot \mathbf{r}}{4\pi\varepsilon_0 r^3} \right) \tag{A11.25}$$

Generally, the gradient at P is needed, the point where V is to be calculated. Since P and Q lie on one and the same vector, $\mathbf{grad}_Q(1/r) = -\mathbf{grad}_P(1/r)$, whence:

$$V_P = \frac{\mu \cdot \mathbf{r}}{4\pi\varepsilon_0 r^3} \tag{A11.26}$$

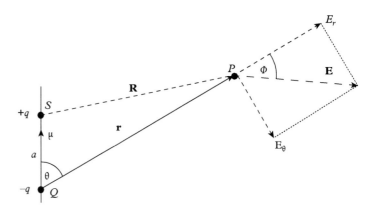

Fig. A11.2 Electrostatic potential V at a point P arising from an ideal dipole of moment μ $(= aqe)$, where a is the length QS of the dipole.

or

$$V_P = \frac{\mu \cos \theta}{4\pi \varepsilon_0 r^2} \quad\quad (A11.27)$$

In molecules, dipoles have a finite length, and so are not ideal. Nevertheless, this equation is used, but with cognizance of possible error. From Fig. A11.3 and using Eq. (A11.23), the potential at a point X lying in the direction of the dipole moment vector and distance a from its end is given by:

$$V_X = \left(\frac{q}{4\pi \varepsilon_0}\right)\left(\frac{1}{a} - \frac{1}{2a}\right) = 0.5\left(\frac{q}{4\pi \varepsilon_0 a}\right) \quad\quad (A11.28)$$

If Eq. (A11.28) be written more generally as

$$V_X = \left(\frac{q}{4\pi \varepsilon_0 a}\right)\left(1 - \frac{1}{m_a}\right) = \left(\frac{\mu}{4\pi \varepsilon_0 a^2}\right)\left(1 - \frac{1}{m_a}\right) \quad\quad (A11.29)$$

where m_a is a multiple of a, then as m_a is increased, then V_X approaches the ideal value.

Fig. A11.3 The error ε' in the electrostatic potential V arising from a molecular dipole of length a, relative to the comparison with the ideal dipole, decreases as its length increases: $\varepsilon' = 11\%$ at X and 4% at Y.

Example A11.1

Three positive charges lie in the straight line; q_1 to q_3 is the direction $+x$:

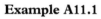

Charges q_1 and q_2 are fixed, with q_3 movable; $r_{12} = x$, $r_{23} = r-x$. It is required to show that, at equilibrium, (a) $q_1 r_{23}^2 = q_3 r_{12}^2$, (b) $x = \dfrac{r}{1+\sqrt{|q_2|/q_1}}$ and (c) the forces on q_1 and q_2 are equal and opposite.

(a) Force on $q_2 = \frac{q_1 q_2}{4\pi \varepsilon_0 r_{12}^2} - \frac{q_1 q_2}{4\pi \varepsilon_0 r_{32}^2}$ in the direction $+x$. At equilibrium this expression is equal to zero. Hence $\frac{q_1 q_2}{4\pi \varepsilon_0 r_{12}^2} = \frac{q_2 q_3}{4\pi \varepsilon_0 r_{23}^2}$, so that $q_1 r_{23}^2 = q_3 r_{12}^2$.

(b) At equilibrium, $q_1 (r - x)^2 = -q_2 x^2$; thus, $\frac{x}{r-x} = \sqrt{\frac{q_1}{|q_2|}}$ so that $x = \dfrac{r}{1+\sqrt{|q_2|/q_1}}$.

(c) From Newton's first law: $\mathbf{F}_{31} = -\mathbf{F}_{13}$, and at equilibrium, $\mathbf{F}_{21} + \mathbf{F}_{23} = 0$. Now, the force on q_1 is $\mathbf{F}_1 = \mathbf{F}_{12} + \mathbf{F}_{13}$. But $\mathbf{F}_{12} = -\mathbf{F}_{32}$ and $\mathbf{F}_{13} = -\mathbf{F}_{31}$, so that the force on q_3 is $\mathbf{F}_3 = \mathbf{F}_{31} + \mathbf{F}_{32}$ and is therefore equal to $-\mathbf{F}_1$.

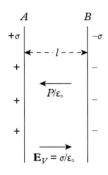

Fig. A11.4 Parallel plate capacitor. *In vacuo*, the applied field $\mathbf{E}_V = \sigma/\varepsilon_0$ and the polarization is P. In the presence of a dielectric, the opposing polarization field $\mathbf{P}/\varepsilon_0$ is set up resulting in a polarization equal to $\varepsilon_0(\varepsilon_r - 1)\mathbf{E}$.

A11.6 Capacitance

Consider a parallel plate capacitor (Fig. A11.4) initially *in vacuo*. If the surface charge density is σ, its electric field strength E_V is:

$$E_V = \sigma/\varepsilon_0 \tag{A11.30}$$

The potential difference between the identical plates A and B is $E_V l$, where l is the perpendicular distance between the plates. The capacitance per unit area C_A is defined by charge/potential difference:

$$C_A = \sigma/V = \varepsilon_0 \times 1/l \tag{A11.31}$$

If the space between the plates of the capacitor be now filled with a medium of dielectric constant ε, then the electric field strength would become:

$$E = \sigma/\varepsilon = \sigma/\varepsilon_r\varepsilon_0 \tag{A11.32}$$

Since $\varepsilon > 1$, a reduction in field strength arises from *polarization* of the dielectric medium. Dipoles are induced in the medium and create an induced field in opposition to the applied field. Since polarization P is *total electrostatic dipole moment per unit volume*:

$$P = \mu al = \mu A$$

where the area of the plate is now deemed to be a general value A. Thus, P has the nature of a surface charge density, and the effective charge density is $(\sigma - P)$. Hence, from Eq. (A11.30):

$$E = (\sigma - P)/\varepsilon_0 = (E\varepsilon/\varepsilon_0) - (P/\varepsilon_0) \tag{A11.33}$$

so that the opposing polarization field is P/ε_0.

The polarization P may be equated as

$$P = \varepsilon_0(\varepsilon_r - 1)E \tag{A11.34}$$

Thus, the presence of the dielectric medium in the capacitor produces a polarization which is $\varepsilon_0(\varepsilon_r - 1)$ times the real field.

This discussion applies strictly to an ideal capacitor having plates of infinite size. In practice, there are fringe effects at the ends of finite plates where the lines of force become non-uniform, but that small effect has been neglected in the results given here (see also Section A11.3.1).

A11.7 Clausius–Mosotti and Lorentz–Lorenz equations

In the condensed state of non-polar molecules, neighbouring molecular interactions with a given molecule are small, so that only electron polarization P_e is significant. If spherical molecules of radius r form, on average, a close-packed, body-centred, cubic array of side a, then the cell volume a^3 is equal to $(4r/\sqrt{3})^3$. Then, the number of species per unit volume N is thus $1/(6.158r^3)$. From Eq. (3.25), $(\varepsilon_r - 1) = 4\pi r^3 N$, then :

$$(\varepsilon_r - 1) = 4\pi r^3/(6.158r^3) \tag{A11.35}$$

so that $\varepsilon_r \approx 3$, and assuming the Maxwell relation for high frequencies Eq. (3.28), the refractive index n of the material is:

$$n \approx 1.73 \tag{A11.36}$$

Many non-polar liquids have values of refractive index close to 1.73: Br_2 (l), 1.76; S (l, 391 K), 1.88; CS_2 (l, 163 K), 1.73; P_4 (s, 298 K), 2.14; glass (s, 298 K), 1.50–1.95. In anisotropic solids, n varies with the direction of observation, which gives rise to *birefringence*.

It has been shown in Chapter 3 that polar molecules in a condensed state interact strongly, and a polarization is induced. If the induced dipole moment is μ', the field E' that it produces is μ'/α, which is commensurate with a field that would produce a polarizability α.

Within a parallel plate capacitor filled with a dielectric, consider a virtual spherical cavity around a molecule at O, the centre of the sphere (Fig. A11.5). The radius r is sufficiently greater than the average molecular separation that there are no molecular interactions within the cavity. In an isotropic material, an atom or molecule is surrounded by other similar species forming a spherical region of nearest neighbours.

The applied field E_a produces a polarization field, and the resultant field E_r on a molecule in the cavity at O depends on the following factors:

- the applied field E_a;
- the field due to the dielectric, excluding the virtual cavity, and comprising both the polarization field $-P/\varepsilon_0$ and the field E_0 arising from the charge density on the surface of the cavity;
- the field E_i arising from the dielectric in the cavity.

Thus, the resultant field is

$$E_r = (E_a - P/\varepsilon_0) + E_0 + E_l = E + E_0 + E_i \tag{A11.37}$$

and E_0 must be determined.

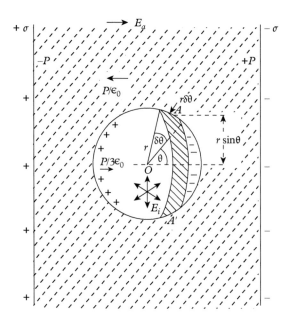

Fig. A11.5 The resultant field $E_0 E_0 (= P/3\varepsilon_0)$ in a condensed polar dielectric under the influence of an applied field E_a.

The dielectric outside the cavity is remote from the molecule at O, and may be regarded as a continuum. Hence, the field at O arising from the bulk dielectric is the same as that arising from the charge density on the walls of the cavity.

Consider an annular region of width $r\delta\theta$ on the sphere and subtending angles of θ and $\theta + \delta\theta$ at O. Since the element $r\delta\theta$ lies at an angle θ to the polarization field P, the apparent surface charge density A is $-P\cos\theta$.

A charge δq on the annulus is $-P\cos\theta 2\pi r^2 \sin\theta\delta\theta$, and its resultant coulombic field δE_0 acting in the direction OA is $\delta q/(4\pi\varepsilon_0 r^2)$. The field may be resolved into components along and perpendicular to the direction of E, as shown in the figure hereunder.

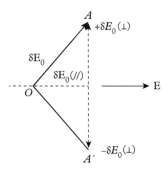

Resolution of the field E into components perpendicular and parallel to the field

The vertical components cancel, but the horizontal components $\delta q/(4\pi\varepsilon_0 r^2)$ combine to give the magnitude in the angular range θ and $\theta + \delta\theta$:

$$\delta E_0 = P(\cos\theta)2\pi r^2 \sin\theta\delta\theta(\cos\theta)/4\pi\varepsilon_0 r^2 \qquad (A11.38)$$

Hence, for the whole sphere:

$$E_0 = P/(2\varepsilon_0)\int_0^\pi \cos^2\theta\sin\theta d\theta = P/(2\varepsilon_0)\left.\frac{-\cos^3\theta}{3}\right|_0^\pi = P/3\varepsilon_0. \qquad (A11.39)$$

From the discussions above:

$$E = \frac{1}{4\pi\varepsilon_0}\mathbf{grad}(\boldsymbol{\mu}\cdot\mathbf{r}/r^3) = \frac{1}{4\pi\varepsilon_0}\{(\boldsymbol{\mu}\cdot\mathbf{r})\,\mathbf{grad}(1/r^3) + (1/r^3)\mathbf{grad}(\boldsymbol{\mu}\cdot\mathbf{r})$$

Setting $\boldsymbol{\mu}$ and $\mathbf{r}$ in Cartesian components, it follows that $\mathbf{grad}(\boldsymbol{\mu}\cdot\mathbf{r}) = \boldsymbol{\mu}$ and $\mathbf{grad}(1/r^3) = -(3/r^4)\mathbf{grad}(r) = -3\mathbf{r}/r^5$. Hence:

$$E = \frac{1}{4\pi\varepsilon_0 r^5}[3(\boldsymbol{\mu}\cdot\mathbf{r})\mathbf{r} - r^2\boldsymbol{\mu}] \qquad (A11.40)$$

All species in the virtual cavity have the same dipole moment μ. Let $\boldsymbol{\mu}$ and $\mathbf{r}$ be referred to Cartesian axes with unit vectors labelled $\mathbf{i}$, $\mathbf{j}$ and $\mathbf{k}$. Then, for the nth dipole in any direction:

$$\mathbf{E}_n = \mathbf{i}E_{n,x} + \mathbf{j}E_{n,y} + \mathbf{k}E_{n,z}$$

which may be given also from Eq. (A11.40) as:

$$\{[3(\mathbf{i}\mu_x + \mathbf{j}\mu_y + \mathbf{k}\mu_z)_n \cdot (\mathbf{i}x + \mathbf{j}y + \mathbf{k}z)_n](\mathbf{i}x + \mathbf{j}y + \mathbf{k}z)_n - r^2(\mathbf{i}\mu_x + \mathbf{j}\mu_y + \mathbf{k}\mu_z)_n\}/(4\pi\varepsilon_0 r^5)$$

If the dipole direction is set along z, then all parameters involving x and y are zero, so that $\mu_z = \mu$ and $E_{n,z} = [1/(4\pi\varepsilon_0 r^5)](3z_n^2 - r^2\mu_n)$; for all particles in the sphere:

$$E_z = [1/(4\pi\varepsilon_0 r^5)]\mu \sum_{n-1}^N 3z_n^2 - x_n^2 - y_n^2 - z_n^2 \qquad (A11.41)$$

In an isotropic medium, all magnitudes x_n, y_n and z_n are equally probable. Thus, $<x_n> = <y_n> = <z_n> = 0$ and $<x_n^2> = <y_n^2> = <z_n^2>$. From the latter term, it follows from Eq. (A11.41) that $E_z = 0$. Similar analyses show that $E_x = E_y = 0$ also. This result means that, because of the high symmetry (effectively spherical) of the isotropic dielectric, the internal field E_i is zero. Thus, from Eqs. (A11.37) and (A11.39), E_i being zero:

$$E_r = E + P/(3\varepsilon_0) \qquad (A11.42)$$

Using Eq. (3.18), the total polarization field is

$$P = N\alpha E_r = N\alpha E + N\alpha P/(3\varepsilon_0)$$

and on rearranging:

$$P/E = N\alpha / \left(1 + \frac{N\alpha}{3\varepsilon_0} \right)$$

Applying Eq. (3.24):

$$(\varepsilon_r - 1) = N\alpha / \left(\varepsilon_0 - \frac{N\alpha}{3} \right) \tag{A11.43}$$

Since $(\varepsilon_r - 1) + 3 = (\varepsilon_r + 2)$,

$$(\varepsilon_r + 2) = (N\alpha + 3\varepsilon_0 - N\alpha)/[\varepsilon_0 - (N\alpha/3)]$$

and further rearrangement leads to:

$$\frac{(\varepsilon_r - 1)}{(\varepsilon_r + 2)} = \frac{N\alpha}{3\varepsilon_0} \tag{A11.44}$$

As $N = LD/M$, where D is density and M molar mass:

$$P_m = \frac{(\varepsilon_r - 1)}{(\varepsilon_r + 2)} \frac{M}{D} = \frac{L\alpha}{3\varepsilon_0} \tag{A11.45}$$

where α may be regarded as the sum $(\alpha_e + \alpha_a + \alpha_o)$ corresponding to the electronic, atomic and orientation polarizabilities respectively; Eq. (A11.45) is the *Clausius–Mosotti equation*, and the left-hand side of the equation is known as the *molar polarization* P_m.

In circumstances where Maxwell relationship Eq. (3.28) is applicable, Eq. (A11.45) becomes:

$$R_m = \frac{(n^2 - 1)}{(n^2 + 2)} \frac{M}{D} = \frac{L\alpha}{3\varepsilon_0} \tag{A11.46}$$

which is the *Lorentz–Lorenz equation*, the left-hand side of which is termed the *molar refraction* R_m.

These two equations are obeyed well for polar liquids, dilute solutions of polar substances in non-polar solvents, amorphous solids and isotropic solids. Where the interaction between species is weak, it may be inferred that $\varepsilon_r \to 1$, whereupon $(\varepsilon_r + 2) \approx 3$, and Eq. (A11.44) degenerates to $\frac{N\alpha}{\varepsilon_0} = \varepsilon_r - 1$, an expression of Eq. (3.25). If $\varepsilon_r \approx 4$, neglect of the strong interaction term gives a value to $(\varepsilon_r - 1)$ that is twice too large; if ε_r is very large, for example, 80 (water at *ca.* 20°C), then the same approximation would make $(\varepsilon_r - 1)$ twenty-five times too large.

A11.8   Volume of molecules in a gas

The constant b in the van der Waals equation of state Eq. (3.30) is related to the volume *occupied* by the gas molecules; more precisely, it is termed an *excluded volume*.

Consider the molecules of a gas as spherical and of radius $\sigma/2$. Two molecules cannot approach each other more closely than the sum of their van der Waals radii, $2 \times \sigma/2$. Figure A11.6 shows the situation of two molecules in close contact. The spherical volume of space in which the centres of other molecules cannot invade is shaded, and corresponds to a sphere of radius σ. Hence, the excluded volume for a *pair* of molecules $\frac{4}{3}\pi\sigma^3$, so that per single molecule it is $\frac{4}{6}\pi\sigma^3$. The actual volume v of a single molecule is $\frac{4}{3}\pi\left(\frac{\sigma}{2}\right)^3$, or $\frac{\pi\sigma^3}{6}$. Thus, the excluded volume for a single molecule is four times its own volume, or $4v$.

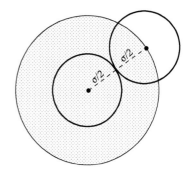

Fig. A11.6 The excluded volume in a gas of diameter σ is $\frac{4}{3}\pi\sigma^3$ *per pair of molecules*, whereas the volume v of a *single* molecule is $\frac{4}{3}\pi\left(\frac{\sigma}{2}\right)^3 = \frac{\pi\sigma}{6}$; hence, the excluded volume is $4v$.

A12
Configurational Energy

Let the probability distribution for N molecules be $P(\mathbf{x}_1, \mathbf{x}_2, \ldots, \mathbf{x}_N)$, where $\mathbf{x}_i$ $(i = 1, 2, \ldots, N)$.[4] The average potential energy is then

$$\int UP\mathrm{d}\mathbf{x}_1\,\mathrm{d}\mathbf{x}_2\ldots\mathrm{d}\mathbf{x}_N$$

Since U is a sum of pairwise interactions:

$$U = \sum_{i>j}^{N} u_{r_{i,j}}$$

There is a total of $N(N-1)/2$ identical integrals. Thus, the average internal energy $<U>$ is given by:

$$<U> \;=\; \frac{1}{2}N(N-1)\int_N Pu_{12}\,\mathrm{d}\mathbf{x}_1\,\mathrm{d}\mathbf{x}_2\ldots\mathrm{d}\mathbf{x}_N$$

Integrating over $\mathbf{x}_3$ to $\mathbf{x}_N$:

$$<U> \;=\; \frac{1}{2}N^2\int P^{(2)}(\mathbf{x}_1, \mathbf{x}_2)u_{12}\mathrm{d}\mathbf{x}_1\mathrm{d}\mathbf{x}_2 \qquad (A12.1)$$

assuming $N-1 \to N$ for large N. Then:

$$P^{(2)}(\mathbf{x}_1, \mathbf{x}_2) = (1/V^2)g(|\mathbf{x}_1 - \mathbf{x}_2|)$$

This expression is one way of defining the radial distribution function g, namely, the difference between the actual distribution and that for random points, which is $1/V^2$ for two particles; g depends only on the distance between points 1 and 2. A change of integration variables to $\mathrm{d}\mathbf{x}_1\mathrm{d}\mathbf{x}_{\{12\}}$, where the subscript $\{12\}$ implies the vector between species 1 and 2, leads to:

$$<U> \;=\; \frac{1}{2}\frac{N^2}{V^2}\int g(|\mathbf{x}_{12}|)U(|\mathbf{x}_{12}|)\mathrm{d}\mathbf{x}_1\mathrm{d}\mathbf{x}_{12}$$

Integrating over $\mathrm{d}\mathbf{x}_1$ leads to the factor V, and over $\mathrm{d}\mathbf{x}_{\{12\}}$ to $4\pi r^2\mathrm{d}r$. Combining the results gives the *configurational energy* equation [1]:

$$<U>/N = 2\pi\frac{N}{V}\int_0^\infty r^2u(r)g(r)\mathrm{d}r \qquad (A12.2)$$

[4] Normalized as given here by virtue of being a probability distribution, that is, $\int Pd\mathbf{x}_1\,\mathrm{d}\mathbf{x}_2\ldots\mathrm{d}\mathbf{x}_N = 1$.

where $u(r)$ is a potential energy function, such as the Lennard-Jones 12-6 function.

Equation (A12.2) can be interpreted through its several parts. Write the multiplier 2 as 4/2, then:

$4\pi r^2 dr$ is the volume of a spherical shell of radius r,

$\frac{N}{V}g(r)\,4\pi r^2 dr$ is the mean number of molecules in the spherical shells,

$\frac{1}{2}\frac{N}{V}u(r)g(r)4\pi r^2 dr$ is the mean energy of a chosen molecule among the others in the shell, and

$\frac{1}{2}\frac{N}{V}\int u(r)g(r)4\pi r^2 dr$ is the sum over all shells, which is the average configurational energy of the system.

REFERENCE A12

[1] Ladd AJC. *Private communication*, 2013.

A13
Numerical Integration

A13.1 Introduction

A numerical integration procedure computes the value of a definite integral from a set of values of the integrand. If these values are ordinates to a curve, the integration defines the area under the curve. If the curve can be fitted by a least-squares or other method, then the integration can be performed analytically. Suppose that the curve in Fig. A13.1 can be fitted exactly by the function

$$y = 2x^2 + 3x + 2$$

then integration to six figures is:

$$\int_0^{0.8} y\mathrm{d}x = \frac{2x^3}{3} + \frac{3x^2}{2} + 2x \mid_0^{0.8} = 2.90133\ldots\ldots$$

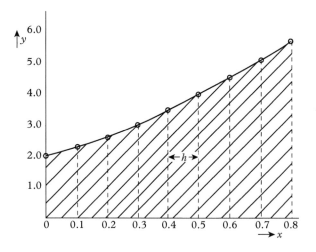

Fig. A13.1 The function $y = 2x^2 + 3x + 2$ from $x = 0$ to $x = 0.8$; the curve is divided into $n(=8)$ areas of equal width h.

A13.2 Numerical methods

Where analytical integration is not possible, a numerical procedure must be adopted. Some numerical integration procedures depend upon the relationship:

$$\int_a^c f(x)\,dx = \int_a^b f(x)\,dx + \int_b^c f(x)\,dx$$

One such procedure is embodied in Simpson's rule [1].

A13.2.1 Simpson's rule

The given curve is divided into an *even* number n of intervals of width h. Then, Simpson's rule states that, for any function $y = f(x)$:

$$\int_{x_0}^{x_n} y\,dx = \int_{x_0}^{x_0+nh} y\,dx$$

$$= (h/3)[y_0 + y_n + 2(y_2 + y_4 + \cdots + y_{n-2})$$

$$+ 4(y_1 + y_3 + \cdots + y_{n-1})]$$

Example A13.1

The curve in Fig. A13.1 is divided into eight intervals of width 0.1 units. From the curve:

x	0.0	0.1	0.2	0.3	0.4	0.5	0.6	0.7	0.8
y	2.0	2.3	2.7	3.1	3.5	4.0	4.5	5.1	5.7

Then, $\displaystyle\int_0^{0.8} y\,dx = (0.1/3)[2.0 + 5.7 + 2(2.7 + 3.5 + 4.5)$

$$+ 4(2.3 + 3.1 + 4.0 + 5.1)]$$

$$= 2.90330$$

The result agrees with the analytical value to 0.067%.

A13.2.2 Gaussian quadrature

A better procedure that also finds the area under the curve, and with which the sub-divisions need not be equally spaced is a Gaussian quadrature [1, 2]. The precision is greater than that obtained with Simpson's rule, and the procedure is embodied in the program QUAD in the Web Program Suite. Using the data set from Example 13.1 the program returns the value 2.9019, to five figures, with a probable error of ± 0.0002. A more extensive Gaussian quadrature procedure [3] returns the value 2.90133... with the same data.

....

REFERENCES A13

[1] Whittaker ET and Robinson G. *The Calculus of Observations*. Dover, 1967.
[2] Arfken G. *Mathematical Methods for Physicists*. Academic Press, 1985.
[3] Casio Computer Co. Ltd. Gaussian quadrature calculator, 2015. http://keisan.casio.com/exec/system/1330940731

A14

X-ray Crystallographic Structure Analyses of Potassium Chloride and Sodium Chloride

The crystal structure analyses of potassium chloride and sodium chloride were determined first in 1913, using the Bragg x-ray spectrometer [1]. The important basis of this instrument is that the collector for measuring the intensity of a diffracted ray was arranged to rotate at twice the angle of rotation of the sample crystal, thus mimicking the geometry of reflection of light from a mirror (Appendix A3). The ensuing years saw a very rapid increase in the determination of the structures of divers compounds by x-ray methods [2].

Potassium and sodium chlorides crystallize as cubic crystals, and x-ray reflections from the planes of Miller indices (Appendix A2) $(h00)$, $(hh0)$ and (hhh) were recorded (Fig. A14.1).

Crystals of the two halides were cut so as to present to the incident x-ray beam each of the three faces indicated in Fig. A14.1. The intensities I of x-rays reflected (diffracted) from these planes were measured as a function of the Bragg angle θ for both KCl and NaCl, and the six sets of results are shown in Fig. A14.2. The results were analyzed in terms of the Bragg equation, in its original form:

$$2d \sin \theta = n\lambda \qquad (A14.1)$$

which is obeyed for the six sets of spectra shown.

Consider the $(h00)$ data for potassium chloride; the three reflections occurred at angles 5.36°, 10.76° and 16.26°. The ratios of the corresponding values of

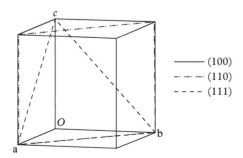

—— (100)
—·—·— (110)
—— —— (111)

Fig. A14.1 Cube, showing the orientation of one plane in each of the crystallographic forms {100}, {110} and {111}.

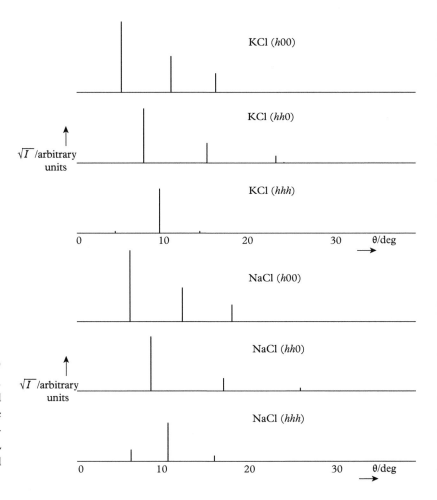

Fig. A14.2 Bar graph of $\sqrt{I}$ against θ for KCl and NaCl; θ ranges from *ca.* 5° to 26° for each compound, and the ordinates are normalized to the value of $\sqrt{I}$ for the first (h00) reflection. [Reproduced from Bragg WL *The Crystalline State*, Vol. I. G Bell and Sons Ltd, 1939.]

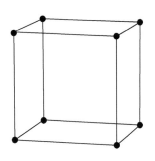

Fig. A14.3 Primitive unit cell in a cubic lattice. The atoms in potassium chloride were first thought to occupy the lattice points at the corners of the unit cell.

$\sin \theta$ are 1:2:3, integral within experimental error, as expected from Eqs. (A3.6) and (A14.1).

Next, the first reflection for potassium chloride in each of the series (h00), (hh0) and (hhh) was examined, and the θ-values were found to be 5.36°, 7.59° and 9.30°. Since $d \propto 1/\sin\theta$, the d-values form the ratios $1:1/\sqrt{2}:1/\sqrt{3}$, which correspond to the interplanar d spacings of the planes shown in Fig. A14.1, and indicate a *primitive* cubic arrangement of atoms in the unit cell (Fig. A14.3).

In the results for sodium chloride, however, the d-ratios for the (hhh) planes were established as $1:1/\sqrt{2}:2/\sqrt{3}$, and these ratios are consistent with a *face-centred* structure (Fig. A14.4); the true spacing was revealed in this (hhh) reflection data, and the totality of results were then explained.

The h00 and hh0 reflections must be relabelled 200, 400, 600 and 220, 440, 660 (n = 2, 4, 6, and $d_{(nh,nk,nl)} = d_{(hkl)}/n$). Reflections such as 100 and 110 are forbidden in a face-centred structure because with interleaving planes of equal scattering power at exactly $d/2$ apart, complete interference takes place, *vide* Eq. (A14.1). The *(hhh)* planes are populated successively by cations and anions, at a spacing $d_{(111)}/2$. Thus, when successive (111) planes are in the reflection position, the interleaving out-of-phase planes will interfere destructively. Thus, reflection 111 will be weaker than reflection 222, which is clear on the graph for the *(hhh)* data of sodium chloride. The 111 and 333 reflections with potassium chloride are so weak that they were first considered negligible at the resolution of the instrument: the K^+ and Cl^- ions are isoelectronic, so that the cancellation of *hhh* with *h* an odd integer is almost complete, significantly more so than with sodium chloride. Thus, the true structure for these halides was completely resolved, and is illustrated in stereoview by Fig. A14.5.

The x-ray powder photograph in Fig. A14.6 is a good illustration of the fact that *hkl* spectra for *h*, *k* and *l* all of *odd* integers are not discernible in KCl, whereas they are clearly evident in the NaCl x-ray spectra.

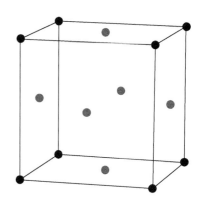

Fig. A14.4 Face-centred unit cell in a cubic lattice, representative of all alkali-metal halides at ambient conditions except CsCl, CsBr and CsI.

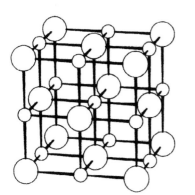 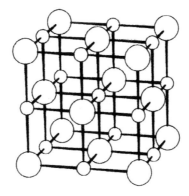

Fig. A14.5 Stereoview of the unit cell and environs of the sodium chloride structure type. The M^+–X^- entity is set down in one and the same orientation at each of the points of the face-centred unit cell shown in Fig. A14.4.

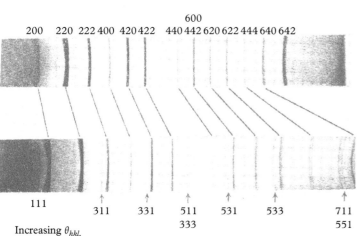

						600						
200	220	222	400	420	422	440	442	620	622	444	640	642

KCl

NaCl

111
311   331   511   531   533   711
333   551

Increasing θ_{hkl},
Decreasing d_{hkl}

Fig. A14.6 X-ray powder photographs of KCl and NaCl. The reflections with *h*, *k* and *l* all odd integers, such as 111 and 311, which are observable on the photograph for NaCl, are not discernible in the x-ray record for KCl; these apparent absences led to the first assignment of KCl as primitive cubic. [Reproduced from Bragg WL *The Crystalline State*, Vol. I. G Bell and Sons Ltd, 1939.]

It is evident from Fig. A14.5 that the unit cell volume contains 4 Na^+ and 4 Cl^- species, and so the total relative mass per unit cell is 4×58.443. The density of sodium chloride is 2165 kg m^{-3}; hence the volume a^3 of the unit cell is $\frac{4 \times 58.433 \times 1.6605 \times 10^{-27} \text{ kg}}{2165 \text{ kg m}^{-3}} = 1.7930 \times 10^{-28}$ m^3; hence, $a = 5.639 \times 10^{-10}$ m, or 0.5639 nm.

The wavelength of the x-rays has not entered into the above calculation. Indeed, the wavelengths of x-rays were not known at the time of these analyses. Taking the value of $6.00°$ for θ_{200}, it follows from Eq. (A14.1) that $2\lambda = 0.5639$ nm $\times \sin 6°$, or $\lambda = 0.0589$ nm, a wavelength now recognized as that of Pd $K\alpha$.

It should be noted that crystal structure analysis is not carried out in this manner today, although several of the principles discussed herein have their counterparts in modern x-ray crystallography. A detailed account of current crystal structure analysis procedures, using both x-rays and neutrons, may be found in the literature [3].

....

REFERENCES A14

[1] Bragg WH and Bragg WL. *Proc. Roy. Soc. (Lond.)* 1913; A88: 428.

[2] Bragg WH and Bragg WL. *X-rays and Crystal Structure*. G. Bell and Sons, 1915.

[3] Ladd M and Palmer R. *Structure Determination by X-ray Crystallography*, 5th ed. Springer Science+Business Media, 2013.

A15
Calculation of Madelung Constants

The Madelung constant was discussed in Section 4.4.1, and some approximate calculations of this parameter were reviewed. The calculation by Ewald [1] has been modified [2] and applied successfully [3, 4] to a wide range of compounds; it has been programmed as MADC in the Web Program Suite. The procedure makes the reasonable assumptions that the ions have a spherical charge distribution, and that their electron densities are linear functions of the radial coordinates in reciprocal space.

The Madelung constant A is given by:

$$A = \frac{(g - Q)d}{rZ} \sum_j q_j^2 - \frac{\pi r^2 d}{V} \sum_{\mathbf{h}} |F_{\mathbf{h}}|^2 \phi(\mathbf{h}) \qquad (A15.1)$$

where the terms have the following meanings:

g	26/35;
Q	A correction for termination of the $\mathbf{h}$ summation (see below);
d	A standard distance, normally the nearest-neighbour distance in the structure;
r	An arbitrary distance less than one half (usually 0.495) of the nearest-neighbour distance;
Z	The number of formula-entities in one unit cell of the structure;
q_j	The charge, including sign, on the jth species in one unit cell;
V	The unit-cell volume;
$\phi(\mathbf{h})$	$288[\alpha \sin \alpha + 2 \cos \alpha - 2]^2 \alpha^{-10}$;
α	$2\pi hr$;
$F_{\mathbf{h}}$	$\sum_j \exp(i2\pi \mathbf{h}\mathbf{x}_j)$, the crystallographic structure factor for point atoms.

The sum over j includes all atoms in one unit cell, and the sum over $\mathbf{h}$ includes all reciprocal lattice vectors hkl in a sphere of radius α; h, k and l are the components of $\mathbf{h}$ with respect to the reciprocal lattice axes. The series termination correction Q depends upon the radius α, according to the following tabulation:

α	Q
2π	0.00030
3π	0.000090
4π	0.000012
5π	0.0000057

Termination of the series at $\alpha = 2\pi$, including the correction term Q, gives results to better than 0.02%, which is often sufficient to match other terms in crystal energy calculations.

....

REFERENCES A15

[1] Ewald PP. *Ann. Phys.* 1921; 64: 253.
[2] Bertaut EF. *J. Phys. Radium.* 1952; 13: 499.
[3] Templeton DH. *J. Chem. Phys.* 1955; 23: 1629.
[4] Jones RE and Templeton DH. *J. Chem. Phys.* 1956; 25: 1062.

A16
Equation of State for a Solid

The *Helmholtz free energy A* is defined by:

$$A = U - TS$$

Since A is an *extensive property*, it follows that:

$$dA = dU - TdS - SdT \qquad\qquad \text{(A16.1)}$$

At constant temperature, and since $TdS = dU + pdV$:

$$dA = dU - (dU + pdV) = -pdV$$

and

$$\left(\frac{\partial A}{\partial V}\right)_T = -p \qquad\qquad \text{(A16.2)}$$

From Eq. (A16.1), at constant temperature:

$$\left(\frac{\partial A}{\partial V}\right)_T = \left(\frac{\partial U}{\partial V}\right)_T - T\left(\frac{\partial S}{\partial V}\right)_T$$

From the Maxwell's relations [1]:

$$\left(\frac{\partial S}{\partial V}\right)_T = \left(\frac{\partial p}{\partial T}\right)_V$$

Furthermore:

$$\left(\frac{\partial p}{\partial T}\right)_V \left(\frac{\partial T}{\partial V}\right)_p \left(\frac{\partial V}{\partial p}\right)_T = -1$$

or

$$\left(\frac{\partial p}{\partial T}\right)_V = -\frac{(\partial V/\partial T)_p}{(\partial V/\partial p)_T} \qquad\qquad \text{(A16.3)}$$

From the thermodynamic definitions of volume expansivity β and compressibility κ, it follows that the right-hand side of Eq. (A16.3) is equal to β/κ. Hence:

$$\left(\frac{\partial A}{\partial V}\right)_T = \left(\frac{\partial U}{\partial V}\right)_T - T\beta/\kappa$$

and with Eq. (A16.2):

$$\left(\frac{\partial U}{\partial V}\right)_T = T\beta/\kappa - p \qquad (A16.4)$$

At 0 K, $(\partial U/\partial V)_{T=0} = -p$, but the term $T\beta/\kappa$ is not negligible at 298.15 K, the temperature at which crystal energy is often considered. A detailed analysis [2] showed that the error in taking $(\partial U/\partial V)$ equal to $-p$ at 298.15 K is *ca.* 1%; in the case of sodium chloride, that would be 8 kJ mol^{-1}.

....

REFERENCES A16

[1] Atkins P and de Paula J. *Atkins' Physical Chemistry*, 9th ed. Oxford University Press, 2009.
[2] Tosi MP. *Solid State Physics*. 1964; 16: 1.

A17
Partial Molar Quantities and Standard States

A17.1 Introduction

In describing the state of a solution, or indeed that of any other system, two kinds of property may be involved: *intensive properties*, such as density, viscosity and refractive index, which are independent of the amount of substance in the system; and *extensive properties*, such as volume, energy and entropy, which depend upon the amount of substance under consideration. While volume, for example, is an extensive property, volume per unit mass (density) is an intensive property. The extensive properties of a given amount of solution can be measured, and in a thermodynamic discussion of solubility the variations with composition of some of these properties are required.

A17.2 Partial molar volume

A system in which the composition may vary is an *open system*, whereas a system of constant composition is a *closed system*.

Consider a solution of n_A mol of a solvent species A and n_B of a solute B. On addition of an amount dn_A mol of A the volume increase will be dV; then:

$$dV/dn_A = \overline{V_A} \qquad (A17.1)$$

where $\overline{V_A}$ is the *partial molar volume* of the solvent in the solution. Similarly, the partial molar volume of the solute in the solution is defined by:

$$dV/dn_B = \overline{V_B} \qquad (A17.2)$$

A partial molar quantity is a property of the solution as a whole; thus, $\overline{V_B}$ is the change in total volume when solute is added. An important part of this change is concerned with the packing of adjacent solvent molecules and, therefore, with the forces between the species in a solution.

In order to assist the appreciation of a partial molar property, Fig. A17.1 is given as an illustration of the partial molar volume of aqueous calcium chloride $\overline{V}(CaCl_2)$ as a function of the number n of moles of calcium chloride, at a temperature of 298.15 K [1].

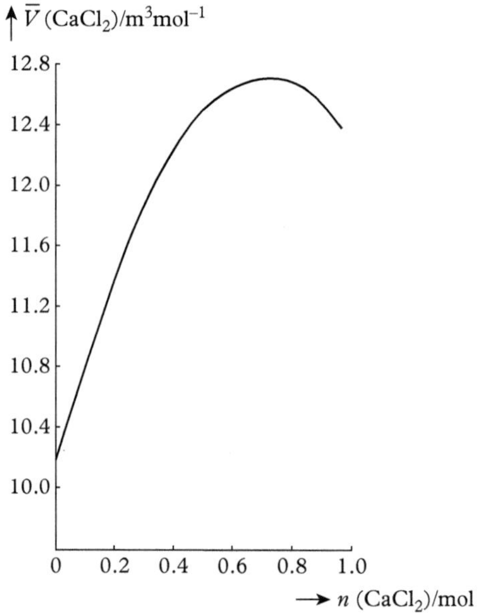

Fig. A17.1 Partial molar volume $\overline{V}$ of calcium chloride in water at 298 K as a function of moles n of the solute $CaCl_2$. The value of $\overline{V}$ is a function of composition because the environment and therefore the interaction of the species changes with the composition.

A17.3   Partial molar entropy

In studying solubility in Section 4.8ff, the defining quantity was determined as $\Delta G_d^{\ominus}$, the standard free energy of dissolution. This quantity was composed of the standard enthalpy change $\Delta H_d^{\ominus}$ and the corresponding change in entropy. The term $\Delta S_d^{\ominus}$ is composed of the entropy of the crystal and the entropy of the hydrated ions. The first of these quantities is obtained from heat capacity data between 0 K and 298 K; the hydrated ion contribution will be considered next.

A17.3.1   Measurement of the partial molar entropy of a hydrated ion

A procedure for determining the *partial molar entropy* of a hydrated ion depends upon measuring the temperature variation of electromotive force (emf) in a suitable galvanic cell. Consider zinc chloride, $ZnCl_2$, as a solute; then, the following thermostatically-controlled cell can be set up:

$$Zn\,|\,ZnCl_2(c)\,\|\,HCl(a_{\pm} = 1)\,|\,H_2(g,\ 1\ atm), Pt \qquad (A17.3)$$

The spontaneous cell reaction may be written as:

$$Zn(s) + 2H_3O^+(aq) \rightarrow Zn^{2+}(aq) + H_2(g) + 2H_2O \qquad (A17.4)$$

The emf of the cell Eq. (A17.3) is measured at several values of concentration c and extrapolated, conveniently against $\sqrt{c}$, to $c = 0$ (infinite dilution). Further measurements of emf are conducted at three or four temperatures between 15°C and 35°C so as to obtain $dE^{\ominus}/dT$. Then, from the equation:

$$\Delta S^{\ominus} = nF(dE^{\ominus}/dT) \qquad \text{(A17.5)}$$

where n is the number of electrons involved in the reaction and F is the Faraday, $\Delta S^{\ominus}$ may be obtained.

But $\Delta S^{\ominus}$ is given also by:

$$\Delta S^{\ominus} = [\overline{S}^{\ominus}(Zn^{2+}) + S^{\ominus}(H_2)] - [\overline{S}^{\ominus}(H_3O^+)] + S^{\ominus}(Zn) \qquad \text{(A17.6)}$$

By convention, $\overline{S}^{\ominus}(H_3O^+) = 0$; hence, the relative partial molar entropy of $Zn^{2+}(aq)$ is:

$$\overline{S}^{\ominus}(Zn^{2+}, \text{ aq}) = \Delta S^{\ominus} + S^{\ominus}(Zn, \text{ s}) - S^{\ominus}(H_2, \text{ g}). \qquad \text{(A17.7)}$$

In a typical experiment, $dE^{\ominus}/dT = -1.00 \times 10^{-4}$ V K^{-1}. Since $S^{\ominus}(Zn, \text{ s}) = 41.6$ J K^{-1} mol^{-1} and $S^{\ominus}(H_2, \text{ g}) = 130.6$ J K^{-1} mol^{-1}, application of Eq. (A17.5) gives $\Delta S^{\ominus} = -2 \times 96485$ C $\times 1.00 \times 10^{-4}$ V K$^{-1} = -19.30$ J K^{-1} mol^{-1}, so that $\overline{S}^{\ominus}(Zn^{2+}, \text{ aq})/\text{J K}^{-1}$ mol$^{-1} = -19.30 + 41.6 - 130.6$; thus, $\overline{S}^{\ominus}(Zn^{2+}, \text{ aq}) = 108.3$ J K^{-1} mol^{-1}.

A17.4 Generalized description of partial molar quantities

An extensive property X is determined by the state of the system and the amount of substances present. Thus, for a property X defined by:

$$X = f(T, \ p, \ n_i), \ (i = 1, 2, \ldots, N) \qquad \text{(A17.8)}$$

$$dX = (\partial X/\partial T)_{p, n_i} dT + (\partial X/\partial p)_{T, n_i} dp + \sum_{i=1}^{N} (\partial X/\partial n_i)_{T, p, n_j} dn_i (j = 1, \ 2, \ldots, j \neq i) \qquad \text{(A17.9)}$$

The derivative $(\partial X/\partial n_i T)_{T, p_i n_j}$ is the partial molar quantity $\overline{X_i}$ for the ith component in the system. If the extensive property is the *Gibbs free energy*, the partial molar property is identical with the *chemical potential*. Thus:

$$(\partial G/\partial n_i)_{T, p_i n_j} = \overline{G_i} = \mu_i. \qquad \text{(A17.10)}$$

Differentiating Eq. (A17.10) with respect to T:

$$(\partial^2 G/\partial n_i \partial T)_{p, n_j} = (\partial \mu_i/\partial T)p, \ n_j \qquad \text{(A17.11)}$$

Since

$$(\partial G/\partial T_i)_p = -S \tag{A17.12}$$

differentiation of Eq. (A17.12) with respect to n_i gives the *partial molar entropy*:

$$(\partial^2 G/\partial T_i \partial n_i)_p = -(\partial S/\partial n_i)_{p,n_j} = \overline{S_i} \tag{A17.13}$$

Since the order of differentiation in Eqs. (A17.11) and (A17.13) is immaterial (G is a state function), it follows that

$$(\partial \mu_i/\partial T)_{p,n_j}[= (\partial^2 G/\partial n_i \partial T)_{p,n_j}] = \overline{S_i} \tag{A17.14}$$

Applying $G = H - TS$ and differentiating with respect to n_i, it follows that

$$\overline{G_i} = \overline{H_i} - T\overline{S_i} \tag{A17.15}$$

Using Eq. (A17.14) and rearranging:

$$\overline{H_i} = \mu_i - T(\partial \mu_i/\partial T)_{p,n_j} \tag{A17.16}$$

Thus, equations for partial molar entropy, partial molar free energy and *partial molar enthalpy* have been established [2].

A17.5   Partial molar enthalpy

In an equilibrium between a solute and its solution, the solid and the saturated solution are at the same chemical potential. Hence, for any species:

$$\mu_i = \mu_i^\ominus + RT \ln a_i \tag{A17.17}$$

or

$$R \ln a_i = \mu_i/T - \mu_i^\ominus/T \tag{A17.18}$$

Quantities such as a_i, the activity of the single ith ionic species, are not measureable experimentally; nevertheless, it is helpful to discuss them as though they were. Under constant pressure and composition:

$$R[\partial(\ln a_i)/\partial T]_{p,n_j} = [\partial(\mu_i/T)/\partial T)]_{p,n_j} - [\partial(\mu_i^\ominus/T)\partial T]_{p,n_j} \tag{A17.19}$$

From Eq. (A17.16), dividing by T^2:

$$\mu_i/T^2 - (1/T)(\partial \mu_i/\partial T)_{p,n_j} = \overline{H_i}/T^2 \tag{A17.20}$$

or

$$[\partial(\mu_i/T)/\partial T]_{p,\,n_j} = -\overline{H}_i/T^2 \qquad (A17.21)$$

Hence, from Eq. (A17.19):

$$(\partial \ln a_i/\partial T)_{p,\,n_j} = (H_i^{\ominus} - \overline{H}_i)/RT^2 \qquad (A17.22)$$

where $\overline{H}_i$ is the *partial molar heat content* of the *i*th constituent in the solution and $H_i^{\ominus}$, equivalent here to $\overline{H}_i^{\ominus}$, is the corresponding value in the pure state of the *i*th constituent. The activity is related to concentration by:

$$a_i = c_i f_i \qquad (A17.23)$$

but concentration in mol dm^{-3}, its normal unit, is not independent of temperature. Therefore, Eq. (A17.23) is formulated as:

$$(\partial \ln f_i/\partial T)_{p,\,n_j} = (H_i^{\ominus} - \overline{H}_i)/RT^2 + [\partial \ln(D_0/D)/\partial T]_{p,\,n_j}, \qquad (A17.24)$$

where D_0 and D are the densities of the solvent and solution respectively; c_i is constant under differentiation with respect to T.

A17.6 Standard states and reference states

In thermochemical work, the *standard state* of a substance, liquid or solid, is its pure form at 1 bar and at a specified temperature, usually 25 °C. The standard change for a reaction, such as enthalpy of fusion, is the difference between the molar enthalpies of the products and reactants each in their standard states at 1 bar and at one and the same temperature, normally 25 °C.

A *standard change* for a reaction, such as the *enthalpy of formation of a substance*, is that change in heat content for the formation of the substance from its elements in their reference state, the *reference state* for elements being their pure, stable states at 1 bar and a specified temperature, usually 25 °C. The enthalpies of all elements in their standard states at 1 bar and 25 °C are defined as zero, since the only reaction to which they could correspond is $X_{T_1} \rightarrow X_{T_2}$.

In the case of ions in solution, since their properties cannot be measured as individuals, a standard state has to be defined: the standard enthalpy of formation of the hydrogen ion $H^+(aq)$ is defined as zero at all temperatures. The value of this convention is that if the standard enthalpy of formation of $HX(aq)$ is determined, that value is then the enthalpy of formation of $X^{\ominus}(aq)$, $\Delta H^{\ominus}(X^-,\ aq)$.

In dealing with solutions, similar concepts are necessarily introduced. Thus, the standard state of a liquid is the pure substance at unit activity a. For the solvent, the standard state is the pure solvent with activity a equal to fx, where f is the activity coefficient and x its mole fraction, under the condition $f \rightarrow 1$ as $x \rightarrow 1$.

In the case of the solute at a molality[5] m, it has an activity $a = fm$, where $a \to 1$ as $m \to 0$. The standard state for a solution is based on the concept of unit activity, and there is freedom in its choice. The standard state should be convenient for the given application, and capable of forming a basis for comparison of different solutions. It is permissible to choose different standard states for a given substance in two phases in equilibrium with each other. The equilibrium constant will be then altered in value, but not in its constancy, at a given temperature. However, when different standard states are chosen for two such substances, their activities are not equal, although their chemical potentials must be identical because they are in equilibrium. Thus, for any species in phase I:

$$\mu_I = \mu_I^\ominus + RT \ln a_I \tag{A17.25}$$

and for phase II, in equilibrium with phase I

$$\mu_{II} = \mu_{II}^\ominus + RT \ln a_{II} \tag{A17.26}$$

Since $\mu_I = \mu_{II}$:

$$\mu_I^\ominus + RT \ln a_I = \mu_{II}^\ominus + RT \ln a_{II} \tag{A17.27}$$

If the standard states are one and the same, then $\mu_I^\ominus = \mu_{II}^\ominus$ and $a_I = a_{II}$.

The standard state for solution is the *infinitely dilute solution*; the activity a of the solute in solution is defined such that the ratio $a/c \to 1$ as $c \to 0$, c being the concentration of the solution in mol dm^{-3} (see also Appendix A18). The standard state is hypothetical: it corresponds to a solution of concentration 1 mol dm^{-3} in which, from Eq, (A17.23) and since D tends to D_0 and f_i tends to 1 as $c \to 0$, the partial molar heat content of the solute in the standard state $\overline{H}_i$ has the same value as in the infinitely dilute solution $\overline{H}_i^\circ$. Thus, in discussing solubility, it is appropriate and convenient to refer the *enthalpy of dissolution* $\Delta H_d^\ominus$ to an infinitely dilute solution, and the corresponding $\Delta G_d^\ominus$ to a hypothetical solution of *unit activity*, knowing that these descriptions apply, in the context, to one and the same state.

In practice, the measureable quantity is the *mean activity* $a_\pm$, but a similar definition of standard state is applicable; Appendix A18 discusses mean activity.

Consider next the equilibrium:

$$NaCl(s) \underset{}{\overset{H_2O}{\rightleftharpoons}} Na^+(aq) + Cl^-(aq) \tag{A17.28}$$

$$\text{(saturated solution)}$$

The standard state is a hypothetical solution of *unit mean concentration* $c_\pm$ and *unit mean activity coefficient* $f_\pm$. This choice has the required property that:

[5] It may be sufficiently accurate to replace molality m by molarity (concentration) c.

$$a_\pm(NaCl) = 1 = c^2 f_\pm^2 \tag{A17.29}$$

where the stoichiometric concentration c has, in this example, the same value as $c_\pm$. The concentration of each ionic species is also unity. Hence, the condition

$$f_\pm = f_+ = f_- \tag{A17.30}$$

holds, and so:

$$\mu^\ominus(\text{NaCl, s}) = \mu^\ominus(\text{Na}^+, \text{aq}) + \mu^\ominus(\text{Cl}^-, \text{aq}) \tag{A17.31}$$

Now consider an unsymmetrical electrolyte, such as $MgCl_2$, in solution of stoichiometric concentration c:

$$MgCl_2(s) \underset{}{\overset{H_2O}{\rightleftharpoons}} Mg^{2+}(aq) + 2Cl^-(aq) \tag{A17.32}$$

$$\text{(saturated solution)}$$

The equation

$$\mu(MgCl_2, s) = \mu^\ominus(MgCl_2, s) + RT \ln a(MgCl_2, aq) \tag{A17.33}$$

requires that, in the standard state,

$$a_\pm(MgCl_2, aq) = 1 = 4c_\pm^3 f_\pm^3 \tag{A17.34}$$

The standard state refers to unit mean concentration $4c^3$ and unit mean activity coefficient. The concentration of magnesium chloride in the standard state is $4^{-1/3}$ mol dm^{-3}, with Mg^{2+} being $4^{-1/3}$ mol dm^{-3} and that of Cl$^-$ ($2 \times 4^{-1/3}$ mol dm^{-3}). Thus, it is clear that $c_\pm^3 = 4^{-1/3}$ mol dm$^{-3} \times (2 \times 4^{-1/3}$ mol dm$^{-3})^2 = 1$. Apparently, the standard state for Cl$^-$ (aq) is different in NaCl(aq) and $MgCl_2$(aq) solutions in the same concentration terms. The following argument may be used to combat this apparent inconsistency.

Let 1 mol of Mg^{2+} ions be concentrated from the hypothetical solution of concentration $4^{-1/3}$ mol dm^{-3} to a new solution of unit concentration, while 2 mol of Cl$^-$ are diluted from the hypothetical solution of concentration $(2 \times 4^{-1/3}$ mol dm$^{-3})$ to a new solution of unit concentration. Both new solutions will be deemed to obey the requirement that $f(Mg^{2+}) = f(Cl^-) = 1$. Hence:

$$\Delta G(Mg^{2+}, aq) = -RT \ln(4^{-1/3}/c_0) = \frac{1}{3} RT \ln 4 \tag{A17.35}$$

and

$$\Delta G(Cl^-, aq) = -2RT \ln(2 \times 4^{-1/3}/c_0) = \frac{2}{3} RT \ln 4 - 2RT \ln 2 \tag{A17.36}$$

where c_0 is a reference concentration of 1 mol dm^{-3} in each case. The total free energy change is zero; hence:

$$\mu^{\ominus}(MgCl_2, s) = \mu^{\ominus}(Mg^{2+}, aq) + 2\mu^{\ominus}(Cl^-, aq) \qquad (A17.37)$$

which compares with Eq. (A17.31). As long as the standard state refers to *unit concentration and unit activity coefficient*, electrolyte solutions can be compared on a common thermodynamic basis.

....

REFERENCES A17

[1] Oakes CS *et al. J. Soln. Chem.* 1995; 24: 9.
[2] Stokes RH and Robinson RA. *Electrolyte Solutions*, 2nd rev. ed. Dover, 2002.

A18
Debye Limiting Law

A strong electrolyte in aqueous solution is fully dissociated, but may not behave as though the concentration of individual ions is equal to the given stoichiometric concentration. On dissolution in water the ions in an electrolyte solution become hydrated, that is, they become attached, albeit loosely, to a number of water molecules in a *hydration sphere*, and some of the ionic charge is distributed over this sphere. Positive and negative ions hydrated ions are attracted one to the other electrostatically, and every hydrated ion may be regarded as being surrounded by oppositely-charged species, so forming an *ionic atmosphere*. The hydrated ions cluster and disperse dynamically but over a period of time, which is long compared to the lifetime of any cluster, there will be a certain fraction of the total stoichiometric concentration which is unavailable as free ions. This effect is expressed as the activity a of a species i:

$$a_i = c_i f_i \qquad (A18.1)$$

where c_i and f_i are, respectively, the concentration and activity coefficient of the ith species. It is a further condition (Appendix A17) that:

$$\lim_{c_i \to 0} f_i = 1 \qquad (A18.2)$$

In many thermodynamic arguments, particularly those involving electrolytes, it is necessary to know the value of f or a. Although single-ion activities cannot be measured, the Debye–Hückel theory of electrolytes leads to an equation for calculating the activity coefficient of a single ion [1]. An approximation derived from the Debye–Hückel theory, and usually termed the *Debye limiting equation* (aka Debye *limiting law*), permits the calculation of activity coefficients in solutions of concentration up to 0.01 molar for 1:1 electrolytes and 0.005 molar for 1:2 electrolytes:

$$\ln f_i = -A z_+ |z_-| \sqrt{I} \qquad (A18.3)$$

where z is the numerical charge on an ion. The term A depends *inter alia* on the relative permittivity and the temperature of the solvent; for pure (conductivity) water at 298.15 K, A is 1.1734; I is the ionic strength of the solution, with concentration in mol dm^{-3}. Equation (A18.3) is often quoted in terms of $\log_{10}$ in which case A, for the given conditions, is $1.1734/\ln 10$, or 0.5096. Extracting the units of A is an unnecessary complicating factor for the application of the limiting law approximation.

For a generalized structure $A^{v_+}_{z_+} B^{v_-}_{z_-}$:

$$f^v_{\pm} = (f^{v_+}_+ f^{v_-}_-)^{1/v} \tag{A18.4}$$

where

$$v = v_+ + v_- \tag{A18.5}$$

The ionic strength I is defined by

$$I = 1/2 \sum_i c_i z_i^2 \tag{A18.6}$$

and includes *all* ionic components in a given electrolyte solution.

Example A18.1

Calculate $f_{\pm}$ for a solution containing magnesium nitrate of concentration (a) 0.002 mol dm^{-3} and (b) 0.0002 mol dm^{-3}.

(a) $I = 1/2[(0.002 \text{ mol dm}^{-3} \times 4) + 2(0.002 \text{ mol dm}^{-3} \times 1)] = 0.006$ mol dm^{-3}. Hence, $\ln f_{\pm} = -1.1734 \text{ mol}^{1/2} \text{ dm}^{-3/2} \times 2\sqrt{0.006 \text{ mol dm}^{-3}} = 0.834$.

(b) Only the ionic strength changes. Hence, $\ln f_{\pm} = -1.1734 \text{ mol}^{1/2}$ dm$^{-3/2} \times 2\sqrt{0.0006 \text{ mol dm}^{-3}} = 0.944$.

Extensions of Eq. (A18.3) for higher ionic strengths have been proposed, that by Davies [2] being one of the more satisfactory, although it is restricted to $I < 0.05$:

$$\ln f_{\pm} = -1.1734 z_+ |z_-| \left(\frac{\sqrt{I}}{(1 + \sqrt{I})} - 0.3I \right) \tag{A18.7}$$

The following data on potassium sulphate at 25 °C illustrate the difference between Eq. (A18.3) and the Davies equation:

c/mol dm^{-3}	I mol dm^{-3}	$f_{\pm}$ Eq. (A18.3)	$f_{\pm}$ Eq. (A18.7)	$f_{\pm}$ (expt)
0.001	0.003	0.88	0.88	0.88
0.005	0.015	0.75	0.77	0.77
0.010	0.030	0.67	0.69	0.69
0.050	0.15	0.40	0.47	0.51
0.100	0.3	0.28	0.35	0.42
0.500	1.5	0.06	0.28	0.37

....

REFERENCES A18

[1] Debye PJW and Hückel E. *Phys. Z.* 1923; 24: 185.
[2] Davies CW. *Ion Association*. Butterworths, 1962.

A19

Quantum Statistics

A19.1 Introduction

Each energy state g_i of energy E_i in a system of electrons in an atom is determined by the four quantum numbers n, l, m_l and m_s, or in the alternative notation n_x, n_y, n_z and m_s used in metallic bonding theory. Electrons are indistinguishable, but any electron of given values of n_x, n_y, n_z and m_s occupies the corresponding energy state; each state can be either empty or filled completely by a single electron. Two states of the same values of n_x, n_y, n_z occupied by electrons with spins $+1/2$ for one and $-1/2$ for the other constitute an electron pair, or fully occupied atomic orbital.

In classical (Maxwell–Boltzmann) statistics, the state of each electron is determined solely by its energy, that is to say, its energy states are non-degenerate. In quantum statistics, energy states are usually degenerate, as the following argument shows.

Consider an electron at 300 K. From Appendix A5 the mean kinetic energy of the electron is $(3/2)kT$. From Eqs. (5.39) and (5.45):

$$E = \frac{h^2}{2m_e a^2}(n_x^2 + n_y^2 + n_z^2) \qquad (A19.1)$$

If the electron be confined to a cubical box of side 1 mm, then equating E to $(3/2)kT$ gives:

$$n_x^2 + n_y^2 + n_z^2 \approx 2.6 \times 10^4 \qquad (A19.2)$$

Thus, an electron possesses the mean classical energy of $(3/2)kT$ if it selects any three values of the quantum numbers n_x, n_y and n_z that satisfy Eq. (A19.2); in other words, the energy levels are highly degenerate.

A19.2 Fermi–Dirac distribution

Consider a set of energy states, or cells, g_i, $(i = 1, 2, \dots, s)$ of similar energy E, and let the number of electrons in the set be N_s such that N_s cells are occupied and $(g_s - N_s)$ are empty. Following Section 4.11.1 and particularly Eq. (4.65), the number W_s of different ways in which N_s electrons can occupy g_s cells, allowing for the indistinguishability of electrons, is

$$W_s = \frac{g_s!}{(g_s - N_s)!N_s!} \tag{A19.3}$$

To a given range of energies, there corresponds for each range an equation like Eq. (A19.3), Hence, the total number W of distinguishable arrangements for an entire system of p sets is given by

$$W = \prod_{i=1}^{p} W_i = \prod_{i=1}^{p} \frac{g_i!}{(g_i - N_i)!N_i!} \tag{A19.4}$$

The most probable distribution is that which maximizes W. However, the maximization is subject to the constraints:

$$\sum_{i=1}^{p} N_i = N \tag{A19.5}$$

where N is the number of electrons in the entire system and

$$\sum_{i=1}^{p} N_i E_i = E \tag{A19.6}$$

where E is the total energy of the entire system. Using Stirling's approximation with Eq. (A19.4):

$$\ln W = \sum_{i=1}^{p} [g_i \ln g_i - (g_i - N_i) \ln(g_i - N_i) - N_i \ln N_i] \tag{A19.7}$$

For a maximum,[6] $d \ln W = 0$; thus:

$$d \ln W = \frac{\partial \ln W}{\partial N_1} dN_1 + \frac{\partial \ln W}{\partial N_2} dN_2 + \ldots + \frac{\partial \ln W}{\partial N_i} dN_i + \ldots + \frac{\partial \ln W}{\partial N_p} dN_p = 0 \tag{A19.8}$$

From Eqs. (A19.5) and (A19.6):

$$\sum_{i=1}^{p} dN_i = 0 \tag{A19.9}$$

and

$$\sum_{i=1}^{p} E_i dN_i = 0 \tag{A19.10}$$

[6] A consideration of entropy *via* Eq. (A19.20) will show that the derivative corresponds to a maximum [3].

In order to solve Eq. (A19.8) subject to the constraints Eqs. (A19.9) and (A19.10), the Lagrange method of undetermined multipliers (Section A19.3) is invoked. Using Eq. (A19.7) in Eq. (A19.8):

$$d \ln W = \sum_{i=1}^{p} [\ln(g_i - N_i) - \ln N_i] \, dN_i = 0 \qquad (A19.11)$$

Multipliers α and β are assigned such that

$$\sum_{i=1}^{p} [\ln(g_i - N_i) - \ln N_i + \alpha + \beta E_i] \, dN_i = 0 \qquad (A19.12)$$

Only two multipliers are required because there are only two constraining equations; Eq. (A19.12) must hold for any small change in N_i and the simplest general variation will be considered. More than one term of N_i must be involved so as to conform to Eq. (A19.9). If only two different terms N_i and N_j are varied, then since $dN_i = -dN_j$ to satisfy Eq. (A19.9), Eq. (A19.10) cannot be satisfied simultaneously because $E_i \neq E_j$. Hence, the simplest general variation involves three N_i terms.

Let dN_i ($i = 1, 2, 3$) be non-zero. Then the coefficients of these quantities must be zero, from Eq. (A19.12). Hence:

$$\ln(g_1 - N_1) - \ln N_1 + \alpha + \beta E_1 = 0 \qquad (A19.13)$$

and

$$\ln(g_2 - N_2) - \ln N_2 + \alpha + \beta E_2 = 0 \qquad (A19.14)$$

Since dN_3 has been chosen to be non-zero

$$\ln(g_3 - N_3) - \ln N_3 + \alpha + \beta E_3 = 0 \qquad (A19.15)$$

The procedure can be repeated for $dN_i \neq 0$ ($i = 2, 3, 4$), giving

$$\ln(g_4 - N_4) - \ln N_4 + \alpha + \beta E_4 = 0 \qquad (A19.16)$$

and so on, and in general for the ith state:

$$\ln(g_i - N_i) - \ln N_i + \alpha + \beta E_i = 0 \qquad (A19.17)$$

and the latter may be written in the form:

$$\frac{g_i}{N_i} - 1 = \exp(\alpha) \exp(\beta E_i) \qquad (A19.18)$$

which is one formulation of the Fermi–Dirac distribution. The multipliers α and β can be determined through Eq. (A19.5), but temporarily $\exp(\alpha)$ is written as A. To determine β it may be noted that at high temperatures the number of states that are energetically available is very large, so that $g_i/N_i > > 1$, whereupon by inverting Eq. (A19.18):

$$N_i = g_i A^{-1} \exp(-\beta E_i) \tag{A19.19}$$

In the high temperature limit, Boltzmann statistics apply (Appendix A4), so that β may be equated to $1/(kT)$. A more rigorous proof may be obtained starting from equation [3]

$$S = k \ln W_{\text{max}} \tag{A19.20}$$

From Eq. (A19.18):

$$N_i/g_i = \frac{1}{A \exp[E_i/(kT)] + 1} \tag{A19.21}$$

The energy E_F, the Fermi energy, is defined such that

$$A = \exp[-E_F/(kT)] \tag{A19.22}$$

then

$$N_i/g_i = \frac{1}{\exp[(E_i - E_F)/(kT)] + 1} \tag{A19.23}$$

Writing $f(E) = N_i/g_i$, where $f(E)$ is the probability that a state of energy E is occupied:

$$f(E) = \frac{1}{\exp[(E - E_F)/(kT)] + 1} \tag{A19.24}$$

which is a convenient formulation of the Fermi–Dirac distribution. The original papers by Fermi and Dirac may be found in the literature [1, 2].

A19.3 Lagrange undetermined multipliers

The method of Lagrange undetermined multipliers is a strategy for finding the local maxima and minima of a function subject to equality constraints.

Consider Fig. A19.1; it shows a portion of the surface $z = xy$ in xyz space. The constraining equation is $g(x, y) = 0$, which is the circle $x^2 + y^2 - 9 = 0$ in the example. The extrema for the function xy are $\pm\infty$ but the result needs to be constrained to lie on $g(x, y)$. Let the position P of the maximum in the positive xy quadrant be $x = \mu$, $y = v$, that is, a position on the curve where $z = c$, a constant.

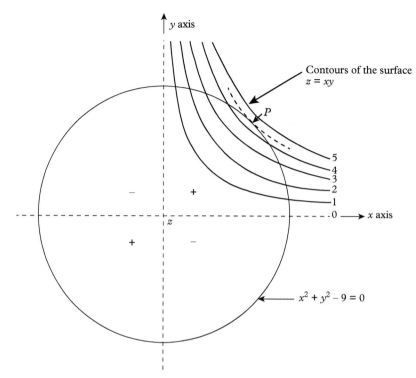

Fig. A19.1 Portion of the surface $z = xy$ shown by contours of unit interval; the constraint $g(xy)$ is the circle $x^2 + y^2 - 9 = 0$. The dashed curve is a portion of one region of the xy surface passing through the required maximum P.

Since two curves $z = c$ and $g = 0$ touch; they have a common tangent. Thus, at the point (μ, ν), $dz = 0$, that is,

$$dz = \frac{\partial z}{\partial x}dx + \frac{\partial z}{\partial y}dy = 0 \qquad (A19.25)$$

or

$$dy/dx = -\frac{\partial z/\partial x}{\partial z/\partial y} \qquad (A19.26)$$

Hence:

$$\frac{\partial z/\partial x}{\partial z/\partial y} = \frac{\partial g/\partial x}{\partial g/\partial y} \qquad (A19.27)$$

Introducing a proportionality constant λ the two equations

$$\partial z/\partial x + \lambda \partial g/\partial x = 0 \qquad (A19.28)$$

and

$$\partial z/\partial y + \lambda \partial g/\partial y = 0 \qquad (A19.29)$$

are satisfied. Also, since

$$g(x, y) = 0 \tag{A19.30}$$

μ, ν and λ can be determined. The function

$$h(x, y, \lambda) = z(x, y) + \lambda g(x, y) \tag{A19.31}$$

can be formed and x and y treated as independent variables. Thus:

$$h = xy + \lambda(x^2 + y^2 - 9) \tag{A19.32}$$

whence

$$\partial h/\partial x = y + 2\lambda x = 0 \tag{A19.33}$$

and

$$\partial h/\partial y = x + 2\lambda y = 0 \tag{A19.34}$$

which, together with

$$x^2 + y^2 - 9 = 0 \tag{A19.35}$$

enables a solution for the four points:

$$
\begin{aligned}
\mu &= 3/\sqrt{2} & \nu &= 3/\sqrt{2} \\
\mu &= -3/\sqrt{2} & \nu &= -3/\sqrt{2} \\
\mu &= 3/\sqrt{2} & \nu &= -3/\sqrt{2} \\
\mu &= -3/\sqrt{2} & \nu &= 3/\sqrt{2}
\end{aligned}
\tag{A19.36}
$$

The first two results give maxima in the positive quadrant of Fig. A19.1, with $xy = 4.5$, whereas the second two results give minima in the negative regions of the figure, with $xy = -4.5$.

The discussion is now extended to its application in the Fermi–Dirac distribution. Let the function to be maximized be

$$z = f(x_1, x_2, \ldots, x_n) \tag{A19.37}$$

subject to

$$g_1(x_1, x_2, \ldots, x_n) = 0 \tag{A19.38}$$

and

$$g_2(x_1, x_2, \ldots, x_n) = 0 \tag{A19.39}$$

At a maximum $dz = 0$, that is,

$$\frac{\partial z}{\partial x_1} dx_1 + \frac{\partial z}{\partial x_2} dx_2 + \ldots, + \frac{\partial z}{\partial x_n} dx_n = 0 \qquad \text{(A19.40)}$$

Following on from the constraining equations:

$$\frac{\partial g_1}{\partial x_1} dx_1 + \frac{\partial g_1}{\partial x_2} dx_2 + \ldots, + \frac{\partial g_1}{\partial x_n} dx_n = 0 \qquad \text{(A19.41)}$$

and

$$\frac{\partial g_2}{\partial x_2} dx_1 + \frac{\partial g_2}{\partial x_2} dx_2 + \ldots, + \frac{\partial g_2}{\partial x_n} dx_n = 0 \qquad \text{(A19.42)}$$

Multiplying Eq. (A19.41) by λ_1 and Eq. (A19.42) by λ_2 and adding to Eq. (A19.40) leads to:

$$\sum_{i=1}^{n} \left(\frac{\partial z}{\partial x_i} + \lambda_1 \frac{\partial g_i}{\partial x_i} + \lambda_2 \frac{\partial g_i}{\partial x_i} \right) dx_i = 0 \qquad \text{(A19.43)}$$

and λ_1 and λ_2 are chosen such that the coefficients of dx_1 and dx_2 in Eq. (A19.43) are zero, that is,

$$\frac{\partial z}{\partial x_1} + \lambda_1 \frac{\partial g_1}{\partial x_1} + \lambda_2 \frac{\partial g_2}{\partial x_1} = 0 \qquad \text{(A19.44)}$$

and

$$\frac{\partial z}{\partial x_2} + \lambda_1 \frac{\partial g_1}{\partial x_2} + \lambda_2 \frac{\partial g_2}{\partial x_2} = 0 \qquad \text{(A19.45)}$$

Since the n variables are subject to two conditions only, $n - 2$ variables are independent. Therefore, the coefficients of dx_i $(i = 3, 4, \ldots n)$ in Eq. (A19.43) must vanish. Hence:

$$\frac{\partial z}{\partial x_i} + \lambda_1 \frac{\partial g_i}{\partial x_i} + \lambda_2 \frac{\partial g_i}{\partial x_i} = 0 \; (i = 1, 2, \ldots, n) \qquad \text{(A19.46)}$$

These n equations and the two constraining equations determine the $n + 2$ unknowns. Further discussions on the Lagrange method of undetermined multipliers are available in the literature [3, 4].

....

REFERENCES A19

[1] Fermi E. *Z. Phys.* 1926; 36: 902.
[2] Dirac PAM. *Proc. Roy. Soc. (Lond.),* 1926; A112: 661.
[3] Reif F. *Fundamentals of Statistical and Thermal Physics.* McGraw-Hill, 1965.
[4] Kittel C. *Introduction to Solid State Physics,* 4th ed. John Wiley & Sons, 1971.

A20
Reciprocal Space and Reciprocal Lattice

A20.1 Reciprocal space

The reciprocal lattice concept was introduced by Ewald in 1913 [1]. He applied it some years later in the interpretation x-ray diffraction photographs [2], although the procedure was conceived independently a year earlier by Bernal [3].

In reciprocal space a lattice, the reciprocal lattice, is defined for each of the fourteen real space Bravais lattices; it is derived here first by the following construction, as applied to a monoclinic lattice in projection on the x, z plane.

A primitive unit cell is outlined by vectors $\mathbf{a}$ and $\mathbf{c}$, with three families of planes indicated by their Miller indices; $\mathbf{b}$ is normal to the plane of projection. Lines are constructed from the origin O that are normal to the families of Bravais lattice planes (hkl) shown (Fig. A20.1a).

Along each of these lines, reciprocal lattice points hkl are defined such that the distances to these points from the origin are the reciprocals of the corresponding interplanar spacings in Bravais space. Thus, in the figure, the families of planes (100), (101) and (001) give rise to reciprocal lattice points at distances from the origin that are proportional to $1/d_{(100)}$, $1/d_{(101)}$ and $1/d_{(001)}$, where $d_{(100)} = OP$, $d_{(101)} = OQ$ and $d_{(001)} = OR$. In general, a reciprocal length d^* is given by

$$d_{hkl}^* = 1/d_{(hkl)} \tag{A20.1}$$

The vectors $\mathbf{d}_{100}^*$, $\mathbf{d}_{010}^*$ and $\mathbf{d}_{001}^*$ may be taken to define the translation vectors $\mathbf{a}^*$, $\mathbf{b}^*$ and $\mathbf{c}^*$ of a unit cell in the reciprocal lattice (aka the reciprocal unit cell). The following equations for the monoclinic reciprocal lattice can now be determined from Fig. A20.1a:

$$d_{100}^* = 1/d_{(100)} = a^* \tag{A20.2}$$

But $d_{(100)}$ is equal to $a \sin \beta$. Hence:

$$a^* = 1/(a \sin \beta) \tag{A20.3}$$

Similarly,

$$c^* = 1/(c \sin \beta) \tag{A20.4}$$

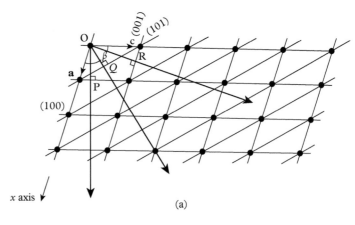

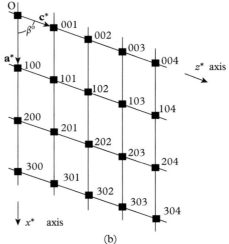

Fig. A20.1 (a) Monoclinic lattice in projection on the *x*, *z* plane in real space, showing primitive unit cells with basic translations *a* and *c*, and traces of the (100), (101) and (001) families of planes; $\beta = \angle O\mathbf{a} - O\mathbf{c}$. (b) Monoclinic reciprocal lattice constructed from the lattice in (a) showing the lengths *a** and *c** of the reciprocal unit cell, in projection on the *x**, *z** plane in reciprocal space; $\beta^* = \angle O\mathbf{a}^* - O\mathbf{b}^*$. [*Structure Determination by X-ray Crystallography*, 5th ed. Springer Science+Business Media, 2013d; reproduced by courtesy of Springer Science+Business Media, NY.]

but

$$b^* = 1/b \qquad (A20.5)$$

because d^*_{010} is normal to the *a, c* plane. The unique β^* angle is given by

$$\beta^* = (180° - \beta) \qquad (A20.6)$$

The *a**, *c** plane from this construction is shown in Fig. A20.1b. Furthermore,

$$\mathbf{a} \cdot \mathbf{a}^* = aa^* \cos \angle \mathbf{aa}^* = a(1/a) \cos(\beta - 90°) = 1 \qquad (A20.7)$$

and similarly for $\mathbf{b} \cdot \mathbf{b}^*$ *and* $\mathbf{c} \cdot \mathbf{c}^*$. For the mixed products:

$$\mathbf{a} \cdot \mathbf{b} = ab^* \angle \cos \mathbf{ab}^* = 0 \qquad (A20.8)$$

and similarly for all other such products. The relationships of Eqs. (A20.7) and (A20.8) apply to all crystal systems.

The reciprocal lattice is considered next in a general manner; knowledge of simple vector manipulations are assumed. In Fig. A20.2, the z^*-axis is normal to the plane a, b. Since $\mathbf{c} \cdot \mathbf{c}^* = 1 = cc^* \cos \angle COR$:

$$c^* = |\mathbf{c}^*| = 1/c \cos \angle COR \qquad \text{(A20.9)}$$

But $\mathbf{c}^*$ is normal to both $\mathbf{a}$ and $\mathbf{b}$, so that it lies in the direction of their vector product. Hence:

$$\mathbf{c}^* = \eta(\mathbf{a} \times \mathbf{b}) \qquad \text{(A20.10)}$$

where η is a constant. Since $V = \mathbf{c}^* \cdot (\mathbf{a} \times \mathbf{b})$, the scalar product of Eq. (A20.10) and $\mathbf{c}$ is

$$\mathbf{c} \cdot \mathbf{c}^* = \eta \mathbf{c} \cdot (\mathbf{a} \times \mathbf{b}) = \eta V = 1$$

so that $\eta = \frac{1}{V}$; hence, from Eq. (A20.10):

$$\mathbf{c}^* = \frac{(\mathbf{a} \times \mathbf{b})}{\mathbf{c} \cdot (\mathbf{a} \times \mathbf{b})} \qquad \text{(A20.11)}$$

and similarly for $\mathbf{a}^*$ and $\mathbf{b}^*$ by cyclic permutation. In scalar form, Eq. (A20.11) becomes:

$$c^* = |\mathbf{c}^*| = \frac{ab \sin \gamma}{V}$$

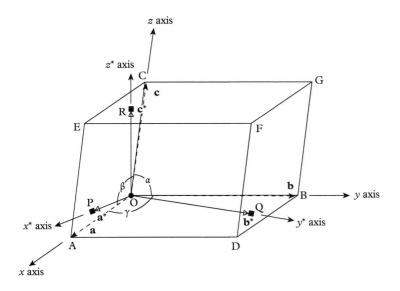

Fig. A20.2 General, triclinic unit cell in real space; **a**, **b** and **c** are the unit cell vectors in real space, and **a***, **b*** and **c*** the corresponding reciprocal space unit cell vectors. Inter-axial angles in real space α, β, γ, where $\alpha = \angle BOC$, and β, γ by cyclic permutation; in reciprocal space the angles are α^*, β^*, γ^*, where $\alpha^* = \angle QOR$, and β^*, γ^* by cyclic permutation. [*Structure Determination by X-ray Crystallography*, 5th ed. Springer Science+Business Media, 2013.]

and

$$V = abc\sqrt{1 - \cos^2\alpha - \cos^2\beta - \cos^2\gamma + 2\cos\alpha\cos\beta\cos\gamma} \qquad \text{(A20.12)}$$

and similarly for a^* and b^* by cyclic permutation.

The angles are given by [1, 2]:

$$\cos\gamma^* = \frac{-\cos\gamma + \cos\alpha\cos\beta}{\sin\alpha\sin\beta} \qquad \text{(A20.13)}$$

with $\cos\alpha^*$ and $\cos\beta^*$ obtained by cyclic permutation.

It remains now to show that the reciprocal lattice points so constructed form a true lattice. The vector normal to the plane (hkl) is $h(\mathbf{b}\times\mathbf{c}) + k(\mathbf{c}\times\mathbf{a}) + l(\mathbf{a}\times\mathbf{b})$[4]. Dividing by V, and denoting the resulting vector $\mathbf{d}^*_{hkl}$:

$$\mathbf{d}^*_{hkl} = \left\{ h\frac{(\mathbf{b}\cdot\mathbf{c})}{\mathbf{a}\cdot(\mathbf{b}\times\mathbf{c})} + k\frac{(\mathbf{c}\times\mathbf{a})}{\mathbf{b}\cdot(\mathbf{c}\times\mathbf{a})} + l\frac{(\mathbf{a}\cdot\mathbf{b})}{\mathbf{c}\cdot(\mathbf{a}\cdot\mathbf{b})} \right\} = h\mathbf{a}^* + k\mathbf{b}^* + l\mathbf{c}^* \quad \text{(A20.14)}$$

Since h, k and l are integers, the vectors $\mathbf{d}^*_{hkl}$ drawn from the common origin form a lattice, the reciprocal lattice, with translation vectors $\mathbf{a}^*$, $\mathbf{b}^*$ and $\mathbf{c}^*$ and interaxial angles:

$$\left. \begin{aligned} \alpha^* &= \angle\mathbf{b}^*\mathbf{c}^* = \angle 010 - 001 \\ \beta^* &= \angle\mathbf{c}^*\mathbf{a}^* = \angle 010 - 001 \\ \gamma^* &= \angle\mathbf{a}^*\mathbf{b}^* = \angle 100 - 010 \end{aligned} \right\} \qquad \text{(A20.15)}$$

It is a standard notation to denote reciprocal lattice points by the Miller indices of the family of planes in the Bravais lattice from which they were derived, but written without parentheses.

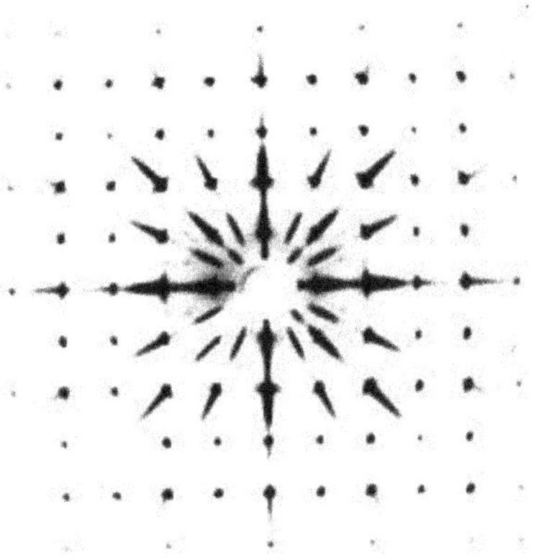

A20.2   Reciprocal lattice and Ewald's construction

The Bragg construction for diffraction (Appendix A3) requires a consideration of the distribution of planes, which is not always easy to visualize. An x-ray photograph of a crystal is a picture of its reciprocal lattice (Fig. A20.3). It is a regular array of spots governed in position by the space group symmetry of the crystal, and in intensity by the nature and positions of the atoms in the crystal structure.

The *Ewald sphere* construction [2, 4–5] discusses the diffraction record in terms of the geometry of the recorded x-ray spot pattern (Fig. A20.4). A sphere of radius unity in reciprocal space is constructed on the x-ray beam as the diameter AQ, where Q is the origin of the reciprocal lattice: the crystal under consideration is at the centre C of the sphere. From the construction, $AQ = 2$ and $\angle APQ = 90°$. Thus, $QP = AQ\sin\theta_{hkl} = 2\sin\theta_{hkl}$, which, from Eqs. (A3.2) and (A20.1) is λ/d_{hkl}. Thus, P is the reciprocal lattice point corresponding to the hkl family of

Fig. A20.3 X-ray precession photograph of the $0kl$ layer of reciprocal space for a crystal of a cubic perovskite, space group $Pm\bar{3}m$. The distances between adjacent spots are the reciprocals of the translation distances b and c in real space; they are equal in this example because this perovskite is a cubic crystal. No systematic absences of spots arise because of the complete absence of translational symmetry in the structure.[a] Streaks attached to the more intense spots indicate a lack of complete monochromatization of the incident beam; the crystal effectively selects the wavelengths from those available in the incident x-ray spectrum that satisfy the Bragg equation for hkl reflections, as in a Laue photograph [5].

[a] Symmetry dependent absences are regular, governed by the space group of the crystal.

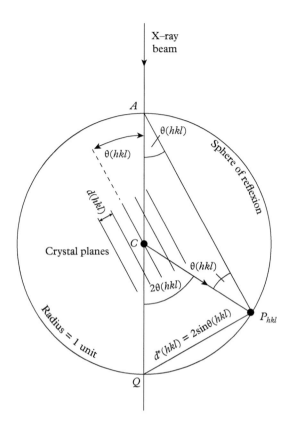

Fig. A20.4 Interpretation of crystal diffraction maxima by the Ewald sphere (aka sphere of reflection) construction, assuming without loss of generality the zero-layer x^*, y^*, 0 of the reciprocal lattice: C crystal, AQ incident x-ray beam and θ_{hkl} its glancing angle of incidence, Q origin of reciprocal lattice, P reciprocal lattice point *hkl* intersecting Ewald sphere and giving rise to a reflected beam in the direction CP. [*Structure Determination by X-ray Crystallography*, 5th ed. Springer Science+Business Media 2013; reproduced by courtesy of Springer Science+Business Media, NY.]

planes and CP is the direction of the reflected x-ray beam. Hence, an x-ray reflection from a family of planes (*hkl*) occurs when the reciprocal lattice point *hkl* intersects the Ewald sphere (aka sphere of reflection), and the direction of the reflected beam is from the crystal C through the point $P(hkl)$. Note that although the term x-ray reflection is used, following the Bragg equation, the process is one of diffraction or, more specifically, of combined diffraction and interference processes.

Another illustration of diffraction through the Ewald sphere concept is shown by Figs. A20.5 and A20.6; the annotations correspond to those of Fig. A20.4. The radius of the Ewald sphere in a practical context is set at $1/\lambda$. A reflection arises at the point P_{hkl} on the Ewald sphere if $\mathbf{QP} = \mathbf{CP} - \mathbf{CQ}$, that is, $\mathbf{QP} = (1/\lambda)(\mathbf{s} - \mathbf{s_0})$, where $\mathbf{s}$ and $\mathbf{s_0}$ are unit vectors in the reflected and incident x-ray beams respectively. Other reciprocal lattice points, such as S, that intersect the Ewald sphere will also give rise to reflections, the direction being CS for the point S. The Ewald sphere is rotated about the origin Q of the reciprocal lattice (in practice the crystal is rotated, taking the reciprocal lattice with it). As reciprocal lattice points intersect successively the Ewald sphere, the corresponding reflections are obtained.

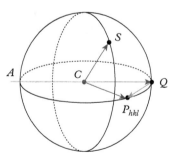

Fig. A20.5 Ewald sphere: reflection is obtained along the direction $|\mathbf{CP}|$ as the reciprocal lattice point P intersects the Ewald sphere; the annotations correspond to those in Fig. A20.4.

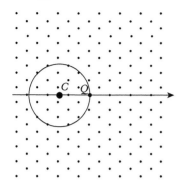

Fig. A20.6 Observable reflections, those for which $|\mathbf{QP}| \leq \lambda/2$, lie within the limiting sphere, radius $2/\lambda$ (= diameter of Ewald sphere). The theoretical maximum d–value ($\sin\theta = 1$) for Cu $K\alpha$ x-radiation is *ca.* 12.97 nm^{-1}.

Only those reflections are observable for which $|\mathbf{QP}| \leq \lambda/2$, the diameter of the Ewald sphere. These reflections lie within a sphere, centre Q and radius $2/\lambda$ that is termed the *limiting sphere*, of which the circle in the figure is one section. From the Bragg equation, it is clear that if $\lambda > d_{max}$ ($\sin\theta > 1$) there can be no reflection. The limiting sphere represents the possible limit of resolution of crystal diffraction. Thus, for Cu $K\alpha$ the maximum radius of the limiting sphere is *ca.* 12.97 nm^{-1}.

Exercise A6

A crystal has the unit-cell dimensions $a = 0.1230$ nm, $b = 0.4560$ nm and $c = 0.7890$ nm, and angles $\alpha = 90.00°$, $\beta = 95.00°$ and $\gamma = 105.00°$. Calculate the constants of the reciprocal unit cell and the volumes of the two unit cells. (Answer at end of Appendices.)

REFERENCES A20

[1] Ewald PP. *Z. Phys.* 1913; 144: 465.
[2] Ewald PP. *Math. Ann.* 1921; 56: 615.
[3] Hodgkin DMC. *Biogr. Mems. Fell. R. Soc.* 1980; 26: 28.
[4] Ladd M and Palmer R. *Structure Determination by X-ray Crystallography*, 5th ed. Springer Science+Business Media, 2013.
[5] Ladd M. *Symmetry of Crystals and Molecules*. Oxford University Press, 2014.

A21
Wave Packets

A21.1 Particle and wave

The de Broglie equation draws together the experimentally observed wave and particle properties of the electron through the well-known equation:

$$p = h/\lambda, \tag{A21.1}$$

whereby momentum, a particle property, and wavelength, a wave property, are reconciled through the Planck constant.

A21.2 Localized particle: superposition of waves

From the de Broglie equation (A21.1) that relates the particle and wave properties of an electron, it follows that:

$$p = h/\lambda = m_e v \; (\text{'vee'})$$

and $\tag{A21.2}$

$$\tfrac{1}{2} m_e v^2 = E = h\nu \; (\text{'nu'})$$

whence

$$\lambda \nu = h/m_e v \times \tfrac{1}{2} m_e v^2 = \tfrac{1}{2} v \tag{A21.3}$$

The speed $\lambda \nu$ of the wave appears to be one half the speed v of the electron, and must be considered further.

Consider an electron moving *in vacuo*. The Born interpretation of the wave-function ψ states that $|\psi|^2 d\tau$ is the probability of finding the electron in a volume element $d\tau$, or dx in one dimension. If this probability is very high at any time t, then there is a high probability that there will be regions of zero electron density where the electron has left or not yet encountered. Given that the electron has energy E, it must possess finite values of momentum and wavelength (or wave number), so that the wave, in one dimension, has the spatially infinite form:

$$\psi(x, t) = A \sin(kx - \omega t) \tag{A21.4}$$

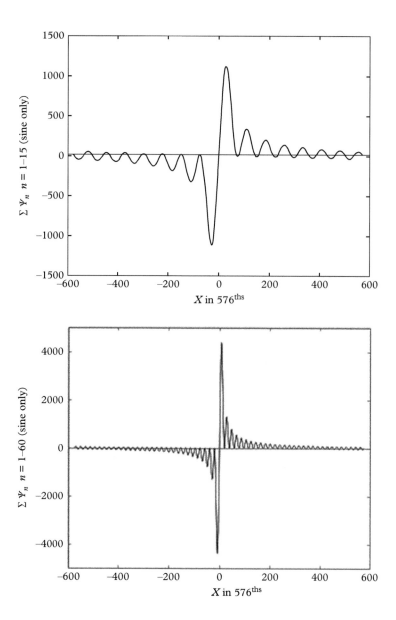

Fig. A21.1 The superposition of sine waves of the form $A\sin(\pi nX/a)$, $X = -576–576$: (a) $n = 1–15$; (b) $n = 1–60$. The succession is moving towards the representation of a localized particle, but the functions are not totally harmonic.

where $k = 2\pi/\lambda$ and $\omega = 2\pi\nu$. In order to begin to represent a localized particle, two waves of the type Eq. (A21.4), differing by the small amounts $\pm\Delta k$ and $\pm\Delta\nu$, must be compounded. By the usual trigonometric manipulation, the resultant wave expression is

$$\psi'(x, t) = 2A\sin(kx - \omega t)\cos(\Delta kx - \Delta\omega t) \qquad (A21.5)$$

The first term on the right-hand side of Eq. (A21.4) is a wave oscillating at the average frequency, but modified by the following cosine term which oscillates by $\pm\pi/\Delta k = \pm\Delta\lambda/2$. Thus, waves originally in phase at the origin become completely out of phase but subsequently come into phase again at the further distance of $2(\pm\Delta\lambda/2)$, and so on.

The beat frequencies decompose the continuous wave into a series of wave packets. A single travelling electron requires a single packet, which is obtained by a superposition of waves of the different orders represented by Δk. Since there is a multitude of wavelengths the out-of-phase waves can be in phase only at the origin.

The superposition of sine waves of the form $A\sin(\pi nX/a)$, where n represents frequency and X/a runs from −576 through zero to 576, is shown in Fig. A21.1, in which the curves represent a succession of 15 waves in (a) and 60 waves in (b). The diagrams show an increasing tendency to form a localized particle. However, it is not a true localized particle since only sine terms have been used and the plots are antisymmetric across the origin: the sine terms are those employed also in the electron-in-a-box equation (Section 2.10) and will be discussed further shortly.

A21.3   Phase and group velocities

Electron waves are unlike electromagnetic waves but resemble water waves in that they differ in phase and group velocities. Referring to Eq. (A21.5), the first term on the right-hand side represents waves with the speed $\tfrac{1}{2}v$ whereas the cosine term has the speed $\Delta\omega/\Delta k$.

From Eq. (A21.2):

$$\mathrm{d}E/\mathrm{d}p = p/m_\mathrm{e} = v \qquad (A21.6)$$

Since $E = h\nu = h\omega/2\pi$ and $p = h/\lambda = hk/2\pi$, it follows that

$$\Delta\omega/\Delta k = \Delta E/\Delta p = v \qquad (A21.7)$$

Thus, the wave packet travels at the same speed v as the electron, whereas the waves within the packets travel at $\tfrac{1}{2}v$.

Example A21.1

A particular glass has the refractive indices for light of $n = 1.49$ at a wavelength of 400 nm, and $n = 1.51$ at 600 nm; then $\delta\lambda = 0.002$ and $<n> = 1.50$. From the foregoing, the phase velocity $v_p = c/<n> = 2.9979 \times 10^8$ m s$^{-1}/1.50 = 1.999 \times 10^8$ m s^{-1}. The group velocity v_g is defined by $v_g = \Delta\omega/\Delta k$, and some simple algebra follows:

$$n = c/v_p = ck/\omega$$

continued

Example A21.1 *continued*

So

$$\mathrm{d}n/\mathrm{d}\lambda = (\mathrm{d}n/\mathrm{d}k)\,(\mathrm{d}k/\mathrm{d}\lambda) = (c/\omega) - (kc/\omega^2)\,(\mathrm{d}\omega/\mathrm{d}k) = (c/\omega) - (kcv_g/\omega^2)$$

Now,

$$\mathrm{d}k/\mathrm{d}\lambda = -2\pi/\lambda^2 = -k/\lambda;$$

Thus,

$$\mathrm{d}n/\mathrm{d}\lambda = -(k/\lambda)(c/\omega - ckv_g/\omega^2\lambda) = -c/(\lambda v_p) + (cv_g/v_p^2\lambda).$$

Re-arranging for v_g :

$$v_g = (v_p^2\lambda/c)\,(c/v_p\lambda + \mathrm{d}n/\mathrm{d}\lambda) = v_p + (v_p\lambda/n)\,(\mathrm{d}n/\mathrm{d}\lambda)$$

$$= v_p \left(1 + (\lambda/n)\frac{\mathrm{d}n}{\mathrm{d}\lambda}\right)$$

Applying this result to the problem in hand: $\lambda/n \approx <\lambda>/<n> = $ 500 nm/1.5, $= 333.33$ nm and $\mathrm{d}n/\mathrm{d}\lambda = 0.02/(200\ \text{nm}) = 10^{-4}\ \text{nm}^{-1}$. Hence, $v_g = 1.999 \times 10^8\ \text{m s}^{-1}(1 + 333.33\ \text{nm} \times 10^{-4}\ \text{nm}^{-1}) = 1.932 \times 10^8\ \text{m s}^{-1}$, very similar to v_p in the given conditions.

Example A21.2

A travelling wave represented in one-dimensional space by the expression $A \exp \mathrm{i}(kx - \omega t)$ when compared with the wave expression $A \exp \mathrm{i}\,(px - Et)$ shows that $k = p/h$ and that $\omega = E/h$. The equation $E = p^2/2m_e$ implies $\hbar\omega = \hbar^2 k^2/2m_e$ or $\omega = \hbar k^2/2m_e$. The phase (wave) velocity $v_p = \omega/k = \hbar k/2m_e$, whereas the group velocity $v_g = \mathrm{d}\omega/\mathrm{d}k = \hbar k/m_e$. Hence, $v_g = 2v_p$. The classical velocity v_c follows from $E = p^2/2m_e = \frac{1}{2}m_e v_c^2$, so that $v_c = p/m = \hbar k/m_e$; thus, the classical velocity is also the group velocity.

A21.4   Localized particle

The situations just discussed may be approached in another way. If a particle is situated at a specific location its representative wavefunction will be of large magnitude, ideally infinite, at that site and zero elsewhere. This function can be created by the superposition of a large number of harmonic functions of the type $A \exp ikx$, or its equivalent in terms of sine and cosine functions.

The wavefunction $A \exp ikx$ describes a particle travelling in the positive x direction with the precise momentum given by $p_x = k\hbar$. However, from the uncertainty principle (Section 2.7) the position of the particle is then

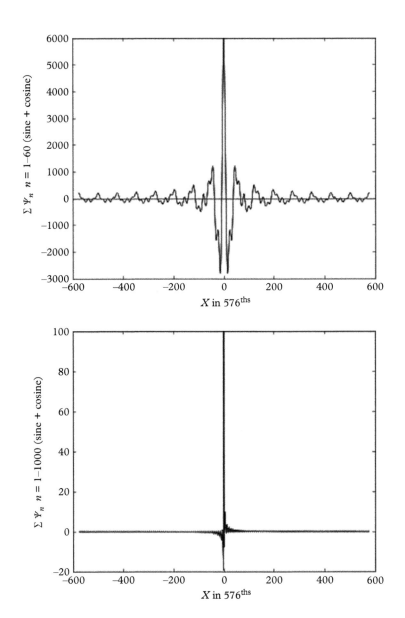

Fig. A21.2 The superimposition of harmonic waves of the form $A[\sin(\pi nX/a) + \cos \sin(\pi nX/a)]$, $X = -576$–576: (a) $n = 1$–60, (b) $n = 1$–10^3, (c) $n = 1$–10^4, (d) $n = 1$–10^6, all of the same range of X as in Fig. A21.1. For the very large (albeit not infinite) number of waves in (d), the particle is completely located within limits of observation but its momentum now is indeterminate.

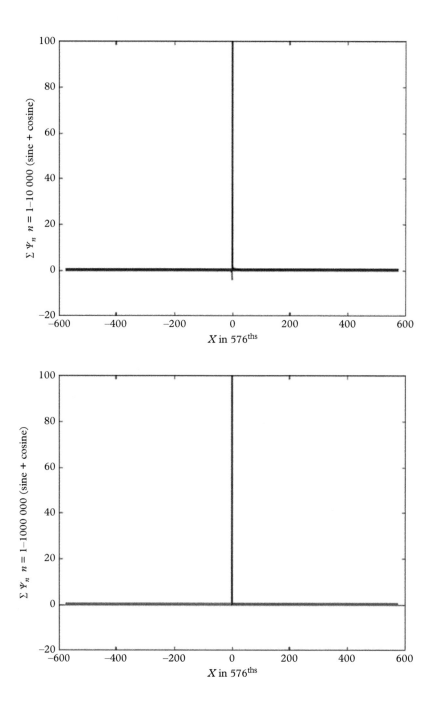

Fig. A21.2 continued

indeterminate. Conversely, if the location of the particle is specified then its momentum cannot be known. The wavefunction of a localized particle can be created by the superposition of exponential functions or, equivalently, a combination of sine and cosine functions: Fig. A21.2 is Fig. A21.1b to which has been added the corresponding cosine functions. Thus, the linear combination of wavefunctions is beginning to form a localized wavefunction or wave packet.

As the position becomes more and more localized by the continued addition of harmonics, so the particle is determined more and more exactly in position and less and less precise in momentum. In the limit, represented here by the superimposition of 10^6 waves, the visible result is a mathematical line of indeterminate height (Fig. A21.2d).

These results may be expressed through the Heisenberg uncertainty principle (Section 2.7) in the form:

$$\Delta p \Delta x \geq \hbar/2 \tag{A21.8}$$

Each uncertainty in Eq. (A21.8) is defined as the rms deviation from its mean, that is, $\Delta p = \sqrt{<p^2> - <p>^2}$ and $\Delta x = \sqrt{<x^2> - <x>^2}$. If, as noted above, the momentum is known precisely, that is, $\Delta p = 0$, then Eq. (A21.8) implies $\Delta x = \infty$, which is an expression of indeterminacy of position. Conversely, if $\Delta x = 0$, as implied by Fig. A21.2d, then $\Delta p = \infty$, and the momentum is totally uncertain.

Although these results have been developed in one-dimensional space for convenience, they apply equally in two- or three-dimensional space.

Example A21.3

What is the least uncertainty in the speed of a standard cricket ball of mass 160 g when it is 1 μm from the bat?

$$\Delta p \Delta x = m \Delta v \Delta x; \text{ hence, } \Delta v = \frac{\hbar}{2m\Delta x} = \frac{6.6261 \times 10^{-34} \text{ J s}}{4 \times \pi \times 0.160 \text{ kg} \times 1 \times 10^{-6} \text{ m}}$$

$$= 3.3 \times 10^{-28} \text{ m s}^{-1}. \tag{A21.9}$$

Exercise A7

A proton travels at a speed of 0.5 Mm s^{-1}. If the uncertainty in its momentum is to be reduced to 0.02%, calculate the uncertainty in its position. (Answer at end of Appendices.)

Answers to Exercises

A1

b (100), e (010), d (001),

f (110), g ($1\bar{1}0$), c (101), a ($10\bar{1}$), p (011), o ($01\bar{1}$),

r (111), m ($1\bar{1}1$), q ($11\bar{1}$), n ($1\bar{1}\bar{1}$).

A2

24.28°.

A3

$r = 2$ nm, $\theta = 135°$, $\phi = 45°$ or $135°$ ($180° - 45°$).

A4

(a) -1.384×10^{-3} J m^{-1} ($\equiv$ N) (b) -1.384×10^{-5} J.

A5

5033 V m^{-1} ($\equiv$ N C^{-1}), *away* from the fly.

A6

$\mathbf{a}^* = 0.4514$ nm^{-1}, $\mathbf{b}^* = 2.2710$ nm^{-1}, $\mathbf{c}^* = 1.2726$ nm^{-1},

$\alpha^* = 88.66°$, $\beta^* = 84.82°$, $\gamma^* = 74.94°$.

$V = 0.04257$ nm^3, $V^* = 23.490$ nm^{-3}.

A7

3.152×10^{-10} m ($\equiv$ 315.2 pm).

Tutorial Solutions

A sum can be put right: but only by going back till you find the error and working it afresh from that point, never by simply going on.

C.S. Lewis, *The Great Divorce*

Introductory note

The solutions to the sets of problems are given here in detail. In some cases it has been found worthwhile to introduce associated material that has not been treated specifically in the text, in the hope that the reader will be encouraged to extend that which has been covered.

SI units are used in the problems, as also in the text; for other commonly-used energy units, a tabulation of conversion factors is given on page xxiv. It is good practice in solving problems to keep a track of the units in problems: the sum total of the units in the working of a problem must agree with that required by the result, otherwise an error probably exists in the organization or working of the solution.

The computer programs in the Web Program Suite are provided as IBM-compatible .EXE files written Fortran 90 (Section 7.1), a language which retains its importance for fast numerical computations. Developed over the past sixty years, it is continuously supported because the number of important scientific programs written in this language is vast. The plotting routines have been prepared in the Python language (Section 7.2). It has a different structure from that of Fortran and is well suited to non-numerical projects, although it can process numerical work, albeit a little slower than Fortran. All programs herein are provided for both 32-bit and 64-bit operating systems. The reader is encouraged to construct small programs in any language, for the preparation of data files for use with the Web Program Suite or for any other appropriate function in the text.

Solutions 1

1.1 The $(NH_4)^+$ ions are either in free rotation or randomly orientated, about their mean positions so that their envelopes of motion, averaged either over time or space, are effectively spherical. (Experiment indicates that the free rotation model is correct. The space group [1] of ammonium nitrate is $Pm\overline{3}m$, with one formula-entity per unit cell, which implies effective spherical symmetry for the nitrate ion in this polymorph.)

1.2   Applying the Bragg equation for a cubic crystal Eq. (1.4).

(100): $a = 0.15418\,\text{nm} \times \left(2\sin\{½\tan^{-1}[(11.05\,\text{mm})/(30.00\,\text{mm})]\}\right)^{-1}$
$= 0.4392\,\text{nm}$.

(110): $a = 0.15418\,\text{nm} \times \sqrt{2}\left(2\sin\{½\tan^{-1}[(16.35\,\text{mm})/(30.00\,\text{mm})]\}\right)^{-1}$
$= 0.4415\,\text{nm}$.

Mean value: $a = 0.440\,\text{nm}$.

1.3   Isomorphous pairs are: NaCl, KCl; $BaSO_4$, $PbSeO_4$; CaF_2, β-PbF_2; NaBr, MgO.

1.4   KCl is mainly an ionic solid and its melt consists of ions. The relative molecular mass of $BeCl_2$ is not very different from that of KCl, so the ΔS value is evidence for molecular association in the melt. ($\Delta G = \Delta H - T\Delta S$: the greater entropy of vaporization indicates greater stability for $BeCl_2$ compared with that for KCl. $BeCl_2$ is a molecular solid of tetrahedrally coordinated Be atoms doubly-bridged by Cl atoms: Be = •; Cl = ○)

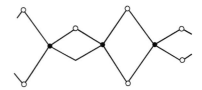

Beryllium chloride $BeCl_2$

1.5   The following enthalpy-level cycle relates the polymorphs of sulphur:

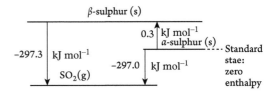

Enthalpy cycle for α-sulphur–β-sulphur

The enthalpy change at the transition temperature is $(300\,\text{J mol}^{-1})/(386\,\text{K})$, so that the entropy of β-sulphur at this temperature is $32.51\,\text{J K}^{-1}\,\text{mol}^{-1}$.

1.6   Following Section 1.8, the following classification is drawn up:

Covalent	Molecular	Ionic	Metallic
AlN	P_4	RbF	Pb
SiC	CO_2	P_2Cl_{10}	Cu_3Au
	C_6H_6	$KClO_3$	
	Ne	Na_2SO_4	

1.7   From the discussion on Miller indices (Appendix A2): (i) $(21\bar{3})$; (ii) (203); (iii) $(\bar{3}00)$; (iv) $(6\bar{9}8)$; (v) $(06\bar{4})$; (vi) $(0\bar{2}1)$.

1.8   From Section 1.4, the only values of $N(= h^2 + k^2 + l^2)$ smaller than that for the 200 ring are 3, 2 and 1, corresponding to planes of type (111), (110) and (100), respectively. From Fig. 1.5, the planes (100) are inter-leaved by identical planes, which will lead to total interference for diffrac-tion from these planes; the same is true for the (110) planes. Hence, the first sharp ring must correspond to (111). From Eq. (1.5), $0.4070\,\text{nm} =$ $0.15418\,\text{nm} \times \sqrt{3}\,(2\sin\{\tfrac{1}{2}\tan^{-1}[r/(30.00\,\text{mm})]\})^{-1}$, whence the required diam-eter r is 23.70 mm.

1.9   The volume V_s of a sphere of radius r is $\tfrac{4}{3}\pi r^3$. In the body-centred, cubic structure there are 2 atoms per unit cell and they are in contact across a body diagonal of the cube, so that $a\sqrt{3} = 4r$; where a is the length of a cube edge; thus the volume occupied per atom in the structure is $\left(\tfrac{4r}{\sqrt{3}}\right)^3 \times \tfrac{1}{2} = \tfrac{32}{3\sqrt{3}}$. Hence, the packing fraction ratio V_s/V_{bcc} is $\tfrac{4}{3}\pi \times \tfrac{3\sqrt{3}}{32} = \sqrt{3}\pi/8 = 0.6802$. In the close-packed, cubic structure there are 4 atoms per unit cell of side a, and close contact takes place across a face diagonal; hence, $V_s/V_{\text{fcc}} = 0.6802 \times \tfrac{4\sqrt{2}}{3\sqrt{3}} = 0.7405$; hence, the change in the packing fraction is +8.87%.

1.10   Density D equals mass of the unit cell over its volume. The relative molar mass of krypton (Periodic Table inside front cover) is 83.798, and there are 4 atoms per cubic unit cell; the atomic mass unit (see table of Physical Constants) is 1.6605×10^{-27} kg. Thus, $D = 4 \times 83.798 \times (1.6605 \times 10^{-27}\,\text{kg})/(0.5648 \times 10^{-9}\,\text{m})^3 = 3089.2\,\text{kg m}^{-3}$.

Solutions 2

2.1   (a) The initial velocity v_0 of $40.00\,\text{m s}^{-1}$ is resolved into horizontal and verti-cal components $v_{0,x} = v_0\cos\theta$ and $v_{0,y} = v_0\sin\theta$, respectively. The y component is needed for the maximum height, and is given by $v_y/\text{m s}^{-1} = v_0\sin 35° = 22.94$. (b) The maximum height $h_{\text{max}} = v_{0,y}t + a_yt^2/2$, where a_y is the acceleration along the y direction and is $-9.812\,\text{m s}^{-2}$. The *time to reach* maximum height is given by $v_{\text{max},y} = v_{0,y} + a_yt = 0$, so that $t = -22.94\,\text{m s}^{-1}/(-9.812\,\text{m s}^{-2}) = 2.338\,\text{s}$. Thus, the maximum height *attained* is $h_{\text{max}} = 22.94\,\text{m s}^{-1} \times 2.338\,\text{s} - 9.812\,\text{m s}^{-2} \times (2.338\,\text{s})^2/2 = 26.82\,\text{m}$. (c) The velocity at maximum height is zero (as in the previous result), so that the kinetic energy at this point is also zero

2.2   Energy $= \tfrac{1}{2}mv^2 = 4.114 \times 10^{-16}$ J; momentum $= mv = 2.745 \times 10^{-23}$ N s. (If special relativity is ignored, 4.094×10^{-16} J and 2.731×10^{-23} N s.)

2.3   Convert Eq. (2.4) in the form $E(\nu)d\nu$ to a function of λ by an appropriate differentiation; remember that $d\nu = -(c/\lambda^2)d\lambda$. At 500 K, $\lambda_{max} = 9.592\,\mu m$.

2.4   Using the Rayleigh–Jeans equation for $E(\lambda)d\lambda$ developed in the solution to Problem 2.3 gives $E(\lambda)d\lambda = 2.14 \times 10^{-2}$ J m^{-3}, whereas the Planck equation, $E(\lambda)d\lambda = (8\pi hc/\lambda^5)\{\exp[hc/(\lambda kT)]-1\}^{-1}d\lambda$, gives $E(\lambda)d\lambda = 1.75 \times 10^{-14}$ J m^{-3}. The Planck equation is essential in this problem, as the exponential term is highly significant under the given conditions.

2.5   (a) By a least squares extrapolation, using program LSLI:

λ/nm	546.2	365.0	312.0
V_0/V	−2.101	−0.9674	−0.3994

   (b) The least squares equation is $V_0 = 4.151 \times 10^{-15}\nu - (-4.379)$; hence, $\eta = 4.151 \times 10^{-15}$ V s $\times$ 1.6022 $\times 10^{-19}$ C $= 6.6507 \times 10^{-34}$ J s, which is an estimate of the Planck constant h. The constant term $\zeta = 4.38$ eV, is a value for the work function of tin, more usually symbolized by ϕ.

2.6   From the de Broglie equation, $p = h/\lambda = mv$; hence, $v = 1.455 \times 10^6$ m s^{-1}.

2.7   Inverting: $1/\lambda = \bar{\nu} = \frac{1}{K}\left(\frac{n^2-4}{n^2}\right)$, and a straightforward rearrangement leads to $\bar{\nu} = \frac{4}{K}\left(\frac{1}{2^2} - \frac{1}{n^2}\right)$, which is Eq. (2.15) with $n_1 = 2$ (Balmer spectral series) and $n_2 = n$; $K = 4/R_H$.

2.8   Using the uncertainty principle, $\Delta p \Delta x = m_e \Delta v \Delta x = \hbar/2$, where $\Delta x = 2a_0$; hence, $\Delta v \approx 547$ km s^{-1}.

2.9   The probability P is proportional to $\sin^2(n\pi m)$, where $n = 1, 2, 3$, and m is a fraction of a. For a maximum, $(n\pi m) = 1$; hence:

n	m
1	1/2
2	1/4, 3/4
3	1/6, 1/2, 5/6

2.10   In three dimensions, the energy for the cubic box is given by $E = (h^2/8m_e)(3/a^2)$, so that $E = 1.807 \times 10^{-7}$ J.

2.11 $P = (1/\pi a_0^3) \int_{1.10a_0}^{1.11a_0} r^2 \exp(-2r)dr \int_{0.20\pi}^{0.21\pi} \sin\theta d\theta \int_{0.60\pi}^{0.61\pi} d\phi$. The integrals over θ and ϕ are straightforward, giving 3.1416×10^{-2} and -1.8862×10^{-2}, respectively.

Integrating over r by parts (Appendix A7, Eq. A7.7) gives $-1/2$. Hence, $P = \frac{1}{2} \times (-0.18862) \times 0.031416 = 2.963 \times 10^{-4}$.

2.12   From the Periodic Table, or otherwise: N: $(He)(2s)^2(2p)^3$; Al: $(Ne)(3s)^2(3p)^1$; Cl⁻: (Ar); Y: $(Kr)(4d)^1(5s)^2$.

2.13   He⁺ is a one-electron, hydrogen-like species: $Z = 2$ and $n = 1$; hence, the energy is four times (ignoring screening) that for hydrogen in its ground state (Table 2.2); $E = -2^2 \times 2.178 \times 10^{-18}$ J $= -8.712 \times 10^{-18}$ J.

2.14   Fig. 2.24 indicates the interparticle interactions: hence,

$$\mathcal{H} = -[\hbar^2/(2M)]\,(\nabla_A^2 + \nabla_B^2) - [\hbar^2/(2m)]\,(\nabla_1^2 + \nabla_2^2)$$
$$- \{[Z_{\text{eff}}^2 e^2/(4\pi\varepsilon_0)][1/r_{A1} + 1/r_{A2} + 1/r_{B1} + 1/r_{B2} + 1/r_{12} + 1/r_e]\},$$

where A and B are nuclei, and 1 and 2 are electrons; r_e is the equilibrium internuclear distance; r_{A1} is the distance between nucleus A and electron 1, and so on; $Z_{\text{eff}} = Z - \sigma$ (Table 2.6).

2.15   By considering overlap integrals (Section 2.17.1), the following values are obtained: (a) $S_{1s+2s} > 0$; (b) $S_{1s+2p_z} > 0$; (c) $S_{1s+2p_{x(y)}} = 0$. Overlap is possible only for orbitals with the same symmetry within the point group of the molecule, C_{2v} [1]: H$(1s)$ and F$(2s$ and $2p_z)$ have the same symmetry (σ^+) in the point group, whereas F$(2p_x$ and $2p_y)$ have symmetry π. In the case of HF, bonding is in the $2p$ level of F, and only $2p_z$ has the requisite symmetry, as the following diagram shows (shaded $= \psi$ negative):

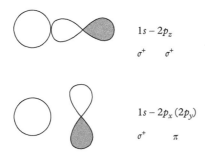

$$1s - 2p_z$$
$$\sigma^+ \qquad \sigma^+$$

$$1s - 2p_x\,(2p_y)$$
$$\sigma^+ \qquad \pi$$

s and *p*-Orbitals for HF

If two orbitals are of different symmetry within the point group of the species, overlap would cancel because of the opposing signs. Hence, bonding is between H$(1s)$ and F$(2p_z)$.

2.16 The three radial distribution function (RDF) plots:

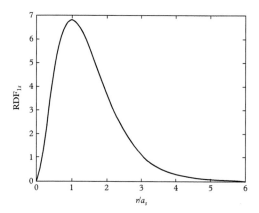

Radial distribution function, 1*s*

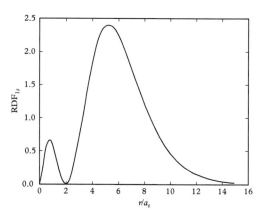

Radial distribution function, 2*s*

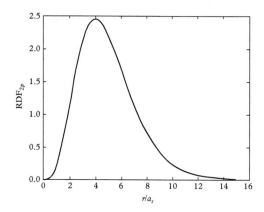

Radial distribution function, 2*p*

(a)   Measurements from the graphs: $1s$, 0.992; $2s$, 0.742 and 5.26; $2p$, 3.97.

(b)   By differentiation: $1s$, 1; $2s$, $3\pm\sqrt{5}$; $2p$, 4. For the $2s$ function, let $r/a_0 = R$; the program ROOT may be useful in solving the derivative equation for RDF-$2s$. (The results in decimal from Bairstow's method, to six-figure accuracy, are 0.763932 and 5.23607.)

2.17   (a) The plots reveal the shape of the function, and the existence of a maximum in $S_{s,p}$. (b) (i) $S_{s,s}(H_2^+) = 0.59$; $S_{s,s}(H_2) = 0.76$. (c) From the graph the maximum is at $\rho = 2.10$, and may be confirmed by differentiating of Eq. (2.104) with respect to r (or ρ). Graphs may be plotted by generating a set of S v. r data for $r = 0$–3 in units of a_0, using the Python program GRFN.

2.18   The dipole moment vector has the magnitude $2|q|e\,d_{O-H}$ $\cos(\angle H-O-H/2)$. Thus, $(1.8\ \text{D} \times 3.3356 \times 10^{-30}\ \text{C m}) / (2 \times 1.6022 \times 10^{-19}\ \text{C} \times 0.96 \times 10^{-10}\ \text{m} \times \cos(104.4/2°)$; hence, $q_0 = -0.32$, so that $q_H = +0.16$.

2.19   Let $\Psi = \psi_s + (c_p/c_s)\psi_p$ where ψ_s and ψ_p are normalized. Normalize Ψ:

$$\int |\Psi|^2 d\tau = \int (\psi_s^2 + x\psi_p^2) d\tau = (1 + x),$$

where $x = c_p^2/c_s^2$, since the cross term of normalized orbitals vanishes. Hence,

$$\Psi = (1 + x)^{\frac{1}{2}}(\psi_s + x^{\frac{1}{2}}\psi_p).$$

2.20   1,3-Butadiene is $CH_2{=}CH{-}CH{=}CH_2$ and a suitable π wavefunction is $\Psi = c_1\psi_1 + c_2\psi_2 + c_3\psi_3 + c_4\psi_4$. Following HMO theory (Section 2.18), the secular equations are:

$$c_1(\alpha - E) + c_2\beta = 0$$

$$c_1\beta + c_2(\alpha - E) + c_3\beta = 0$$

$$c_2\beta + c_3(\alpha - E) + c_4\beta = 0$$

$$c_3\beta + c_4(\alpha - E) = 0$$

leading to the secular determinant, with $\alpha - E = y$:

$$\begin{vmatrix} y & 1 & 0 & 0 \\ 1 & y & 1 & 0 \\ 0 & 1 & y & 1 \\ 0 & 0 & 1 & y \end{vmatrix}.$$

Expansion of the determinant (Appendix 10) gives $y^4 - 3y^2 + 1 = 0$, whence by the substitution $p = y^2$, $y = \pm 1.618$ and ± 0.618, so that $E = \alpha \pm 1.618\beta$, $\alpha \pm 0.618\beta$, and the energy level diagram is below:

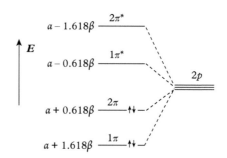

π-**Energy level diagram for 1,3-butadiene**

$E\pi = 4\alpha + 4.472\beta$. Two ethenic double bonds would have π energy $4\alpha + 4\beta$; E_π (butadiene) $= 0.472\beta$, or *ca.* -36 kJ mol^{-1}. The c_i coefficients are calculated as described in Appendix 10, and are listed below.

Ψ_i	$c_{i,1}$	$c_{i,2}$	$c_{i,3}$	$c_{i,4}$
1	0.3717	0.6015	0.6015	0.3717
2	0.6015	0.3717	−0.3717	−0.6015
3	0.6015	−0.3717	−0.3717	0.6015
4	0.3717	−0.6015	0.6015	−0.3717

Bond orders:

$C_{1,2} = C_{3,4} = (2 \times 0.3717 \times 0.6015) + (2 \times 0.3717 \times 0.6015) = 0.894$

$C_{2,3} = (0.6015 \times 0.3717) + (0.6015 \times 0.3717) = 0.447$

$P_{i,j}$ values are 1.894 and 1.447.

Free valence parameters:

$\mathscr{F}_1 = \mathscr{F}_4 = 4.732 - (2 + 1 + 0.894) = 0.838$

$\mathscr{F}_2 = \mathscr{F}_3 = 4.732 - (2 + 1 + 0.447 + 0.894) = 0.391$

Charge distribution:

$q_1 = q_2 = q_3 = q_4 = 1 - [2(\pm 0.6015)^2 + 2(\pm 0.3717)^2] = 0$; the species is neutral.

2.21   The secular determinant is

$$\begin{vmatrix} \alpha - E & \beta & 0 & \beta \\ \beta & \alpha - E & \beta & 0 \\ 0 & \beta & \alpha - E & \beta \\ \beta & 0 & \beta & \alpha - E \end{vmatrix} = \begin{vmatrix} y & 1 & 0 & 1 \\ 1 & y & 1 & 0 \\ 0 & 1 & y & 1 \\ 1 & 0 & 1 & y \end{vmatrix},$$

where $y = (\alpha - E)/\beta$. Expansion of the determinant by cofactors, as shown in Appendix 10, gives:

$$y^4 = 4y^2 = y^2(y^2 - 4) = y^2(y + 2)(y - 2).$$

The roots and energy level diagram are hereunder:

$$E_1 = \alpha + 2\beta \quad E_2 = \alpha \quad E_3 = \alpha \quad E_4 = \alpha - 2\beta.$$

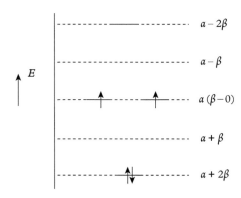

π-**Energy level diagram for 1,3-cyclobutadiene**

The frontier orbitals are the degenerate pair of energy α (HOMO), and $\alpha - 2\beta$ (LUMO). The total energy is $4\alpha + 4\beta$, which the same as for two isolated ethenic double bonds; hence, $D_\pi = 0$. The ground state is a di-radical and, therefore, highly reactive (actually self-reactive). (It distorts readily (Jahn–Teller effect [2]) to a more stable rectangular conformation.)

2.22   Express ΔE from Eq. (2.37); then, $\lambda = hc/\Delta E$. Evaluating λ as a function of n and l: $\lambda = 3.2972 \times 10^{-6} \text{ m}^{-2} \times l^2/(2n + 1)$. Since $l = 1.12$ nm, $n = 5$ and the first transition is from energy level 5 to level 6, with $\lambda = 3.2972 \times 10^{-6} \text{ m}^{-1} \times 1.12^2 \text{ m}^2/11 = 3.760 \times 10^{-7}$ m, or 376 nm; its colour is violet.

2.23

	Weak field		Strong field			Weak field		Strong field	
n in d^n	Config.	N	Config.	N	n in d^n	Config.	N	Config.	N
1	t^1	1	t^1	1	6	$t^4 e^2$	4	t^6	0
2	t^2	2	t^2	2	7	$t^5 e^2$	3	$t^6 e^1$	1
3	t^3	3	t^3	3	8	$t^6 e^2$	2	$t^6 e^2$	2
4	$t^3 e^1$	4	t^4	2	9	$t^5 e^3$	1	$t^6 e^3$	1
5	$t^3 e^2$	5	t^5	1					

2.24 The C_4 and S_4 *axial directions* coincide in a cube and a tetrahedron (Fig. 2.34). Lines from points a, c, g and e to the centre of the cube represent directions of the ligand orbitals. In a cube of side a, the e molecular orbitals (involving d_{xy}, d_{zx} and d_{yz}) point to the mid-points of edges of the cube (Fig. 2.21), whereas the t_2 molecular orbitals (involving d_{z^2} and $d_{x^2-y^2}$) point to the mid-points of faces. Thus, the t approach distance is $a/2$, whereas the e approach is $a\sqrt{2}/2$, or 1.414 times that of the t. The ligand field splits the d levels, but oppositely to the octahedral case; the t_2 level now lies above the e. Hence the favourable interaction is with the t_2, leading to a tetrahedral conformation. The Zn(II) configuration is $(\mathrm{Ar})(3d)^{10}$, with no unpaired spins, and the complex is diamagnetic.

2.25 Inserting values of the constants into Eq. (2.154) gives $\mu = 0.8943\sqrt{\chi_m T}$, so that the magnetic moment is $5.19\,\mu_B$. This value corresponds to four unpaired electrons; thus, the oxidation state of iron is Fe(II) in the given compound.

2.26 Draw Lewis electron-pair diagrams to show the total number of electron around the central atom.

(a) F_2O: eight valence electrons around oxygen indicate a tetrahedral configuration. Two electron pairs bond to the two fluorine atoms, and repulsion by the remaining lone pairs leads to a bent molecule, but less bent than in H_2O because the larger electronegativity of fluorine decreases the repulsion effect of the oxygen lone pairs.

(b) NF_3: eight valence electrons around nitrogen indicate a tetrahedral configuration. Three pairs bond with fluorine, and the single lone pair repulsion distorts the tetrahedral configuration to trigonal-pyramidal, but nearer to tetrahedral that in the water molecule.

(c) SiF_4: eight valence electrons around silicon again indicate a tetrahedral shape. Four bonds are made, satisfying fully the valence requirements and the tetrahedral configuration remains.

(d) SF_4: ten valence electrons around sulphur indicate a trigonal bi-pyramidal configuration. The repulsion from the remaining lone pair produces a see-saw shaped molecule, so that there are two types of F–S–F angles, three equal and < tetrahedral, and one angle < linear.

(e) SF_6: twelve valence electrons around sulphur form a regular octahedral configuration, with all F–S–F angles 90° (See also Section 2.23.1.)

(f) XeF_4: twelve valence electrons around xenon indicate a potentially octahedral shape. Only four bonds are formed. The remaining two lone pairs are set in opposition, resulting in a square-planar configuration.

2.27 The NH_3 pyramidal molecular configuration is shown below. The N atom is $(1s)^2(2s)^2(2p_z)(2p_x)(2p_y)$, but combining three H $1s$ with three N $2p$ orbitals would not lead to the correct bond angles. The C_3 symmetry element passing through the nitrogen atom (and the coincident z reference axis) is normal to the plane passing through the three hydrogen atoms; the σ_v, σ'_v and σ''_v vertical symmetry plane elements pass through the N–H$_1$, N–H$_2$ and N–H$_3$ bond directions, respectively. However, since the three σ_v-type operators arise from the interaction of C_3 and C_2 in this group, the problem can be solved in the simpler point group symmetry, C_3. An LCAO procedure uses H $1s$ and N $2s$ and $2p$, and a representation can be generated on the three N–H bond vectors:

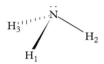

The NH$_3$ molecule: the z-axis passes through N and is normal to the plane of the three hydrogen atoms

The unshifted atom procedure on the three hydrogen $1s$ orbitals $\psi_i (i = 1 - 3)$ of the molecule has the following effects:

Symmetry	H atoms unshifted	Contribution
E	None	3
C_3	All	0
C_3^2	All	0

generating the representation

C_3	E	C_3	C_3^2
Γ	3	0	0

and an inspection of the character table for C_3, or otherwise, show that Γ reduces to $A + E$. The projection operators are $P_\Gamma A_1(\psi_1)$ and $P_\Gamma E(\psi_1)$; the symmetry operators in the character table are already shown in full, as required by the projection operator procedure:

C_3	E	C_3	C_3^2
Symmetry effect on ψ_1	ψ_1	ψ_2	ψ_3
Characters of E	$\left\{\begin{array}{c} 1 \\ 1 \end{array}\right.$	$\begin{array}{c} \varepsilon \\ \varepsilon^* \end{array}$	$\left.\begin{array}{c} \varepsilon^* \\ \varepsilon \end{array}\right\}$
$P_\Gamma E$	$\left\{\begin{array}{c} \psi_1 \\ \psi_1 \end{array}\right.$	$\begin{array}{c} \varepsilon\psi_2 \\ \varepsilon^*\psi_2 \end{array}$	$\left.\begin{array}{c} \varepsilon^*\psi_3 \\ \varepsilon\psi_3 \end{array}\right\}$

The action of $P_\Gamma A'_1(\psi_1)$ is straightforward since all character are 1; the result is

$$P_\Gamma A_1(\psi_1) = \psi_1 + \psi_2 + \psi_3.$$

Normalizing:

$$\Psi_1 = \sqrt{1/3}(\psi_1 + \psi_2 + \psi_3). \tag{1}$$

The same result would be obtained with ψ_2 and ψ_3 because they are symmetry related. From the above table,

$$P_\Gamma E(\psi_1) = \psi_1 + \varepsilon\psi_2 + \varepsilon^*\psi_3$$
$$P_\Gamma E(\psi_1) = \psi_1 + \varepsilon^*\psi_2 + \varepsilon\psi_3 \tag{2}$$

These two SALCs are orthogonal, and are better expressed by the linear combinations of addition and subtraction in Eq. (2). Thus, remembering that, for an angle of 120°,

$$(\varepsilon + \varepsilon^*) = -1 \text{ and } (\varepsilon - \varepsilon^*) = i\sqrt{3},$$

combining, neglecting any common factors and normalizing gives:

$$\Psi_2 = (1/\sqrt{6})(2\psi_1 - \psi_2 - \psi_3)$$
$$\Psi_2 = (1/\sqrt{2})(\psi_2 - \psi_3) \tag{3}$$

(common multipliers can always be eliminated before normalizing). The same results would be obtained, albeit with a little more labour, by working in C_{3v}. The form of Eqs. (3) shows that they are normalized orbitals.

The orbitals Ψ_1, Ψ_2 and Ψ_3 are obtained by the matched symmetry-adapted orbitals of $2s$ and $2p$ on nitrogen with $1s$ on hydrogen. The energy level diagram is shown below; the MO levels have been marked in accord with point group C_{3v}. (Lying between the H($1s$) atomic orbital levels and the molecular orbitals are the H($1s$) SALC energy levels, and between the N($2s$)/($2p$) atomic orbitals and the molecular orbitals are the N($2s$)-($2p$)-mixing energy levels.)

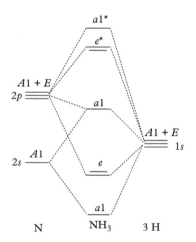

Energy level diagram for the NH₃ molecule

2.28 The water molecule has the following symmetry elements:

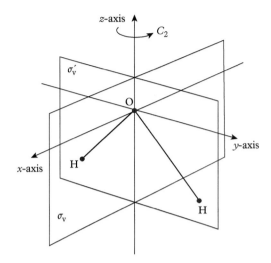

Symmetry elements for the H₂O molecule

The shifts on the O–H bond vectors under the symmetry C_{2v} lead to the following representation:

C_{2v}	E	C_2	σ_v	σ_v'
Γ	2	0	0	2

which reduces to $A + B_2$. Applying the projection operators:

C_{2v}	E	C_2	σ_v	σ_v'
Symmetry effect on ψ_1	ψ_1	ψ_2	ψ_2	ψ_1
Characters of B_2	1	−1	−1	1
Product	ψ_1	$-\psi_2$	$-\psi_2$	ψ_1

The action of $P_\Gamma A_1$ on ψ_1 is self-evident; hence,

$$\text{For O–H}_1: P_\Gamma A_1(\psi_1) = \psi_1 + \psi_2 + \psi_2 + \psi_1 = 2\psi_1 + 2\psi_2 \equiv \psi_1 + \psi_2$$
$$P_\Gamma B_2(\psi_1) = \psi_1 - \psi_2 - \psi_2 + \psi_1 = 2\psi_1 - 2\psi_2 \equiv \psi_1 - \psi_2$$

The same results would be obtained with O–H₂. Normalizing:

$$\Psi_1 = (1/\sqrt{2}(\psi_1 + \psi_2)$$
$$\Psi_2 = (1/\sqrt{2}(\psi_1 - \psi_2)$$

That they are orthogonal follows from $\int \tfrac{1}{2}(\psi_1 + \psi_2)(\psi_1 - \psi_2)d\tau = 0$.

The Walsh diagram applies to AH_2 compounds: the B_1 interaction is a maximum at $180°$, whereas that for A_1 has a maximum at a smaller angle. The compromise for the water molecule occurs at $104.5°$, as indicated on the diagram [2, 3].

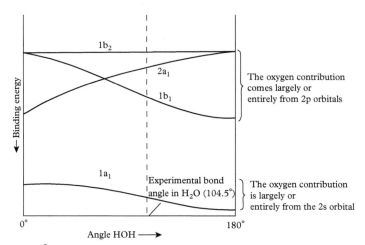

At $\widehat{HOH} = 180°$, $2a_1$ and $1b_2$ become degenerate (linear triatomic)
At $\widehat{HOH} = 0°$, $1b_1$ and $1b_2$ become degenerate (linear diatomic)

Walsh diagram for AH_2 compounds

2.29   Using the unshifted atom technique on the P–F σ-bond vectors, the following reducible representation is obtained:

D_{3h}	E	C_3	C_2	σ_h	S_3	σ_v
Γ_σ	5	2	1	2	0	3

From Table 2.14, reduction of Γ_σ by inspection, or analytically, leads to the irreducible representations $2A_1' + A_2'' + E'$. The character table shows that A_1' spans s and d_{z^2}, A_2'' spans p_z, and E' spans d_{xy} and $d_{x^2-y^2}$. The configuration of phosphorus is $(Ne)(3s)^2(3p)^3(3d)^0$, so that possible combinations are:

$$
\begin{array}{ccc}
\quad 3p_x,\ 3p_y & \quad 3p_x,\ 3p_y & \quad 3p_x,\ 3p_y \\
\diagup & \diagup & \diagup \\
3s,\ 4s,\ 3p_z & 3d_{z^2},\ 4d_{z^2},\ 3p_z & 3s,\ 3d_{z^2},\ 3p_z \\
\diagdown & \diagdown & \diagdown \\
3d_{xy},\ 3d_{x^2-y^2} & 3d_{xy},\ 3d_{x^2-y^2} & 3d_{xy},\ 3d_{x^2-y^2} \\
\text{I} & \text{II} & \text{III}
\end{array}
$$

On energetic grounds, the most likely configuration is III. (VSEPR would predict a trigonal bi-pyramidal conformation (Table 2.10), but could go no further with the description.)

2.30   (a) The first six energy levels of the box will be filled with electrons ($n = 6$). The length a of the box is 11×0.140 nm. (a) $\Delta E_{6\to7} = 13h^2/(8m_e a^2)$; hence, $\Delta E_{6\to7} = 3.302 \times 10^{-19}$ J. (b) The emission wavelength $\lambda = 7 \to 6$ is 601.5 nm. (c) If the box is lengthened by a single bond, λ becomes 715.8 nm, and if by a double bond, 840.1 nm, and conversely if the chain is shortened.

2.31 (a) For $(H_3)^+$, the determinant is $\begin{vmatrix} y & 1 & 0 \\ 1 & y & 1 \\ 0 & 1 & y \end{vmatrix} = 0$, or $y^3 - 2y = 0$, so that

$y = 0, \pm\sqrt{2}$. Hence, the energies are $\alpha + \sqrt{2}\beta$, α, $\alpha - \sqrt{2}\beta$.

—————— $\alpha - \sqrt{2}\,\beta$

—————— α

——↑↓—— $\alpha + \sqrt{2}\,\beta$

Energy level diagram for linear $(H_3)^+$

$E_{(H_3)^+} = 2\alpha + 2\sqrt{2}\beta$. In (H_3), one electron is added to the α level: $E_{(H_3)} = 3\alpha + 2\sqrt{2}\beta$.

For $(H_3)^-$, yet one more electron is added, pairing in the α level: $E_{(H_3)^-} = 4\alpha + 2\sqrt{2}\beta$.

(b) In the cyclic form for $(H_3)^+$, the determinant is $\begin{vmatrix} y & 1 & 1 \\ 1 & y & 1 \\ 1 & 1 & y \end{vmatrix} = 0$, or $y^3 -$

$3y + 2 = 0$, so that $y = 1, 1, -2$. Hence, the energy levels are $\alpha - \beta$, $\alpha - \beta$, $2\alpha + 4\beta$.

———— ———— $\alpha - \beta$ [a]

——↑↓—— $\alpha + 2\beta$

[a]Doubly degenerate

Energy level diagram for cyclic $(H_3)^+$

$E_{(H_3)^+} = 2\alpha + 4\beta$; $E_{(H_3)} = 3\alpha + 3\beta$; $E_{(H_3)^-} = 4\alpha + 2\beta$.

2.32   The Hückel determinant for 1,3,5-hexatriene solves to $y^6 - 5y^4 + 6y^2 - 1 = 0$, giving $y = \pm1.8019, \pm1.2470, \pm0.44504$; total energy: $E = 6\alpha + 6.988\beta$, and delocalization energy: $D_\pi = 0.988\beta$. The program HUCK provides also the following data:

Coefficients and wavefunctions:

$\Psi_1 = 0.2319\psi_1 + 0.4179\psi_2 + 0.5211\psi_3 + 0.5211\psi_4 + 0.4179\psi_5 + 0.2319\psi_6$
$\Psi_2 = 0.4179\psi_1 + 0.5211\psi_2 + 0.2319\psi_3 - 0.2319\psi_4 - 0.5211\psi_5 - 0.4179\psi_6$
$\Psi_3 = 0.5211\psi_1 + 0.2319\psi_2 - 0.4179\psi_3 - 0.4179\psi_4 + 0.2319\psi_5 + 0.5211\psi_6$
$\Psi_4 = 0.5211\psi_1 - 0.2319\psi_2 - 0.4179\psi_3 + 0.4179\psi_4 + 0.2319\psi_5 - 0.5211\psi_6$
$\Psi_5 = 0.4179\psi_1 - 0.5211\psi_2 + 0.2319\psi_3 + 0.2319\psi_4 - 0.5211\psi_5 + 0.4179\psi_6$
$\Psi_6 = 0.2319\psi_1 - 0.4179\psi_v + 0.5211\psi_3 - 0.5211\psi_4 + 0.4179\psi_5 - 0.2319\psi_6$

Bond orders:

The three orbitals of lowest energy are fully occupied:

$p_{1,2} = p_{5,6} = 2(0.2319 \times 0.4179) + 2(0.4179 \times 0.5211) + 2(0.5211 \times 0.2319) = 0.871.$

Similarly,

$p_{2,3} = p_{4,5} = 0.483$

$p_{3,4} = 0.785$

Charge distributions:

$q_1 = 1 - [2(0.2319)^2 + 2(0.4179)^2 + 2(0.5211)^2] = 7 \times 10^{-5} \equiv 0 = q_2 = q_3 = q_4 = q_5 = q_6.$

The difference from zero arises from rounding errors; the molecule is neutral. Note that the program HUCK does not subtract $\sum_i Nc_i^2$ from 1.

Free valence indexes:

$\mathfrak{F}_1 = \mathfrak{F}_6 = 4.732 - (2p_{\text{C--H}} + p_{1,2}\sigma + p_{1,2}\pi) = 4.732 - (2 + 1 + 0.871) = 0.861$
$\mathfrak{F}_2 = \mathfrak{F}_5 = 4.732 - (3 - p_{1,2}\pi - p_{2,3}\pi) = 0.378$
$\mathfrak{F}_3 = \mathfrak{F}_4 = 4.732 - (3 - p_{2,3}\pi - p_{3,4}\pi) = 0.464$

2.33

	Pyridine			Pyrrole	
Atom	Charge*	Free valence index	Atom	Charge*	Free valence index
1	1.195	0.425	1	1.720	0.853
2	0.923	0.409	2	1.035	0.502
3	1.004	0.398	3	1.106	0.389
4	0.950	0.402	4	1.106	0.389
5	1.004	0.398	5	1.035	0.502
6	0.923	0.409			

*The charge distributions q_i above should be subtracted from 1, as per Eq. (2.153).

From the charge distribution $(1 - q_i)$ results: (a) Sodamide is a nucleophilic agent and attacks at the more positive position 2 (or 6) to form 2-aminopyridine. (b) The Br radical from gaseous bromine is attracted to the same atom site producing 2-bromopyridine. (c) The nitrating agent is the nitronium ion $[NO_2]^+$, an electrophile, which seeks the negative 3- (or 5-) position giving 3-nitropyridine.

The HMO results should be studied. Note particularly, (i) $\sum_i q_i = N$, where N is the number of π-electrons; (ii) the relation of the free valence index to the site of reactivity, and (iii) the similar delocalization energies D_π for pyridine and pyrrole, approximately 0.25β and 0.54β, respectively—significantly less than for benzene.

2.34   Consider the two structural formulae:

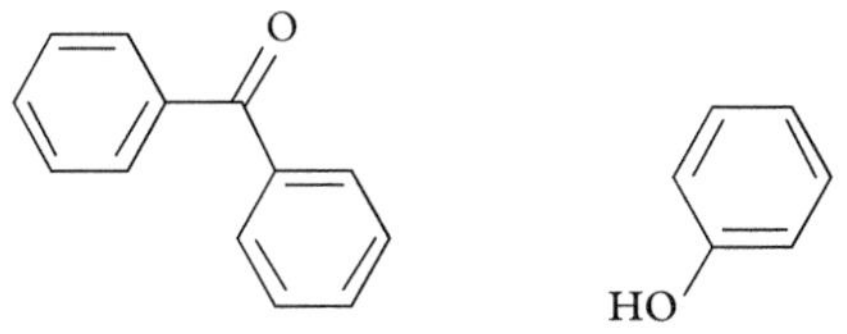

Benzophenone   **Phenol**

In benzophenone, both lone-pair electrons on oxygen are in non-bonding molecular orbitals, whereas in phenol one lone pair is non-bonding while the other takes part in the conjugation of the ring system.

2.35   The lone-pair electrons in pyridine are non-bonding and, therefore, readily accept H_3O^+ to form a quaternary salt. Pyrrole is less basic because the lone-pair electrons are part of the conjugated ring system, so that it does not readily take part in quaternary salt formation. (The basicity of a base *BN* is judged frequently by the acidity of the conjugate acid BNH^+: the base equilibrium is $BN + H_2O \rightleftharpoons BNH^+ + OH$, with $K_a = \frac{[BNH^+][OH^-]}{[BN]}$. The stronger the acid, the weaker the base: K_a (pyridine) $= 5.9 \times 10^{-6}$, K_a (pyrrole) $= 0.40$.)

2.36 (a) From the HMO calculation on anthracene:

$p_{1,2} = 0.737$, $p_{2,3} = 0.586$, $p_{4,12} = 0.535$, $p_{9,11} = 0.606$, $p_{11,12} = 0.485$.

(b) From the Coulson formula with the above bond orders (Note: the Coulson formula gives results in Å; 1 Å $\equiv$ 0.1 nm):

Atom pair	1, 2	2, 3	4, 12	9, 11	11, 12
d/nm	0.1384	0.1412	0.1421	0.1408	0.1430
(Ref. [4]d/nm)	0.1366	0.1419	0.1433	0.1399	0.1436

(These values should be compared with the literature references given in the text of the problem; the references form an interesting comment on progress in x-ray crystallographic structure determination.) The graph in Fig. 2.41 shows an almost linear correlation between d and P.

(c)   The 9, 10 sites with a free valence index of 0.52 are the most reactive; see, for example, the Diels–Alder reaction, and note the two $>C=O$ electron withdrawing groups on the attacking electrophile.

2.37 (a) There are four geometrical isomers, (i)–(iv):

(i)   (ii)   (iii)   (iv)

Geometrical isomers of $[Co(NH_3)_2H_2O(NH_2)_2Cl]$

(b)   The point groups are (i) C_s, (ii) C_s, (iii) C_1, (iv) C_1, (c) (iii) and (iv) are optical isomers.

2.38   The angular frequency ω depends on the wave number k, and $\omega = 2\pi\nu$. Using the given equation for λ,

$$\omega(k) = \frac{2\pi c}{\lambda} = 2\pi\sqrt{\nu^2 - \nu_0^2} = 2\pi\sqrt{\frac{c^2}{\lambda^2} - \nu_0^2} = 2\pi\sqrt{\frac{c^2 k^2}{4\pi^2} - \nu_0^2}$$

The group velocity (Appendix A21) is given by

$$v_g = \frac{d\omega(k)}{dk} = \frac{2\pi}{2\sqrt{\frac{c^2 k^2}{4\pi^2} - \nu_0^2}} \times \frac{2kc^2}{4\pi^2} = \frac{2\pi}{2\nu} \times \frac{2kc^2}{4\pi^2} = \frac{kc^2}{2\pi\nu}$$

$$= \frac{2\pi}{\lambda} \frac{c^2}{2\pi\nu} = \frac{c^2}{\lambda\nu} = \frac{c^2}{v_p}$$

2.39   (a)

	n	l	m_l	m_s
	4	1	$\pm 1, 0$	$\pm\frac{1}{2}$

Therefore 6 electrons can possess $n = 4$, $l = 1$.

(b)   With $n = 4$ and $l = 0$; only two electrons can have the given values of n and l.

2.40 (a) For a 4*p* electron, $n = 4$, $l = 1$, $m_l = \pm 1$, 0 and $m_s = -\tfrac{1}{2}$.

(b) Mg has the configuration [Ne] $3s^2$. Thus, the valence electrons have $m_s = \pm \tfrac{1}{2}$.

Solutions 3

3.1 With the carbon atom at the origin and C=O along the $+x$-axis, the following data set results:

	x	y	z	q
C	0	0	0	+0.45
H_1	−0.5829	0.9329	0	+0.13
H_2	−0.5829	−0.9329	0	+0.13
O	1.22	0	0	−0.38

Evidently the y-components of the hydrogen atoms cancel, so that

$$|\mu| = \sqrt{\mu_x^2} = \sum_i |q_i| \, ex_i = 1.6022 \times 10^{-19}\,\text{C}\,[(0.018 \times 2 \times 0.110 \times 10^{-9}\,\text{m} \times \cos 58°)$$
$$+ (0.38 \times 0.122 \times 10^{-9}\,\text{m})]$$

$$= 7.764 \times 10^{-30}\,\text{C m},$$

or 2.33 D. Compare the result with literature values [5].

3.2 From Newton's laws, at equilibrium the force on q_3 is $\frac{q_1 q_3}{4\pi\varepsilon_0 r_{13}^2} + \frac{q_2 q_3}{4\pi\varepsilon_0 r_{13}^2}$ along $+x$. Hence, (a) $q_1 r_{23}^2 = -q_2 r_{13}^2$. (b) From (a), $q_1(x-r)^2 = -q_2 x^2 = |q^2|x^2$; then, $|q_2|/q_1 = (x-r)^2/x^2 = [1 - (r/x)]^2$. Thus, $x = \frac{r}{(1-\sqrt{|q_2|/q_1})}$. (c) No equilibrium position exists between the positions of q_1 and q_3 because q_3 would be attracted to q_1.

3.3 $\mu = qed$. Hence, $|q| = (2.50\,\text{D} \times 3.3356 \times 10^{-30}\,\text{C m})/(1.6022 \times 10^{-19}\,\text{C} \times 0.121 \times 10^{-9}\,\text{m}) = 0.430$; thus, the ionic character is 43%. The difference in electronegativities is 0.89; thus, the % ionic character would be $= 0.16 \times 0.89 + 0.035 \times 0.89^2 = 0.170$, that is, 17% by this method. This example shows that results based on electronegativities need independent confirmation (see Example 3.3a).

3.4 From the periodic table, $M_r(\text{Se}) = 78.96$. Using next the Clausius–Mosotti equation, $\alpha = \frac{(6.701-1)}{(6.701+2)} \times \frac{0.07896\,\text{kg mol}^{-1}}{4290\,\text{kg m}^{-3}} \times \frac{3\times8.8542\times10^{-12}\,\text{F m}^{-1}}{6.0221\times10^{23}\,\text{mol}^{-1}} = 5.319 \times 10^{-40}\,\text{C m}^2\,\text{V}^{-1}$. From Section 3.3.2.1, $\alpha' = 5.319 \times 10^{-40}\,\text{F m}^{-1}/(10^{-6} \times 4\pi \times 8.8542 \times 10^{-12}\,\text{C m}^2\,\text{V}^{-1}) = 4.780 \times 10^{-24}\,\text{cm}^{-3}$. (b) The effective radius is given by $\alpha = 4\pi\varepsilon_0 r^3$; hence, $r = \sqrt[3]{\frac{5.319\times10^{-40}\,\text{C m}}{4\pi\times8.8542\times10^{-12}\,\text{F m}^{-1}}} = 0.168\,\text{nm}$, which is the effective radius of the argon atom. (A literature value for the collision diameter of argon is *ca.* 0.36 nm.)

3.5 (a) The density is deemed constant over the given temperature range; thus, $M/D = 0.15223$ kg mol^{-1}/ 992 kg m^{-3} = 1.5346×10^{-4} m^3 mol^{-1}. Using the Clausius–Mosotti equation, the required data are:

$(10^3/T)/$K^{-1}	3.663	3.413	3.195	3.003	2.833	2.681
$(\varepsilon_r - 1)/(\varepsilon_r + 2)$	0.7931	0.7761	0.7656	0.7500	0.7391	0.7248
$10^{4Pm}P_m$/m^3 mol^{-1}	1.217	1.191	1.175	1.151	1.134	1.112

By least squares of P_m against the independent variable $1/T$, the slope is 1.0458×10^{-2} and this parameter is equal to $L\mu^2/(9\varepsilon_0 k)$. *It is instructive* to check the units of the slope: mol^{-1} C^2 m^2/F m^{-1} J K^{-1} = mol^{-1} C^2 m^2/[(C V^{-1}) m^{-1} (C V) K^{-1}] = m^3 K mol^{-1}, which is correct. On inserting values for the constants, μ evaluates as 4.37×10^{-30} C m, or 1.31 D.

(b) Since P_a is deemed negligible, the total polarization P_m is $\{L\mu^2/(9\varepsilon_0 kT) + L\alpha_e/(3\varepsilon_0)\}$. The intercept on the graph at $(10^3/T) = 0$ is 8.3588×10^{-5} m^3 mol^{-1}, which is $L\alpha/(3\varepsilon_0)$; hence, $\alpha = 3.687 \times 10^{-39}$ C m^2 V^{-1}, and the volume polarizability $\alpha' = 10^6\alpha/(4\pi\varepsilon_0)$. The units of this expression are C m^2 V^{-1} / F m^{-1} = m^3, and the factor of 10^6 ensures that the result $\alpha' = 33.1 \times 10^{-24}$ is in the usual (volume) units of cm^{-3}.

3.6 (a) From the discussion in Appendix A11, Sections A11.4 and A11.5, it follows that the potential V_P at the point P is $\mu r \cos 30°/(4\pi\varepsilon_0 r^3) = \mu(\sqrt{3}/2)/(4\pi\varepsilon_0 r^2)$, where $r = 50$ nm; hence, $V_P = 1.890 \times 10^{-5}$ V.

(b) From the same discussion, the field strength **E** is $\frac{1}{4\pi\varepsilon_0 r^5}\{3(\boldsymbol{\mu}\cdot\mathbf{r})\mathbf{r} - r^2\boldsymbol{\mu}\}$. The vector expression reduces to $a\mathbf{r} - b\boldsymbol{\mu}$, where $a = 3\mu r \cos 30°$ and $b = r^2$. Evaluating the vector expression as the dot product of itself followed by multiplication by $1/(4\pi\varepsilon_0 r^5)$ leads to the value $|\mathbf{E}| = 697$ V m^{-1}.

3.7 From vector addition of C–Cl bond moments:

	μ/D
1,2,3,4-tetrachlorobenzene	2.65
1,2,3,5-tetrachlorobenzene	1.53
1,2,4,5-tetrachlorobenzene	0
1,2,3,4,5-pentachlorobenzene	1.53

3.8 In dicyanoethyne, partial overlap of the π-orbitals of the carbon and nitrogen atoms of adjacent molecules facilitates the packing adopted by this substance in the crystalline state; the C...N contact distance is less than the sum of the van der Waals radii sum by 0.02 nm. In hexane, there is no interaction other than the London dispersion energy, and the structure maintains the normal van der Waals contact distance between molecules.

3.9   The data required are:

$(10^3 1/T)/\mathrm{K}^{-1}$	3.422	3.236	3.003	2.584	3.421	2.242
$10^6 P_m/\mathrm{m}^3\,\mathrm{mol}^{-1}$	57.57	55.01	51.22	44.99	42.51	39.59

(a) By least squares as before, the slope $= (1.5096 \times 10^{-2})\mu^2$ m C^{-2} K mol^{-1}, and the intercept at $1/T = 0$ is 6.0090×10^{-6} m^3 K mol^{-1}. Hence, $\mu = 1.5096 \times 10^{-2} \times 9\varepsilon_0 k/L = 5.251 \times 10^{-30}$ C m $= 1.57$ D.

(b) The polarization at 0 °C is the value of the intercept at $1/T = 0$ K^{-1}, that is, 6.009 m^3 mol^{-1}.

(c) The volume polarizability α' is $6.009 \times 10^{-6} \times 3\varepsilon_0/L \times (10^6/4\pi\varepsilon_0) = 2.38 \times 10^{-24}$ cm^3.

3.10 (a) Differentiate the expression $(\sigma/r)^{12} - (\sigma/r)^6$ with respect to r and equate to zero, whereupon $r\ (= r_e) = (2)^{1/6}\sigma = 0.427$ nm.

(b) $V(r)_{r=r_e} = 4\varepsilon\left\{\left(\frac{\sigma^{12}}{4\sigma^{12}}\right) - \left(\frac{\sigma^6}{2\sigma^6}\right)\right\} = -\varepsilon$

(c) $V(r)_{r=r_e} = -214\,\mathrm{K} \times 1.3807 \times 10^{-23}\,\mathrm{JK}^{-1} = -2.955 \times 10^{-21}\,\mathrm{J} \equiv -1779\,\mathrm{J\,mol}^{-1}$

3.11   The electric field strength $|\mathbf{E}|$ of the Na$^+$ ion at the distance r is $e/(4\pi\varepsilon_0 r^2) = 9.000 \times 10^9$ V m^{-1}. The dipole μ_i induced in the methane molecule is $\alpha E = 2.601 \times 10^{-30}$ C m. This value corresponds to a movement Δx of electron density in the methane molecule given by $\Delta x = \mu_{\mathrm{ind}}/e$, which is 0.016 nm. The fractional change in $d_{\mathrm{CH_4}}$ is 0.04. However, if it is assumed that all C–H bonds are affected similarly, the fractional reduction would be 0.16.

3.12   Apply Eq. (3.43) for a single species. In J per atom, $I_{\mathrm{Kr}} = 2080.6 \times 10^3$ J mol$^{-1}/6.0221 \times 10^{23}$ mol$^{-1} = 3.455 \times 10^{-18}$ J. For a non-polar gas, the polarizability is $4\pi\varepsilon_0\zeta^3$, where ζ is the atomic radius. Hence, α' (in m^3) is ζ^3. Hence, from Eq. (3.43), the London energy:

$$V_{\mathrm{id;id}} = -\frac{3}{4}\frac{(\alpha'^2)^2}{(r)^6}I = -\frac{3}{4}\frac{(\zeta^3)^2}{(2\zeta)^6}I = -\frac{3}{4}\frac{I}{64}.$$

Thus, $V_{\mathrm{id,id}} = -4.05 \times 10^{-20}$ J, or -24.4 kJ mol^{-1}. The negative sign indicates an attractive interactional energy.

3.13   (a) The equilibrium interatomic distance r_e is first obtained by differentiating $V(r)$ with respect to r and setting the derivative to zero; thus, r_e is $(2)^{1/6}\sigma$, which is 0.438 nm for krypton. The Lennard-Jones equation is then used to obtain $V(r)$ for krypton:

$$V(r) = 4 \times 155\,\mathrm{K} \times 1.3807 \times 10^{-23}\,\mathrm{JK}^{-1}\left[\left(\frac{\sigma}{2^{1/6}\sigma}\right)^{12} - \left(\frac{\sigma}{2^{1/6}\sigma}\right)^6\right]$$

$$= -8.5603 \times 10^{-21}\,\mathrm{J}/4 = 2.14 \times 10^{-21}\,\mathrm{J} \equiv -1.29\,\mathrm{kJ\,mol}^{-1}.$$

(b) The attractive force F between two atoms is the negative slope of the potential, $-\frac{dV(r)}{dr}$, the expression which was used in part (a) to obtain r_e; the curvature of $V(r)$, $\frac{d}{dr}\left(-\frac{dV(r)}{dr}\right)$, evolves as $24\varepsilon\left[\left(\frac{26\sigma^{12}}{r^{14}}\right)-\left(\frac{7\sigma^6}{r^8}\right)\right]$.

(c) If the second derivative is set to zero, $r = (26/7)^{1/6}\sigma$. This value represents a point of inflection on the $V(r)$ curve at $r = 1.244\sigma$; in the case of krypton $r = 0.485 \times 10^{-9}$ nm. Its only significance is that it represents the maximum of the curvature, the value of which changes continuously as r increases from σ, where it is zero, to $r = \infty$.

3.14 (a) $\frac{dV(r)}{dr} = \frac{6A}{r^7} - \frac{12B}{r^{13}}$. Setting this derivative to zero gives $r = r_e = \left(\frac{2B}{A}\right)^{1/6}$. Since $r_e = 2^{1/6}\sigma$, $\sigma = B/A$.

(b) The force F is given by $\frac{dF}{dr} = \frac{d}{dr}\left(-\frac{dV(r)}{dr}\right) = \frac{42A}{r^8} - \frac{156B}{r^{14}}$. The maximum is obtained by setting this second derivative to zero; thus, $r_m = (26B/7A)^{1/6} = 3.9353 \times 10^{-10}$ m. The maximum force F_{max} is obtained by substituting $r = (26B/7A)^{1/6}$ into the expression for $-\frac{dV(r)}{dr}$. Thus, $F_{max} = -\frac{6.0\times10^{-77}\,\mathrm{J\,m^6}}{(3.93531\times10^{-10}\,\mathrm{m})^7} + \frac{12.0\times10^{-134}\,\mathrm{J\,m^{12}}}{(3.93531\times10^{-10}\,\mathrm{m})^{13}} = -\frac{6.0\times10^{-7}\,\mathrm{J\,m^6}}{(3.93531\,\mathrm{m})^7} + \frac{12.0\times10^{-1}\,\mathrm{J\,m^{12}}}{(3.93531\,\mathrm{m})^{13}} = 1.895 \times 10^{-11}\,\mathrm{N}\,(\equiv \mathrm{J\,m^{-1}})$

(c) If $V(r) = 0$ at $r = r_0$, then $-\frac{A}{r^6} + \frac{B}{r^{12}} = 0$, so that $r = (B/A)^{1/6}$. Hence, $r_e/r_0 = \frac{(2B/A)^{1/6}}{(B/A)^{1/6}} = 2^{1/6} = 1.122$.

(d) Reading from the graph, $r_0 = 1.00 \times 10^{-10}$ m and $r_e = 1.12 \times 10^{-10}$ m. It is hardly practicable to read r_m since the maximum is not well defined on the plot:

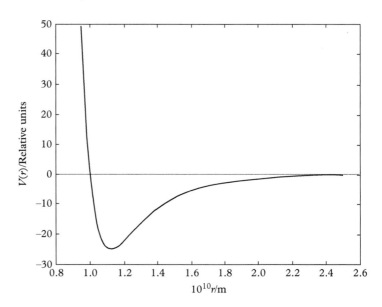

Lennard-Jones potential energy function

3.15 (a) The interactional energy is given on Fig. 3.7: $E = -\frac{Q^2 p^2}{6(4\pi\varepsilon_0)^2 kT r^4}$; $Q = e$
and kT at 25 °C is 0.4114×10^{-20} J. $E = -0.09341 \times 10^{-20}$ J.

(b) The assumption is valid because $|E| < kT$.

(c) Calculating at the energy level $10^{20}E$ and plotting, with the line $y = 0.4114$ drawn, gives:

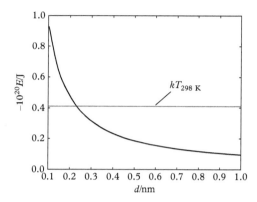

Ion–dipole: energy vs. distance

With the mouse pointer over the plot, the critical value of d is $\leq$ *ca.* 0.23 nm.

3.16 T data required are as follows:

$10^3(1/T)/\text{K}^{-1}$	2.604	2.381	2.247	2.066	1.916
$10^5 P'_\text{m}/\text{m}^3\,\text{mol}^{-1}$	5.74	5.35	5.01	4.69	4.31

From least squares of P_m *v.* $(1/T)$, the slope $= 2.078 \times 10^{-2}$ m C^{-2} K mol^{-1}, and the intercept at $1/T = 0$ is 3.571×10^{-6} m^3 K mol^{-1}. Hence, $\mu = 6.161 \times 10^{-30}$ C m $\equiv 1.85$ D, and $\alpha = 1.575 \times 10^{-40}$ C m^2 V^{-1} or as a polarizability volume $\alpha' = 1.42 \times 10^{-24}$ cm^3.

Solutions 4

4.1 (a) From $\varepsilon = h\nu = hc/\lambda = hc\bar{\nu}$, $\varepsilon/\text{eV} = hc\bar{\nu} \times 10^2/e$. Evaluating, $\varepsilon = 6.1062$ eV.

(b) $\varepsilon/\text{eV} \times Le = \varepsilon/\text{kJ mol}^{-1}$. Evaluating, $\varepsilon = 589.16$ kJ mol^{-1}.

4.2 The following data are required a least squares calculation of $1/T$ *v.* $\ln(p/p_0)$, where p_0 is a reference pressure of 1 N m^{-2}:

$10^3(1/T)/\text{K}^{-1}$	1.1168	1.0845	1.0368	0.99039	0.95648
$\ln(p/p_0)$	−2.5472	−1.5847	−0.61804	0.33647	1.2238

By least squares, $\ln(p/p_0) = -2.2722 \times 10^4(1/T)K + 22.925$. The slope of the line is $-\Delta H_{vap}R$; hence, $\Delta H_{vap} = -22722\,\text{K} \times 8.3145\,/10^3 = 188.9\text{ kJ mol}^{-1}$.

4.3   The required thermodynamic cycle is:

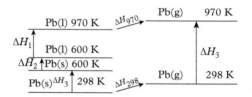

Cycle for the enthalpy of sublimation of lead

4.4   Applying the Born–Haber cycle (Section 4.3.1), the crystal energy for MgCl is *ca.* −717.3 kJ mol⁻¹, whereas that for (MgCl₂) is *ca.* −2057.2 kJ mol⁻¹. Although more energy is required to form the Mg^{2+} cation compared to Mg^+, it is more than compensated by the formation of crystalline $MgCl_2$ rather than MgCl; hence $MgCl_2$ is formed from the reaction of magnesium and chlorine.

4.5   The required thermodynamic cycle for the reactions leading to NH_4Cl is:

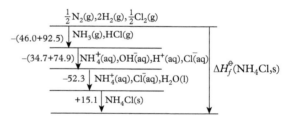

Cycle for the enthalpy of formation of NH₄Cl(s)

Since $\sum \Delta H_f^{\ominus} = 0$ when summed cyclically, taking note of the directions of the arrows,

$$\Delta H_f^{\ominus}(NH_4Cl, s) = -(46.0\text{ kJ mol}^{-1} + 92.5\text{ kJ mol}^{-1}) - (34.7\text{ kJ mol}^{-1} + 74.9\text{ kJ mol}^{-1})$$
$$- 52.3\text{ kJ mol}^{-1} + 15.1\text{ kJ mol}^{-1}$$
$$= -285.3\text{ kJ mol}^{-1}$$

4.6   (a)   The energy lost per ion in the spherical shell when the shell is earthed $= 0.806e^2/(4\pi\varepsilon_0 V^{1/3})$. The volume V occupied *per ion* (Na^+ or Cl^-) in the sodium chloride structure is $\frac{a^3}{4\times 2} = \frac{(2r)^3}{8} = r^3$. Hence, taking the conducting shell on any ion of radius r equal to r_e, the energy lost per ion is $0.806e^2/(4\pi\varepsilon_0 a/2)$. (The units are correct: C^2/F $m^{-1} \times m = $ J.) Hence, the Madelung constant (for a pair of ions) is 2×0.806, or 1.612.

(b) Take Cl^- as an origin. Then, ions totally inside the cube have their nominal value; those on a face are 1/2, along an edge 1/4, at a corner 1/8, outside the cube of side $2a$, zero. Then the following scheme can be set up for the ions within or in contact with the structure:

Number/type of ion	Distance from origin	Sign and Contribution to cube
6 Na^+	1	-1
12 Cl^-	$\sqrt{2}$	$+1$
8 Na^+	$\sqrt{3}$	-1
6 Cl^-	$\sqrt{4}$	$+\frac{1}{2}$
24 Na^+	$\sqrt{5}$	$-\frac{1}{2}$
24 Cl^-	$\sqrt{6}$	$+\frac{1}{2}$
12 Cl^-	$\sqrt{8}$	$+\frac{1}{4}$
24 Na^+	$\sqrt{9}$	$-\frac{1}{4}$
8 Cl^-	$\sqrt{12}$	$+\frac{1}{8}$

The appropriate sum is:

$$- (6 \times 1) + (12 \times 1)/\sqrt{2} - (8 \times 1)/\sqrt{3} + (6 \times \tfrac{1}{2})/\sqrt{4} - (24 \times \tfrac{1}{2})/\sqrt{5}$$
$$+ (24 \times \tfrac{1}{2})/\sqrt{6} + (12 \times \tfrac{1}{4})/\sqrt{8} - (24 \times \tfrac{1}{4})/\sqrt{9} + (8 \times 1/8)/12 = -1.752$$

The Madelung constant is 1.752 (precise value 1.74756)

(c) The Evjen result reflects Pauling's rule that *the charge on an ion tends to be neutralized by its nearest neighbours.*

4.7 Following the discussion in Section 4.4ff, $U(r_e) = -3952.1$ kJ mol^{-1}. Using now the Born–Haber cycle equation (4.16): -3952.1 kJ $mol^{-1} = -601.8$ kJ mol^{-1} $- 737.7$ kJ $mol^{-1} - 1451.$ kJ $mol^{-1} - 148.1$ kJ $mol^{-1} - 244.95$ kJ $mol^{-1} - E(O^{2-})$ /kJ $mol^{-1} + 5.0$ kJ $mol^{-1} = 773.6$. Hence, $E(O^{2-}) = +774$ kJ mol^{-1}. (A more exact calculation, including polarization terms in the crystal energy calculation, led to $E(O^{2-}) = 748.9 \pm 33.0$ (4%) kJ mol^{-1} [6].)

4.8 Obtain $\Delta H_c^{\ominus}$ from a Born–Haber cycle (including $+3RT$ in the cycle) for each halide. Then, $\Delta H_c^{\ominus}$ $(SrX_2) + \Delta H_d^{\ominus}$ $(SrX_2) = \Delta H_h^{\ominus}$ (SrX_2). Finally, $\Delta H_h^{\ominus}$ $(SrX_2) - 2\Delta H_h^{\ominus}$ $(X^-) = \Delta H_h^{\ominus}$ (Sr^{2+}).

	SrF_2	$SrCl_2$	$SrBr_2$	SrI_2
$-U_c$ /kJ mol^{-1}	2479.5	2142.3	2059.8	1961.6

Hence, an average value for $\Delta H_h^{\ominus}(Sr^{2+}, aq)$ is -1456 kJ mol^{-1}. A sketch of the cycle of operations is given below:

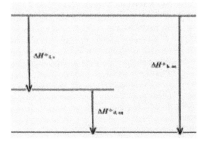

Cycle for enthalpy of hydration of the Sr^{2+}(aq)

4.9 There are 4 S–O equal contacts and 6 O–O equal contacts to consider. Therefore, the electrostatic self-energy is

$$U_S = -\sum \frac{e^2}{4\pi\varepsilon_0}[(-4 \times 0.7 \times 0.8)/0.13\,\text{nm} + (6 \times 0.7^2)/0.2123\,\text{nm}]$$

$$= \frac{(1.6022 \times 10^{-19}\,\text{C})^2 \times (-3.3824 \times 10^9\,\text{nm}^{-1})}{4\pi \times 8.8542 \times 10^{-12}\,\text{F}\,\text{m}^{-1}} = -7.804 \times 10^{-19}\,\text{J} \equiv -469.9\,\text{kJ}\,\text{mol}^{-1}.$$

4.10 Allowing $+3\%$ on the radii, and then summing them gives $r_e = 0.3924$ nm. The, density is the mass of the unit cell (it contains 1 CsCl entity) divided by its volume: $D = \frac{259.81 \times 1.6605 \times 10^{-27}\,\text{kg}}{8 \times \sqrt{3} \times (0.3924 \times 10^{-9}\,\text{m})^3/9} = 4.638$ kg m^3.

4.11 From the diagram: $\mathbf{p} = (x_{Ti}-x_O)\mathbf{a} + (y_{Ti}-y_O)\mathbf{b} + (z_{Ti}-z_O)\mathbf{c}$, whence $p^2 = (x_{Ti}-x_O)^2 a^2 + (y_{Ti}-y_O)^2 b^2 + (z_{Ti}-z_O)^2 c^2$; the cross terms such as $2(x_{Ti}-x_O)a(y_{Ti}-y_O)b\cos\gamma$ are zero because the inter-axial angles α, β and γ are all $90°$ in the tetragonal system. Thus, $d_p^2 = [(0.5-0.3056)^2 + (0.5-0.3056)^2] \times (0.4593\,\text{nm})^2 + 0.5^2 \times (0.2959\,\text{nm})^2$, since $a = b$ and $z_O = 0$. Hence, $d_p = 0.1945_1$ nm. Similarly, $d_q = 0.1985_0$ nm. Check the diagrams to decide upon the correct values for x and y; z will be zero for q.

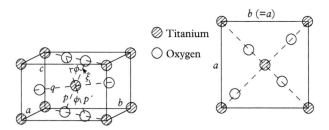

TiO$_2$ (rutile): Clinographic drawing and projection on to (001)

For the angles, $\cos\phi = \frac{\mathbf{p}\cdot\mathbf{p}'}{pp'}$; note that 10^{-18} in the numerator and denominator cancel. Thus,

$$pp'\cos\phi = [(x_{Ti}-x_O)\mathbf{a} + (y_{Ti}-y_O)\mathbf{b} + (z_{Ti}-z_O)\mathbf{c}]\cdot$$
$$[(x_{Ti}-x_{O'})\mathbf{a} + (y_{Ti}-y_{O'})\mathbf{b} + (z_{Ti}-z_{O'})\mathbf{c}]$$
$$= [(x_{Ti}-x_O)(x_{Ti}-x_{O'}) + (y_{Ti}-y_O)(y_{Ti}-y_{O'})]a^2 + (z_{Ti}-z_O)(z_{Ti}-z_{O'})c^2$$
$$= [0.1944\times(-0.1944) + 0.1944\times(-0.1944)]\times0.4593^2 + 0.25\times0.2959$$
$$= 5.9445\times10^{-3}$$

Since $p = p'$, $\cos\phi = 5.5445 \times 10^{-3}/0.19451^2 = 0.15712$. Hence, $\phi = \cos^{-1} 0.15712 = 80.96°$. Getting the angles correct hinges on selecting the correct co-ordinates for the three atoms involved. The angle ξ can be found in similar manner (90° exact). Other O–Ti–O angles exist which are not quite 180° because of the small distortion of the $[TiO_6]$ octahedral grouping. A discussion on the use of vector methods for bond lengths, bond angles and torsion angles, and their precision, may be found in the literature [7].

4.12 (a) With the atoms are in close contact, $a = 2r_-$ and $r_- + r_+ = (3/8)c = (3/8)2\sqrt{2/3}\,a$. Eliminating a between the two equations and dividing throughout by r_-, $r_+/r_- = \sqrt{3/2} - 1 = 0.225 = R_4$.

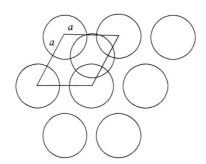

Projection of the unit cell (outlined) of wurtzite ZnS on to (0001)

(b) The shortest Zn–S distance = $0.375 \times .6261$ nm = 0.235 nm. (The sum of the ionic radii is 0.244 nm, indicating a small polarization component in the bonding.)

(c) Following the above procedure in two dimensions, the radius ratio is 0.155.

4.13 The equations are:

		ΔH_{reac}
(1) $AgI(s) + K^+(aq) + I^-(aq)$	$\rightarrow K^+(aq) + [AgI_2]^-(aq)$	$-9.54\,kJ\,mol^{-1}$
(2) $AgI(s) + Na^+(aq) + I^-(aq)$	$\rightarrow Na^+(aq) + [AgI_2]^-(aq)$	$-7.46\,kJ\,mol^{-1}$
(3) $Ag(s) + I_2(s) + K^+(aq) + I^-(aq)$	$\rightarrow K^+(aq) + [AgI_2]^-(aq)$	$-72.0\,kJ\,mol^{-1}$
(4) $Ag(s) + I_2(s) + Na^+(aq) + I^-(aq)$	$\rightarrow Na^+(aq) + [AgI_2]^-(aq)$	$-70.3\,kJ\,mol^{-1}$

From (3) − (1), $Ag(s) + \frac{1}{2}I_2 \rightarrow AgI(s)$ $\Delta H_f(AgI, s) = -62.46$ kJ mol^{-1}, and from (4) − (2) $\Delta H_f(AgI, s) = -62.84$ kJ mol^{-1}. Hence, the average $\Delta H_f(AgI, s) = -62.7$ kJ mol^{-1}.

4.14 (a) $\Delta G_d^{\ominus} = \Delta H_d^{\ominus} - T\Delta S_d^{\ominus} = 39.6$ kJ mol^{-1}. The negative $\Delta H_d^{\ominus}$ promotes solubility, but this effect is negated by the large negative $\Delta S_d^{\ominus}$. (The small ions F$^-$ and more particularly Mg^{2+} are structure-making in water; this effect is enhanced by F$\cdots$H$\cdots$F hydrogen bonds. The entropy (ions + water) system is less than that of the sum of its separate components.)

(b) Since $-RT\ln(4c_{\pm}^3 f_{\pm}^3) = 39600$ J mol^{-1}, $c_{\pm} \times f_{\pm} = (2.8630 \times 10^{-8})^{1/3} = 3.059 \times 10^{-3}$, so that $f_{\pm} = 0.0408$. (The Debye limiting law gives $f_{\pm} = 0.329$, which is far from the correct value at this concentration.)

4.15   Following the analysis in the text for Schottky defects (Section 4.11.1), W_1 (for the number of defects) is $N!/[(N-n)!]$ and W_2 (for the number of interstices) is $N'!/[(N'-n)!n!]$. The total probability $W_1 W_2$ is W:

$$W = \frac{N!N'!}{(N-n)!\,n!\,(N'-n)!\,n!}.$$

Since the initial ideal state has a probability of unity, $\Delta S = k\ln(W/1)$, and the change in the Helmholtz free energy is

$$\Delta A = n\Delta u - kT\ln W = n\Delta u - kT\ln\frac{N!}{(N-n)!n!} + \ln\frac{N'!}{(N'-n)!n!}.$$

Applying Stirling's equation to the factorial expressions:

$$[N\ln N - (N-n)\ln(N-n) - n\ln n] + [N'\ln N' - (N'-n)\ln(N'-n)! - n\ln n].$$

At equilibrium, $(\partial\Delta A/\partial n)_T = 0 = \Delta u - T(\partial\Delta S/\partial n)_T$; hence,

$$\Delta u/kT = \left\{\ln\frac{N-n}{n} + \ln\frac{N'-n}{n}\right\} = \left\{\ln\frac{(N-n)(N'-n)}{n^2}\right\}.$$

Making the reasonable assumption that n is very much smaller than N and N',

$$\Delta u = kT\left\{\ln\frac{NN'}{n^2}\right\}.$$

Hence,

$$n = \sqrt{NN'}\exp(-\Delta u/2kT).$$

For the creation of a defect (ion + hole) in silver chloride at 500 K, $n/\sqrt{NN'} = \exp[-1.5\,\mathrm{eV} \times 1.6022 \times 10^{-19}\,\mathrm{C}/(2 \times 500\,\mathrm{K} \times 1.3807 \times 10^{-23}\,\mathrm{J\,K^{-1}})] = 2.76 \times 10^{-8}$.

4.16 Following through the analysis in Section 4.11.1 for the Schottky defect shows that $n/N = \exp[-\Delta u/(3kT)]$, since two anions must be lost per cation vacancy in order to preserve electrical neutrality.

4.17 Taking ln of each side: $\ln \beta_t = \ln A - \frac{x^2}{Dt}$. By least squares of $\ln \beta_t$ *v.* x^2, the least-squares equation is $\ln \beta_t = -17.79x^2 + 6.419$, whence $D = 1/(17.79\,\text{mm}^{-2} \times 6 \times 10^4\,\text{s}) = 9.368 \times 10^{-7}\,\text{mm}^2\,\text{s}^{-1}$; $A = 613.4$ (arbitrary units).

4.18 By least squares, $\ln D = -2.296 \times 10^5(1/T) - 14.89$; hence, $U = -2.296 \times 10^5\,\text{K} \times R\,\text{J}\,\text{K}^{-1}\,\text{mol}^{-1} \times 10^{-3} = 190.9\,\text{kJ}\,\text{mol}^{-1}$; $\sigma(U)$, (from the least-squares fit) $= 1.7\,\text{kJ}\,\text{mol}^{-1}$.

4.19 The data for the least squares plot of C_p/T against T are:

T/K	15.05	25.20	47.10	67.13	82.11	133.4	204.1	256.5	283.0	298.0
$10^2(C_p/T)/\text{JK}^{-2}\text{mol}^{-1}$	1.2924	2.3786	7.4989	11.379	12.301	13.402	11.132	9.6725	9.2191	8.8054

The area under the curve between 15.05 K and 283.0 K is obtained by quadrature (program QUAD). Between 15.05 K and 0 K, the Debye approximation may be used: $C_p = aT^3$. $C_{p(15.05\,\text{K}-0\,\text{K})}/T = \int_0^{15.05} aT^2 dT = aT^3/3 = C_p/3$, where C_p is the minimum value recorded. Thus, C_p (0–15.05 K) = 0.1945/3 J $\text{K}^{-1}\,\text{mol}^{-1} = 0.065\,\text{J}\,\text{K}^{-1}\,\text{mol}^{-1}$. The area under the curve is 29.64 J K^{-1} mol^{-1}, to which is added 0.065 J K^{-1} mol^{-1} to give 29.71 J K^{-1} mol^{-1} as the standard molar entropy of nickel.

4.20 From Appendix A3, $\sin \theta_{300} = \frac{0.15418\,\text{nm} \times \sqrt{9}}{2 \times 0.4562\,\text{nm}}$; hence, $\theta_{300} = 30.46°$. The intensity I_{300} would be weak as equal electron densities of Cs$^+$ and I$^-$ ions lie on planes interleaving (300) at $d_{300}/2$ and the isoelectronic Cs$^+$ and I$^-$ species have scattering factors f that are very nearly equal in magnitude. (The phase difference for d_{300} with respect to the origin is given by $\{f_{\text{Cs}^+} + f_{\text{I}^-} \cos 2\pi\,(3 \times \frac{1}{2})\}$, so that the two contributions are clearly in opposition.)

4.21 The density D of copper is given by $D = \frac{4 \times M_r \times m_u}{V} = \frac{4 \times 63.546 \times 1.6605 \times 10^{-27}\,\text{kg}}{(0.3615 \times 10^{-9}\,\text{m})^3} = 8934.3\,\text{kg}\,\text{m}^{-3}$. Then, the number of lattice sites per m^3 is given by $N = D \times L/(10^{-3}M_r)$. Thus, $N = 8934.3\,\text{kg}\,\text{m}^{-3} \times 6.0221 \times 10^{23}\,\text{mol}^{-1}/0.063546\,\text{kg}\,\text{mol}^{-1} = 8.467 \times 10^{28}\,\text{m}^{-3}$. Hence, the number n of defects in 1 m^3 is: $n = N \exp(-U/kT) = 8.647 \times 10^{28}\,\text{m}^{-3} \times \exp\left(-\frac{0.901\,\text{eV} \times 1.6022 \times 10^{-19}\,\text{C}}{2 \times 1.3807 \times 10^{-23}\,\text{JK}^{-1} \times 1273\,\text{K} \times 1\,\text{m}^3}\right) = 1.42 \times 10^{27}$ (for the given m^3).

Solutions 5

5.1   Lithium is body-centred cubic in a unit cell with $Z = 2$. (a) The mobility μ is given by $\mu = e\tau/m_e$ and the conductivity $\sigma = (N/V)e^2\tau/m_e$, or $\sigma = (N/V)e\mu$, and $1/\mu = (N/V)e\rho$, where τ is the relaxation time and N/V, the electron concentration for 2 atoms per unit cell, is given by $N/V = 2/(0.3510 \times 10^{-9}\,\text{m})^3 = 4.6250 \times 10^{28}\,\text{m}^3$, so that $\mu = 1.578 \times 10^{-3}\,\text{m}^2\,\text{s}^{-1}\,\text{V}^{-1}$.

(b) Also, $1/\tau = (N/V)e^2\rho/m_e$. so that $\tau = 8.974 \times 10^{-15}$ s, and the mean free path λ is given by $\lambda = \tau\sqrt{\overline{v^2}} = 8.976 \times 10^{-15} \times s \times 1.30 \times 10^6\,\text{m s}^{-1} = 1.17 \times 10^{-8}$ m.

5.2   The mean value $<x>$ of a function $f(x)$ is $\int xf(x)dx/\int f(x)dx$. Thus, the mean displacement $<x>$ of the function $V(x)$ is given by $<x> = \frac{\int_{-\infty}^{\infty} x\exp(-ax^2+bx^3)\beta dx}{\int_{-\infty}^{\infty}\exp(-ax^2+bx^3)\beta dx}$, where $\beta = 1/(kT)$. Applying the approximation $\exp(bx^3) = (1 + bx^3)$, $<x> = \frac{\int_{-\infty}^{\infty}[x\exp(-ax^2)\beta+bx^4\beta\exp(-ax^2)\beta]dx}{\int_{-\infty}^{\infty}[\exp(-ax^2)\beta+bx^3\beta\exp(-ax^2)\beta]dx}$. The integrals involving odd power of x are zero, while those in even powers are twice the integral from 0 to infinity. These integrals are solved readily by use of the gamma function, Appendix 7: $<x> = \frac{(b/a^{5/2})\beta^{-3/2}3(\pi^{1/2})/4}{(\pi/a\beta)^{1/2}} = \frac{3b}{4a^2\beta} = \frac{3bkT}{4a^2}$. Thus, $<x> \propto T$, as expected for thermal expansion

5.3   (a)   The Fermi energy E_F is given by $E_F = \frac{\hbar^2}{2m_e}[3\pi^2(N/V)]^{2/3}$, with $N/V = 4.6250 \times 10^{28}\,\text{m}^3$ (from Solution 5.1); thus, $E_F = 7.257 \times 10^{-19}$ J $= 4.529$ eV. (Recall: J $\equiv$ kg m^2 s^{-2})

(b)   The Fermi temperature E_F/k is 7.257×10^{-19} J/ 1.3807×10^{-23} J K^{-1} $= 52{,}560$ K.

5.4   From Eq. (5.49), it is straightforward to show that $g(E_F)g(E_F) = 3.33 \times 10^9$.

5.5 $<E> = \frac{\int_0^{E_F} Eg(E)dE}{\int_0^{E_F} g(E)dE} = \frac{\int_0^{E_F} \phi E^{3/2}dE}{\int_0^{E_F} \phi E^{1/2}dE}$, where ϕ includes the constant terms in $g(E)$. Integration shows that $<E> = (3/5)E_F$, a result used in Section 5.5 of the text.

5.6   (a)

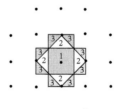

The first three Brillouin zones in a square, primitive, two-dimensional lattice of spacing a

Each Brillouin zone has the same total area as each of the others.

(b) By definition, a cube of side $2\pi/a$ (Section 5.5).

5.7 From the geometry of the structure, the appropriate construction can be drawn:

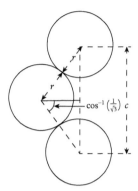

Vertical section through three successive layers of a close-packed hexagonal array of equal spheres

In the close-packed hexagonal structure, $c = a\sqrt{8/3}$, and $a = 2r$, the volume of the unit cell (containing 2 spheres) is $a^2 c\sqrt{3}/2 = 4r^2 \times 2r\sqrt{8/3} \times \sqrt{3}/2 = 8r^3\sqrt{2}$. The volume of 2 spheres is $8\pi r^3/3$. Hence, the packing fraction is $\frac{8\pi r^3}{3} \times \frac{1}{8r^3\sqrt{2}} = \pi/(3\sqrt{2}) = 0.740$.

5.8 (a) The fractional coordinates are as follow:

Eight tetrahedral holes: ¼, ¼, ¼; ¾, ¼, ¼; ¼, ¾, ¼; ¼, ¼, ¾

¾, ¾, ¾; ¼, ¾, ¾; ¾, ¼, ¾; ¾, ¾, ¼

Four octahedral holes: ½, 0, 0; 0, ½, 0; 0, 0, ½; ½, ½, ½

(A model of a close-packed cubic structure helps—failing that, Fig. 5.28a.)

(b) Since an atom at ¼, ¼¼ in contact with 8 other atoms, the radius of the atom in the hole (or radius of the hole) is $\sqrt{3}-1$, or 0.732. (Cf. the radius ratio for the CsCl structure type.)

5.9 (a) The radius of the sphere is the distance between the fractional coordinates 0, 0, 0 and ½, ½, ½, or $a\sqrt{3}/2$. Hence, the volume of V_I the sphere is $\frac{4\pi}{3} a^3 \frac{3\sqrt{3}}{8} = \pi a^3 \sqrt{3}/2$.

(b) The appropriate distance now (FCC) is between fractional coordinates 0, 0, 0 and ½, ½, 0, or $\sqrt{2}/2$, and the volume V_F is $\frac{4\pi}{3} a^3 \frac{2\sqrt{2}}{8} = \pi a^3 \sqrt{2}/3$. Hence, $V_I/V_F = (3/2)\sqrt{3/2}$.

5.10   (a) The conduction band is partly filled; the crystal is metallic and will conduct electricity. (b) If the band gap ΔE is large, the crystal would be an insulator, whereas if ΔE is sufficiently small ($\leq kT$), it would behave as an intrinsic semiconductor. Removal of the electrons from the upper band could be accomplished by a chemical oxidant. (c) Adding electrons to fill the partially occupied band will make the crystal behave as an insulator; it could be accomplished by a chemical reductant. However, if a higher unoccupied band with a small energy gap is present, semiconduction could arise.

5.11   (a) With a partially filled conduction band and unoccupied valence band, the solid will conduct electricity. (b) The solid would be an insulator; an oxidant would bring about this change. (c) The solid would be again either an insulator or an intrinsic semiconductor if the band gap is small.

5.12   (a)

Number of unit cells in the cube	Surface atoms (S)	Interior atoms (I)	S/I
1	8	1	8
8	26	9	2.89
27	6	35	1.60
64	98	91	1.08
125	152	189	0.80

So 5^3 unit cells are needed. (b) Since the cell side for sodium is 0.4291 nm, the necessary crystal is a cube of side 2.146 nm.

5.13   From $E = hc/\lambda$, $E = 7.946 \times 10^{-19}$ J, or 4.959 eV. Hence, the kinetic energy of emitted electrons is 0.709 eV, and the average electron speed is, therefore, 499.6×10^5 m s^{-1}.

5.14   A photon wavelength just greater than that given by $hc/\Delta E$, which evaluates to 799.9 nm.

5.15   The wavelengths in white light range from *ca.* 700 nm (red) to 400 nm (violet). Since the crystal transmits red, it will absorb in the yellow–blue region of the visible spectrum. As the material is pure and has a finite band gap it acts as a semiconductor. In order to promote electron charge carriers, photons of energy greater than *ca.* 700 nm (red) are needed, say, 650 nm. Taking $\lambda = 650$ nm as the optical absorption edge for this material, $E = hc/(650 \times 10^{-9}$ nm$) = 3.056 \times 10^{-19}$ J $\equiv 1.907$ eV, so that the band gap ΔE is *ca.* 1.9 eV.

5.16 A possible band structure is

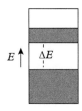

Band structure for a metal showing a partially-filled valence band

where the upper and lower bands are the conduction and valence bands, respectively. The material is transparent to photons of energy less than $h \times 1.25 \times 10^{14}/e = 0.516$ eV, which value is an approximate value for the energy gap ε (ΔE).

5.17 (a) From the uncertainty principle, $\Delta E \Delta t = \hbar/2 = 1.05457 \times 10^{-34}$ J s/2; hence, the uncertainty in the energy ΔE is 1.05457×10^{-34} J s/$(2 \times 20 \times 10^{-9}$ s$) = 2.636 \times 10^{-27}$ J. (b) $\Delta \nu = \Delta E/h = 3.978$ MHz. The spread width of the line $\Delta \lambda$ is given by $\Delta \lambda/m = \Delta(c/\nu) = -c\Delta\nu/\nu^2 = \lambda^2 \Delta\nu/c = -4.611 \times 10^{-15}$, so that the spread width $|\Delta \lambda| = 4.611$ fm. (c) The photon length l is given by $c\Delta t$, which evaluates to 5.996 m.

5.18 (a) The energy per unit volume $= 1360$ W m$^{-2}/2.9979 \times 10^8$ m s$^{-1} = 4.536 \times 10^{-6}$ m^{-3}. (Recall; W $=$ kg m^2 s$^{-3} \equiv$ J s^{-1}.) (b) The luminous power from the light is 0.15×60 W, or 9 W. Thus, the light falling on the table surface is 0.75×9 W, or 6.75 W. The energy of a single photon of wavelength 600 nm is hc/λ: 6.6261×10^{-34} J s $\times 2.9979 \times 10^8$ m s$^{-1}/600 \times 10^{-9}$ m $= 3.311 \times 10^{-19}$ J. Hence, the number of photons reaching the table surface in 1 s is 6.75 W/3.311×10^{-19} J $= 2.039 \times 10^{19}$.

5.19 For CdS, $\Delta E = 3.829 \times 10^{-19}$ J and for α-ZnS 5.784×10^{-19} J. The minimum absorption frequencies determined by these energies are 5.779×10^{14} Hz and 8.729×10^{14} Hz, respectively. Then, the corresponding wavelengths are: CdS: 518.8 nm, which is in the green region of the visible spectrum, so that the compound exhibits the complementary colour, orange; lower frequencies are not absorbed. For α-ZnS, the wavelength of 343.4 nm is in the ultraviolet region of the electromagnetic spectrum and appears colourless.

5.20 (a) $\Delta E = hc/\lambda = 5.842 \times 10^{-19}$ J $\equiv 3.65$ eV. (b) (i) n-type (ii) p-type. (c) p-type (forward bias).

5.21 (a) The mole fraction of lead x_{Pb} in the alloy is $20.76/M_r(\text{Pb}) = 0.1002$, or 10.02%. Thus, the zinc content of the alloy is 89.98%. (b) The number of moles of the two components in the silver/cadmium alloy are $n_{Ag} = 37.49/107.87 = 0.3475$ and $n_{Cd} = 62.51/112.41 = 0.5561$; thus, the corresponding mole fractions are $x_{Ag} = 0.3846$ and $x_{Cd} = 0.6154$, and the alloy composition is therefore $Ag_{1.0}Cd_{1.6}Ag_{1.0}Cd_{1.6}$ or, in integral formulation, Ag_5Cd_8.

5.22   The upward shift of 0.1 eV means that $(E - E_F) = (1.11\ \text{eV}/2) - 0.1\ \text{eV} = 0.455\ \text{eV}$. Then, the probability of promotion is given by

$$f(E) = \frac{1}{\exp[0.455\ \text{eV}/(8.6173 \times 10^{-5}\ \text{eV K}^{-1} \times 293\ \text{K})] + 1}$$

$$= 2.02 \times 10^{-8}$$

5.23   From Section 5.3, an equation for the mean collision time is developed as $<\tau> = \frac{V m_e}{N e^2 \rho}$. The face-centred unit cell contains 4 atoms; hence, $N/V = 4/(0.3615 \times 10^{-9}\ \text{nm})^3$, so that $<\tau> = \frac{(0.3615 \times 10^{-9}\ \text{nm})^3 \times 9.109410^{-31}\ \text{kg}}{4 \times (1.6022 \times 10^{-19}\ \text{C})^2 \times 1.68 \times 10^{-8}\ \Omega\,\text{m}} = 2.495 \times 10^{-14}\ \text{s}$.

5.24   Recall Section 5.5ff as necessary. Since elemental sodium is body-centred cubic, the electron concentration n_c is $2/(0.42906\ \text{nm})^3 = 2.5321 \times 10^{28}\ \text{nm}^{-3}$. Then, $k_F = \sqrt[3]{3\pi^2 n_c} = 9.0845 \times 10^9\ \text{nm}^{-1}$. The energy E_F is given by $\frac{\hbar^2 k_F^2}{2 m_e} = 5.0358 \times 10^{-19}\ \text{J} \equiv 3.144\ \text{eV}$. The Fermi velocity may be derived from the equation $E = \frac{1}{2} m v^2$. Thus, $v_F = \sqrt{2 E_F/m_e} = 1.05148 \times 10^6\ \text{m s}^{-1}$. Summarizing: $E_F = 3.144\ \text{eV}$, $v_F = 1.051 \times 10^6\ \text{m s}^{-1}$, $k_F = 9.085 \times 10^9\ \text{nm}^{-1}$.

5.25   The resistivity follows the relationship: $\rho = \rho_0 \exp(E_g/2kT)$, where ρ_0 is a constant and E_g is the band gap. Since resistance is proportional to resistivity, the equation $\ln R = m\left(\frac{1}{T}\right) + c$ should be a linear relationship, with the slope m equal to $E_g/2k$. The least-squares line is $\ln R = 3442.3 \left(\frac{1}{T}\right) - 2.3000$; Pearson's $r = 0.996$. The plot is shown below, and from the slope: $E_g = 3442.3\ \text{K} \times 2 \times 1.3807 \times 10^{-23}\ \text{J K}^{-1}/1.6022 \times 10^{-19}\ \text{C} = 0.59\ \text{eV}$.

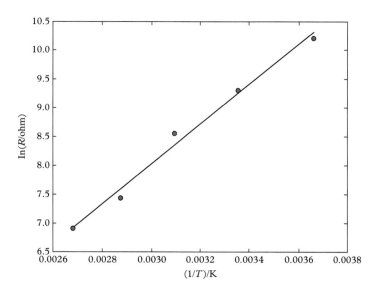

ln R vs. (1/T) for a germanium thermistor

Solutions 6

6.1   From the Periodic table (inside front cover), the relative molecular mass of silicon is 28.085. Hence, the volume of the atom is $28.085 \times 10^{-3} \, \text{kg mol}^{-1}/(6.0221 \times 10^{23} \, \text{mol}^{-1} \times 2329.0 \, \text{kg m}^{-3}) = 2.002 \times 10^{-29} \, \text{m}^3$. Thus, the radius r of a silicon atom is $\sqrt[3]{\frac{3 \times 2.002 \times 10^{-29} \, \text{m}^3}{4 \times \pi}} = 1.684 \times 10^{-10}$ m, or 1.684×10^{-7} mm. The size of the transistor is $20r^2$, so that the number of transistors that will fit on a chip 250 mm square is $(250 \, \text{mm})^2/[20 \times (1.684 \times 10^{-7} \, \text{mm})^2] = 5.510 \times 10^{15}$.

6.2   The irradiating wavelength λ is given by $\lambda = hc/E = 6.6261 \times 10^{-34} \, \text{J s} \times 2.9979 \times 10^8 \, \text{m s}^{-1}/(1.31 \, \text{eV} \times 1.6022 \times 10^{-19}) \, \text{C} = 9.464 \times 10^{-7}$ m. The wavelength of 946.4 nm would be found in the far infrared region of the spectrum.

6.3   Since the energy ($h\nu$) of the quantum dot is inversely proportional to its size, *vide* Eq. (2.37), the energy, and hence the frequency, of the fluorescence falls to one quarter of its original value.

6.4   For a fixed value of the electric field, mobility is proportional to velocity. Since holes have a lesser velocity than electrons, typically one-half, their mobility is less. If the applied field is reduced to zero, both electrons and holes move randomly; mobility is not defined in the absence of an applied field.

[Electrons have free movement in the conduction band, whereas holes have 'obstacles' in the form of atomic sites which restrict their free movement. A hole is the conceptual opposite of an electron—the lack of an electron where one could exist in the valence band.]

6.5   (a) The volume of the cantilever is $0.125 \, \text{mm} \times 0.03 \, \text{mm} \times 4.0 \, \mu\text{m}$ $(1 \, \text{mm} = 10^3 \, \mu\text{m}) = 15,000 \, \mu\text{m}^3$. Hence, the mass m of the cantilever is $1500 \times 10^{-18} \times 2329 = 3.4935 \times 10^{-11}$ g. (b) The resonant frequency ν is given by $\nu = \frac{1}{2\pi}\sqrt{\frac{f}{m}} = \frac{1}{2\pi}\sqrt{\frac{41.0 \, \text{N m}^{-1}}{3.4935 \times 10^{-11} \, \text{kg}}} = 172.4 \, \text{kHz}$.

6.6   Refer to Section 6.3.5 if necessary. The standard designations of the three tube forms are: (a) chiral, moderate semiconductor; (b) and (c) armchair, metallic; (d) zigzag, good semiconductor; (e) chiral, good semiconductor; (f) zigzag, moderate semiconductor.

6.7   (a) The nanotube is obtained by wrapping the sheet around the (4, 3) vector until the ends of the two vectors $4\mathbf{a}_1$ and $3\mathbf{a}_2$ on the sheet meet.

(b) The diameter of the nanotube is $\frac{0.2461 \, \text{nm} \times \sqrt{4^2 + 3^2 + (4 \times 3)}}{\pi} = 0.476 \, \text{nm}$.

6.8   From the above reference for Problem 6.7, $\theta = \tan^{-1}\left[\frac{\sqrt{3}m}{(2n+m)}\right]$. The chiral angles for the vectors in Problem 6.6 are, in order: $26.3°, 30°, 30°, 0°, 8.2°, 0°$.

6.9   (a) Let P and H be the numbers of pentagons and hexagons respectively. Then, $P + H = F$, the total number of faces on the polyhedron. Each edge is shared by two faces, so that $E = (5P + 6H)/2$. Also, each vertex is shared by three faces, whereupon $V = (5P + 6H)/3$. Since $F = P + H$, applying Euler's formula gives

$$\frac{5P + 6H}{3} + (P + H) = \frac{5P + 6H}{2} + 2,$$

whence $P = 12$.

(b) The smallest fullerene molecule has $H = 0$; it is dodecahedral, C_{20}.

(c) The fullerene map for C_{20} is

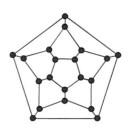

C_{20} molecule

(*Note.* The complexity of fullerene maps increases with the number of atoms; that for C_{60} follows:

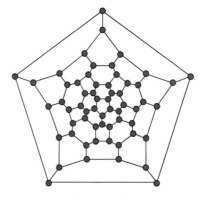

C_{60} molecule

Fullerene maps are drawn normally as plane projections along a symmetry axis, fivefold in the two examples here.)

(d) From the C_{20} figure in (c) above, or by Euler's formula, $E = 30$.

6.10   From a central hexagon, the addition of six pentagons sharing edges with the hexagon and with one another produces figure (a). However, all carbon atoms in fullerenes are of sp^2 configuration, so that the valence requirements are not satisfied by (a). Linking the outer bonds produces figure (b), but that now comprises two hexagons. A fullerene can have any number of hexagons except 1.

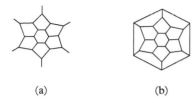

(a)   (b)

(a) Valence requirements unsatisfied   **(b) Valence requirements satisfied**

6.11   The required data are as follows:

T/K	273	298	323	348	373
G/S	0.1198	0.04450	0.1411	0.3489	0.7942
$10^3(1/T)/K^{-1}$	3.663	3.356	3.096	2.874	2.681
$\ln(G/G^*)$	−4.425	−3.112	−1.958	−1.053	−0.2304

The graph of $\ln(G/G^*)$, where G^* is a reference value equal to 1 S, against $10^3/T$ follows the equation $\ln(G/G^*) = -4.293(10^3/T) + 11.30$, and the least-squares line is shown below:

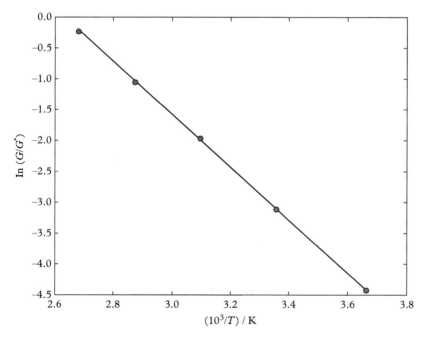

Graph of ln G (conductance) vs. $1/T$

The slope of the line is $-\Delta E/2k$, $\Delta E = 2 \times 1.3807 \times 10^{-23}\,\mathrm{J\,K^{-1}} \times 4293\,\mathrm{K^{-1}} = 1.1855 \times 10^{-19}\,\mathrm{J} \equiv 0.740$ eV.

6.12 $K_c = \frac{c_{ADP} \times c_{[iP]}}{c_{ATP}} = 5.8 \times 10^5$, since the value of c_{H_2O} may be taken as unity. From the equation for the hydrolysis, $c_{ADP} = c_{[iP]}$, and if the fractional degree of dissociation of ATP is α, then $\frac{\alpha^2}{1-\alpha} = 5.8 \times 10^5$, whence $1 - \alpha$, which is c_{ATP}, is 1.72×10^{-6} and is also the concentration in mmol dm^{-3} at equilibrium. For the forward reaction, $\Delta G = -RT\ln(1.72 \times 10^{-6}) = -34.2$ kJ mol^{-1}. (The large negative value of ΔG is then the driving force for ATP hydrolysis.)

6.13   Reduced mass: $m_r^* = \frac{m_e^* m_h^*}{m_e^* + m_h^*} = \frac{0.26 \times 0.36 \times m_e^2}{(0.26 + 0.36)m_e} = 0.1510 m_e$. Thus,

$$r_{ex} = \frac{\varepsilon_r \varepsilon_0 a_0}{m_r^* m_e} = \frac{11.9 \times 5.29177 \times 10^{-11}\,\mathrm{nm}}{0.1510} = 4.17\,\mathrm{nm}.$$

6.14   In a strip of conducting material (a), the resistance R is given by the conventional formula: $R = \rho l/A$, where ρ is the electrical resistivity, and l and A are, respectively, the length and cross-sectional area of the narrow bar conductor.

   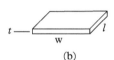

(a)   (b)

Conductor of small   Monomolecular semiconductor
cross-section A semiconductor sheet of thickness t

In a layered or sheet semiconductor material (b), it is usual to work with a parameter R_S, the *sheet resistance*, where $R = \rho l/(wt) = (\rho/t)(l/w) = R_S(l/w)$, the area being unity by definition. Thus: $R_S = \rho/t$, where t is given in the text as 0.335 nm. Hence: $\sigma = 1/(tR_S) = \frac{1}{5900\,\Omega \times 0.335 \times 10^{-9}\,\mathrm{m}} = 5.06 \times 10^5\,\mathrm{S\,m^{-1}}$.

6.15   (a)   The correct sequence is that of (iii) [8].

  (b)   The energy of the signal is hc/λ. Introducing the data: $E = \frac{6.6261 \times 10^{-34}\,\mathrm{J\,s} \times 2.9979 \times 10^8\,\mathrm{m\,s^{-1}}}{2250 \times 10^{-9}\,\mathrm{m}} = 8.8286 \times 10^{-20}\,\mathrm{J} \equiv \frac{8.8286 \times 10^{-20}\,\mathrm{J}}{1.6022 \times 10^{-19}\,\mathrm{C}} = 0.551$ eV. Hence, the energy of the signal is insufficient to activate the p–n junction of band gap 1.73 eV.

6.16   (a)   From Eq. (5.49)

$$g(E) = (8 \times \pi \times \sqrt{2} \times 1.0 \times 10^{-22}\,\mathrm{m^3})\left(\frac{1.05 \times 9.1094 \times 10^{-31}\,\mathrm{kg}}{(6.62607 \times 10^{-34}\,\mathrm{J\,s})^2}\right)^{3/2}\sqrt{0.10\,\mathrm{eV} \times 1.6022 \times 10^{-19}\,\mathrm{C}}$$

$$= 1.45 \times 10^{26}\,\mathrm{J^{-1}}$$

  (b)   Units of expression: $\frac{(\mathrm{m^3})(\mathrm{kg^{3/2}})(\mathrm{V^{1/2}C^{1/2}})}{\mathrm{J^3}} = \frac{(\mathrm{J^{3/2}})(\mathrm{J^{1/2}})}{\mathrm{J^3}} = \mathrm{J^{-1}}$, which is correct for $g(E) \equiv \frac{dN}{dE}$. [$\mathrm{J} = \mathrm{kg\,m^2\,s^{-2}} = (\mathrm{eV})\mathrm{C}$]

Solutions 7

7.1 The graph below is the function $y = x^3 - 3x^2$ for $x = -2$ to $+4$ and the following approximate results were determined from it:

	Maximum	Minimum
x	0	2
y	0	−4

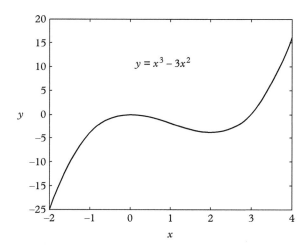

The function $f(x) = x^3 - 3x^2$

$f' = 3x^2 - 6x$; $f'' = 6x - 6$, from which the above values of the maximum and minimum may be confirmed. Note that if $x = 0$, then $y = 0$, whence $0 = x^3 - 3x^3 = 0$, so that $x = 0$ and 3 are the x-intercepts of $f(x)$ with the line $y = 0$. This is shown clearly by adding the command **plt.plot([–2, 4], [0, 0])** which draws the line $y = 0$.

7.2 The number of (non-trivial) nodes is $n - l - 1$, so that the number of nodes in each case is 4. Check with 'Orbitron' (see Section 7.8).

7.3

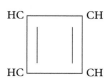

1,3-Cyclobutadiene

β-Energies: −2, 0, 0, 2
Bond orders: 0.5 (all)
Free valence indexes: 0.732 (all)

7.4   From the form of the equation, a plot of $\ln D$ v. $1/T$ should be linear with a slope of $-U/R$:

$\ln D/D^*$	$(10^3\,T)/\text{K}^{-1}$
−46.800	3.333
−33.330	2.000
−27.554	1.429
−24.350	1.111
−22.308	0.909

The plot is linear; D^* is a reference value of unity:

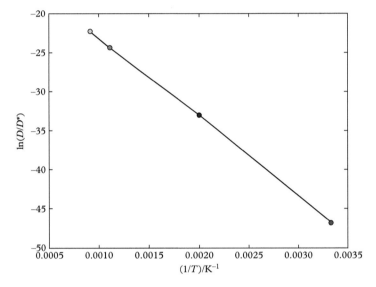

Graph of $f(D)$ vs. $1/T$

From program LSLI the slope $(-U/R)$ is -1.0083×10^4 K; hence, $U = 83.8$ kJ mol^{-1}.

7.5   Enter data at the keyboard, using the sample data in the program MADC as a guide: $\beta = 90.0°$; $Z = 4$; $d = a/2 = 0.28201$ nm; System Number $= 1$; $N = 8$; $\alpha = 5$.

Coordinates	0.0 0.0 0.0 1	0.5 0.0 0.0 −1
and charges:	0.0 0.5 0.5 1	0.0 0.5 0.0 −1
	0.5 0.0 0.5 1	0.0 0.0 0.5 −1
	0.5 0.5 0.0 1	0.5 0.5 0.5 −1

Madelung constant calculated to 6 decimal places $= 1.747564$

7.6 (a) $|\mathbf{A}| = 5$. (b) $\text{Cof}(\mathbf{A}) = \begin{pmatrix} 3 & \bar{4} & 2 \\ 1 & 2 & \bar{1} \\ \bar{1} & 3 & 1 \end{pmatrix}$. (c) $\mathbf{A}^{-1} = \frac{1}{\det \mathbf{A}} \text{Cof}(\mathbf{A}^{\mathrm{T}}) =$

$\frac{1}{5}\begin{pmatrix} 3 & 1 & \bar{1} \\ \bar{4} & 2 & 3 \\ 2 & \bar{1} & 1 \end{pmatrix} = \begin{pmatrix} 0.6 & 0.2 & \overline{0.2} \\ \overline{0.8} & 0.4 & 0.6 \\ 0.4 & \overline{0.2} & 0.2 \end{pmatrix}$.

7.7 The plot of RDF_{2p} $v.$ r/a_0 is

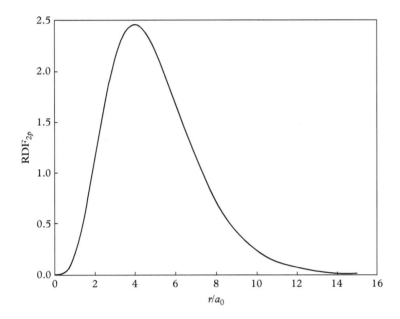

Radial distribution function $4\pi r^2 |\psi(r)|^2$ $v.$ r/a_0

from which the maximum, using the mouse pointer, ≈ 4. From Table 2.2, $\text{RDF}_{2p} = C\{r^4 \exp(-r/a_0)\}$, where C is a constant.

$$\frac{d(\text{RDF}_{2p})}{dr} = r^4 \exp(-r/a_0)(-1/a_0) + 4r^3 \exp(-r/a_0), \text{ and} = 0 \text{ (for the maximum)};$$

hence, $r_{\max}/a_0 = 4$.

7.8 From QUAD the area is 293.319 with an error of –0.001. By integration, the area = 293.333.....

7.9 The equation solves for three real roots: $x = 1.11491, -1.25410$ and -2.86081, to a precision of 10^{-7}.

7.10   The sum of 500 harmonic functions

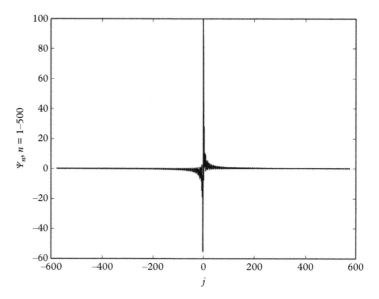

$C\{\sin(n\pi j/576) + \cos(n\pi j/576)\}, j = -576\text{--}576, n = 1\text{--}500$

approaches the ideal wave packet shown by Fig. A21.2d; C is a normalizing constant.

7.11   The electron-in-a-box functions ψ_n of period a for $n = 1$–5 are

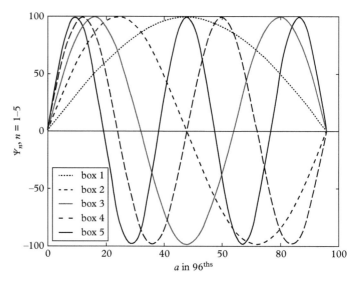

Electron-in-a-box functions ψ_n, $n = 1$–5

They are the same as those discussed in the text and illustrated separately by Fig. 2.7, albeit there of smaller amplitude:

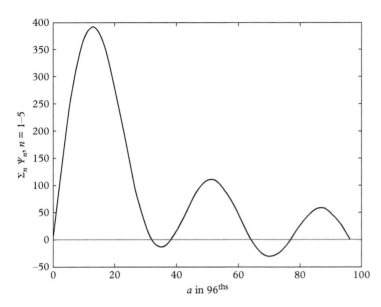

Sum of electron-in-a-box functions $\Sigma \psi_n$, $n = 1\text{--}5$

The sum $\sum\limits_{n=1}^{5} \psi_n$ point by point shows an early stage in the formation of a wave packet (sine functions only shown).

7.12 $\bar{\nu} = 43327.1 \text{ cm}^{-1}$.

....

REFERENCES

[1] Ladd M. *Symmetry of Crystals and Molecules*. Oxford University Press, 2014.

[2] Kettle SFA. *Symmetry and Structure*, 3rd ed. John Wiley & Sons, 2007.

[3] Walsh AD. *J. Chem. Soc.* 1963; p. 2250.

[4] Cruickshank DWJ. *Acta Crystallogr.* 1956; 9: 915.

[5] Nelson RD *et al. Selected Values of Electric Dipole Moments in the Gas Phase*, National Bureau of Standards Circular No. 537, 1957 (online at http://www.nist.gov/data/nsrds/NSRDS-NBS-10.pdf/)

[6] Ladd MFC and Lee WH. *Acta Crystallogr.* 1960; 13: 959.

[7] Ladd MFC and Palmer RA. *Structure Determination by X-ray Crystallography*, 5th ed. Springer Science+Business Media, 2013.

[8] http://www.nist.gov/data/PDFfiles/jpcrd22.pdf/

Index